Ocular Transporters in Ophthalmic Diseases and Drug Delivery

OPHTHALMOLOGY RESEARCH

JOYCE TOMBRAN-TINK, PhD, AND COLIN J. BARNSTABLE, DPhil
SERIES EDITORS

Ocular Transporters in Ophthalmic Diseases and Drug Delivery, edited by Joyce Tombran-Tink, PhD and Colin J. Barnstable, DPhil, 2008.

Visual Prosthesis and Ophthalmic Devices: New Hope in Sight, edited by Joseph F. Rizzo, MD, Joyce Tombran-Tink, PhD, and Colin J. Barnstable, DPhil, 2007

Retinal Degenerations: Biology, Diagnostics, and Therapeutics, edited by Joyce Tombran-Tink, PhD and Colin J. Barnstable, DPhil, 2007

Ocular Angiogenesis: Diseases, Mechanisms, and Therapeutics, edited by Joyce Tombran-Tink, PhD and Colin J. Barnstable, DPhil, 2006

Ocular Transporters in Ophthalmic Diseases and Drug Delivery

Edited by

Joyce Tombran-Tink, PhD
Department of Ophthalmology and Visual Science
Yale University School of Medicine, New Haven, CT

Colin J. Barnstable, DPhil
Department of Neural and Behavioral Sciences
Penn State University College of Medicine, Hershey, PA

Editors and Series Editors

Joyce Tombran-Tink, PhD
Department of Neural and
   Behavioral Sciences
Penn State University College of Medicine
500 University Drive
Hershey, PA 17033, USA
jttink@aol.com

Colin J. Barnstable, DPhil
Department of Neural and
   Behavioral Sciences
Penn State University College of Medicine
500 University Drive
Hershey, PA 17033, USA
cbarnstable@hmc.psu.edu

ISBN: 978-1-58829-958-1   e-ISBN: 978-1-59745-375-2

Library of Congress Control Number: 2007932582

© 2008 Humana Press, a part of Springer Science+Business Media, LLC
All rights reserved. This work may not be translated or copied in whole or in part without the written permission of the publisher (Humana Press, 999 Riverview Drive, Suite 208, Totowa, NJ 07512 USA), except for brief excerpts in connection with reviews or scholarly analysis. Use in connection with any form of information storage and retrieval, electronic adaptation, computer software, or by similar or dissimilar methodology now known or hereafter developed is forbidden.
The use in this publication of trade names, trademarks, service marks, and similar terms, even if they are not identified as such, is not to be taken as an expression of opinion as to whether or not they are subject to proprietary rights.

While the advice and information in this book are believed to be true and accurate at the date of going to press, neither the authors nor the editors nor the publisher can accept any legal responsibility for any errors or omissions that may be made. The publisher makes no warranty, express or implied, with respect to the material contained herein.

Printed on acid-free paper.

9 8 7 6 5 4 3 2 1

Cover illustration: Schematic representation of the ABCR protein (*see* complete Fig. 1 in Chapter 17 and caption on p. 319)

springer.com

Preface

Detection and responses to light are common features found throughout the plant and animal kingdoms. In most primitive life forms, a patch of light-sensitive cells make up a region containing a cell sheet devoid of any specialized anatomical structure. With the development of the eyes in more advanced life forms, light-sensing structures became more complex but primitive eyes are still in contiguity with other body tissues and fluids. The evolution of the eyeball promoted an increase in visual acuity and visual processing that, in turn, allowed vision to become the dominant sensory system for many species, including humans. The formation of a totally enclosed structure, however, required a unique set of solutions to enable the eye to control its environment.

Like most organs, the eye evolved a series of homeostatic mechanisms to regulate its environment within tightly controlled limits. Unlike most organs, however, this advanced light-sensing structure has a series of requirements that place a tremendous burden on molecules that are responsible for controlling ocular homeostasis. There are many signaling molecules and pathways that work in parallel or through crosstalk to maintain the normal ocular environment required for visual function. Perhaps none are so critical as the group of membrane molecules that are collectively termed transporters. These molecules are responsible for the controlled and selective movements of ions, nutrients, and fluid across various ocular layers necessary to optimize the internal milieu to preserve visual function.

One of the most critical functions of the eye is to maintain a clear optical path to the retina. The cornea is composed of just two cell types with a stromal layer between them. This delicate structure acts as a barrier to the external environment without the advantage of a protective layer of skin. It must also remain transparent and maintain the right level of curvature to allow refraction of light. Maintaining the correct hydration of the stroma requires coordinated ion and water transport by both the corneal epithelium and corneal endothelium. In many tissues, transport involves regulated movement of molecules from adjacent blood vessels. The cornea must maintain its structure in the absence of any blood supply. In the first three chapters of this volume, the authors explore these transport mechanisms and the ways in which they not only control corneal volume and transparency, but promote epithelial renewal, endothelial migration, and wound healing.

A second structure that must maintain optical clarity over many decades is the lens. Lens epithelial cells undergo a complex terminal differentiation into fiber cells that involves extensive elongation, loss of nuclei and other organelles, and expression of large quantities of specific lens proteins, including the crystallins. Maintenance of this complex tissue organization depends on tightly controlled levels of hydration and carefully controlled fluxes of ions. The roles of specific lens transporters in normal lens physiology, and in the development of cataracts, are discussed in several chapters of this volume.

Because the cornea and lens are avascular structures, they depend on fluid and nutrients transported across the ciliary epithelium. Materials flow from the ciliary processes, into the posterior chamber, forward into the anterior chamber via the pupil, and finally out through the trabecular meshwork. There is a dynamic balance between the inflow and the outflow of this fluid pathway and the difference gives rise to an intraocular pressure of approximately 15 mm Hg. Too little pressure and the optical path can be compromised; too much pressure and a pathological cascade can be activated leading to glaucoma.

Unlike the cornea and lens, the retina, a thin sheet of CNS tissue that lines the back of the eye, is highly vascularized. Indeed, unlike most regions of the CNS, the retina receives two separate blood supplies. The inner retina of most mammals contains a capillary network that arises from vessels entering the eye through the optic nerve. These vessels spread over the inner surface of the retina and then ramify this stratified structure at the level of the outer plexiform layer. The vasculature and associated glia form a blood-retinal barrier that is similar to the barrier found elsewhere in the CNS. The photoreceptor layer of the retina has an alternative source for oxygen and nutrients, the choroidal blood supply at the back of the eye that allows free passage of molecules across the fenestrated capillaries. Before reaching the retina, molecules have to traverse the retinal pigment epithelium (RPE). The tight junctions between the RPE cells form the second type of blood-retinal barrier. Because of this barrier the RPE cells play a critical role in providing glucose and other nutrients to the retina, removing retinal waste products, regenerating visual pigments as part of the visual cycle, and generally regulating the extracellular environment of the retina. Many of the transporters and channels expressed by the RPE cells show a remarkable asymmetric distribution between the apical and basal surfaces. This results in an ability to separately control the subretinal and choroidal extracellular spaces. It also results in vectorial transport of many molecules into and away from the retina. A group of five chapters in this volume discusses a wide array of RPE transporters and their functions. It is clear that the catalog of molecules is not complete and that there is much to find out about the molecules already identified. For example, we are only beginning to understand the many ways in which circulating molecules, trophic factors, and neuromodulators can affect transport of particular molecules across the RPE.

As with other neural tissues, the function of neurons that comprise the retina requires tight control of levels of ion and neurotransmitters surrounding synaptic terminals. Ion exchangers are important molecules in maintaining synaptic terminal ion gradients, which are essential for normal function. Removal of neurotransmitters from the synaptic cleft regulates the magnitude and time course of synaptic transmission. The transporters that carry out this removal are important molecules that can shape synaptic responses.

One measure of the importance of transporters in the eye is the range of pathological changes seen when these molecules are mutated or dysfunctional. One of the best examples of such dysfunction is found in mutations of the ABCA4 transporter, which can lead to some forms of Stargardt's disease and retinal degenerations. Altered glutamate transport is also associated with various retinal pathologies, a topic discussed in detail in two chapters of this volume.

Ocular transporters are also important targets for drug therapy to the eye and good conduits for drug delivery, a topic explored in three chapters in this volume. Because of its unique properties, the eye is a well-studied experimental model to understand how drugs penetrate tissues and how enzymatic alterations can affect their bioavailability. It is rapidly becoming a popular model to explore new avenues to deliver drugs by nanotechnology. Drugs such as Timolol™ and its derivatives are potent agents that regulate intraocular pressure and aqueous flow. Delivered as eyedrops, such drugs must diffuse across ocular tissues and work against the direction of ocular fluid flow. Although the exact mechanism of action of the adrenergic drugs at the ciliary epithelium remains poorly understood, they have proved to be effective agents for treatment of many forms of glaucoma. There is less progress in devising ways to deliver drugs to the back of the eye to treat retinal diseases. Given the numbers of patients worldwide suffering from macular degeneration, glaucoma, and diabetic retinopathy, easy and effective delivery systems are urgently needed. A better understanding of ocular transport would certainly facilitate the development of better drug delivery systems to the eye.

Molecules involved in the transport of fluid, ions, micronutrients, metals, and neurotransmitters have been studied for many decades. It is only in the last few year, however, that we have identified the structure of some of these molecules and have only now begun to understand their structure–function relationships. Most of the discussions on transporters have previously been restricted to a specific type of molecule or cell. It is our hope in putting together this volume that researchers working on the front of the eye will read about work going on in the back of the eye, and vice versa. We also hope that clinicians and pharmacologists will also benefit from the excellent reviews in this text by those who have worked diligently in the field to provide information that shows how altering transport of molecules in one part of the eye might affect the physiology of other ocular structures. As molecular, pharmacological, and genetic approaches establish the importance of ocular transporters in physiological and pathological functions, we may have better insight into their regulation and potential for exploitation in delivering therapeutics to the eye.

Joyce Tombran-Tink, PhD
Colin J. Barnstable, DPhil

Contents

Preface v
Contributors xiii
Companion CD xvii

I Transport in the Anterior Segment

1. Aquaporins and Water Transport in the Cornea 3
   Alan S. Verkman

2. Roles of Corneal Epithelial Ion Transport Mechanisms
   in Mediating Responses to Cytokines and Osmotic Stress 17
   *Peter S. Reinach, José E. Capó-Aponte,
   Stefan Mergler, and Kathryn S. Pokorny*

3. Vitamin C Transport, Delivery, and Function in the Anterior
   Segment of the Eye 47
   Ram Kannan and Hovhannes J. Gukasyan

II Transporters of the Ciliary Epithelium

4. Mechanisms of Aqueous Humor Formation: *Cellular Model
   of Aqueous Inflow* 61
   *Chi-wai Do, Chi-wing Kong, Chu-yan Chan,
   Mortimer M. Civan, and Chi-ho To*

III Lens Transporters

5. Membrane Transporters: *New Roles in Lens Cataract* 89
   Paul J. Donaldson and Julie Lim

6. Lens Na^+, K^+-ATPase 111
   Nicholas A. Delamere and Shigeo Tamiya

IV Transport Across the Blood–Retinal Barrier

7. Pathophysiology of Pericyte-containing Retinal Microvessels:
   Roles of Ion Channels and Transporters 127
   Donald G. Puro

8. Molecular Mechanisms of the Inner Blood-Retinal
   Barrier Transporters 139
   Masatoshi Tomi and Ken-ichi Hosoya

V Transport Across the Retinal Pigment Epithelium

9 Regulation of Transport in the RPE.. 157
   Adnan Dibas and Thomas Yorio

10 Glucose Transporters in Retinal Pigment Epithelium
   Development... 185
   Lawrence J. Rizzolo

11 Ca^{2+} Channels in the Retinal Pigment Epithelium:
   *Modulators of Retinal Pigment Epithelium
   Function and Communication with Neighboring Tissues* 201
   Olaf Strauss

12 Taurine Transport Pathways in the Outer Retina
   in Relation to Aging and Disease 217
   Ali A. Hussain and John Marshall

13 P-Glycoprotein Expression and Function
   in the Retinal Pigment Epithelium 235
   Paul A. Constable, John G. Lawrenson, and N. Joan Abbott

VI Transporters in the Retina

14 The Retinal Rod NCKX1 and Cone/Ganglion Cell NCKX2
   Na^+/Ca^{2+}-K^+ Exchangers.. 257
   *Paul P. M. Schnetkamp, Yoskiyuki Shibukawa, Haider F. Altimimi,
   Tashi G. Kinjo, Pratikhya Pratikhya, Kyeong Jing Kang,
   and Robert T. Szerencsei*

15 Excitatory Amino Acid Transporters in the Retina 275
   Vijay Sarthy and David Pow

16 Localization and Function of Gamma Aminobutyric Acid
   Transporter 1 in the Retina 293
   Giovanni Casini

VII Genetic Variants of Ocular Transporters:
Implication in Drug Metabolism and Eye Diseases

17 Biochemical Defects Associated with Genetic Mutations
   in the Retina-Specific ABC Transporter, ABCR,
   and Macular Degenerative Diseases... 317
   Esther E. Biswas-Fiss

18 Glutamate Transporters and Retinal Disease and Regulation 333
   Nigel L. Barnett and Natalie D. Bull

19 Glutamate Transport in Retinal
   Glial Cells during Diabetes 355
   Erica L. Fletcher and Michelle M. Ward

VIII Ocular Drug Delivery

20 The Emerging Significance of Drug Transporters and Metabolizing Enzymes to Ophthalmic Drug Design 375
    Mayssa Attar and Jie Shen

21 Barriers in Ocular Drug Delivery 399
    Sriram Gunda, Sudharshan Hariharan, Nanda Mandava, and Ashim K. Mitra

22 Ophthalmic Applications of Nanotechnology 415
    Swita Raghava, Gaurav Goel, and Uday B. Kompella

23 Vitamin C Transporters in the Retina 437
    Vadivel Ganapathy, Sudha Ananth, Sylvia B. Smith, and Pamela M. Martin

24 The Plasma Membrane Transporters and Channels of Corneal Endothelium 451
    Jorge Fischbarg

Index 459

CONTRIBUTORS

N. JOAN ABBOTT • *Pharmaceutical Science Research Division, School of Biomedical and Health Sciences, King's College, London UK*

HAIDER F. ALTIMIMI • *Department of Physiology and Biophysics, Faculty of Medicine, University of Calgary, Alberta, Canada*

SUDHA ANANTH • *Department of Biochemistry and Molecular Biology, Medical College of Georgia, Augusta, GA*

MAYSSA ATTAR • *Department of Pharmacokinetics and Drug Metabolism, Allergan Inc., Irvine, CA*

NIGEL L. BARNETT • *Vision, Touch and Hearing Research Centre, School of Biomedical Sciences, University of Queensland, Australia*

COLIN J. BARNSTABLE • *Department of Neural and Behavioral Sciences, Penn State University College of Medicine, Hershey, PA*

ESTHER E. BISWAS-FISS • *Department of Bioscience Technologies, Jefferson College of Health Professions, Thomas Jefferson University, Philadelphia, PA*

NATALIE D. BULL • *Cambridge Centre for Brain Repair, University of Cambridge, UK*

JOSÉ E. CAPÓ-APONTE • *Department of Biological Sciences, State University of New York, State College of Optometry, New York, NY*

GIOVANNI CASINI • *Dipartimento di Scienze Ambientali, Università della Tuscia, Largo dell'Università snc, Viterbo, Italy*

CHU-YAN CHAN • *Laboratory of Experimental Optometry, School of Optometry, The Hong Kong Polytechnic University, Hong Kong, PR China*

MORTIMER M. CIVAN • *Department of Physiology, University of Pennsylvania School of Medicine, Philadelphia, PA*

PAUL A. CONSTABLE • *Department of Optometry and Visual Science, City University, Henry Wellcome Laboratories for Visual Sciences, City University, London, UK*

NICHOLAS A. DELAMERE • *Department of Physiology, College of Medicine, University of Arizona, Tucson, AZ*

ADNAN DIBAS • *Department of Pharmacology and Neuroscience, University of North Texas Health Science Center, Fort Worth, TX*

CHI-WAI DO • *Laboratory of Experimental Optometry, School of Optometry, The Hong Kong Polytechnic University, Hong Kong, PR China*

PAUL J. DONALDSON • *Department of Physiology, University of Auckland, Auckland, New Zealand*

JORGE FISCHBARG • *Department of Physiology and Cellular Biophysics, and Ophthalmology, Columbia University, New York, NY*

ERICA L. FLETCHER • *Department of Anatomy and Cell Biology, The University of Melbourne, Australia*

VADIVEL GANAPATHY • *Department of Biochemistry and Molecular Biology, Medical College of Georgia, Augusta, GA*

GAURAV GOEL • *Department of Pharmaceutical Sciences and Department of Ophthalmology, University of Nebraska Medical Center, Omaha, NE*

HOVHANNES J. GUKASYAN • *Pfizer Global Research and Development, Pfizer Inc. La Jolla Laboratories, San Diego, CA*

SRIRAM GUNDA • *Division of Pharmaceutical Sciences, School of Pharmacy, University of Missouri, Kansas City, MO*

SUDHARSHAN HARIHARAN • *Division of Pharmaceutical Sciences, School of Pharmacy, University of Missouri, Kansas City, MO*

KEN-ICHI HOSOYA • *Department of Pharmaceutics, Graduate School of Medical and Pharmaceutical Sciences, University of Toyama, Japan*

ALI A. HUSSAIN • *Department of Ophthalmology, The Rayne Institute, King's College London, St. Thomas' Hospital, London, UK*

KYEONG JING KANG • *Department of Physiology and Biophysics, Faculty of Medicine, University of Calgary, Alberta, Canada*

RAM KANNAN • *Doheny Eye Institute, University Of Southern California, Doheny Vision Research Center, Los Angeles, CA*

TASHI G. KINJO • *Department of Physiology and Biophysics, Faculty of Medicine, University of Calgary, Alberta, Canada*

UDAY B. KOMPELLA • *Department of Pharmaceutical Sciences and Department of Ophthalmology, University of Nebraska Medical Center, Omaha, NE*

CHI-WING KONG • *Laboratory of Experimental Optometry, School of Optometry, The Hong Kong Polytechnic University, Hong Kong, PR China*

JOHN G. LAWRENSON • *Department of Optometry and Visual Science, City University, Henry Wellcome Laboratories for Visual Sciences, City University, London, UK*

JULIE LIM • *Department of Physiology, University of Auckland, Auckland, New Zealand*

NANDA MANDAVA • *Division of Pharmaceutical Sciences, School of Pharmacy, University of Missouri, Kansas City, MO*

JOHN MARSHALL • *Department of Ophthalmology, The Rayne Institute, King's College London, St. Thomas' Hospital, London, UK.*

PAMELA M. MARTIN • *Department of Biochemistry and Molecular Biology, Medical College of Georgia, Augusta, GA*

STEFAN MERGLER • *Eye Clinic, Charité University Medicine, Berlin, Germany*

ASHIM K. MITRA • *Division of Pharmaceutical Sciences, School of Pharmacy, University of Missouri, Kansas City, MO*

KATHRYN S. POKORNY • *The Institute of Ophthalmology and Visual Science, New Jersey Medical School, University of Medicine and Dentistry, Newark, NJ*

DAVID POW • *Department of Ophthalmology, Northwestern University Feinberg School of Medicine, Chicago, IL*

PRATIKHYA PRATIKHYA • *Department of Physiology and Biophysics, Faculty of Medicine, University of Calgary, Alberta, Canada*

DONALD G. PURO • *Department of Ophthalmology and Visual Sciences and Department of Molecular and Integrative Physiology, University of Michigan, Ann Arbor, MI*

Contributors

SWITA RAGHAVA • *Department of Pharmaceutical Sciences and Department of Ophthalmology, University of Nebraska Medical Center, Omaha, NE*

PETER S. REINACH • *Department of Biological Sciences, State University of New York, State College of Optometry, New York, NY*

LAWRENCE J. RIZZOLO • *Departments of Surgery and of Ophthalmology and Visual Science, Yale University School of Medicine, New Haven, CT*

VIJAY SARTHY • *Department of Ophthalmology, Northwestern University Feinberg School of Medicine, Chicago, IL*

PAUL P. M. SCHNETKAMP • *Department of Physiology and Biophysics, Faculty of Medicine, University of Calgary, Alberta, Canada*

JIE SHEN • *Department of Pharmacokinetics and Drug Metabolism, Allergan Inc., Irvine, CA*

YOSKIYUKI SHIBUKAWA • *Department of Physiology, Tokyo Dental College, Chiba, Japan*

SYLVIA B. SMITH • *Department of Cellular Biology and Anatomy, Medical College of Georgia, Augusta, GA*

OLAF STRAUSS • *Experimentelle Ophthalmologie, Klinik und Poliklinik für Augenheilkunde, Klinikum der Universität Regensburg, Regensburg, Germany*

ROBERT T. SZERENCSEI • *Department of Physiology and Biophysics, Faculty of Medicine, University of Calgary, Alberta, Canada*

SHIGEO TAMIYA • *Department of Ophthalmology and Visual Sciences and the Department of Pharmacology and Toxicology, University of Louisville, School of Medicine, Louisville, KT*

CHI-HO TO • *Department of Medicine, University of Pennsylvania School of Medicine, Philadelphia, PA*

JOYCE TOMBRAN-TINK • *Department of Neural and Behavioral Sciences Penn State University College of Medicine, Hershey, PA*

MASATOSHI TOMI • *Department of Pharmaceutics, Graduate School of Medical and Pharmaceutical Sciences, University of Toyama, Japan*

ALAN S. VERKMAN • *Cardiovascular Research Institute, University of California, San Francisco*

MICHELLE M. WARD • *Department of Anatomy and Cell Biology, The University of Melbourne, Australia*

THOMAS YORIO • *Department of Pharmacology and Neuroscience, University of North Texas Health Science Center, Fort Worth, TX*

COMPANION CD

Illustrations listed here may be found on the Companion CD attached to the inside back cover. The image files are organized into folders by chapter number and are viewable in most Web browsers. The CD is compatible with both Mac and PC operating systems.

CHAPTER 1, FIG. 1, P. 5
CHAPTER 1, FIG. 2, P. 6
CHAPTER 1, FIG. 3, P. 9
CHAPTER 1, FIG. 4, P. 11
CHAPTER 1, FIG. 5, P. 12
CHAPTER 1, FIG. 6, P. 12
CHAPTER 2, FIG. 1, P. 18
CHAPTER 2, FIG. 2, P. 21
CHAPTER 2, FIG. 3, P. 25
CHAPTER 2, FIG. 4, P. 29
CHAPTER 2, FIG. 5, P. 35
CHAPTER 2, FIG. 6, P. 36
CHAPTER 2, FIG. 7, P. 36
CHAPTER 2, FIG. 8, P. 38
CHAPTER 4, FIG. 1, P. 62
CHAPTER 4, FIG. 2, P. 64
CHAPTER 4, FIG. 3, P. 65
CHAPTER 5, FIG. 1, P. 90
CHAPTER 5, FIG. 2, P. 94
CHAPTER 5, FIG. 3, P. 96
CHAPTER 5, FIG. 4, P. 99
CHAPTER 5, FIG. 5, P. 100
CHAPTER 5, FIG. 6, P. 103
CHAPTER 5, FIG. 7, P. 105
CHAPTER 6, FIG. 1, P. 112
CHAPTER 6, FIG. 2, P. 113
CHAPTER 6, FIG. 3, P. 115
CHAPTER 6, FIG. 4, P. 117
CHAPTER 6, FIG. 5, P. 118
CHAPTER 6, FIG. 6, P. 119
CHAPTER 6, FIG. 7, P. 120
CHAPTER 7, FIG. 1, P. 128
CHAPTER 7, FIG. 2, P. 128

CHAPTER 7, FIG. 3, P. 130
CHAPTER 7, FIG. 4, P. 133
CHAPTER 8, FIG. 1, P. 140
CHAPTER 8, FIG. 2, P. 142
CHAPTER 8, FIG. 3, P. 145
CHAPTER 8, FIG. 4, P. 147
CHAPTER 8, FIG. 5, P. 150
CHAPTER 9, FIG. 1, P. 160
CHAPTER 9, FIG. 2, P. 173
CHAPTER 9, FIG. 3, P. 174
CHAPTER 9, FIG. 4, P. 175
CHAPTER 10, FIG. 1, P. 187
CHAPTER 10, FIG. 2, P. 190
CHAPTER 10, FIG. 3, P. 192
CHAPTER 10, FIG. 4, P. 193
CHAPTER 11, FIG. 1, P. 203
CHAPTER 12, FIG. 1, P. 221
CHAPTER 12, FIG. 2, P. 223
CHAPTER 12, FIG. 3, P. 225
CHAPTER 12, FIG. 4, P. 226
CHAPTER 12, FIG. 5, P. 228
CHAPTER 12, FIG. 6, P. 229
CHAPTER 13, FIG. 1, P. 237
CHAPTER 13, FIG. 2, P. 239
CHAPTER 13, FIG. 3, P. 244
CHAPTER 14, FIG. 1, P. 263
CHAPTER 14, FIG. 2, P. 264
CHAPTER 14, FIG. 3, P. 268
CHAPTER 15, FIG. 1, P. 277
CHAPTER 15, FIG. 2, P. 278
CHAPTER 15, FIG. 3, P. 278
CHAPTER 15, FIG. 4, P. 285
CHAPTER 15, FIG. 5, P. 287

CHAPTER 16, FIG. 1, P. 297
CHAPTER 16, FIG. 2, P. 300
CHAPTER 16, FIG. 3, P. 301
CHAPTER 16, FIG. 4, P. 302
CHAPTER 16, FIG. 5, P. 303
CHAPTER 16, FIG. 6, P. 303
CHAPTER 16, FIG. 7, P. 304
CHAPTER 17, FIG. 1, P. 319
CHAPTER 17, FIG. 2, P. 320
CHAPTER 17, FIG. 3, P. 324
CHAPTER 18, FIG. 1, P. 337
CHAPTER 18, FIG. 2, P. 343
CHAPTER 19, FIG. 1, P. 357
CHAPTER 19, FIG. 2, P. 358
CHAPTER 19, FIG. 3, P. 359
CHAPTER 19, FIG. 4, P. 360

CHAPTER 19, FIG. 5, P. 361
CHAPTER 19, FIG. 6, P. 364
CHAPTER 20, FIG. 1, P. 388
CHAPTER 20, FIG. 2, P. 389
CHAPTER 21, FIG. 1, P. 400
CHAPTER 21, FIG. 2, P. 402
CHAPTER 21, FIG. 3, P. 409
CHAPTER 22, FIG. 1, P. 417
CHAPTER 22, FIG. 2, P. 428
CHAPTER 22, FIG. 3, P. 430
CHAPTER 23, FIG. 1, P. 438
CHAPTER 23, FIG. 2, P. 443
CHAPTER 23, FIG. 3, P. 445
CHAPTER 24, FIG. 1, P. 453
CHAPTER 24, FIG. 2, P. 456
CHAPTER 24, FIG. 3, P. 457

I
Transport in the Anterior Segment

1
Aquaporins and Water Transport in the Cornea

Alan S. Verkman

CONTENTS

INTRODUCTION
AQUAPORIN STRUCTURE AND TRANSPORT FUNCTION
AQUAPORIN EXPRESSION IN OCULAR TISSUES–INDIRECT EVIDENCE
  FOR A ROLE IN EYE PHYSIOLOGY
ROLES OF AQUAPORINS IN MAMMALIAN PHYSIOLOGY DEDUCED
  FROM PHENOTYPES OF AQP-NULL MICE
OCULAR ROLES OF AQUAPORINS OUTSIDE OF THE CORNEA
ROLES OF AQUAPORINS IN THE CORNEA
SUMMARY AND PERSPECTIVE
ACKNOWLEDGMENTS
REFERENCES

INTRODUCTION

The eye contains specialized avascular tissues for corneal and lens transparency, secretory epithelia for regulation of aqueous fluid volume and pressure, and electrically excitable cells for retinal signal transduction. Regulated fluid transport between extravascular spaces and adjacent tissues or microvessels supports these specialized functions. The general paradigm in mammalian tissues, including the eye, is that water movement follows osmotic gradients generated by active and secondary active solute transport. Although all cell membranes have significant water permeability, AQP-type water cells are found in some cell types where they increase plasma membrane water permeability by a few up to approximately 100-fold compared to membranes without AQPs. This chapter provides a brief description of AQP structure and function, followed by evidence about the physiological role of AQPs in extraocular and ocular tissues, with focus on cornea.

From: *Ophthalmology Research: Ocular Transporters in Ophthalmic Diseases and Drug Delivery*
Edited by: J. Tombran-Tink and C. J. Barnstable © Humana Press, Totowa, NJ

AQUAPORIN STRUCTURE AND TRANSPORT FUNCTION

The AQPs are a family of small, hydrophobic integral membrane proteins (approximately 30 kDa/monomer) that are expressed widely in the animal and plant kingdoms, with 13 members identified to date in mammals. Aquaporins are expressed in many epithelia and endothelia involved in fluid transport, as well as in cell types that are thought not to carry out fluid transport, such as skin, fat, and urinary bladder cells. In most cell types, the AQPs reside constitutively at the plasma membrane. A notable exception is kidney AQP2, which undergoes vasopressin-regulated exo- and endocytosis in a manner similar to insulin-regulated GLUT4 targeting. High-resolution X-ray crystal structures exist for AQP1 *(1)*, the bacterial glycerol-transporter GlpF *(2)*, and the major intrinsic protein (formerly called MIP) of lens fiber, AQP0 *(3)*. Aquaporin 1 monomers contain six tilted alpha-helical domains, forming a barrel-like structure in which the first and last three helices exhibit inverted symmetry (reviewed in *4,5*). Two conserved Asn-Pro-Ala (NPA) motifs reside on opposite sides of the AQP monomer, which permits water but not small solutes to pass through the pore. Monomeric AQP units contain independently-functioning pores, though they are assembled in membranes as tetramers *(6)*. Molecular dynamics simulations based on the AQP1 crystal structure suggest tortuous, single-file passage of water through a narrow, less than 0.3 nm diameter pore, in which steric and electrostatic factors prevent transport of protons and other small molecules *(7)*. Aquaporins 1, 2, 4, 5 and 8 are primarily water-selective, whereas AQP3, AQP7 and AQP9 (called aquaglyceroporins) also transport glycerol and possibly other small solutes. Water and glycerol transport via some AQPs can be inhibited by nonspecific, mercurial sulfhydral-reactive compounds such as $HgCl_2$, though there is considerable interest in the identification of non-toxic, AQP-selective inhibitors *(8, 9)*.

AQUAPORIN EXPRESSION IN OCULAR TISSUES–INDIRECT EVIDENCE FOR A ROLE IN EYE PHYSIOLOGY

Figure 1 summarizes the sites of fluid transport (panel A) and AQP protein expression (panel B) in the eye. Expression of AQP0 (MIP) in lens fiber has been known for many years, where its involvement in lens transparency and cataracts is well established. Mutations in AQP0 are associated with congenital cataracts in humans *(10)*. However, the mechanisms by which AQP0 deficiency produces cataracts remain unknown. Aquaporin 1 is expressed strongly in endothelial cells and keratocytes in cornea, as well as in nonpigmented ciliary epithelium and in the epithelium at the anterior surface of the cornea. Aquaporin 3 is expressed throughout the corneal and conjunctival epithelium at the ocular surface. Aquaporin 5 is strongly expressed at the corneal epithelium as well. Aquaporin 4 is expressed in Müller cells in retina, and colocalizes with AQP1 in nonpigmented ciliary epithelium. The ocular expression pattern of AQPs provides indirect evidence for their possible involvement in intraocular pressure regulation (AQP1 and AQP4), corneal and lens transparency (AQP0, AQP1, and AQP5), visual signal transduction and retinal swelling following injury (AQP4), corneal and conjunctival barrier function (AQP3), and tear formation by lacrimal glands (AQP5). Our lab has generated transgenic mice individually lacking the major eye AQPs to establish their physiological functions. As described below, several predicted functions, such as involvement of AQP1

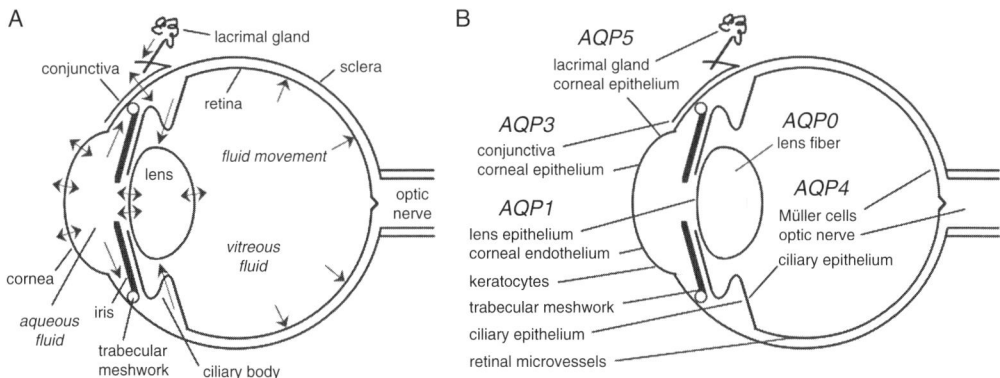

Fig. 1. Sites of water movement and aquaporin (AQP) expression in ocular tissues. **(A)** Schematic showing physiological movements of water across various cell barriers in the eye. **(B)** Sites of AQP protein expression in the eye.

in aqueous fluid secretion, were confirmed, and a number of unexpected functions were discovered.

ROLES OF AQUAPORINS IN MAMMALIAN PHYSIOLOGY DEDUCED FROM PHENOTYPES OF AQP-NULL MICE

Analysis of the extraocular phenotype of transgenic mice lacking specific AQPs has provided considerable insight into their physiological roles with respect to the eye (reviewed in ref. *11*). Mice lacking AQPs 1–4 manifest a defect in urinary concentrating ability *(12)*. Near-isosmolar fluid secretion is impaired in salivary and airway submucosal glands in AQP5 deficiency *(13)*. From these findings, and associated mechanism studies, it follows that high transepithelial water permeability facilitates rapid water transport in response to active transepithelial salt transport. As shown in Fig. 2A, AQP deletion impairs osmotic equilibration, resulting in secretion of a reduced volume of relatively hypertonic fluid, as found for saliva secretion in AQP5 null mice. Involvement of AQPs in fluid secretion is relevant to aqueous fluid secretion by ciliary epithelium in the eye *(14)*. Figure 2B depicts a second mechanism for involvement of AQPs in mammalian physiology, in which they facilitate passive, osmotically driven water transport, as in osmotic extraction of water in kidney collecting duct. A related AQP role is in the pathophysiology of tissue fluid accumulation, as found for AQP4 in brain edema *(15)*. Aquaporin 4 facilitates water entry into and exit from the brain in response to clinically relevant stimuli, such as altered cellular ionic homeostasis in cytotoxic brain edema *(16)*. Involvement of AQP4 in tissue swelling has relevance to retinal edema and damage following injury *(17)*. Not shown in the figure is a separate role for AQP4 in neural signal transduction, which is relevant in retina and may involve AQP4-dependent extracellular space dynamics and Kir4.1-facilitated K^+ buffering *(18, 19)*. Mice lacking AQP4 have altered seizure threshold and duration *(20)*, and as described below, abnormal electroretinograms *(21)*.

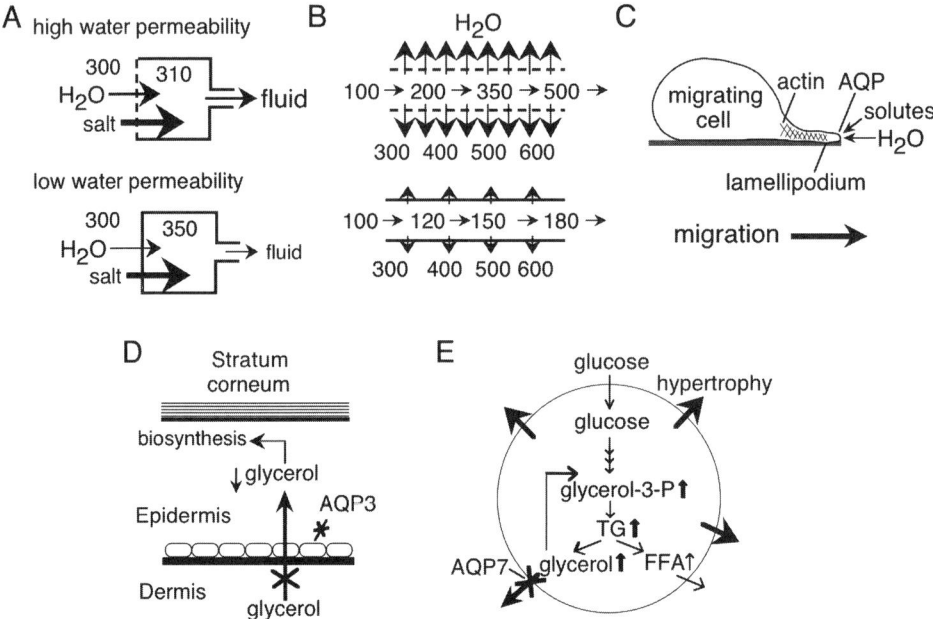

Fig. 2. Physiological functions of aquaporins (AQPs) discovered from phenotype analysis of AQP knockout mice. (**A**) Reduced water permeability in glandular epithelium impairs active, near-isosmolar fluid transport by slowing osmotic water transport into the acinar lumen, producing hypertonic secretion. (**B**) Reduced transepithelial water permeability in kidney collecting duct impairs urinary concentrating ability by preventing osmotic equilibration of luminal fluid. (**C**) AQP-facilitated water entry into protruding lamellipodia in migrating cells, accounting for AQP-dependent cell migration. (**D**) Reduced steady-state glycerol content in epidermis and stratum corneum in skin following AQP3 deletion, accounting for reduced skin hydration in AQP3 deficiency. (**E**) Impaired AQP7-dependent glycerol escape from adipocytes resulting in intracellular glycerol accumulation and increased triglyceride content, accounting for progressive adipocyte hypertrophy in AQP7 deficiency. See text for further explanations.

Recently, a novel cellular role for AQPs in cell migration was discovered, as originally demonstrated in endothelial cells and various transfected cells *(22)*, and subsequently in brain astroglial cells *(23)*, kidney proximal tubule cells *(24)* and tumor cells *(25)*. Figure 2C shows the proposed mechanism for AQP involvement in cell migration, in which actin cleavage and ion uptake at the tip of a lamellipodium create local osmotic gradients that drive water influx. As described below, AQP3-dependent cell migration is relevant in corneal epithelial wound healing *(26)*. Also, AQP-facilitated cell migration may be important in microvascular proliferation in retinopathies and in corneal keratocyte migration during the healing of corneal stromal wounds.

The aquaglyceroporins, of which airway AQP3 is an example, have unique biological roles that are related to their glycerol-transporting function. Aquaporin 3-facilitated glycerol transport in skin is an important determinant of epidermal and stratum corneum hydration (Fig. 2D). Mice lacking AQP3, which is normally expressed in the basal layer of keratinocytes in the epidermis, have reduced stratum corneum hydration and skin elasticity, as well as impaired stratum corneum biosynthesis and wound healing *(27)*. The mechanism responsible for the skin phenotype in AQP3 deficiency involves reduced epidermal cell-skin glycerol permeability, resulting in reduced glycerol content in the stratum corneum and epidermis. Interesting recent data suggest involvement of AQP3 in epidermal-cell proliferation by a mechanism that may involve reduced cellular ATP-content and impaired mitogen-activated protein (MAP) kinase signaling *(28)*. As described below, AQP3-dependent cell proliferation has relevance to repair of the corneal epithelium following injury. Another aquaglyceroporin, AQP7, is expressed in the plasma membrane of adipocytes. Aquaporin 7-null mice have a greater fat-mass than wild-type mice as they age, with remarkable adipocyte hypertrophy and accumulation of glycerol and triglycerides *(29)*. As shown in Fig. 2E, hypertrophy of AQP7-deficient adipocytes probably results from reduced plasma membrane glycerol permeability, and consequent increased glycerol accumulation and triglyceride biosynthesis.

OCULAR ROLES OF AQUAPORINS OUTSIDE OF THE CORNEA

Aquaporins and Intraocular Pressure

The ciliary epithelium is a tissue bilayer consisting of pigmented ciliary epithelia (PCE) and non-pigmented ciliary epithelia, whose apical surfaces are juxtaposed, and basolateral surfaces face the ciliary body and aqueous humor, respectively. The principal determinants of intraocular pressure (IOP) are the rates of aqueous fluid production by the ciliary epithelium and aqueous fluid drainage (outflow) in the canal of Schlemm. Aqueous fluid production involves near-isosmolar water secretion across the ciliary epithelium into the posterior aqueous chamber. Aqueous fluid is drained by pressure-driven bulk fluid flow into the canal of Schlemm and across the sclera. Non-pigmented ciliary epithelial cells coexpress AQP1 and AQP4 *(30–33)*, suggesting their involvement in aqueous fluid production. An initial study on human ciliary epithelial cell cultures reported AQP1 protein expression and partial sensitivity of fluid transport to Hg^{2+} and AQP1 small interfering RNA (siRNA), suggesting AQP1-dependent aqueous inflow *(34)*. Intraocular pressure measurements in mice using a fluid-filled microneedle inserted into the anterior chamber showed a modest reduction in pressure by 2–3 mm Hg in mice lacking AQP1 and/or AQP4 compared to wild-type mice *(14)*. Aquaporin 1 is also expressed in trabecular meshwork endothelium in the canal of Schlemm, where a role in cellular volume regulation has been proposed *(35)*. However, direct measurement of aqueous fluid outflow in mice by a pulsed infusion method showed no effect of AQP1 deletion *(14)*. Together with measurements of aqueous fluid production by a fluorescein iontophoresis-confocal detection method, it was concluded that reduced IOP in AQP-deficient mice was due to reduced aqueous fluid production by the ciliary epithelium. Whether AQP1/AQP4 inhibition will be therapeutically useful in the treatment of glaucoma remains to be determined.

Aquaporin 1 and Cataract Formation

Like the AQP1-expressing corneal endothelium covering the corneal inner surface as discussed further in below, the anterior surface of the lens is covered by an AQP1-

expressing epithelium. Based on immunohistochemical evidence showing AQP1 expression in epithelial cells at the anterior pole of the lens, the involvement of AQP1 in lens epithelial water permeability and cataract formation was tested *(36)*. Osmotic water permeability, measured in calcein-stained epithelial cells in intact lenses from fluorescence changes in response to osmotic gradients, was reduced approximately threefold in lenses from AQP1-null mice. Aquaporin 1 deletion did not alter baseline lens morphology or transparency, though basal water content was significantly increased in AQP1-null mice as measured by gravimetry using kerosene-bromobenzene gradients and wet/dry weight ratios. Cataract formation was induced in vitro by incubation of lenses in a high-glucose solution. Loss of lens transparency was greatly accelerated in AQP1-null lenses bathed in a 55 mM glucose solution for 18 hours, as measured by optical contrast analysis of transmitted grid images. Cataract formation in vivo was significantly accelerated in a mouse model of acetaminophen toxicity. Aquaporin 1 thus facilitates the maintenance of lens transparency and opposes cataract formation, suggesting the possibility of AQP1 induction to retard cataractogenesis. As mentioned above, an association between AQP0 mutations and cataracts is well established. Recent data suggests AQP0-functioning in lens fiber cells as a pH- and calcium-related water channel *(37)*, though the link between lens fiber water permeability and lens opacification is not known.

Aquaporin 4 and Retinal Signal Transduction and Edema Following Injury

Aquaporin 4 is strongly expressed in Müller cells, especially in perivascular and end-feet processes (facing the retinal capillaries and vitreous body), where it has been proposed to form a multiprotein complex involving the inwardly rectifying K^+ channel, Kir4.1 *(38, 39)*. Analogous to its roles in brain astroglial cells and cochlear supportive cells *(40)*, Müller-cell AQP4 has been proposed to maintain extracellular space volume and K^+ concentration during bipolar cell excitation. Aquaporin 4-null mice exhibit mildly altered retinal signal transduction as evidenced by reduced electroretinogram (ERG) b-wave amplitude and latency *(21)*, suggesting functional coupling between water and K^+ clearance. Aquaporin 4 deletion in Müller cells also provides protection against edema and ganglion cell death following retinal ischemia *(17)*. Aquaporin 4 inhibitors might therefore limit inner retinal pathology following vascular-occlusive and other ischemic diseases causing cytotoxic (cellular) edema.

ROLES OF AQUAPORINS IN THE CORNEA
Aquaporin 1 in Corneal Endothelium – its Role in Corneal Transparency

Maintenance of corneal transparency requires precise regulation of stromal water content *(41, 42)*. Aquaporin 1 is expressed in corneal endothelial cells and AQPs 3 and 5 in epithelial cells. To test the possible involvement of AQPs in the maintenance of corneal volume, transparency and thickness, water permeability and response to experimental swelling was measured in wild-type versus AQP1-null mice *(43)*. Compared to wild-type mice, which have a corneal thickness of 123 µm, corneal thickness was reduced in AQP1-null mice (101 µm). Thickness measurements were made in fixed eyes (Fig. 3A), as well as

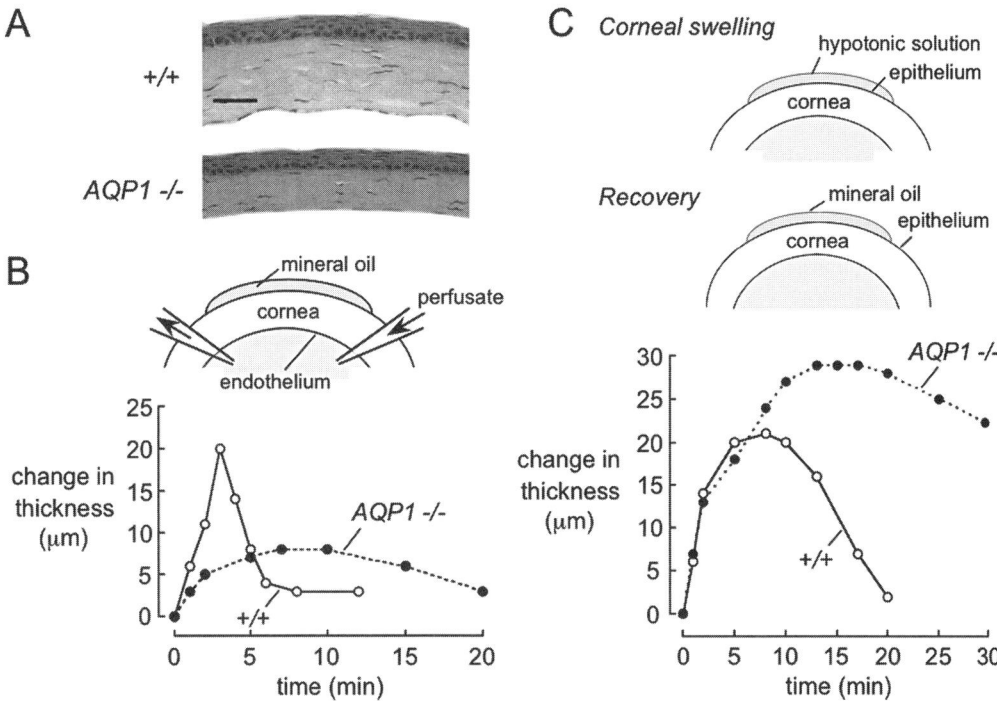

Fig. 3. Involvement of aquaporin 1 in the maintenance of corneal transparency. (**A**) Reduced thickness in AQP1-deficient adult mouse corneas in paraffin-embedded central corneal sections. Bar = 50 μm. (**B**) Osmotic water-transport across the corneal endothelium. Top: schematic showing micropipette placement for anterior chamber perfusion. Bottom: time course of corneal thickness following corneal endothelial exposure to hypotonic saline in wild-type (*open circles*) and AQP1-null (*filled circles*) mice. (**C**) Restoration of corneal thickness after osmotic swelling. Top: transient exposure of the corneal surface to hypotonic saline to induce corneal swelling and to follow recovery of thickness. Bottom: time course of corneal thickness following exposure of the corneal surface to hypotonic saline. Corneal thickness measured *in vivo* by z-scanning confocal microscopy. Adapted from *(43)*.

in vivo by brightfield scanning confocal microscopy. Aquaporin 1 water-transport function in corneal endothelium in vivo was demonstrated by slowed corneal swelling upon hypotonic challenge at the endothelial surface, utilizing an anterior chamber microperfusion method (Fig. 3B). An important role for AQP1 in the maintenance of corneal transparency was demonstrated in an experimental model of corneal edema produced by transient exposure of the corneal surface to hypotonic solution, in which AQP1 deficiency was associated with remarkably impaired recovery of corneal transparency and thickness. Although baseline corneal transparency was not impaired by AQP1 deletion, the return of corneal transparency and thickness after hypotonic swelling (10 minutes exposure of corneal surface to distilled water) was remarkably delayed in AQP1-null mice, with approximately 75% recovery at seven minutes in wild-type mice compared to 5% recovery in AQP1-null mice (Fig. 3C). The impaired recovery of corneal transparency in AQP1-null mice provides evidence for the involvement of AQP1 in active extrusion of fluid from the corneal stroma across the corneal endothelium.

The mechanisms by which AQP1 deletion impairs maintenance of corneal transparency remain unclear. In primary corneal endothelial cell cultures, AQP1 deficiency reduced osmotically driven cell membrane water permeability, but did not impair active near-isosmolar transcellular fluid transport *(44)*. The generally assumed mechanism of transcellular, AQP-facilitated fluid transport has been questioned in relation to the corneal endothelium, with Fischbarg and colleagues proposing a central role for electro-osmotic coupling of fluid transport to recirculating currents in intercellular junctions *(45)*. This model, which is described more fully in another chapter in this volume, posits that AQP1 contributes primarily to cell volume regulation, a role that remains difficult to reconcile with the dramatic corneal swelling phenotype of AQP1-null mice and with the substantially slower rate of cell volume regulation versus osmotic equilibration.

Aquaporins and Ocular Surface Fluid Secretion

The ocular surface is lined by stratified corneal and conjunctival epithelia, which lie in contact with the tear film. The water permeability of the ocular surface, together with the rates of evaporative water-loss and tear-fluid production and drainage, determine tear-film volume and osmolality, as well as corneal stromal water content. Active Cl^- secretion and Na^+ absorption drive net water secretion into tears across both corneal and conjunctival epithelia (reviewed in ref. *46*). The ocular surface, and the conjunctival epithelium in particular (covering 17 times more area than the cornea in humans), contributes to active tear-fluid secretion under basal conditions, and even more so upon stimulation. A computational model of tear-film balance was developed to investigate the theoretical importance of ocular surface water permeability on tear-film osmolality. The model demonstrated the sensitivity of tear film osmolarity to both excessive tear evaporation and inadequate tear-fluid secretion, the two principal causes of dry-eye syndrome (Fig. 4; ref. *47*). In this model, tear fluid generated by osmotic flux (J_v) and isosmolar active secretion (J_s) is removed by evaporation (J_e) and isotonic drainage (J_d), such that in the steady state, $J_s + J_v = J_e + J_d$. Predicted tear-film osmolarity depended strongly on both passive water permeation and active fluid secretion at the ocular surface.

A series of measurements indicated the involvement of AQPs in osmotically-driven water transport at the ocular surface *(47)*. Mice lacking AQP5 have increased corneal thickness (144 versus 123 µm, from ref. *43*). Plasma membrane osmotic water permeability of corneal epithelial cells was measured in mice utilizing an ocular surface perfusion method (Fig. 5A) involving microfluorimetric measurement of calcein quenching in surface cells. Water permeability was high (0.045 cm/s), and reduced approximately twofold in AQP5 deficiency (Fig. 5B). Membrane water permeability was AQP3 dependent in conjunctiva when measured similarly, with approximately fourfold slowing of osmotic equilibration in AQP3 deficiency. Water permeability across the intact cornea and conjunctiva (P_f), the relevant parameters describing water movement into the hyperosmolar tear film in vivo, was measured by a dye-dilution method from the fluorescence of Texas red dextran in an anisosmolar solution in a microchamber at the ocular surface. The P_f of whole cornea was 0.0017 cm/s, and

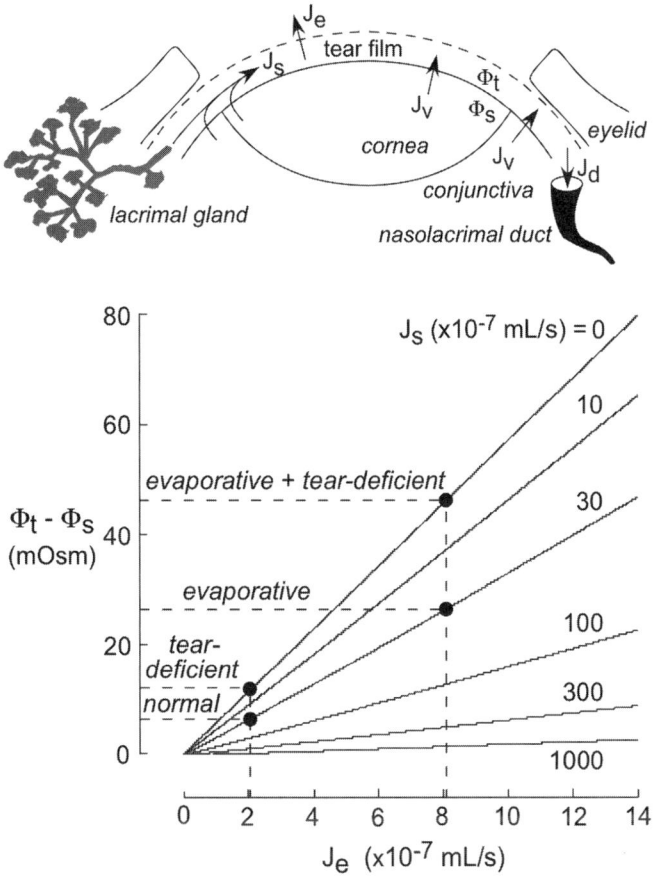

Fig. 4. Theoretical role of ocular surface water permeability in tear-film dynamics. Top: Schematic of ocular surface geometry and contributors of tear fluid balance: J_e, water evaporation at the corneal/conjunctival surface; J_s, fluid secretion by the lacrimal gland and ocular surface epithelia; J_v, osmotic water flow across the corneal and conjunctival surfaces; J_d, tear fluid removal by nasolacrimal drainage; Φ_t and Φ_s, osmolarities of tear film and surface tissue, respectively. Bottom: theoretical dependence of tear-film hyperosmolarity, plotted as $(\Phi_t - \Phi_s)$, on tear-fluid evaporation and secretion rates. Physiology of evaporative and tear-deficient dry eye syndromes shown. See text for explanations (adapted from *47*).

was reduced by greater than fivefold in AQP5-deficient mice. P_f in AQP5-null mice was restored to 0.0015 cm/s after epithelial removal, implicating the corneal epithelium as the major barrier for transcorneal osmosis. Together, these studies suggest the theoretical importance of AQP-dependent water transport across the ocular surface in tear-film osmolality. Measurements of tear-film properties in AQP-deficient mice are needed to test these predictions.

Aquaporin 3 and Healing of Corneal Epithelial Wounds

Aquaporin 3 is expressed in corneal and conjunctival epithelia, and in various extraocular tissues, including skin epidermis and colonic crypt epithelia. Figure 6A shows AQP3

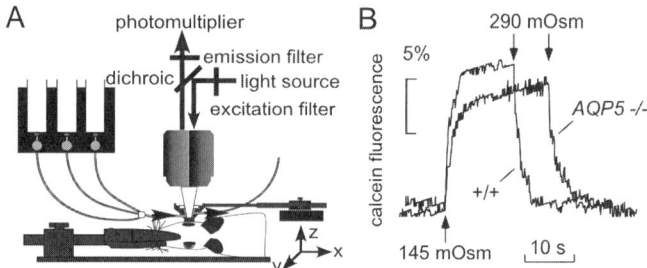

Fig. 5. Aquaporin 5 (AQP5-) dependent water transport across corneal epithelium. (**A**) Ocular surface perfusion for fluorescence measurements of corneal epithelial cell volume changes, containing a microchamber positioned on the corneal surface, optical elements for calcein fluorescence measurement, rapid exchange perfusion system, and stereotaxic platform. (**B**) Time course of corneal epithelial cell calcein fluorescence in response to osmotic gradients in wild-type and AQP5-deficient mice (adapted from ref. *47*).

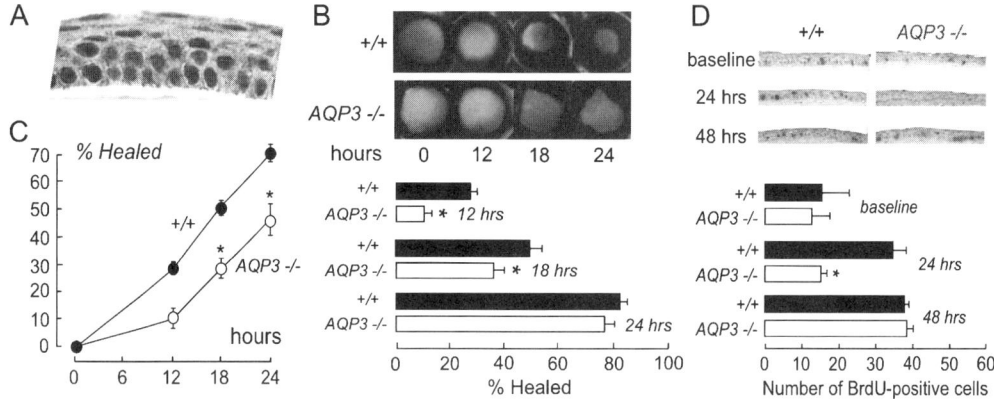

Fig. 6. Aquaporin 3- (AQP3-) dependent corneal epithelial cell proliferation. (**A**) Antibody staining of AQP3 protein expression in plasma membrane in mouse corneal epithelium. (**B**) AQP3-dependent corneal re-epithelialization following wounding. Top: epithelial defects (diameter 2.3 mm) were created in central corneas and resurfacing was followed by fluorescein staining. Bottom: percent remaining epithelial defect (± standard error, $^*p < 0.01$). (**C**) Corneal epithelial cell migration after wound healing in organ culture. Eyes were enucleated after corneal scraping and incubated in serum-containing medium. (**D**) AQP3-dependent corneal epithelial cell proliferation. Top: Mid-peripheral corneas showing bromodeoxyuridine- (BrdU-) immunoreactive cells (dark nuclei) before and at 24 or 48 hours after corneal scraping. Bottom: BrdU-positive cells in uninjured corneas and during healing (adapted from ref. *26*).

protein expression in plasma membranes of corneal epithelial cells. A role for AQP3 in cell proliferation was suggested in studies in skin where delayed recovery of epidermal barrier function was found following removal of the stratum corneum, as well as delayed epidermal wound closure *(27)*.

The possible involvement of AQP3 in corneal epithelial cell proliferation was recently investigated *(26)*. Osmotic water permeability of corneal epithelial cells was slowed by 2.4-fold in AQP3-null mice and ^{14}C-glycerol uptake was slowed by more than tenfold, confirming functional AQP3 expression. Following creation of a 2.3-mm diameter corneal wound (sparing limbal cells), re-epithelialization, which was quantified as defect area by fluorescein pooling, was significantly slowed in AQP3-null mice at 12 and 18 hours (Fig. 6B). Evidence was found for distinct defects in corneal epithelial cell migration and proliferation. Slowed migration of mouse corneal epithelia lacking AQP3 was found in organ culture of enucleated globes following epithelial scraping, in which initial corneal resurfacing results from epithelial cell migration (as shown by bromodeoxyuridine (BrdU) analysis and 5-fluorouracil insensitivity; Fig. 6C), and in primary corneal epithelial cultures by scratch-wound assay. A separate defect in epithelial cell proliferation was shown by delayed restoration of full-thickness corneas over days following scraping, and reduction in the density of proliferating BrdU-positive cells in AQP3-deficient mice, as well as reduced proliferation in primary cultures of corneal epithelial cells from AQP3-null mice. BrdU incorporation studies of proliferation were done in nonwounded mice, and at different times after wounding. Figure 6D (top) shows representative images of mid-peripheral cornea stained for BrdU. No significant difference in BrdU staining was observed among the genotypes in the absence of wounding. As summarized in Fig. 6D (bottom), at 24 hours AQP3-null corneas showed greatly diminished basal-cell proliferation compared to wild-type corneas.

Evidence for involvement of AQP3 in corneal epithelial cell proliferation was also seen in mouse primary cultures in which corneal-limbal epithelial sheets from wild-type and AQP3-null mice were isolated by dispase II digestion, and cells were cultured on plastic. Cells from both genotypes grew similarly as monolayers of mostly large squamous epithelial cells that stratified at areas of confluence. Cell proliferation, measured after five days in culture, was reduced approximately fourfold in cultures from AQP3-null versus wild-type cornea. These findings provided evidence for AQP3-dependent corneal epithelial cell proliferation. Based on recent work on AQP3-dependent epidermal cell proliferation, the AQP3-dependent corneal epithelial cell proliferation may involve AQP3-facilitated glycerol transport and consequent metabolic alterations involving MAP kinase signaling and cellular energy balance. However, these mechanisms identified in skin cells will require direct testing in corneal epithelial cell cultures.

SUMMARY AND PERSPECTIVE

Multiple AQPs are strongly expressed in ocular tissues, where they appear to serve a variety of functions related to epithelial fluid secretion, regulation of tissue water content, cell migration and proliferation, and neural signal transduction. Small-molecule modulators of AQP function or expression might thus be exploited clinically. At the ocular surface, AQP3 or AQP5 upregulation could accelerate wound healing and reduce corneal edema. Corneal endothelial AQP1-inducers might also reduce corneal edema and associated opacity. Induction of lens AQPs might slow cataract-related opacification. Inhibition of AQP1/AQP4 represents a possible strategy for reducing intraocular pressure associated with glaucoma. In the retina, AQP4 inhibitors might offer neuroprotection following retinal ischemia. These possibilities will require experimental verification in large animal model experiments when nontoxic AQP-selective modulators become available. Aquaporin modulators will also be useful in confirming conclusions from phenotype analysis of knock-out mice.

ACKNOWLEDGMENTS

Supported by grants EY13574, HL73856, DK72517, DK35124, HL59198, and EB00415 from the National Institutes of Health, and drug discovery and research development program grants from the Cystic Fibrosis Foundation.

REFERENCES

1. Sui H, Han BG, Lee JK, Walian P, Jap BK. Structural basis of water-specific transport through the AQP1 water channel. Nature 2001;414:872–878.
2. Fu D, Libson A, Miercke LJ, Weitzman C, Nollert P, Krucinski J Stroud RM. Structure of a glycerol-conducting channel and the basis for its selectivity. Science 2000;290:481–486.
3. Harries WE, Akhavan D, Miercke LJ, Khademi S, Stroud RM. The channel architecture of aquaporin 0 at a 2.2-Å resolution. Proc Natl Acad Sci USA 2004;101:14045–14050.
4. Fujiyoshi Y, Mitsuoka K, de Groot BL, Philippsen A, Grubmuller H, Agre P, Engel A Structure and function of water channels. Curr Opin Struct Biol 2002;12:509–515.
5. Stroud RM, Nollert P, Miercke L. The glycerol facilitator GlpF its aquaporin family of channels, and their selectivity. Adv Protein Chem 2003;63:291–316.
6. Verbavatz JM, Brown D, Sabolic I, Valenti G, Ausiello DA, Van Hoek AN, Ma T, Verkman AS. Tetrameric assembly of CHIP28 water channels in liposomes and cell membranes: a freeze-fracture study. J Cell Biol 1991;23:605–618.
7. Tajkhorshid E, Nollert P, Jensen MO, Miercke LJ, O'Connell J, Stroud RM, Schulten K. Control of the selectivity of the aquaporin water channel family by global orientational tuning. Science 2002;296:525–530.
8. Castle NA. Aquaporins as targets for drug discovery. Drug Discov Today 2005;10:485–493.
9. Verkman AS. Applications of aquaporin inhibitors. Drug News Perspect 2001;14:412–320.
10. Francis P, Berry V, Bhattacharya S, Moore A. Congenital progressive polymorphic cataract caused by a mutation in the major intrinsic protein of the lens, MIP (AQP0). Br J Ophthalmol 2000;84:1376–1379.
11. Verkman AS. Novel roles of aquaporins revealed by phenotype analysis of knockout mice. Rev Physiol Biochem Pharmacol 2005;155:31–55.
12. Verkman AS. Roles of aquaporins in kidney revealed by transgenic mice. Semin Nephrol 2006;26:200–208.
13. Ma T, Song Y, Gillespie A, Carlson EJ, Epstein CJ, Verkman AS. Defective secretion of saliva in transgenic mice lacking aquaporin-5 water channels. J Biol Chem 1999;274:20071–20074.
14. Zhang D, Vetrivel L, Verkman AS. Aquaporin deletion in mice reduces intraocular pressure and aqueous fluid production. J Gen Physiol 2002;119:561–569.
15. Verkman AS, Binder DK, Bloch O, Auguste K, Papadopoulos MC. Three distinct roles of aquaporin-4 in brain function revealed by knockout mice. Biochim Biophys Acta 2006;1758:1085–1093.
16. Manley GT, Fujimura M, Ma T, Noshita N, Filiz F, Bollen A, Chan P, Verkman AS. Aquaporin-4 deletion in mice reduces brain edema following acute water intoxication and ischemic stroke. Nature Med 2000;6:159–163.
17. Da T, Verkman AS. Aquaporin-4 gene disruption in mice protects against impaired retinal function and cell death after ischemia. Invest Ophthalmol Vis Sci 2004;45:4477–4483.

18. Binder D, Papadopolous MC, Haggie PM, Verkman AS. In vivo measurement of brain extracellular space diffusion by cortical surface photobleaching. J Neurosci 2004;24:8049–8056.
19. Padmawar P, Yao X, Bloch O, Manley GT, Verkman AS. K^+ waves in brain cortex visualized using a long-wavelength K^+-sensing fluorescent indicator. Nat Methods 2005;2:825–827.
20. Binder DK, Yao X, Sick TJ, Verkman AS, Manley GT. Increased seizure duration and slowed potassium kinetics in mice lacking aquaporin-4 water channels. Glia 2006;53:631–636.
21. Li J, Patil RV, Verkman AS. Mildly abnormal retinal function in transgenic mice without Müller cell aquaporin-4 water channels. Invest Ophthalmol Vis Sci 2002;43:573–579.
22. Saadoun M, Papadopoulos MC, Hara-Chikuma M, Verkman AS. Impairment of angiogenesis and cell migration by targeted of aquaporin-1 gene disruption. Nature 2005;434:786–792.
23. Saadoun M, Papadopoulos MC, Watanabe H, Yan D, Manley GT, Verkman AS. Involvement of aquaporin-4 in astroglial cell migration and glial scar formation. J Cell Sci 2005;118: 5691–5698.
24. Hara-Chikuma M, Verkman AS. Aquaporin-1 facilitates epithelial cell migration in kidney proximal tubule. J Am Soc Nephrol 2006;17:39–45.
25. Hu J, Verkman AS. Increased migration and metastatic potential of tumor cells expressing aquaporin water channels. FASEB J 2006;20:1892–1894.
26. Levin MH, Verkman AS. Aquaporin-3-dependent cell migration and proliferation during corneal re-epithelialization. Invest Ophthalmol Vis Sci 2006;47:4365–4372.
27. Hara M, Ma T, Verkman AS. Selectively reduced glycerol in skin of aquaporin-3-deficient mice may account for impaired skin hydration, elasticity and barrier recovery. J Biol Chem 2002;277:46616–46621.
28. Hara-Chikuma M, Verkman AS. Aquaporin-3 facilitates epidermal cell migration and proliferation during wound healing. J Mol Med in press.
29. Hara-Chikuma M, Sohara E, Rai T, Ikawa M, Okabe M, Sasaki S, Uchida S, Verkman AS. Progressive adipocyte hypertrophy in aquaporin-7 deficient mice: adipocyte glycerol permeability as a novel regulator of fat accumulation. J Biol Chem 2005;280:15493–15496.
30. Nielsen S, Smith BI, Christensen EI, Agre PA. Distribution of the aquaporin CHIP in secretory and resorptive epithelia and capillary endothelia. Proc Natl Acad Sci USA 1993;90:7275–7279.
31. Hamann S, Zeuthen T, La Cour M, Nagelhus EA, Ottersen OP, Agre P, Nielsen S. Aquaporins in complex tissues: distribution of aquaporins 1–5 in human and rat eye. Am J Physiol 1998;274: C1332–1345.
32. Frigeri A, Gropper M, Turck CW, Verkman AS. Immunolocalization of the mercurial-insensitive water channel and glycerol intrinsic protein in epithelial cell plasma membranes. Proc Natl Acad Sci USA 1995;92:4328–4331.
33. Hasegawa H, Lian SC, Finkbeiner WB, Verkman AS. Extrarenal tissue distribution of CHIP28 water channels by in situ hybridization and antibody staining. Am J Physiol 1994;266:C893–C903.
34. Patil RV, Saito I, Yang X, Wax MB. Expression of aquaporins in the rat ocular tissue. Exp Eye Res 1997;64:203–209.
35. Stamer WD, Peppel K, O'Donnell ME, Roberts BC, Wu F, Epstein DS. Expression of aquaporin-1 in human trabecular meshwork cells: role in resting cell volume. Invest Ophthalmol Vis Sci 2001;42:1803–1811.
36. Ruiz-Ederra J, Verkman AS. Accelerated cataract formation and reduced lens epithelial water permeability in aquaporin-1deficient mice. Invest Ophthalmol Vis Sci 2006;47:3960–3967.
37. Varadaraj K, Kumari S, Shiels A, Mathias RT. Regulation of aquaporin water permeability in the lens. Invest Ophthalmol Vis Sci 2005;46:1393–1402.
38. Connors NC, Kofuji P. Potassium channel Kir4.1 macromolecular complex in retinal glial cells. Glia 2006;53:124–131.

39. Nagelhus EA, Horio A, Inanobe Al, Fujita FM, Haug S, Nielson S, Kurachi Y, Ottersen OP. Immungold evidence suggests that coupling of K^+ siphoning and water transport in rat retinal Müller cells is mediated by a coenrichment of Kir4.1 and AQP4 in specific membrane domains. Glia 1999;26:47–54.
40. Li J, Verkman AS. Impaired hearing in mice lacking aquaporin-4 water channels. J Biol Chem 2001;276:31233–31237.
41. Freegard TJ. The physical basis of transparency of the normal cornea. Eye 1997;11:465–471.
42. Maurice DM. The structure and transparency of the cornea. J Physiol 1957;136:263–286.
43. Thiagarajah JR, Verkman AS. Aquaporin deletion in mice reduces corneal water permeability and delays restoration of transparency after swelling. J Biol Chem 2002;277:19139–19144.
44. Kuang K, Yiming M, Wen Q, Li Y, Ma L, Iserovich P, Verkman AS, Fischbarg J. Fluid transport across cultured layers of corneal endothelium from aquaporin-1 null mice. Exp Eye Res 2004;78:791–798.
45. Fischbarg J, Diecke FP, Iserovich P, Rubashkin A. The role of the tight junction in paracellular fluid transport across corneal endothelium. Electro-osmosis as a driving force. J Membr Biol 2005;210:117–130.
46. Candia OA. Electrolyte and fluid transport across corneal, conjunctival, and lens epithelia. Exp Eye Res 2004;78:527–535.
47. Levin MH, Verkman AS. Aquaporin-dependent water permeation at the mouse ocular surface: in vivo microfluorimetric measurements in cornea and conjunctiva. Invest Ophthalmol Vis Sci 2004;45:4423–4432.

2
Roles of Corneal Epithelial Ion Transport Mechanisms in Mediating Responses to Cytokines and Osmotic Stress

Peter S. Reinach, José E. Capó-Aponte, Stefan Mergler, and Kathryn S. Pokorny

CONTENTS

OVERVIEW

INVOLVEMENT OF ION TRANSPORT MECHANISMS IN MEDIATED RECEPTOR
  CONTROL OF CORNEAL EPITHELIAL CELL RENEWAL
  AND VOLUME REGULATION

ROLES OF OSMOLYTE TRANSPORTERS IN MEDIATING CONTROL OF BARRIER,
  DETURGESCENCE, AND EPITHELIAL CELL RENEWAL

ROLES OF ION CHANNELS IN MEDIATING CONTROL OF BARRIER,
  DETURGESCENCE, AND EPITHELIAL CELL RENEWAL

ACKNOWLEDGEMENTS

REFERENCES

OVERVIEW

Normal vision depends, in part, on the combined refractive powers of the cornea and crystalline lens to permit adequate focusing of light onto the retina. Such refractive function requires that the cornea remain transparent, a requirement that is met provided that corneal hydration, i.e., deturgescence, is maintained within specific physiological limits. Maintenance of corneal deturgescence is reliant upon coupled ion and fluid transport activities within the epithelial and endothelial layers. Net ion transport activity offsets the natural tendency of the corneal stroma to imbibe fluid from the anterior chamber, thus keeping the cornea transparent (1–5). Although most of the ion transport activity involved in maintaining corneal deturgescence is contingent upon ion transport processes localized in the corneal endothelial layer, corneal epithelial ion transport activity plays a fine-tuning role in maintaining corneal deturgescence during exposure to environmental challenges (6) (Fig. 1). Only under maximally stimulated conditions is the epithelial-side fluid transport rate able to increase sufficiently, i.e., to approximately 25% of the endothelial-side fluid transport rate (7). This realization has prompted a host of studies concentrated on characterizing receptor-mediated regulation of corneal epithelial active ion transport.

From: *Ophthalmology Research: Ocular Transporters in Ophthalmic Diseases and Drug Delivery*
Edited by: J. Tombran-Tink and C. J. Barnstable © Humana Press, Totowa, NJ

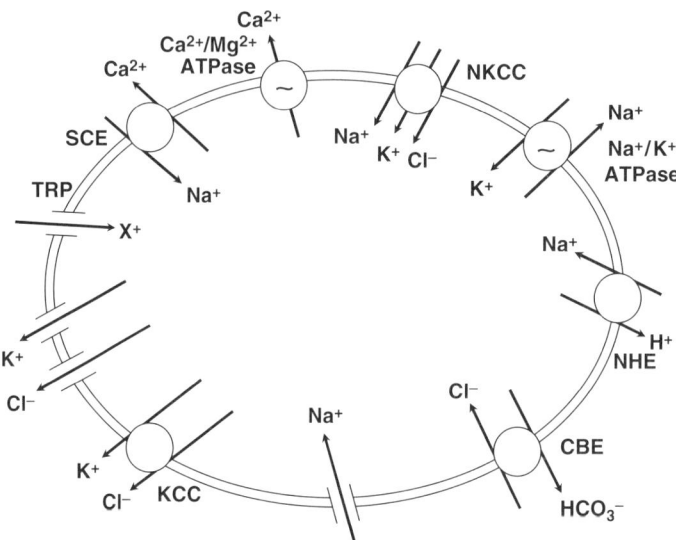

Fig. 1. Plasma membrane inorganic osmolyte transport mechanisms. Diagram depicting the ion pumps, coupled transporters, and channels mediating control of corneal epithelial cell functions. Arrows indicate net fluxes of ions for each specific pathway.

Results of these investigations are relevant for identifying clinical strategies for improving corneal renewal and transparency, and for optimizing the cornea's refractive properties.

The cornea comprises three very different tissue types. In humans, there is an approximately 50 μm-thick outer epithelial layer that faces the tears; an approximately 10 μm-thick inner endothelial layer that is exposed to the aqueous humor contained in the anterior chamber *(8)*; and an intermediate 450 μm-thick stromal layer, comprising orthogonally arranged collagen bundles within a glycosaminoglycan-containing ground substance. Distributed within this matrix are keratocytes, which secrete the macromolecules that form the ground substance *(9)*. Not only the organization, but also the embryonic origins of these tissues, differ greatly from one another. The epithelium is derived from ectoderm and is 5 to 7 cell-layers thick. The uppermost epithelial layers, which are adjacent to the tears, form tight junctions of relatively high electrical resistance, thus providing the crucial barrier function of the cornea. The endothelium and keratocytes are derived from neural crest cells and mesenchymal tissue, respectively. The endothelium comprises a single-cell layer, with tight junctions of low electrical resistance—a property reflected by the endothelium's much higher solute and greater ionic permeabilities than exhibited by the epithelial layer *(10)*. Epithelial layer ion transporters elicit coupled net salt and fluid movement outwards from the stroma into the tears, whereas endothelial transporters mediate salt and fluid translocation inwards to the anterior chamber. The endothelial layer fluid-transport rate exceeds that of the epithelial side, as evidenced by the fact that net solute transport into the anterior chamber is greater than that into the tears. Therefore, under normal conditions, the contribution of the corneal epithelium to maintenance of transparency and deturgescence is much less than that of the endothelium. Nevertheless, the barrier function of the epithelium is critical for protecting the cornea and other intraocular tissues from the damage imposed by exposure to environmental insults.

This review describes the involvement of ion transporter and channel activity in mediating cytokine receptor control of responses required for corneal epithelial function and renewal. Accordingly, a description is first provided of the mechanisms mediating in this tissue osmolyte transport and cell-volume regulation. With this background, it is then possible to appreciate how changes in ion transport and channel activity contribute to cytokine-mediated control of corneal epithelial cell renewal and transparency. Finally, we discuss new evidence that transient receptor potential (TRP) protein superfamily expression and function activate cytokine receptor-linked signaling events inducing responses needed for the maintenance of these corneal epithelial functions.

INVOLVEMENT OF ION TRANSPORT MECHANISMS IN MEDIATED RECEPTOR CONTROL OF CORNEAL EPITHELIAL CELL RENEWAL AND VOLUME REGULATION

Identification of receptors that mediate control of corneal epithelial ion transporter function has been partially elucidated. There are adrenergic *(11)*, serotonergic *(12)*, and cholinergic *(13)* receptors that contribute to the control of ion transport activity. Regulation of ionic transport in frog corneal epithelia has been demonstrated to be under adrenergic control *(14–16)*. Serotonergic control is shown by the finding that neural serotonin stimulates chloride transport in rabbit corneal epithelia *(12)*. Endogenous cholinergic agonists are able to stimulate active ionic transport in rabbit corneal epithelia. Similarly, cholinergic receptors mediate regulation of epithelial cell proliferation *(13, 17)*. Vital for perpetuation of corneal epithelial function are a host of cytokines that are released into the tears and anterior chamber from the epithelium and accessory tissues of the anterior ocular surface. Cytokine expression is critical for inducing control of proliferation and cell migration through stimulation of cognate receptors *(18)*. Accordingly, control of cell proliferation and migration by cytokines and neuronal agonists is required for synchronization of the corneal epithelial renewal process and preservation of corneal epithelial functions. The uppermost epithelial layers are continuously being lost (into the tears during their terminal differentiation) and replaced, the latter process being necessary for maintenance of corneal transparency. Active ion transporters and channels are components of a myriad of cell signaling pathways that mediate cytokine receptor control of this renewal process *(19, 20)*. Understanding of these events, coupled with the requirement for ion transport activity in maintaining corneal epithelial transparency, has prompted further studies into the mechanisms by which epithelial receptors elicit control of corneal epithelial renewal through stimulation of ion transport activity.

Synchronized epithelial renewal is essential for the maintenance of the cornea's multilayer integrity and function. Renewal of the upper epithelial layer preserves tight junctional resistance. In addition, epithelial renewal sustains receptor-mediated control of net ion transport, which is required for receptor modulation of epithelial membrane and tight junctional permeability *(21)*. Should epithelial layer renewal become compromised, net ionic fluxes will decline as result of decreased ionic pump function, loss of membrane permselectivity, and decreased tight junctional electrical resistance *(22)*.

Under physiological conditions, cellular losses due to terminal differentiation are countered by replacement with younger cells located in the inner cell layers. This renewal process ensures continuation of normal epithelial ion transport activity. Consequently, epithelial receptor-mediated control of renewal is required for the maintenance of corneal transparency and barrier function.

Even though the contribution of epithelial ion transport activity to deturgescence is much less than that of its endothelial counterpart, the epithelial component is required for the preservation of the integrity of the epithelial layers during exposure to anisosmotic stresses. Anisosmotic insults to the cornea frequently occur during activities of daily life, e.g., swimming or bathing, as well as from contact lens wear and ocular diseases such as dry eye syndrome (DES), as result of which, afflicted individuals experience chronic exposure to hypertonic tears. The physiological disturbances induced by such anisosmotic stresses are, very likely, counteracted by regulatory volume-response activation, such activity having been described in cultured corneal epithelial cells. Two different types of regulatory volume activations have been described for such an in vitro system. Exposure to a hypotonic challenge induces regulatory volume decrease (RVD) behavior, which acts to restore isotonic cell volume *(23, 24)*. Such cell volume restoration is partially due to increases in K^+ and Cl^- membrane permeability, which results in KCl egress coupled to fluid loss. In human corneal epithelial cells, this regulatory response brings about complete recovery of isotonic cell volume within minutes of onset of the hypotonic stress. In contrast, cell exposure to hypertonic challenges, which simulate increased tear-film osmolarity in DES *(25)*, induces another type of regulatory volume response, referred to as the regulatory volume increase (RVI) *(24, 26)*. Regulatory volume increase behavior restores the cell's isotonic volume through stimulation of ion and solute influx transport mechanisms, which mediate a rise in intracellular osmolyte content coupled with net fluid influx. Even though the RVI response is somewhat slower than the RVD response, corneal epithelial cells are able to achieve complete restoration of their isotonic volume following anisosmotic stress. In corneal epithelial cells, there is some evidence that specific receptors sense changes in osmolarity and activate a unique array of signaling pathways. These signaling pathways have some features in common to those linked to the cytokine receptors that mediate control of the corneal epithelial responses—proliferation and migration—required for renewal *(27)*. This commonality—coupled with results from recent studies of other tissues, showing that cell volume regulation is requisite for proliferation and migration—suggests the importance of ion transport regulation in the maintenance of corneal epithelial function *(28)*.

There is emerging evidence that receptor-mediated control of ion transport activity is requisite for corneal epithelial renewal *(27, 29)*. This requirement is self-evident, since parent cell volume must increase to accommodate rises in genomic and cytoplasmic content prior to karyokinesis and cytokinesis. Similarly, changes in cell volume are requisite for cell migration, as this process involves repeated, coordinated leading-edge cytoplasmic volume extension, with retraction at the opposite pole. Because changes in ion transport activity underlie cell volume regulation, cytokine-induced control of renewal is, thus, dependent upon modulation of ion transport mechanisms. For this control to occur, corneal epithelial-induced cytokine expression of the corneal epithelium and accessory ocular tissues must be synchronized and manifested at appropriate times to alter cell volume, and to allow cell-cycle progression and migration to occur unperturbed. Numerous cytokines are involved in regulating these processes *(18)*, some of which are needed to elicit control of proliferation and migration, while others affect rates of differentiation. Such controls occur through a host of cell-signaling pathways, each of which is specific for one of the cognate receptors. Any particular cytokine can elicit control of a variety of responses by activating different pathways within a cell-signaling network. The mitogen-activated protein kinase (MAPK) cascade is a superfamily of cytokines that mediates this type of exquisite control. Mitogen-activated protein kinase cascades have three different parallel pathways: the extracellular signal-regulated kinase (ERK); the p38; and the c-jun *N*-terminal/stress-activated protein kinase (JNK/SAPK) pathways. In corneal epithelial cells, different stimuli selectively activate specific receptors linked to one or more of these three different pathways. Cytokines that mediate increases in proliferation induce a response through stimulation of the ERK (i.e., p44/42) pathway, whereas those involved in increasing cell migration act through the p38 MAPK pathway *(30–34)*. Stressors, such as anisosmotic challenges or apoptosis-inducing agents, e.g., ultraviolet light, activate the JNK/SAPK pathway *(29, 35)*. Furthermore, the ability of a cytokine to elicit a particular response can be modulated by crosstalk between different branches of a cell-signaling network that is linked to different responses. Such

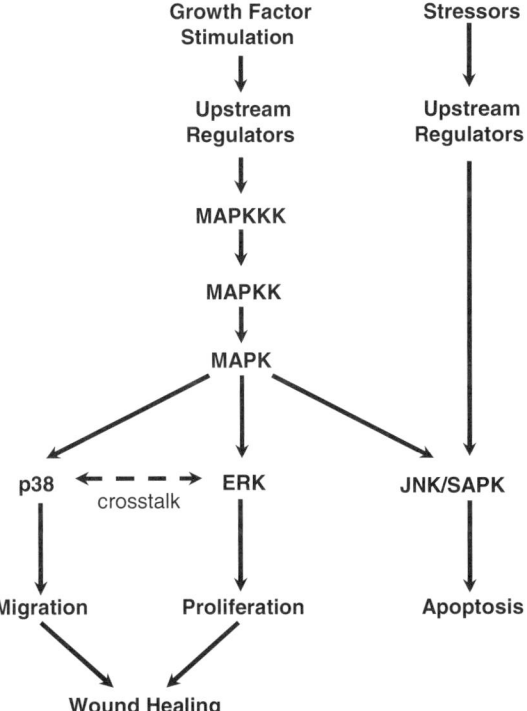

Fig. 2. Mitogen-activated protein kinase (MAPK) pathways in corneal epithelial cells. Diagram depicting the three MAPK pathways (p38, extracellular signal-regulated kinase [ERK], and c-jun *N*-terminal/stress-activated protein kinase [JNK/SAPK]) and their respective cellular function in corneal epithelial cells. The broken arrow indicates phosphatase-mediated crosstalk control of growth factors between p38 and ERK MAPK signaling. This control modulates the magnitude of growth factor-induced increases in corneal epithelial cell migration and proliferation, both of which are required for wound healing or apoptosis.

crosstalk depends on MAPK-induced phosphatase activation by both the ERK and p38 MAPK pathways *(31, 34)* (Fig. 2).

The direct effects of mitogens on epithelial ion transport rates and downstream signaling have also been determined for corneal epithelial cells *(19, 20)*. Stimulation by numerous mitogen-containing serums, including epidermal growth factor (EGF), is dependent upon increases in the activity of specific ion transporters and channels. The involvement of ion transporters and channels in mediating responses to mitogens indicates that their activation is an important component of the second-messenger cascade linking receptor stimulation to increases in proliferation. Therefore, receptor-control of ion transport activity is of critical importance in the maintenance of corneal epithelial renewal, transparency, and refractive properties.

ROLES OF OSMOLYTE TRANSPORTERS IN MEDIATING CONTROL OF BARRIER, DETURGESCENCE, AND EPITHELIAL CELL RENEWAL

Na^+/K^+-Adenosine Triphosphatase (ATPase) and Na:K:2Cl Cotransporter

The specific ionic fluxes that account for active ion transport in corneal epithelia vary among species. In amphibians, this process is essentially a result of net Cl^- transport from the stroma into the tears *(36)*. In mammals, this mechanism is accounted for by approximately equal contributions by inwardly directed net Na^+ transport towards the stroma and outwardly directed active Cl^- transport into the tears *(1)*. Active Cl^- transport is described as a secondary active process, because Cl^- uptake from the stroma into the epithelial layer is dependent upon basolateral membrane-coordinated Na:K:2Cl cotransporter (NKCC) and Na^+/K^+-ATPase (i.e., Na^+/K^+ pump) activity. Pump coupling to NKCC function is a result of the Na^+/K^+ pump establishing a bath-to-cell inwardly directed Na^+ gradient, which provides the chemical driving force for NKCC-mediated uphill intracellular Cl^- accumulation. Na^+/K^+-ATPase activity has a variable load-dependent Na^+/K^+ stoichiometry *(37)*. Such function results in intracellular K^+ accumulation above its predicted electrochemical value, leading to net K^+ transport from cell to bath across the basolateral membrane. Na^+/K^+-ATPase function in concert with K^+ egress across the basolateral membrane is described in terms of the pump-leak model for ion transport *(38)*. Net K^+ transport, in turn, establishes the repelling electrical driving force for Cl^- electro-diffusion across the epithelial tear-side-facing membrane. In amphibians, this membrane is essentially Cl^- permselective, and electroneutrality is maintained as result of passive Na^+ efflux across the paracellular route. In mammals, this outfacing membrane is also Na^+ permeable, resulting in inwardly directed net Na^+ transport towards the stroma—driven by the Na^+/K^+ pump opposed by net Cl^- efflux into the tears.

A number of different corneal sequelae, all of which lead to impaired corneal epithelial renewal, are caused by dysfunctional control of ion transport function, perhaps as a result of disruption of cytokine-mediated control of ion transport function. The probable underlying causes include changes in cytokine expression profiles, changes in levels of cognate receptor expression, loss of receptor coupling to signal control of ion transport function, alterations in ion transporters, and various channel pathophysiologies. Examples of corneal disease resulting in losses in Na^+/K^+-ATPase expression are aphakic bullous keratopathy (ABK) and pseudophakic bullous keratopathy

(PBK) following cataract surgery, and Fuchs' dystrophy. These conditions are characterized by persistent corneal edema, epithelial blisters (bullae), and loss of transparency *(39)*. Such Na$^+$/K$^+$-ATPase losses could also directly or indirectly affect the epithelial renewal process. Consideration of these factors could be of great value in the search for novel drugs to ameliorate the symptomatic problems associated with the aforementioned pathologies.

Further implicating Na$^+$/K$^+$-ATPase as important in maintaining cell volume homeostasis is the disrupted pump function and corneal edema that occur during the inflammatory response. Swelling is typically observed during acute corneal injuries, bacterial infection, chemical burns, and extended contact lens wear-induced hypoxia. Under these in vivo conditions, stimulation of epithelial 12(R)-hydroxy-5,8,10,14-eicosatetraenoic acid [12(R)-HETE] synthesis occurs *(40–44)*. It is recognized that 12(R)-HETE is a potent inflammatory eicosanoid (one of the major arachidonate metabolites of the corneal epithelial cytochrome P450 system) as well as an endogenous inhibitor of Na$^+$/K$^+$-ATPase, in which role it leads to perturbed hydration control and edema during the inflammatory process. Increased 12(R)-HETE may act in the pathogenesis of diseases such as PBK, ABK and Fuchs' dystrophy, since these conditions are characterized by decreased epithelial Na$^+$/K$^+$-pump expression. On the other hand, the enantiomer of 12(R)-HETE, 12(S)-HETE, mediates EGF-induced increases in corneal epithelial proliferation *(45)*. However, the effect of 12(S)-HETE on Na$^+$/K$^+$-ATPase has not been described for the corneal epithelium. Hence, another potential approach to treatment of corneal inflammation may be to identify strategies that selectively inhibit 12(R)-HETE synthesis.

In the corneal epithelium, the functions of Na$^+$/K$^+$-ATPase and NKCC are modulated by exposure to a hypertonic challenge or to a mitogen, such as EGF. Responses to these challenges mediate increases in net osmolyte uptake through NKCC stimulation, resulting in cell swelling. EGF has a similar effect on NKCC activity, as shown by the fact that bumetanide inhibition of NKCC suppresses both RVI and the mitogenic response to EGF. An increase in cell volume is thus a component of the signaling pathways that mediate an increase in cell proliferation *(19, 46)*. More evidence in support of the notion that an increase in cell volume is a component of EGF receptor (EGFR)-induced signaling is provided by the fact that selective inhibition of the ERK branch of the MAPK superfamily suppresses both NKCC1 activity and proliferation *(19, 34)*. A further indication that changes in ion transport activity affect the mitogen-induced signaling process is that chemical inhibition of K$^+$ channel activity suppresses both RVD- and growth-factor-mediated increases in proliferation *(20, 47)*. This correspondence indicates that cell shrinkage is an additional component of the volume change involved in EGF signaling. Taken together, these results suggest that the mitogenic response to EGF is dependent on a transient cell-volume change rather than a steady-state change.

Cytokine receptor-induced interaction with the p38 MAPK pathway for mediating cell migratory control has been described for the corneal epithelium. The p38 MAPK branch is coupled to various cytokine receptors, including hepatocyte growth factor (HGF), keratinocyte growth factor (KGF), transforming growth factor-β (TGFβ), and EGF *(30, 31, 34)*. Selective stimulation of each of these cognate receptors results in p38 MAPK stimulation. Because, in some tissues, cell migration is dependent upon alternate activation of ion transporters that mediate cell volume regulatory responses (RVI and RVD) *(48, 49)*, it was of interest to determine whether the RVI response resulting from exposure to a hypertonic challenge induces NKCC activation through interaction with the p38 MAPK pathway *(27, 29)*. Results of such studies indicate that there is protein–protein interaction between these macromolecules, thus suggesting that NKCC is requisite for phosphorylation by p38 MAPK triggering of RVI and for recovery of barrier function. Other evidence suggesting dependence of NKCC activation on p38 MAPK stimulation is that EGF-induced stimulation of NKCC activity is inhibited by exposure of cells to EGF in the presence of the selective p38 MAPK inhibitor, SB203580 *(27)*. Therefore, cytokine-mediated control of the responses involved in corneal epithelial renewal occurs through stimulation of the ERK and p38 MAPK pathways, followed by NKCC activation and cell-volume modulation.

Tear hyperosmolarity is frequently detected in individuals afflicted with DES *(25)*. This stress induces epithelial cell shrinkage and losses in the protective barrier (tight-junction) function against noxious agents *(27)*. Compromising the integrity of the corneal epithelium increases the rate of corneal epithelial erosion *(50)* and infections *(51)*, as well as decreases in corneal sensitivity *(52)* and transparency *(53)*. Hypertonicity-induced cell shrinkage can be reversed in corneal epithelial cells through activation of the RVI response *(24, 46)*. NKCC plays a major role during the RVI process, and RVI activation stimulates intracellular chloride accumulation. Although there is protein expression of both NKCC1 (secretory) and NKCC2 (absorptive) isoforms in bovine corneal epithelium, NKCC1 expression predominates in this tissue *(54)*.

The importance of regulating NKCC expression for permitting adaptation to a chronic hypertonic challenge has been demonstrated in comparative studies of human and rabbit corneal epithelial cells. Rabbit cells adapt better to chronic hypertonic stress than human cells do, as shown by the fact that the proliferative capacity of the rabbit cells was far less inhibited during such stress. The greater adaptive capacity of the rabbit cells is attributable to their unique ability to upregulate NKCC gene and protein expression during hypertonic stress, and to sustain functional NKCC and proliferative activity *(26, 46)*. Another indication of the importance of NKCC protein expression upregulation to rabbit corneal epithelial cell survival during hypertonic challenge is that these cells are better able to restore their translayer electrical resistance (i.e., barrier function) *(27)* than their human counterparts. Results of regulatory volume studies on corneal epithelial cells, coupled with results of studies showing that EGF stimulates NKCC functional activity through MAPK activation, further support the view that volume modulation is a component of the signaling pathway that mediates EGFR-induced mitogenesis *(19)*. Taken together, corneal epithelial cell proliferative capacity is associated with the ability of these cells to modulate NKCC activity, and NKCC gene and protein expression.

Recent studies suggest that survival of hypertonicity-stressed corneal epithelial cells depends on the their capacity to activate the p38 MAPK pathway, and on the ability of this pathway to stimulate NKCC1 activity through protein–protein interactions *(29)*. The level of NKCC1 activation has an effect on the extent of cell-volume recovery and, in turn, epithelial survival capacity. The interaction between p38 MAPK and NKCC is related to p38 MAPK phosphorylation status; hypertonicity-induced NKCC activation is, in turn, dependent on this association. Rapid and substantial p38 MAPK activation leads

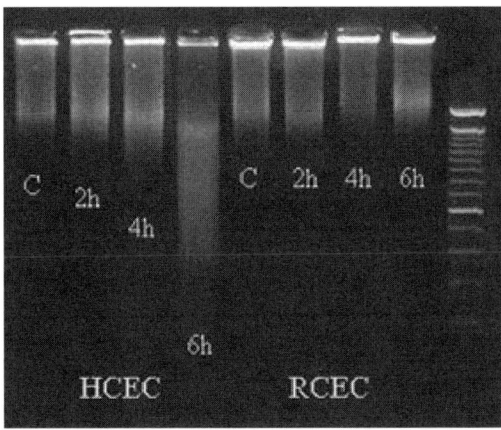

Fig. 3. Comparison of hypertonicity-induced DNA fragmentation in human and rabbit corneal epithelial cells. Representative DNA fragmentation pattern showing DNA integrity analyzed by high-resolution 2% agarose gel electrophoresis following exposure of human (HCEC) and rabbit (RCEC) corneal epithelial cells to 600 mOsm medium for up to 6 h.

to increases in NKCC1 activity, which determines the kinetics and extent of RVI-mediated recovery of the cell to isotonic volume conditions. These events determine whether corneal epithelial cells will be able, under conditions of hyperosmolar adversity, to recover their p44/42 MAPK activity, proliferate, adapt to stress, and avoid SAPK/JNK-induced apoptosis. The greater resilience of rabbit versus human corneal epithelial cells to withstand hypertonic challenges indicates that maintenance of cell-volume regulatory capacity correlates with resistance to apoptotic cell death during hypertonic stress, since only cells that exhibit a loss in cell volume degrade their DNA *(55)* (Fig. 3). The earlier onset of apoptosis in human versus rabbit corneal epithelia is consistent with the lower capacity of the human cells to undergo RVI and to proliferate in a hypertonic medium *(29, 46)*. Characterization of stress-activated signaling pathways that induce regulatory volume responses or apoptosis is complex for a variety of reasons. For one thing, there may be crosstalk between the various pathways. Coupled with this notion is the idea that the same pathway may mediate variable responses in different species.

The inability of human corneal epithelia to adapt to chronic hypertonicity suggests that exposure to hypertonic tears in DES patients leads to decreases in NKCC activity. This would result in both disruption of barrier function and impairment of wound healing in these patients *(56)*.

K^+/Cl^- Cotransporter

Corneal epithelial cells are capable of adapting to hypotonic stress. They do so by restoring their volumes to isotonic levels through activation of specific ion transport mechanisms that mediate net ion efflux. The potassium-chloride cotransporter (KCC)—in parallel with K^+ and Cl^- channel-coupled conductance—is a mechanism contributing to ion efflux in human corneal epithelial cells *(24)*. Although four different isoforms—KCC1 to 4—have been identified in mammalian cells, Capó-Aponte et al. found that human, rabbit, and bovine corneal epithelial cells express only KCC1, 3, and 4 *(57)*.

Comparative studies of human and rabbit corneal epithelial cells show that human cells are capable of more rapid and complete RVD than are rabbit cells *(57)*. This finding is in accord with the fact that isotonic KCC1 membrane expression is twofold higher in human than in rabbit cells, and that the amount of membrane KCC1 increases during hypotonic stress in human cells. In contrast, the expression of KCC3 is higher in rabbit than in human corneal epithelial cells, although KCC4 expression levels are similar for both species, and levels of KCC3 and KCC4 are not altered during exposure to hypotonic stress. These results indicate that, as in other cell types, KCC1 is the house-keeping isoform, i.e., involved in maintenance of steady-state ionic homeostasis as well as volume regulation, of corneal epithelial cells

In addition to the well-known role of KCC in ionic and osmotic homeostasis, recent studies demonstrate KCC's involvement in regulating cell growth and proliferation in ovarian and cervical cancer cells *(58, 59)*. Furthermore, in these latter cell types, expression of KCC1, 3, and 4—regulated during the cell cycle—is greater than in their nonmalignant counterparts. KCC activity of these cancer cells could also be stimulated by certain growth factors (particularly insulin-like growth factor-1), which act via specific kinases (P13K/Akt and p44/42 MAPK) to promote gene transcription and synthesis of KCC *(60)*. Similarly, in rabbit and human corneal epithelia, hypotonicity-induced activation of RVD is accompanied by transient increases in p44/42 and p38 MAPK activity *(57)*. Activation of p44/42 and p38 MAPK are key elements for cell proliferation and migration, which in turn are essential for corneal epithelial renewal. Apart from the functional role of KCC in maintaining volume homeostasis and restoring isotonic cell volume during hypotonic stress in corneal epithelial cells, KCC1 is involved in the hypotonicity-induced activation of the p44/42 MAPK pathway. In human corneal epithelial cells, KCC's involvement in p44/42 MAPK activation has been demonstrated by: (i) the higher membrane content of this isoform during hypotonic stress, (ii) inhibition of hypotonicity-induced phosphorylation of p44/42 MAPK by the KCC inhibitor [(dihydroindenyl)oxy] alkanoic acid (DIOA), and (iii) increased phosphorylation of p44/42 MAPK activation during isotonic conditions induced by the KCC activator *N*-ethylmaleimide (NEM). In rabbit corneal epithelia, on the other hand, such pharmacological manipulation has no significant effect, a finding that is in accord with the low level of KCC1 protein expression in this latter cell type.

Thus, KCC—specifically KCC1—appears to be an upstream regulator of hypotonicity-induced p44/42 MAPK activation in human corneal epithelial cells. In contrast, pharmacological inhibition or stimulation of KCC does not affect hypotonicity-induced p38 MAPK phosphorylation, suggesting that this signaling pathway is largely independent of KCC1. The functional roles of KCC3 and KCC4 isoforms in corneal epithelial cells remain to be elucidated.

Na^+/H^+ and Cl^-/HCO_3^- Exchangers

Normal metabolic functions of cells are largely influenced by changes in intracellular pH (pH_i). Corneal epithelial cells are often exposed to hypoxic stress (e.g., by contact-lens wear), which leads to intracellular acidification *(61)*. Early functional studies of rabbit and bovine corneal epithelial cells showed that the Na$^+$/H$^+$ exchanger (NHE) (amiloride-sensitive subtype) is involved in the maintenance of steady-state pH$_i$ and restoration of pH$_i$ after intracellular acidification *(62, 63)*. It appears that the corneal epithelium is under constant acid load, partially counteracted by NHE, as evidenced by the fact that inhibition of this transporter induces pH$_i$ acidification *(64)*. Na$^+$/H$^+$ exchanger protein makes use of the large downhill Na$^+$ gradient generated by Na$^+$/K$^+$-ATPase to extrude excess protons (1:1 stoichiometry), which either leak into or are generated by acid-forming reactions within the intracellular space.

Although there is limited information regarding isoform-specific involvement of NHE in corneal epithelial cells, its gene expression has been described for rabbit corneal epithelia *(65)*. In addition to its involvement in pH$_i$ regulation, NHE—specifically, the amiloride-insensitive subtype NHE-2—contributes to mediating regulatory volume responses induced by either hypertonic or hypotonic challenges *(24, 66, 67)*. Although it is well established that NHE is activated by low pH$_i$ and shrinkage, the signaling mechanism by which this antiporter is activated in corneal epithelial cells is not well understood, particularly the signal transduction pathways mediating steady-state pH$_i$ maintenance and hypertonic stress-induced NHE activation. The signaling pathway activating pH$_i$ recovery has been partially elucidated for acidic loading conditions in bovine corneal epithelia *(66)*. This clarification stemmed from studies of the effects of the cytokine endothelin (ET) on this process. ET is a mitogen in corneal epithelial cells, where it potentiates mitogenic responses to EGF *(68–70)*. In corneal epithelial cells, ET 1 completely suppresses pH$_i$ recovery by stimulating ET$_A$ receptor-induced signaling, an inhibitory effect that occurs as a consequence of protein kinase C (PKC) and Ca^{2+}-dependent calmodulin (CaM) II kinase-mediated stimulation of protein phosphatase (PP) activity. This view is supported by the fact that pharmacological inhibition of PP-1, PP-2A, or PP-2B activity prevents ET 1-induced pH$_i$ recovery suppression. Increases in the activity of any one of these PPs could lead to NHE dephosphorylation and inactivation. On the other hand, ET$_A$ receptor stimulation has no effect on pH$_i$ under steady-state conditions. Although NHE activation by several other growth factors has been described for other tissues, their involvement in corneal epithelia as mediators of receptor control of proliferation through pH$_i$ transients has not been described.

Na$^+$/H$^+$ exchangers produce a net influx of ions by coupling with the Cl$^-$/HCO$_3^-$ anion exchanger (CBE). In the corneal endothelium, sodium-dependent and sodium-independent CBEs have been described *(71, 72)*. Both export HCO$_3^-$ from the cell to mediate pH$_i$ recovery and maintain steady-state pH$_i$. In human corneal epithelial cells, there is evidence of CBE involvement in mediating the RVI response *(24)*. In addition to the aforementioned pH$_i$–regulation transporters, the functional involvement of a K$^+$/H$^+$ exchanger was described for rabbit corneal epithelial cells. This transporter was proposed to be involved in pH$_i$ control and to play a role in cell volume regulation *(64)*.

Plasma Membrane and Endoplasmic Reticulum (ER) Ca^{2+} Pumps

As in numerous other cell types, receptor-induced control of net Cl$^-$ transport and proliferation of corneal epithelial cells is mediated through transient changes in intracellular calcium, [Ca^{2+}]$_i$ concentration *(68, 73–75)*. In order for calcium to mediate a signaling function, its intracellular concentration must be regulated within the sub-micromolar range. Corneal epithelial cells accomplish such regulation through a variety of ATP-dependent transporters, localized in the plasma membrane and intracellular calcium stores (ICS). Ca^{2+} Mg^{2+}-ATPase plasma membrane localization was identified in a plasma membrane-enriched subcellular fraction, in which calmodulin stimulated vesicular calcium uptake, which was sensitive to inhibition by trifluoperzine *(76, 77)*.

In human corneal epithelia, four genes—ATP2B1 to ATP2B4—encode plasma membrane Ca^{2+}-ATPase (PMCA) transporter function, and are designated PMCA1 to PMCA4, respectively *(78)*. In humans, these multiple PMCA isoforms are differentially expressed and localized among the corneal epithelial layers *(79, 80)*. PMCA4 is the isoform that is primarily expressed, except that it is notably absent in the membranes of the basal cells adjacent to the stroma. PMCA1 and PMCA2 are located mainly in basal- and wing-cell cytoplasm and membranes. In contrast to PMCA4, PMCA1 immunoreactivity is evident in the basal cell plasma membranes adjacent to the stroma. PMCA2 immunoreactivity is detectable in the cytoplasm of basal and wing cells in the central cornea and limbus. PMCA3 is enclosed in basal-cell nuclei in the central cornea. Such selective corneal epithelial localization of PMCA expression suggests that each of these isoforms has a specialized function, contributing to the requisite flexibility of the cornea to respond to the different Ca^{2+} requirements needed for renewal and regeneration of its multiple cell types. In addition to ATP-dependent transporters, the Na^+/Ca^{2+} exchanger (SCE) was described as a Ca^{2+} efflux pathway for rabbit corneal epithelial cells *(81)*.

ICS calcium transporter activity exists, as evidenced by results showing that agonist-induced calcium transients are consistent with capacitative calcium entry (CCE). This was demonstrated by showing that adding back calcium to a calcium-free medium, following depletion of ICS with cyclopiazonic acid, induces a calcium transient. This add-back response reflects calcium influx through store-operated channels (SOC) at the plasma membrane, whose activation is dependent upon depletion (using cyclopiazonic acid) of ICS *(75, 82)*. The CCE mechanism is a component of the signaling pathways mediating control of proliferation by EGF, ET_A, and ET_B receptors *(68, 75, 82)*. Cholinergic receptor-mediated control of corneal epithelial cell proliferation has also been described, and such control also entails calcium signaling *(17, 83)*. EGF-induced mitogenesis is dependent upon calcium signaling in corneal epithelial cells, as shown by the fact that EGF fails to induce such a response in a calcium-free medium *(84)*. EGF-induced stimulation of SOC and CCE activity depends upon selective increases of PKCβ and PKCδ isoforms *(85)*. Therefore, studies targeted at identifying mechanisms of Ca^{2+} transporter regulation are relevant for revealing novel drugs targeted to hastening proliferation and wound healing of the corneal epithelia (Fig. 4).

Organic Ion Transporters

During acute anisosmotic stress, corneal epithelial cells are able to mediate rapid regulatory volume behavior through activation of inorganic plasma membrane transporters that mediate net osmolyte influx or efflux. In contrast, chronic anisosmotic stress requires additional protective mechanisms, e.g., stimulation of transcription of genes encoding for osmolyte transport mechanisms and activation of organic osmolyte transporters (Table 1). These responses enhance cell survival by preventing the intracellular

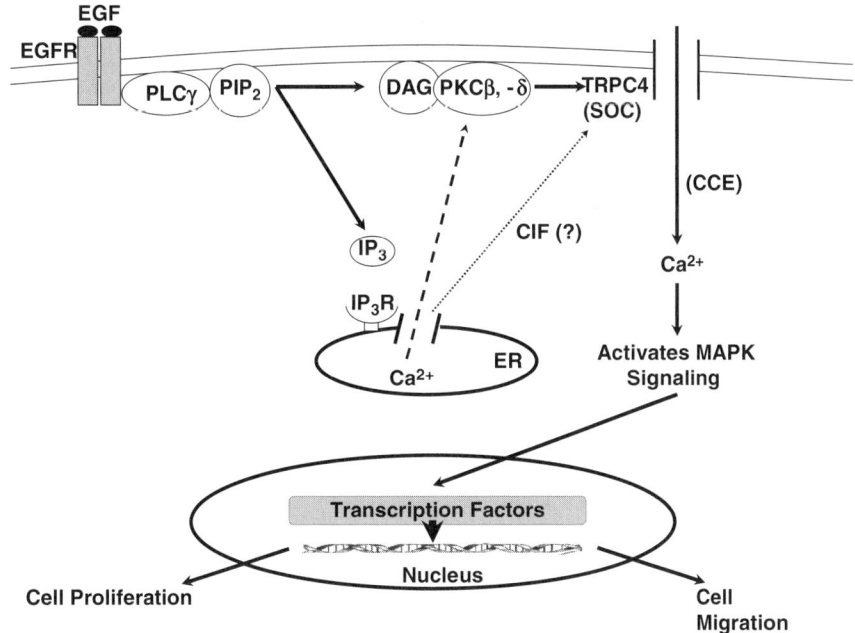

Fig. 4. Epidermal growth factor (EGF)-induced Ca^{2+} signaling mechanisms. EGF receptor (EGFR) activation stimulates phospholipase C (PLC)γ, which hydrolyzes phosphatidylinositol 4,5-bisphosphate (PIP_2) to release diacylglycerol (DAG) and inositol 1,4,5-triphosphate (IP_3). IP_3 activates its receptor, which is located at the endoplasmic reticulum (ER) membrane, releasing Ca^{2+} to the intracellular space. Activation of PKCβ and PKCδ by DAG and Ca^{2+} release from the ER, and/or capacitative influx factor (CIF) activates capacitative calcium entry (CCE) through the store-operated channel (SOC) TRPC4. CCE activates the mitogen-activated protein kinase (MAPK) signaling cascade and nuclear transcription factors leading to cell migration and proliferation, both of which are required for corneal epithelial wound healing.

Table 1 Organic osmolyte transporters in corneal epithelial cells

Transporter	Main organic substrates
Lactate-H^+ cotransporter	Lactate
Glucose transporter (GLUT)1	Glucose
Organic anion transporting peptide (OATP)-E	Thyroid hormone
Type amino acid transporter (LAT)1	Phenylalanine
Cation and neutral amino acid transporter ($B^{0,+}$)	L-Arginine
Ala-, Ser-, and Cys-preferring transport (ASCT)1	L-Alanine
Carrier-mediated riboflavin transport system	Riboflavin (vitamin B_2)
Sodium-dependent vitamin C transporter (SVCT)1	L-ascorbic acid (vitamin C)
Sodium-dependent multivitamin transporter (SMVT)	Biotin (vitamin H); pantothenic acid (vitamin B_5); lipoic acid
Taurine transporter	Taurine

ionic composition from realizing nonphysiological conditions. Moreover, the normal corneal epithelium is avascular, being partially nourished by nutrients (e.g. glucose, peptides, amino acids, vitamins) delivered via inorganic osmolyte-coupled transporters located primarily at the apical side of the epithelium. There are no studies describing whether such transport function is modulated by the cytokines that mediate control of corneal epithelial renewal. Such a mechanism of modulation is conceivable, however, since epithelial renewal is dependent upon adequate provision of substrates to meet the energy requirements of this process. Nevertheless, the presence of these membrane transporters may provide new opportunities for the design of transporter-targeted drugs with enhanced corneal epithelial permeability.

One of the first organic osmolyte transport systems to be identified in corneal epithelial cells was the lactate-H^+ cotransporter *(86)*. Under aerobic conditions, most of the glucose consumed by the corneal epithelium is metabolized to lactate. During hypoxic conditions (e.g., as in contact lens wear), lactate production (in rabbit corneal epithelial cells) increases, and an equivalent amount of protons are generated, inducing stromal acidification *(61, 87)*. Accumulation of lactate leads to local fluid accumulation and increases in the light-scattering effects commonly observed during epithelial hypoxia *(88)*. Metabolic waste products are removed from the corneal epithelium by the lactate-coupled proton transporter, whose function is essential for pH_i regulation during increased anaerobic glycolysis.

Glucose availability is critical in the maintenance of numerous corneal epithelial functions. In the corneal epithelium, it is well established that glucose uptake occurs through a facilitative transport process involving the glucose transporter GLUT1 *(89)*. After corneal epithelial wounding, GLUT1 mRNA and protein expression are rapidly enhanced *(90)*. This effect reflects the fact that increases in intracellular glucose levels are needed to meet the energy requirements of cell proliferation and migration. In addition, several GLUTs may be involved in the maintenance of cell homeostasis, based on the fact that facilitative GLUTs act as minor transmembrane pathways for water flow in the presence of an osmotic gradient *(91)*.

The organic anion transporting peptide (OATP)-E gene and protein expression have been characterized in several rat ocular tissues, including the corneal epithelium *(92)*. Expression of OATP-E tends to be greater in the basal cells than in the superficial cells of the corneal epithelium. This pattern may be related to the fact that basal cells are more metabolically active than are their superficial layer counterparts. OATP-E activity is thought to function in transporting thyroid hormone into the cornea for the purpose of maintaining optical transparency; this hormone plays a role in corneal dehydration and transparency *(93)*. It has also been proposed that this sodium-independent exchanger is involved in the transcellular movement of other organic anions and amphipathic compounds.

Three different types of neutral amino acid transporters have been identified in the corneal epithelium. The first is the L-type amino acid transporter (LAT)1, a large neutral amino acid and a Na^+-, energy-, and pH-independent facilitative transport system, first described in rabbit corneal epithelial cells *(94)*. LAT 1 preferentially transports the large amino acid phenylalanine (Phe) across the cornea. The second is the cationic and neutral amino acid transporter, $B^{0,+}$ *(95)*, which actively transports L-arginine (L-Arg) across rabbit and human corneal epithelia. Unlike LAT1, this transporter is Na$^+$, Cl$^-$, and energy dependent. The third and most recently discovered is the neutral amino acid transport system ASCT1 (ASC for Ala-, Ser-, and Cys-preferring), which was characterized in intact rabbit cornea and epithelial cell culture *(96)*. The ASCT1 isoform transports the neutral amino acid L-alanine (L-Ala) in a sodium-dependent manner. Although ASCT1 transport function is highly sodium dependent, neither substrate uptake into corneal epithelial cells nor transport across rabbit cornea is altered in this system in the presence of ouabain. Such an inconsistency suggests that uphill L-Ala influx is not driven by Na$^+$ entry, but instead mediates obligatory amino acid exchange. In this system, the sodium pump is not required for the transporter to function, since it can mediate Na$^+$ flux in either an inward or outward direction in exchange for amino acids.

Vitamins comprise another group of nutrients that play critical roles in maintaining normal metabolism, differentiation, and growth of mammalian cells. Riboflavin (vitamin B$_2$) is an essential water-soluble vitamin and a precursor for flavin mononucleotide (FMN) and flavinadenine dinucleotide (FAD), both of which are coenzymes that metabolize carbohydrates, amino acids, lipids, and proteins. Riboflavin, which is present in the tears, has a role in the development and maintenance of the surface structures of epithelial and goblet cells. Riboflavin deficiency leads to poor development of the glycocalyx layer of the conjunctival surface, and contributes to DES and other corneal changes. Microscopic evaluation of corneas from riboflavin-deficient rats display hemidesmosomal damage to the basal cells, decreases in microvilli and microplicae, and presence of cellular debris in the superficial epithelial cells *(97)*. These morphological alterations correlate with the clinical ocular signs and symptoms that are described for riboflavin-deficient patients: tearing, foreign-body sensation, photophobia, decreased vision, and keratoconjunctivitis sicca *(98, 99)*. Such clinical and structural corneal changes are reversible with riboflavin restitution. Recent findings for rabbit corneal epithelia suggest that riboflavin is translocated by a highly specific carrier-mediated system, a mechanism described as pH, Na$^+$, and Cl$^-$ independent, but temperature and energy dependent *(100)*.

The protective properties of ascorbic acid (vitamin C) are attributed to its antioxidant action against damage caused by ultraviolet exposure and ocular infections. Abundant amounts of ascorbic acid are present in the corneal epithelium *(101)* and tears *(102, 103)*. Moreover, through inhibition of corneal lipoxygenase, the antioxidative effect of this water-soluble vitamin acid reduces free hydroxyl radicals and eliminates subepithelial deposition of extracellular matter, as well as the haze produced by excimer laser ablation *(104, 105)*. The expression of the sodium-dependent vitamin C transporter (SVCT)1 was detected in the deeper layers of the rabbit corneal epithelium *(106)*. Recent evidence demonstrates that SVCT1 is pH and Na$^+$ dependent, with high affinity for L-ascorbic acid in rabbit corneal epithelial cells *(107)*.

The functional and molecular expression of the sodium-dependent multivitamin transporter (SMVT) has been characterized for rabbit corneal epithelial cells *(108)*. SMVT has broad substrate specificity and is, thus, able to translocate biotin (vitamin H) and several of its analogues, e.g., pantothenic acid (vitamin B$_5$) and the cofactor, lipoic acid. Biotin, which is found in tears, is necessary for the proper metabolism of certain amino acids, fatty acids, and oxaloacetic acid. Biotin uptake, and that of its analogues, is inhibited in the presence of the calmodulin antagonist, calmidazolium. This suggests the involvement of Ca^{2+}/calmodulin-mediated pathways in biotin's intracellular regulation. Biotin transport via SMVT is coupled to the Na^+ electrochemical gradient, but not to H^+. As biotin uptake is an energy (ATP)-dependent process, it is also dependent on the normal functioning of the of Na^+/K^+ pump.

An osmosensitive taurine transporter gene and its protein expression were also described for human corneal epithelial cells *(109)*. This transporter's activity is sodium dependent, and is manifested under isotonic conditions. Exposure to a hypertonic challenge increases its function in a time-dependent manner proportional to the magnitude of the stress. Such challenges induce a biphasic effect, i.e., an initial elevation, followed by declines in gene and protein expression, and return to isotonic levels after 48 hours, suggesting that these changes are under feedback control. Since the increases in intracellular taurine accumulation are less than those predicted as being necessary for restoring isotonic cell volume, the role of taurine appears, instead, to be that of improving cell viability.

ROLES OF ION CHANNELS IN MEDIATING CONTROL OF BARRIER, DETURGESCENCE, AND EPITHELIAL CELL RENEWAL

There is emerging evidence that the regulation of the different types of corneal epithelial ion channels is crucial for inducing the cytokine-mediated responses essential for corneal epithelial function and renewal. These pathways have diverse modes of selectivity and are regulated in different ways. Modulation of these pathways permits transient increases in ionic influx, which activate the downstream signaling events that are linked to disparate receptor-controlled responses. Receptor-operated and voltage-gated activities have already been identified. Furthermore, there is leak channel activity, whose ionic activity is dependent on electronegative intracellular membrane voltage. Various other channels can be stimulated by mechanical-stress- or receptor-induced second messenger formation. In the corneal epithelium, activation of ion channels, in response to receptor stimulation by a host of different stimuli or stressors, induces the responses needed for corneal epithelial renewal, as well as for regulation of fluid and electrolyte homeostasis. In the corneal epithelium, there is heterogeneous ion channel expression.

Ion channels are widely regarded to be causally involved in many diseases (the so-called channelopathies) and to contribute indirectly to the genesis of several diseases *(110)*. However, a presumptive role for these channels vis-à-vis specific diseases of the cornea is still undetermined. So far, there are only limited studies concerning the mechanisms by which ion channels, especially Ca^{2+}-permeable ion channels, function in the human cornea. This void leaves unclear the relationships between defective ion channel control, cell death, and disease.

K^+ Channels

Potassium-selective ion channels are essential to and widely distributed in all cell types. Basically, K^+ channel activity sets the resting membrane potential. In different cell types, these channels generate electrical signals (in excitable cells), and regulate cell volume, proliferation, and movement. There are a variety of K^+ channels types: voltage-gated K^+ channels ($K_v1.1–K_v1.6$) (delayed rectifiers, A-type voltage-gated K^+

channels); inward rectifier K⁺ channels ($K_{ir}1.1$–$K_{ir}3.1$) (these preferentially pass K⁺ ions inward; steep voltage dependence) *(111)*; and ligand-gated K⁺ channels, e.g., the ATP-dependent K⁺ channels (K_{ATP}) or Ca²⁺-dependent K⁺ channels ($K_{Ca}1$) *(112, 113)*. The various K⁺ channel types differ in their functions *(114–119)*. More is known about K⁺ channels in corneal epithelial cells than in corneal endothelial cells. Early studies showed that K⁺ channel activity changes modulate the essential corneal epithelial functions needed to mediate adrenoceptor stimulation of net Cl⁻ transport *(120)*. In rabbit corneal epithclia, cholinergic receptor-mediated increases in cyclic GMP are also modulators of some types of K⁺ channel activity, and are sensitive to diltiazem *(121)*. Another type of external Ca²⁺-dependent K⁺ channel activity, which is outwardly rectifying and active at hyperpolarized voltages, but insensitive to the L-type Ca²⁺ channel blocker diltiazem, has also been detected. These, on the other hand, are activated by flufenamic acid and blocked by Ba²⁺. In addition, an inwardly rectifying K⁺ current has also been uncovered *(122)*. Additional studies of freshly isolated bovine corneal epithelial cells reveal an inactivating voltage-gated K⁺ channel. Arachidonic acid and some fatty acids may directly activate the large-conductance K⁺ channel to augment its housekeeping functions in corneal epithelial cells. There is also another noisy, sustained K⁺ current, which resembles the large-conductance K⁺ current in the rabbit *(123)*. Recently, some important functions of K⁺ channels in the corneal epithelium were reviewed *(35)*. Specifically, ultraviolet-irradiation-induced K⁺ channel (Kv3.4) hyperactivation, resulting in rapid intracellular K⁺ content losses and cell-volume shrinkage followed by apoptosis, was described for rabbit corneal epithelia. As well, growth factor cognate receptor stimulation induced the activation of other types of K⁺ channels through induction of corneal epithelial cell proliferation *(20)*. Therefore, selective K⁺ channel activation via diverse receptor-activated events is essential for determining whether this activation induces cell death or promotes cell proliferation.

Na⁺ Channels

There is very limited information regarding the importance of selective sodium channel modulation in corneal epithelial function. Watsky et al. (1991) reported the occurrence of sodium currents in ocular epithelial cells from three different species. These currents have a current/voltage (I/V) relationship consistent with traditional voltage-gated Na⁺ currents, which are tetrodotoxin (TTX)-blockable *(124)*. The presence of these currents is surprising, since TTX-sensitive Na⁺ channels are normally not found in nonexcitable cells such as corneal epithelial cells. Na⁺ channels of this type are usually inactive in membrane voltages of less than −60 mV. In corneal epithelial cells, the membrane voltage under Cl⁻-transporting conditions is less than −60 mV with respect to the bathing solution *(125)*. Therefore, the importance of the functional activity of these currents in cytokine- and anisosmotic-mediated signaling is unclear.

Cl⁻ Channels

Chloride channel activity modulation in the corneal epithelium occurs via adrenergic signaling pathways that elicit increases in fluid secretion into the tears through transient cytosolic increases in cyclic adenosine monophosphate (cAMP) and $[Ca^{2+}]_i$ *(16, 21, 73)*. Little is known about the effects of cytokines on net Cl⁻ transport, with the exception of results from a single study showing that stimulation of the ET-1 receptor subtype, ET_A, inhibited Cl⁻-orginated fluid transport across epithelial layers in the rabbit cornea *(126)*. Cl⁻ channel activity within the so-called ClC family of voltage-dependent Cl⁻ channels is, accordingly, delineated: (i) the cAMP-activated transmembrane conductance regulator (CFTR); (ii) Ca^{2+}-activated Cl⁻ channels (CaCC); and (iii) volume-regulated anion channels (VRAC) *(127–130)*; (iv) Voltage-dependent Ca^{2+} channels are also referred to as voltage-operated Ca^{2+} channels (VOCC). Corneal epithelial cells express low-conductance Cl⁻ channels, comparable to the anion channels of other vertebrate epithelia *(131)*. cAMP-dependent chloride currents have been identified in rabbit corneal epithelia. This chloride conductance is mediated by CFTR *(132)*. In the same study, functional and molecular properties of VRAC were identified. Specifically, Northern blot analysis verified expression of ClC-3 gene transcripts, suggesting that ClC-3 activation underlies the effects of known ClC-3 inhibitors on hypotonicity-induced cell swelling *(47)*.

Voltage-Dependent Ca^{2+} Channels

VOCCs are divided into L-type channels ($Ca_V1.1–1.4$; dihydropyridine sensitive); P/Q-type channels ($Ca_V2.1$, ω-agatoxin sensitive); N-type channels ($Ca_V2.2$, ω-conotoxin GVIA sensitive); R-type channels ($Ca_V2.3$, former blocker resistant channel, SNX-482 sensitive); and T-type channels ($Ca_V3.1–3.3$, low-voltage activated) *(133, 134)*. All are expressed in a variety of tissues, including skeletal muscle, cardiac muscle, endocrine cells, neurons, dendrites, and nerve terminals *(133–135)*. Pharmacological and electrophysiological properties of VOCCs are determined by their pore-forming α_1 subunits. Ten members, all of which differ in their unique α_1 subunits, of the VOCC family are known (Ca_V subunits) *(133, 134)*.

There is variable L-type dihydropyridine-sensitive Ca^{2+} channel expression in corneal epithelial cells. In the rabbit, Ca^{2+} influx is mediated through nonselective non-voltage-gated channels driven by membrane voltage electronegativity, which is dihydropyridine insensitive *(81)*. However, in human corneal epithelial cells, L-type Ca^{2+} channel ($Ca_V1.2$) expression is detectable in every second patched cell. When detected, the current amplitudes are less than 50 pA and have typical characteristics of VOCCs of the L-type (Mergler et al., unpublished data) (Fig. 5).

Extracellular application of the specific L type channel blocker nifedipine, reduces plasma membrane Ca^{2+} influx, causing $[Ca^{2+}]_i$ to decline (Mergler et al., unpublished data) (Fig. 6); washout fails to restore the decreased levels to normal.

Ca^{2+} transients are components of some signaling pathways, which mediate innate immunity in response to bacterial exposure through activation of toll-like receptors (TLRs). There are 11 different TLR subtypes (TLR1-11), all of which respond to bacterial challenge. In corneal epithelial cells, TLR5 activation by a Gram-negative bacterial component agonist (i.e., flagellin) mediates increases in pro-inflammatory cytokines, e.g., interleukin-8 (IL-8). As corroborated by several other studies, Ca^{2+} influx is a component of such signaling, provided that the TLR5 coreceptor, asialoGM1, is activated. TLR5 activation is dependent on asialoGM1 activation, as indicated by the fact that the TLR5 agonist binding site is not accessible to the external medium. TLR5 stimulation in turn elicits an increase in Ca^{2+} influx through L-type Ca^{2+} channels. Stimulation of

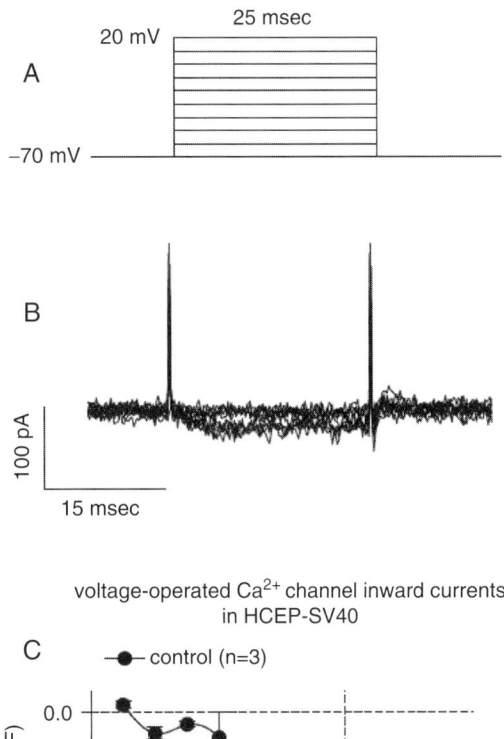

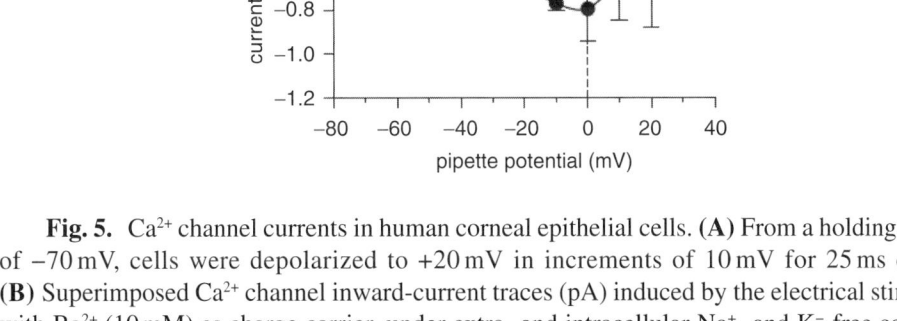

Fig. 5. Ca^{2+} channel currents in human corneal epithelial cells. (**A**) From a holding potential of −70 mV, cells were depolarized to +20 mV in increments of 10 mV for 25 ms duration. (**B**) Superimposed Ca^{2+} channel inward-current traces (pA) induced by the electrical stimulation, with Ba^{2+} (10 mM) as charge carrier, under extra- and intracellular Na^+- and K^--free conditions. (**C**) Ca^{2+} channel peak current amplitudes (pA) were plotted against the pipette potentials (mV) used in the stimulation protocol (standard current/voltage plot). Characteristics of the *I/V* line nearly correspond to voltage-operated Ca^{2+} channels (VOCCs) of the L-type.

such a Ca^{2+} transient is the result of TLR5-mediated ATP release and, in turn, ionotropic P2Y receptor stimulation *(137)* (Fig. 7). L-type Ca^{2+} channel activity is not the sole mechanism for plasma membrane Ca^{2+} influx in human corneal epithelial cells. Other Ca^{2+} influx pathways participating in receptor-mediated Ca^{2+} signaling include SOCs, of tetrameric structure, comprising disparate monomeric transient receptor potential (TRP) protein isoform subunits.

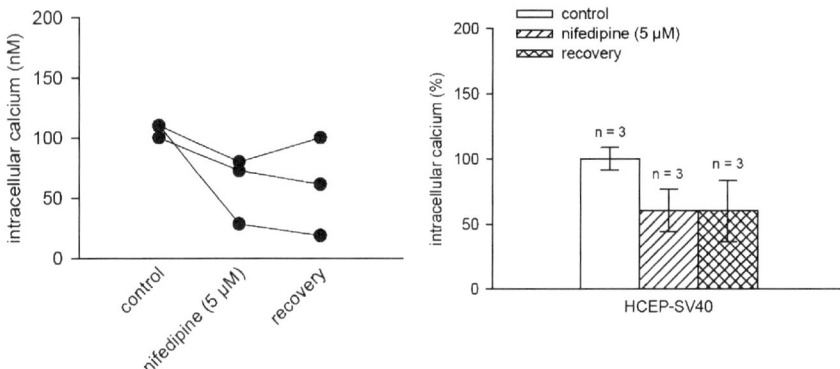

Fig. 6. L-type channel activity in non-stimulated human corneal epithelial cells. **Left panel:** Effect of extracellular application of nifedipine (5 µM) on cytosolic, free Ca^{2+}. Changes in cytosolic, free Ca^{2+} are depicted as the ratio of the fluorescence values induced by excitation wavelengths 340 and 380 nm. Intracellular Ca^{2+} concentration was calculated by the equation from Grynkiewicz et al. *(136)*. **Right panel:** Summary of nifedipine experiments.

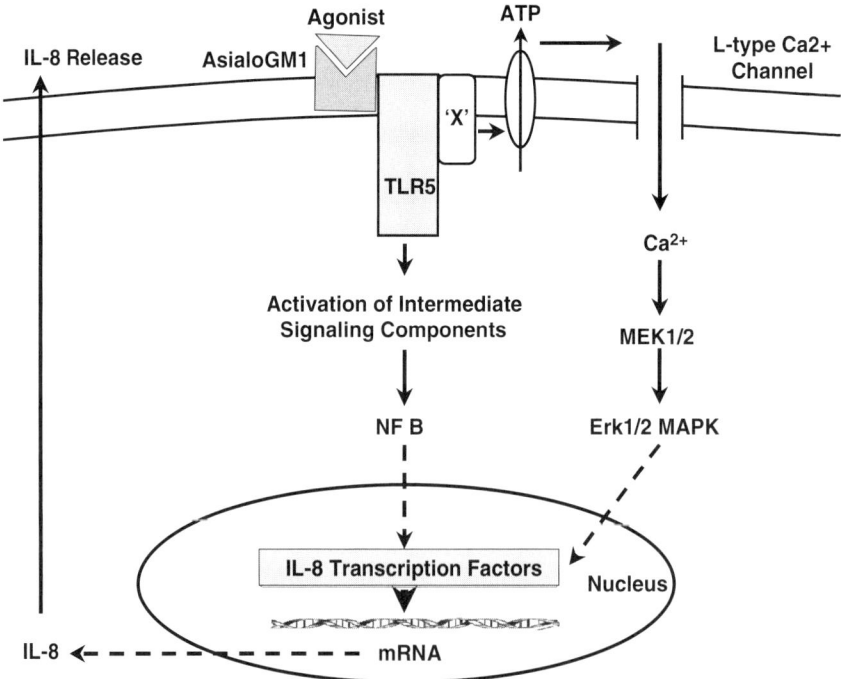

Fig. 7. AsialoGM1 interaction with Toll-like receptor 5 (TLR5) induces interleukin-8 (IL-8) release. Agonist activation of a coreceptor, asialoGM1, results in protein-protein interaction with TLR5 followed by extracellular release of adenosine triphosphate and purinoceptor-induced transient increases in Ca^{2+} influx through L-type voltage-dependent Ca^{2+} channels. This response activates Ca^{2+}-dependent kinase-mediated MEK1/2 and Erk1/2 mitogen-activated protein kinase (MAPK) phosphorylation and IL-8 release. There is also parallel NFκB activation.

Table 2 Transient receptor potential (TRP) channel expression in corneal epithelial cells

Subtype	Function
Canonical	
TRPC4	Component of SOC mediating Ca^{2+} influx induced by EGF
TRPC1,3,5,7	Unknown
Vanilloid	
TRPV1	Ca^{2+} influx pathway activated by capsaicin mediating IL-6 and IL-8 release
TRPV4	Alleged osmosensor inducing RVD responses to a hypotonic challenge
Melastatin	
TRPM8	Induces Ca^{2+} influx increases of in response to temperature changes

TRP Channels

Within the TRP superfamily are seven subfamilies, containing a total of 27 different TRP genes *(110, 138–142)*. Their gene products form homomeric and heteromeric complexes that are nonselective cation channels. Channel activity control is receptor mediated through kinases and influx factors. Influx factor formation may be responsive to the Ca^{2+} filling state of the ICS. The various receptor types that activate Ca^{2+} influx through these nonselective pathways act as cellular sensors with diverse functions. Recent studies in human corneal epithelia reveal TRP expression from three of the seven subfamilies. The TRP subfamilies represented are: (i) canonical (TRPC), (ii) vanilloid (TRPV), and (iii) melastatin (TRPM) (Table 2).

Seven different isoforms—TRPC1 to 7— exist. In different tissues, these can form heteromers or homomers with one another to give rise to a four-member grouping that constitutes a channel. On the basis of sequence homology and functional similarities, members of the mammalian TRPC family are grouped accordingly: TRPC1, TRPC2, TRPC3/6/7, and TRPC4/5 *(142)*. Specific TRPC channels from different cell types are each regulated by different signaling mechanisms. Canonical TRP activation can be induced either through changes in the Ca^{2+} filling states of ICS or by direct receptor-mediated control *(142)*. The isoform TRPC4 is expressed in human corneal epithelial cells by mediating Ca^{2+} influx in response to EGFR stimulation *(84)*. The mitogenic response to EGFR stimulation is dependent upon TRPC4 expression, based on the knowledge that, following knockdown of its expression, EGF-induced increases in human corneal epithelial cell proliferation are suppressed. Similarly, EGFR activation of Ca^{2+} influx declines as result of the failure to empty Ca^{2+} from the ICS, which is needed to activate the TRP channels. This is controlled by the filling state of these compartments. Even though TRPC1,3,5,7 gene expression has been detected in human corneal epithelia, there is no information regarding their roles in mediating Ca^{2+} influx induced by activation of other receptor types that induce control of cell proliferation. Nevertheless, targeting drugs to mediate control of TRPC4 expression and function may provide an alternative approach to hastening corneal epithelial proliferation and wound healing.

The TRPV subfamily, on the basis of structure and function, consists of two groups in mammals: TRPV1–4, and TRPV5/6 *(142)*. TRPV1–4 are nonselective thermosensitive cation channels, although TRPV1 and 4 can also be activated by numerous other stimuli. TRPV1–4 are modestly permeable to Ca^{2+}. TRPV1 is activated by vanilloid

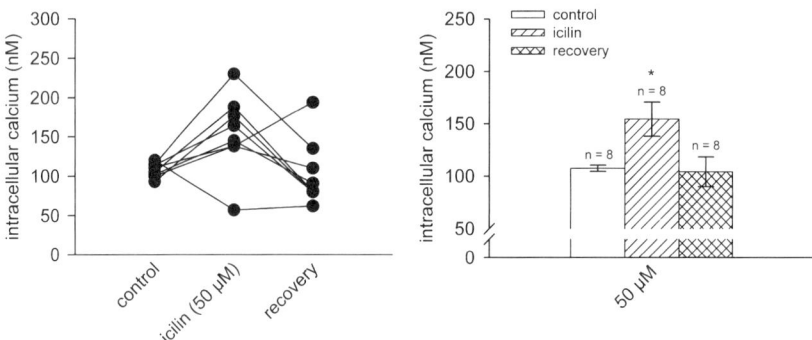

Fig. 8. Melastatin transient receptor potential 8 (TRPM8) activity in human corneal epithelial cells (HCECs). **Left panel:** Effect of extracellular application of the cooling agent icilin (50 μM) on cytosolic free Ca^{2+}. Changes in cytosolic, free Ca^{2+} are depicted as the ratio of the fluorescence values induced by excitation wavelengths 340 and 380 nm. Intracellular Ca^{2+} concentration was calculated by the equation from Grynkiewicz et al. *(136)*. **Right panel:** Summary of icilin experiments.

compounds, such as capsaicin, resiniferatoxin, and olvanil, as well as by multiple other stimuli, including moderate heat (≥43°C) *(143–145)*.

TRPV1 was first identified in the dorsal root ganglion (DRG) and trigeminal ganglion neurons. It is also highly expressed in spinal and peripheral nerve terminals, and in multiple non-neuronal cell types (reviewed in detail in *142*). Interestingly, there is emerging evidence of TRPV1 expression in various epithelial cell types. TRPV1 has also been detected in human corneal epithelial cells. Its activation results in increases in pro-inflammatory cytokine expression through activation of Ca^{2+} influx (Reinach et al., unpublished data).

Another isoform of the TRPV subfamily is TRPV4, which acts as an osmosensor to induce the RVD response *(146)*. It is widely expressed in brain, DRG neurons, and multiple excitable and nonexcitable peripheral cell types *(147, 148)*. Interestingly, TRPV4 expression is also detected in human corneal epithelial cells, and appears to be involved in mediating the RVD response during a hypotonic challenge (Reinach et al., unpublished data).

On the basis of sequence homology, members of the TRPM family fall into four groups: TRPM1/3, TRPM2/8, TRPM4/5, and TRPM6/7 *(142)*. TRPM channels exhibit highly varying permeabilities to Ca^{2+} and Mg^{2+}. Mergler et al. (2005) showed that icilin, the agonist for the temperature-sensitive transient receptor potential protein, TRPM8, induced transient increases in $[Ca^{2+}]_i$ in human corneal epithelia *(149)* (Fig. 8).

There are several reports indicating temperature-dependent effects on corneal viability during storage *(150, 151)*. In this context, it is possible that the activity of putative temperature-sensitive TRP channels, e.g., TRPM8, may be increased at lower storage temperatures than are possible without TRPM8. This could result in graded shifts of voltage-dependent activation curves of these TRP channels *(144)*. Another important point is that corneal epithelial cells may be sensitive to hypothermic storage, because they may be susceptible to a similar cold-induced injury, a phenomenon that could provide an explanation for putative corneal epithelial sensitivity to hypothermic storage.

ACKNOWLEDGEMENTS

This work was support by grants EY04795 (PR) and by an unrestricted grant from Research to Prevent Blindness, Inc., NY (KP).

REFERENCES

1. Klyce SD. Transport of Na, Cl, and water by the rabbit corneal epithelium at resting potential. Am J Physiol 1975;228:1446–1452.
2. Maurice DM. The permeability to sodium ions of the living rabbit's cornea. J Physiol 1951;112:367–391.
3. Maurice DM. Influence on corneal permeability of bathing with solutions of differing reaction and tonicity. Br J Ophthalmol 1955;39:463–473.
4. Maurice DM. The location of the fluid pump in the cornea. J Physiol 1972;221:43–54.
5. Zucker BB. Hydration and transparency of corneal stroma. Arch Ophthalmol 1966;75:228–231.
6. Klyce SD, Wong RK. Site and mode of adrenaline action on chloride transport across the rabbit corneal epithelium. J Physiol 1977;266:777–799.
7. Klyce SD. Enhancing fluid secretion by the corneal epithelium. Invest Ophthalmol Vis Sci 1977;16:968–973.
8. Li HF, Petroll WM, Moller-Pedersen T, Maurer JK, Cavanagh HD, Jester JV. Epithelial and corneal thickness measurements by in vivo confocal microscopy through focusing (CMTF). Curr Eye Res 1997;16:214–221.
9. Jakus M. The fine structure of the hyman cornea. In: Smelser G, ed. The structure of the eye. New York: Academia, 1969:344.
10. Baum JP, Maurice DM, McCarey BE. The active and passive transport of water across the corneal endothelium. Exp Eye Res 1984;39:335–342.
11. Zadunaisky JA, Lande MA, Chalfie M, Neufeld AH. Ion pumps in the cornea and their stimulation by epinephrine and cyclic-AMP. Exp Eye Res 1973;15:577–584.
12. Klyce SD, Palkama KA, Harkonen M, Marshall WS, Huhtaniitty S, Mann KP, Neufeld AH. Neural serotonin stimulates chloride transport in the rabbit corneal epithelium. Invest Ophthalmol Vis Sci 1982;23:181–192.
13. Pesin SR, Candia OA. Acetylcholine concentration and its role in ionic transport by the corneal epithelium. Invest Ophthalmol Vis Sci 1982;22:651–659.
14. Candia OA, Podos SM, Neufeld AH. Modification by timolol of catecholamine stimulation of chloride transport in isolated corneas. Invest Ophthalmol Vis Sci 1979;18:691–695.
15. Chu TC, Candia OA. Role of alpha 1- and alpha 2-adrenergic receptors in Cl- transport across frog corneal epithelium. Am J Physiol 1988;255:C724–730.
16. Montoreano R, Candia OA, Cook P. alpha- and beta-adrenergic receptors in regulation of ionic transport in frog cornea. Am J Physiol 1976;230:1487–1493.
17. Cavanagh HD, Colley AM. Cholinergic, adrenergic, and PGE1 effects on cyclic nucleotides and growth in cultured corneal epithelium. Metab Pediatr Syst Ophthalmol 1982;6:63–74.
18. Lu L, Reinach PS, Kao WW. Corneal epithelial wound healing. Exp Biol Med (Maywood) 2001;226:653–664.
19. Yang H, Wang Z, Miyamoto Y, Reinach PS. Cell signaling pathways mediating epidermal growth factor stimulation of Na:K:2Cl cotransport activity in rabbit corneal epithelial cells. J Membr Biol 2001;183:93–101.
20. Roderick C, Reinach PS, Wang L, Lu L. Modulation of rabbit corneal epithelial cell proliferation by growth factor-regulated K(+) channel activity. J Membr Biol 2003;196:41–50.

21. Candia OA, Grillone LR, Chu TC. Forskolin effects on frog and rabbit corneal epithelium ion transport. Am J Physiol 1986;251:C448–454.
22. Candia OA. Ouabain and sodium effects on chloride fluxes across the isolated bullfrog cornea. Am J Physiol 1972;223:1053–1057.
23. Wu X, Yang H, Iserovich P, Fischbarg J, Reinach PS. Regulatory volume decrease by SV40-transformed rabbit corneal epithelial cells requires ryanodine-sensitive Ca2+-induced Ca2+ release. J Membr Biol 1997;158:127–136.
24. Capo-Aponte JE, Iserovich P, Reinach PS. Characterization of regulatory volume behavior by fluorescence quenching in human corneal epithelial cells. J Membr Biol 2005;207:11–22.
25. Farris RL. Tear osmolarity–a new gold standard? Adv Exp Med Biol 1994;350:495–503.
26. Bildin VN, Yang H, Fischbarg J, Reinach PS. Effects of chronic hypertonic stress on regulatory volume increase and Na-K-2Cl cotransporter expression in cultured corneal epithelial cells. Adv Exp Med Biol 1998;438:637–642.
27. Bildin VN, Wang Z, Iserovich P, Reinach PS. Hypertonicity-induced p38MAPK activation elicits recovery of corneal epithelial cell volume and layer integrity. J Membr Biol 2003;193:1–13.
28. Lang F, Busch GL, Ritter M, Volkl H, Waldegger S, Gulbins E, Haussinger D. Functional significance of cell volume regulatory mechanisms. Physiol Rev 1998;78:247–306.
29. Capo-Aponte JE, Wang Z, Bildin VN, Pokorny KS, Reinach PS. Fate of Hypertonicity-Stressed Corneal Epithelial Cells Depends on Differential MAPK Activation and p38MAPK/Na-K-2Cl Cotransporter1 Interaction. Exp Eye Res 2007;84:361–372.
30. Saika S, Okada Y, Miyamoto T, Yamanaka O, Ohnishi Y, Ooshima A, Liu CY, Weng D, Kao WW. Role of p38 MAP kinase in regulation of cell migration and proliferation in healing corneal epithelium. Invest Ophthalmol Vis Sci 2004;45:100–109.
31. Sharma GD, He J, Bazan HE. p38 and ERK1/2 coordinate cellular migration and proliferation in epithelial wound healing: evidence of cross-talk activation between MAP kinase cascades. J Biol Chem 2003;278:21989–21997.
32. Kang SS, Li T, Xu D, Reinach PS, Lu L. Inhibitory effect of PGE2 on EGF-induced MAP kinase activity and rabbit corneal epithelial proliferation. Invest Ophthalmol Vis Sci 2000;41:2164–2169.
33. Kang SS, Wang L, Kao WW, Reinach PS, Lu L. Control of SV-40 transformed RCE cell proliferation by growth-factor-induced cell cycle progression. Curr Eye Res 2001;23:397–405.
34. Wang Z, Yang H, Tachado SD, Capo-Aponte JE, Bildin VN, Koziel H, Reinach PS. Phosphatase-Mediated Crosstalk Control of ERK and p38 MAPK Signaling in Corneal Epithelial Cells. Invest Ophthalmol Vis Sci 2006;47:5267–5275.
35. Lu L. Stress-induced corneal epithelial apoptosis mediated by K(+) channel activation. Prog Retin Eye Res 2006;25:515–538.
36. Zadunaisky JA, Lande MA. Active chloride transport and control of corneal transparency. Am J Physiol 1971;221:1837–1844.
37. Candia OA, Reinach PS, Alvarez L. Amphotericin B-induced active transport of K+ and the Na+-K+ flux ratio in frog corneal epithelium. Am J Physiol 1984;247:C454–461.
38. Davson H. The influence of the lyotropic series of anions on cation permeability. Biochem J 1940;34:917–925.
39. Ljubimov AV, Atilano SR, Garner MH, Maguen E, Nesburn AB, Kenney MC. Extracellular matrix and Na+,K+-ATPase in human corneas following cataract surgery: comparison with bullous keratopathy and Fuchs' dystrophy corneas. Cornea 2002;21:74–80.
40. Cejkova J, Lojda Z, Brunova B, Vacik J, Michalek J. Disturbances in the rabbit cornea after short-term and long-term wear of hydrogel contact lenses. Usefulness of histochemical methods. Histochemistry 1988;89:91–97.

41. Conners MS, Urbano F, Vafeas C, Stoltz RA, Dunn MW, Schwartzman ML. Alkali burn-induced synthesis of inflammatory eicosanoids in rabbit corneal epithelium. Invest Ophthalmol Vis Sci 1997;38:1963–1971.
42. Masferrer JL, Rios AP, Schwartzman ML. Inhibition of renal, cardiac and corneal (Na(+)-K+)ATPase by 12(R)-hydroxyeicosatetraenoic acid. Biochem Pharmacol 1990;39:1971–1974.
43. Schwartzman ML, Balazy M, Masferrer J, Abraham NG, McGiff JC, Murphy RC. 12(R)-hydroxyicosatetraenoic acid: a cytochrome-P450-dependent arachidonate metabolite that inhibits Na+,K+-ATPase in the cornea. Proc Natl Acad Sci USA 1987;84:8125–8129.
44. Vafeas C, Mieyal PA, Urbano F, Falck JR, Chauhan K, Berman M, Schwartzman ML. Hypoxia stimulates the synthesis of cytochrome P450-derived inflammatory eicosanoids in rabbit corneal epithelium. J Pharmacol Exp Ther 1998;287:903–910.
45. Ottino P TF, Bazan HE. Growth factor-induced proliferation in corneal epithelial cells is mediated by 12(S)-HETE. Exp Eye Res 2003;76:613–622.
46. Bildin VN, Yang H, Crook RB, Fischbarg J, Reinach PS. Adaptation by corneal epithelial cells to chronic hypertonic stress depends on upregulation of Na:K:2Cl cotransporter gene and protein expression and ion transport activity. J Membr Biol 2000;177:41–50.
47. Al-Nakkash L, Iserovich P, Coca-Prados M, Yang H, Reinach PS. Functional and molecular characterization of a volume-activated chloride channel in rabbit corneal epithelial cells. J Membr Biol 2004;201:41–49.
48. Wehner F, Olsen H, Tinel H, Kinne-Saffran E, Kinne RK. Cell volume regulation: osmolytes, osmolyte transport, and signal transduction. Rev Physiol Biochem Pharmacol 2003;148:1–80.
49. Jakab M, Ritter M. Cell volume regulatory ion transport in the regulation of cell migration. Contrib Nephrol 2006;152:161–180.
50. Gobbels M, Spitznas M. Corneal epithelial permeability of dry eyes before and after treatment with artificial tears. Ophthalmology 1992;99:873–878.
51. Fleiszig SM, Zaidi TS, Pier GB. Mucus and Pseudomonas aeruginosa adherence to the cornea. Adv Exp Med Biol 1994;350:359–362.
52. Xu KP, Yagi Y, Tsubota K. Decrease in corneal sensitivity and change in tear function in dry eye. Cornea 1996;15:235–239.
53. Imanishi J, Kamiyama K, Iguchi I, Kita M, Sotozono C, Kinoshita S. Growth factors: importance in wound healing and maintenance of transparency of the cornea. Prog Retin Eye Res 2000;19:113–129.
54. Bildin VN, Iserovich P, Fischbarg J, Reinach PS. Differential expression of Na:K:2Cl cotransporter, glucose transporter 1, and aquaporin 1 in freshly isolated and cultured bovine corneal tissues. Exp Biol Med (Maywood) 2001;226:919–926.
55. Bortner CD, Cidlowski JA. Absence of volume regulatory mechanisms contributes to the rapid activation of apoptosis in thymocytes. Am J Physiol 1996;271:C950–961.
56. Wilson SE, Mohan RR, Hong J, Lee J, Choi R, Liu JJ. Apoptosis in the cornea in response to epithelial injury: significance to wound healing and dry eye. Adv Exp Med Biol 2002;506:821–826.
57. Capo-Aponte JE, Wang Z, Bildin VN, Iserovich P, Pan Z, Zhang F, Pokorny KS, Reinach PS. Functional and molecular characterization of multiple K-Cl cotransporter isoforms in corneal epithelial cells. Exp Eye Res 2007;84:1090–1103.
58. Shen MR, Chou CY, Hsu KF, Liu HS, Dunham PB, Holtzman EJ, Ellory JC. The KCl cotransporter isoform KCC3 can play an important role in cell growth regulation. Proc Natl Acad Sci USA 2001;98:14714–14719.
59. Shen MR, Chou CY, Hsu KF, Hsu YM, Chiu WT, Tang MJ, Alper SL, Ellory JC. KCl cotransport is an important modulator of human cervical cancer growth and invasion. J Biol Chem 2003;278:39941–39950.

60. Shen MR, Lin AC, Hsu YM, Chang TJ, Tang MJ, Alper SL, Ellory JC, Chou CY. Insulin-like growth factor 1 stimulates KCl cotransport, which is necessary for invasion and proliferation of cervical cancer and ovarian cancer cells. J Biol Chem 2004;279:40017–40025.
61. Bonanno JA, Polse KA. Corneal acidosis during contact lens wear: effects of hypoxia and CO_2. Invest Ophthalmol Vis Sci 1987;28:1514–1520.
62. Korbmacher C, Helbig H, Forster C, Wiederholt M. Evidence for Na+/H+ exchange and pH sensitive membrane voltage in cultured bovine corneal epithelial cells. Curr Eye Res 1988;7:619–626.
63. Korbmacher C, Helbig H, Forster C, Wiederholt M. Characterization of Na+/H+ exchange in a rabbit corneal epithelial cell line (SIRC). Biochim Biophys Acta 1988;943:405–410.
64. Bonanno JA. K(+)-H+ exchange, a fundamental cell acidifier in corneal epithelium. Am J Physiol 1991;260:C618–625.
65. Shepard AR, Rae JL. Ion transporters and receptors in cDNA libraries from lens and cornea epithelia. Curr Eye Res 1998;17:708–719.
66. Wu X, Torres-zamorano V, Yang H, Reinach PS. ETA receptor mediated inhibition of intracellular pH regulation in cultured bovine corneal epithelial cells. Exp Eye Res 1998;66:699–708.
67. Reinach P, Ganapathy V, Torres-Zamorano V. A Na:H exchanger subtype mediates volume regulation in bovine corneal epithelial cells. Adv Exp Med Biol 1994;350:105–110.
68. Takagi H, Reinach PS, Tachado SD, Yoshimura N. Endothelin-mediated cell signaling and proliferation in cultured rabbit corneal epithelial cells. Invest Ophthalmol Vis Sci 1994;35:134–142.
69. Takagi H, Reinach PS, Yoshimura N, Honda Y. Endothelin-1 promotes corneal epithelial wound healing in rabbits. Curr Eye Res 1994;13:625–628.
70. Tao W, Liou GI, Wu X, Abney TO, Reinach PS. ETB and epidermal growth factor receptor stimulation of wound closure in bovine corneal epithelial cells. Invest Ophthalmol Vis Sci 1995;36:2614–2622.
71. Fischbarg J, Hernandez J, Liebovitch LS, Koniarek JP. The mechanism of fluid and electrolyte transport across corneal endothelium: critical revision and update of a model. Curr Eye Res 1985;4:351–360.
72. Bonanno JA. Identity and regulation of ion transport mechanisms in the corneal endothelium. Prog Retin Eye Res 2003;22:69–94.
73. Candia OA, Montoreano R, Podos SM. Effect of the ionophore A23187 on chloride transport across isolated frog cornea. Am J Physiol 1977;233:F94–101.
74. Leiper LJ, Walczysko P, Kucerova R, Ou J, Shanley LJ, Lawson D, Forrester JV, McCaig CD, Zhao M, Collinson JM. The roles of calcium signaling and ERK1/2 phosphorylation in a Pax6+/− mouse model of epithelial wound-healing delay. BMC Biol 2006;4:27.
75. Yang H, Sun X, Wang Z, Ning G, Zhang F, Kong J, Lu L, Reinach PS. EGF stimulates growth by enhancing capacitative calcium entry in corneal epithelial cells. J Membr Biol 2003;194:47–58.
76. Reinach P, Holmberg N. Ca-stimulated Mg dependent ATPase activity in a plasma membrane enriched fraction of bovine corneal epithelium. Curr Eye Res 1987;6:399–405.
77. Reinach PS, Holmberg N, Chiesa R. Identification of calmodulin-sensitive Ca(2+)-transporting ATPase in the plasma membrane of bovine corneal epithelial cell. Biochim Biophys Acta 1991;1068:1–8.
78. Verma AK, Filoteo AG, Stanford DR, Wieben ED, Penniston JT, Strehler EE, Fischer R, Heim R, Vogel G, Mathews S, et al. Complete primary structure of a human plasma membrane Ca2+ pump. J Biol Chem 1988;263:14152–14159.

79. Johnson JA, Grande JP, Roche PC, Campbell RJ, Kumar R. Immuno-localization of the calcitriol receptor, calbindin-D28k and the plasma membrane calcium pump in the human eye. Curr Eye Res 1995;14:101–108.
80. Talarico EF, Jr., Kennedy BG, Marfurt CF, Loeffler KU, Mangini NJ. Expression and immunolocalization of plasma membrane calcium ATPase isoforms in human corneal epithelium. Mol Vis 2005;11:169–178.
81. Rich A, Rae JL. Calcium entry in rabbit corneal epithelial cells: evidence for a nonvoltage dependent pathway. J Membr Biol 1995;144:177–184.
82. Tao W, Wu X, Liou GI, Abney TO, Reinach PS. Endothelin receptor-mediated Ca2+ signaling and isoform expression in bovine corneal epithelial cells. Invest Ophthalmol Vis Sci 1997;38:130–141.
83. Socci RR, Tachado SD, Aronstam RS, Reinach PS. Characterization of the muscarinic receptor subtypes in the bovine corneal epithelial cells. J Ocul Pharmacol Ther 1996;12:259–269.
84. Yang H, Mergler S, Sun X, Wang Z, Lu L, Bonanno JA, Pleyer U, Reinach PS. TRPC4 knockdown suppresses epidermal growth factor-induced store-operated channel activation and growth in human corneal epithelial cells. J Biol Chem 2005;280:32230–32237.
85. Zhang F, Wen Q, Mergler S, Yang H, Wang Z, Bildin VN, Reinach PS. PKC isoform-specific enhancement of capacitative calcium entry in human corneal epithelial cells. Invest Ophthalmol Vis Sci 2006;47:3989–4000.
86. Bonanno JA. Lactate-proton cotransport in rabbit corneal epithelium. Curr Eye Res 1990;9:707–712.
87. Klyce SD. Stromal lactate accumulation can account for corneal oedema osmotically following epithelial hypoxia in the rabbit. J Physiol 1981;321:49–64.
88. Lambert SR, Klyce SD. The origins of Sattler's veil. Am J Ophthalmol 1981;91:51–56.
89. Kumagai AK, Glasgow BJ, Pardridge WM. GLUT1 glucose transporter expression in the diabetic and nondiabetic human eye. Invest Ophthalmol Vis Sci 1994;35:2887–2894.
90. Takahashi H, Kaminski AE, Zieske JD. Glucose transporter 1 expression is enhanced during corneal epithelial wound repair. Exp Eye Res 1996;63:649–659.
91. Loike JD, Cao L, Kuang K, Vera JC, Silverstein SC, Fischbarg J. Role of facilitative glucose transporters in diffusional water permeability through J774 cells. J Gen Physiol 1993;102:897–906.
92. Ito A, Yamaguchi K, Tomita H, Suzuki T, Onogawa T, Sato T, Mizutamari H, Mikkaichi T, Nishio T, Unno M, Sasano H, Abe T, Tamai M. Distribution of rat organic anion transporting polypeptide-E (oatp-E) in the rat eye. Invest Ophthalmol Vis Sci 2003;44:4877–4884.
93. Coulombre AJ, Coulombre JL. Corneal Development. 3. The Role of the Thyroid in Dehydration and the Development of Transparency. Exp Eye Res 1964;75:105–114.
94. Jain-Vakkalagadda B, Dey S, Pal D, Mitra AK. Identification and functional characterization of a Na+-independent large neutral amino acid transporter, LAT1, in human and rabbit cornea. Invest Ophthalmol Vis Sci 2003;44:2919–2927.
95. Jain-Vakkalagadda B, Pal D, Gunda S, Nashed Y, Ganapathy V, Mitra AK. Identification of a Na+-dependent cationic and neutral amino acid transporter, B(0,+), in human and rabbit cornea. Mol Pharm 2004;1:338–346.
96. Katragadda S, Talluri RS, Pal D, Mitra AK. Identification and characterization of a Na+-dependent neutral amino acid transporter, ASCT1, in rabbit corneal epithelial cell culture and rabbit cornea. Curr Eye Res 2005;30:989–1002.
97. Takami Y, Gong H, Amemiya T. Riboflavin deficiency induces ocular surface damage. Ophthalmic Res 2004;36:156–165.

98. Stern JJ. The ocular manifestations of riboflavin deficiency. Am J Ophthalmol 1950;33:1127–1136.
99. Jackson CR. Riboflavin deficiency with ocular signs: report of a case. Br J Ophthalmol 1950;34: 259–260.
100. Hariharan S, Janoria KG, Gunda S, Zhu X, Pal D, Mitra AK. Identification and functional expression of a carrier-mediated riboflavin transport system on rabbit corneal epithelium. Curr Eye Res 2006;31:811–824.
101. Ringvold A, Anderssen E, Kjonniksen I. Impact of the environment on the mammalian corneal epithelium. Invest Ophthalmol Vis Sci 2003;44:10–15.
102. Choy CK, Benzie IF, Cho P. Is ascorbate in human tears from corneal leakage or from lacrimal secretion? Clin Exp Optom 2004;87:24–27.
103. Brubaker RF, Bourne WM, Bachman LA, McLaren JW. Ascorbic acid content of human corneal epithelium. Invest Ophthalmol Vis Sci 2000;41:1681–1683.
104. Williams RN, Paterson CA. Modulation of corneal lipoxygenase by ascorbic acid. Exp Eye Res 1986;43:7–13.
105. Shimmura S, Masumizu T, Nakai Y, Urayama K, Shimazaki J, Bissen-Miyajima H, Kohno M, Tsubota K. Excimer laser-induced hydroxyl radical formation and keratocyte death in vitro. Invest Ophthalmol Vis Sci 1999;40:1245–1249.
106. Tsukaguchi H, Tokui T, Mackenzie B, Berger UV, Chen XZ, Wang Y, Brubaker RF, Hediger MA. A family of mammalian Na+-dependent L-ascorbic acid transporters. Nature 1999;399:70–75.
107. Talluri RS, Katragadda S, Pal D, Mitra AK. Mechanism of L-ascorbic acid uptake by rabbit corneal epithelial cells: evidence for the involvement of sodium-dependent vitamin C transporter 2. Curr Eye Res 2006;31:481–489.
108. Janoria KG, Hariharan S, Paturi D, Pal D, Mitra AK. Biotin uptake by rabbit corneal epithelial cells: role of sodium-dependent multivitamin transporter (SMVT). Curr Eye Res 2006;31:797–809.
109. Shioda R, Reinach PS, Hisatsune T, Miyamoto Y. Osmosensitive taurine transporter expression and activity in human corneal epithelial cells. Invest Ophthalmol Vis Sci 2002;43:2916–2922.
110. Nilius B, Voets T. TRP channels: a TR(I)P through a world of multifunctional cation channels. Pflugers Arch 2005;451:1–10.
111. Kubo Y, Adelman JP, Clapham DE, Jan LY, Karschin A, Kurachi Y, Lazdunski M, Nichols CG, Seino S, Vandenberg CA. International Union of Pharmacology. LIV. Nomenclature and molecular relationships of inwardly rectifying potassium channels. Pharmacol Rev 2005;57:509–526.
112. Goldstein SA, Wang KW, Ilan N, Pausch MH. Sequence and function of the two P domain potassium channels: implications of an emerging superfamily. J Mol Med 1998;76:13–20.
113. Chandy KG, Gutman GA. Nomenclature for mammalian potassium channel genes. Trends Pharmacol Sci 1993;14:434.
114. Desir G. Molecular physiology of renal potassium channels. Semin Nephrol 1992;12:531–540.
115. Faber ES, Sah P. Calcium-activated potassium channels: multiple contributions to neuronal function. Neuroscientist 2003;9:181–194.
116. Giebisch G. Renal potassium channels: function, regulation, and structure. Kidney Int 2001;60:436–445.
117. Hebert SC, Desir G, Giebisch G, Wang W. Molecular diversity and regulation of renal potassium channels. Physiol Rev 2005;85:319–371.

118. Korn SJ, Trapani JG. Potassium channels. IEEE Trans Nanobiosci 2005;4:21–33.
119. MacKinnon R. Potassium channels. FEBS Lett 2003;555:62–65.
120. Wolosin JM, Candia OA. Cl- secretagogues increase basolateral K+ conductance of frog corneal epithelium. Am J Physiol 1987;253:C555–560.
121. Farrugia G, Rae JL. Regulation of a potassium-selective current in rabbit corneal epithelium by cyclic GMP, carbachol and diltiazem. J Membr Biol 1992;129:99–107.
122. Bockman CS, Griffith M, Watsky MA. Properties of whole-cell ionic currents in cultured human corneal epithelial cells. Invest Ophthalmol Vis Sci 1998;39:1143–1151.
123. Takahira M, Sakurada N, Segawa Y, Shirao Y. Two types of K+ currents modulated by arachidonic acid in bovine corneal epithelial cells. Invest Ophthalmol Vis Sci 2001;42:1847–1854.
124. Watsky MA, Cooper K, Rae JL. Sodium channels in ocular epithelia. Pflugers Arch 1991;419:454–459.
125. Nagel W, Reinach P. Mechanism of stimulation by epinephrine of active transepithelial Cl transport in isolated frog cornea. J Membr Biol 1980;56:73–79.
126. Yang H, Reinach PS, Koniarek JP, Wang Z, Iserovich P, Fischbarg J. Fluid transport by cultured corneal epithelial cell layers. Br J Ophthalmol 2000;84:199–204.
127. Hartzell C, Putzier I, Arreola J. Calcium-activated chloride channels. Annu Rev Physiol 2005;67:719–758.
128. Jentsch TJ, Stein V, Weinreich F, Zdebik AA. Molecular structure and physiological function of chloride channels. Physiol Rev 2002;82:503–568.
129. Nilius B, Droogmans G. Amazing chloride channels: an overview. Acta Physiol Scand 2003;177:119–147.
130. Pusch M. Structural insights into chloride and proton-mediated gating of CLC chloride channels. Biochemistry 2004;43:1135–1144.
131. Marshall WS, Hanrahan JW. Anion channels in the apical membrane of mammalian corneal epithelium primary cultures. Invest Ophthalmol Vis Sci 1991;32:1562–1568.
132. Al-Nakkash L, Reinach PS. Activation of a CFTR-mediated chloride current in a rabbit corneal epithelial cell line. Invest Ophthalmol Vis Sci 2001;42:2364–2370.
133. Catterall WA, Perez-Reyes E, Snutch TP, Striessnig J. International Union of Pharmacology. XLVIII. Nomenclature and structure-function relationships of voltage-gated calcium channels. Pharmacol Rev 2005;57:411–425.
134. Catterall WA. Structure and regulation of voltage-gated Ca2+ channels. Annu Rev Cell Dev Biol 2000;16:521–555.
135. McDonald TF, Pelzer S, Trautwein W, Pelzer DJ. Regulation and modulation of calcium channels in cardiac, skeletal, and smooth muscle cells. Physiol Rev 1994;74:365–507.
136. Grynkiewicz G, Poenie M, Tsien RY. A new generation of Ca2+ indicators with greatly improved fluorescence properties. J Biol Chem 1985;260:3440–3450.
137. Du JW, Zhang F, Capo-Aponte JE, Tachado SD, Zhang J, Yu FS, Sack RA, Koziel H, Reinach PS. AsialoGM1-mediated IL-8 release by human corneal epithelial cells requires coexpression of TLR5. Invest Ophthalmol Vis Sci 2006;47:4810–4818.
138. Clapham DE. TRP channels as cellular sensors. Nature 2003;426:517–524.
139. Clapham DE, Julius D, Montell C, Schultz G. International Union of Pharmacology. XLIX. Nomenclature and structure-function relationships of transient receptor potential channels. Pharmacol Rev 2005;57:427–450.
140. Montell C. Physiology, phylogeny, and functions of the TRP superfamily of cation channels. Sci STKE 2001;2001:re1.
141. Montell C. The TRP superfamily of cation channels. Sci STKE 2005;2005:re3.

142. Pedersen SF, Owsianik G, Nilius B. TRP channels: an overview. Cell Calcium 2005;38: 233–252.
143. Sanchez MG, Sanchez AM, Collado B, Malagarie-Cazenave S, Olea N, Carmena MJ, Prieto JC, Diaz-Laviada II. Expression of the transient receptor potential vanilloid 1 (TRPV1) in LNCaP and PC-3 prostate cancer cells and in human prostate tissue. Eur J Pharmacol 2005;515:20–27.
144. Voets T, Droogmans G, Wissenbach U, Janssens A, Flockerzi V, Nilius B. The principle of temperature-dependent gating in cold- and heat-sensitive TRP channels. Nature 2004;430:748–754.
145. Weil A, Moore SE, Waite NJ, Randall A, Gunthorpe MJ. Conservation of functional and pharmacological properties in the distantly related temperature sensors TRVP1 and TRPM8. Mol Pharmacol 2005;68:518–527.
146. Liedtke W. TRPV4 as osmosensor: a transgenic approach. Pflugers Arch 2005;451:176–180.
147. Kochukov MY, McNearney TA, Fu Y, Westlund KN. Thermosensitive TRP ion channels mediate cytosolic calcium response in human synoviocytes. Am J Physiol Cell Physiol 2006;291:C424–432.
148. Zhang L, Jones S, Brody K, Costa M, Brookes SJ. Thermosensitive transient receptor potential channels in vagal afferent neurons of the mouse. Am J Physiol Gastrointest Liver Physiol 2004;286:G983–991.
149. Mergler S, Pleyer U, Reinach P, Bednarz J, Dannowski H, Engelmann K, Hartmann C, Yousif T. EGF suppresses hydrogen peroxide induced Ca2+ influx by inhibiting L-type channel activity in cultured human corneal endothelial cells. Exp Eye Res 2005;80:285–293.
150. Hsu JK, Cavanagh HD, Jester JV, Ma L, Petroll WM. Changes in corneal endothelial apical junctional protein organization after corneal cold storage. Cornea 1999;18:712–720.
151. Lindstrom RL. Advances in corneal preservation. Trans Am Ophthalmol Soc 1990;88:555–648.

3
Vitamin C Transport, Delivery, and Function in the Anterior Segment of the Eye

Ram Kannan and Hovhannes J. Gukasyan

CONTENTS

INTRODUCTION
LENS
CORNEA
OCULAR SURFACE AND TEAR FILM
FUTURE DIRECTIONS
REFERENCES

INTRODUCTION

It is well known that ocular tissues and fluids contain significant amounts of ascorbic acid (AA) *(1)*. High and nearly equal amounts of AA were found in both the aqueous humor and vitreous humor *(1, 2)*. Due to the higher levels of AA in these two compartments relative to other intraocular tissues, aqueous and vitreous humors are considered as reservoirs of AA in the eye. Comprehensive reviews on transport dynamics of AA in the iris-cilliary body, aqueous humor, and vitreous humor have been published. Furthermore, AA appears to be secreted onto the ocular epithelial surface and is found in tear fluid *(3, 4)*. Multiple transmembrane carrier proteins of facilitative diffusion and secondary active type have also been identified as transporters of vitamin C across various ocular tissue layers and cell membranes. In this chapter, we will restrict our discussion to the function and transport of vitamin C in the anterior segment of the eye, particularly the lens, cornea and the ocular surface.

LENS

Function

Rats and mice possess biosynthetic pathways to synthesize L-ascorbic acid (vitamin C) endogenously via the glucuronic acid pathway. In contrast, vitamin C is an essential nutrient for humans and primates since these species lack the ability to biosynthesize it from precursors *(5)*. Guinea pigs must obtain vitamin C from their diets, in a manner similar to humans and primates. Absorption by the gastrointestinal tract in these species is a specialized, critical, rate-limiting step following which vitamin C is transported to target tissues via the circulation.

Ascorbic acid is one of the most important cellular antioxidants and participates in a number of critical functions *(6)*. The finding that the human lens and intraocular fluids maintain a high AA level has led investigators to propose that AA plays a special role in the eye *(3)*. It is generally found that the eye of diurnal animals has higher levels of AA than ocular tissues of nocturnal animals, which have lower levels *(7)*. Ascorbic acid of ocular origin could compensate for the lack of antioxidant protection that comes normally with the blood supply, or alternatively serve the special need for transparent tissues to protect from injury due to light exposure. High levels of AA in aqueous humor have protective effects against ultraviolet (UV)-induced damage to the lens epithelium *(8)*. Hegde and Varma (2004) found that the presence of high ascorbate was highly beneficial in protecting the lens against oxidative damage and cataract formation in aldose reductase-deficient mice, implying that it is an important antioxidant. Our laboratory found that restricting AA in the diet increased the propensity of cataract formation in the guinea pig and proposed that this could be a suitable model for studying mechanisms of galactosemia-induced cataract *(9)*. However, AA can also act as a pro-oxidant in the aqueous humor and in the lens. The metal-catalyzed reaction of AA produces reactive ascorbate free radicals, dehydroascorbic acid (DHAA) and hydrogen peroxide, which must be reduced by a mechanism such as GSH redox cycle because they are toxic to the lens *(10)*. In fact DHA and H_2O_2 have been linked to the formation of senile cataract *(11–13)*. Thus, it is important to understand the specific transport mechanisms of AA when dissociated from those of DHAA to better assess the role played by AA in the protection of the lens.

Transport

Cellular transport of AA and DHAA have been shown to be mediated by separate mechanisms *(14)*. Studies with *Xenopus* oocytes showed that AA is transported by a Na^+-dependent mechanism that transports its structural analog isoascorbic acid, but not D-glucose *(15)*. In contrast, DHAA is a transport substrate for the mammalian Na^+-independent hexose transporters *(16)*. It should be recognized that once transported, DHAA is immediately reduced in the intracellular space to AA *(14)*. Results from studies on AA uptake in bovine lens epithelial cells led to the conclusion that the aqueous humor pool of AA upon oxidative stress serves as a source of DHA, which is readily taken up from the aqueous humor and converted to AA intracellularly in lens epithelial cells, thereby sparing the pool of cellular AA *(17)*. Early in vivo and in vitro transport studies have led to conflicting results. Both AA and DHAA can be taken up by the lens in vitro *(18)*. However, subsequent functional studies suggested that DHAA is the predominant form transported *(19)*. Radiolabeled AA uptake by lens and its compartments in vivo revealed that AA accumulated in guinea-pig lens epithelium by a concentrative mechanism but did not accumulate in the lens cortex *(20, 21)*.

More recently, two isoforms of the Na$^+$-dependent vitamin C transporters (SVCTs) have been cloned from rats and humans and have been named SVCT1 and SVCT2 *(22–26)*. These transporters are encoded by the genes Slc23a1 and Slc23a2, respectively *(27, 28)*. In the considered species, at least one SVCT isoform has been found in every tissue examined. SVCT1 is expressed predominantly in the liver, intestine, and kidney, while SVCT2 is found in the brain and the eye *(24)*. The two isoforms have 84% sequence homology but differ in their substrate affinity and capacity for AA transport *(29)*. Heterozygous SVCT2 knockout mice have been generated that can serve as a valuable model for the study of the physiological contribution of this transporter. In contrast to the homozygous Slc23a2(–/–) mice that die shortly after birth, the heterozygous mice survive to adulthood *(30)*. It is also noteworthy that the two SVCT isoforms appear to function independently of each other. For example, SVCT1 expression and AA levels in SVCT1-predominant organs were illustrated as not being influenced by the suppression of SVCT2 *(31)*. Recent studies have suggested that SVCT1 and SVCT2 can be regulated by protein kinase C in COS-1 cells *(32)*. An age-dependent decrease in gene expression of SVCT1 and SVCT2 has also been shown in rat hepatocytes *(33)*, while analogous studies in the lens or other ocular tissues are lacking. Carrier-mediated transport of AA and the expression of AA transporters in human lens epithelial cells have been investigated. Remarkably, these gene expression studies in human lens epithelia revealed the presence of SVCT2, and not SVCT1. Furthermore, regulation of the specific human lens epithelial AA transporter gene expression by oxidant stress in addition to various messenger signals was also be demonstrated *(34)*.

A recent report attempted to explain reasons behind for low levels of AA in rodent lenses as compared to the human tissue by examining comparative uptake of AA in cultured human and mouse lens epithelial cells. A significant (5- to 10-fold) suppression of vitamin C uptake and SVCT2 expression in mouse 17EM 15 cells as compared to human HLE-B3 cells was found *(35)*. The mechanism causing this difference may need to be investigated by further studies and may involve a wide variety of factors such as changes in glycosylation, phosphorylation and other regulatory processes. At any rate, the finding is consistent with the metabolic studies of Mody et al. who found a marked resistance to AA uptake in the guinea-pig lens *(36)*. It is of interest that an alternate-spliced variant of SVCT2 was identified recently, and that it behaves as a dominant negative by inhibiting vitamin C transport through protein–protein type interaction *(37)*. Studies on the role of the spliced variants as well as other identified isoforms are likely to provide valuable information on regulation of AA levels in the lens *(31)*.

CORNEA

Function

As mentioned above, AA is highly concentrated in ocular tissues and fluids compared with most other biosystems. The guinea-pig aqueous humor, for example contains 20 times more AA than plasma. Vitamin C content also varies significantly in different sections of the eye, with maximal values in the aqueous humor and the corneal epithelium, and with the lowest reported values in the retina *(38, 39)*. Ringvold et al. found that the distribution pattern of vitamin C levels in the bovine eye followed the order of, from greatest to least, central corneal epithelium, peripheral epithelium, aquous humor, sclera/conjunctiva, serum *(40)*. Additionally, the ocular vitamin C content of diurnal animals is higher than that of nocturnal animals, supporting the documented role for vitamin C in protection from UV radiation.

A number of physiological functions have been attributed to vitamin C in the cornea, particularly in the context of its role in modulating wound healing and inflammatory responses. There are several reports on the ascorbic acid content in mammalian corneal tissue *(38, 41–43)*. The levels in the corneal tissue in various species studied were higher than in plasma and in general similar to in aqueous humor.

Ascorbic acid regulates collagen biosynthesis, serving as a cofactor for the hydroxylation of proline and lysine residues on precollagen *(44, 45)*. Since greater than 80% of corneal dry weight is made up of collagen, the relationship between AA concentration and collagen production has been studied in corneal stroma. In a series of studies, Saika et al. found that AA and AA phosphate augmented the proliferation of corneal keratocytes and increased the synthesis of type I and type III collagen *(46, 47)*. Evidence for the beneficiary role of AA in wound healing was provided by studies by Pfister and his group who found that wound healing was promoted by topical application of AA *(48, 49)*. The mechanism is likely to be through inhibition of the effect of oxygen radicals produced by inflammatory process. Within this context, ascorbic acid was shown to also modulate the metabolism of arachidonic acid in cornea *(50, 51)*.

Other mechanisms of wound healing involve the participation of the electrophysiological currents that are modulated by AA. The corneal epithelium contains an active Na^+ transport system (Na^+/K^+-ATPase) displaying a net inward flow of sodium ions *(52, 53)*. On the other hand, Cl^- ions are actively transported outward, from aqueous humor, across stroma and epithelium, to the tear side *(54, 55)*. Ascorbic acid increases Na^+ and Cl^- transport across amphibian cornea *(56, 57)*. Recently Reid et al. described a technique for pharmacologically manipulating the wound-induced electric fields, and found that AA caused a significant increase in the wound-edge current compared to the wound-center current *(58)*.

Transport

DiMattio studied the transport of vitamin C in cornea using similar approaches to those in lens tissue mentioned above *(59)*. He concluded that, in guinea-pig and rat cornea, vitamin C enters the tissue from the aqueous humor passing through the endothelium into the stroma. DiMattio further found that there was an impairment in AA entry in streptozotocin-treated diabetic guinea pigs and rats *(60)*. Using bovine corneal endothelial cells, Bode et al. showed that DHAA is taken up seven times more than AA and suggested that cellular uptake of DHAA served as a source for AA for corneal protection *(61)*. Tsukaguchi et al. in their characterization of two Na^+-dependent vitamin C transporters, SVCT1 and SVCT2, found that SVCT2 was localized in deeper layers of rabbit corneal epithelial tissue *(24)*. This was confirmed by very recent work in which Na⁺-dependent, carrier-mediated transport of AA was localized on the apical domain of rabbit corneal epithelium. Further, reverse-transcriptase polymerase chain reaction (RT-PCR) identified the presence of SVCT1 and SVCT2 message in the tissue *(62)*. Studies so far support the hypothesis that L-ascorbic acid transport may represent its uptake by lacrimal gland from plasma and subsequent secretion into tears. This is consistent with original findings that AA (and DHAA) were taken up by the porcine lacrimal gland in in vitro studies with isolated tissue slices *(63)*.

OCULAR SURFACE AND TEAR FILM

The conjunctiva is a thin, mucus-secreting, vascular tissue that covers most of the inner surface of the eyelids, and is part of the anterior sclera where the cornea begins. The conjunctiva is thought to function as a passive physical protective barrier and to participate in the maintenance of tear-film stability due to the mucus secreted by the resident goblet cells *(64)*. Our laboratory as well as others have provided ample evidence for additional functional features of the conjunctiva, viz. acting as a conduit for drug delivery to the posterior segment of the eye following ocular drug instillation and contributing to the regulation of electrolyte and fluid balance in the microenvironment of its mucosal surface *(65)*. The dynamic nature of conjunctiva was demonstrated from the identification of several transport mechanisms for Na⁺ absorption: Na⁺-glucose *(66, 67)*, Na⁺-amino acid *(68)* and Na⁺-nucleoside cotransporters *(69)*, in addition to active Cl⁻ secretion *(70)*. The permeability of the conjunctiva to a wide variety of hydrophilic and lipophilic molecules was also reported *(71–73)*.

The ocular epithelium is very different in structure and function from the rest of the dead epidermis covering the mammalian body, which serves as a primary barrier from the environment. This small area (compared to the rest of the body surface) is a combination of two unique tissues, the conjunctiva, covering more than 85% of the surface, and the cornea. These epithelia are relatively unprotected and constantly exposed to light radiation, atmospheric oxygen, environmental chemicals, and physical abrasion. Each one of these forms of stress has the potential to generate reactive oxygen species (ROS) that contribute to ocular damage and disease if left unchecked. While the eye as an organ contains natural protective components, viz. water-soluble antioxidants such as vitamin C, cysteine, GSH, uric acid, pyruvate, and tyrosine; lipid-soluble antioxidants such as tocopherols and retinols; and highly specialized enzymes such as superoxide dismutase, catalase, and glutathione peroxidase, the ocular epithelium serves as the front line for protection equipped rather moderately with proper ammunition. Studies employing advanced analytical methods have demonstrated the presence of several unambiguous water-soluble antioxidants in human tear fluid collected at normal and stimulated flow rates. The mechanisms of action or the glandular sources of these antioxidants have not been addressed in any detail so far.

Hydrogen peroxide has been detected in aqueous humor and other ocular fluids; however reports of basal H_2O_2 levels within mammalian tear fluid are scarce. Concentrations of H_2O_2 present in aqueous humor range from 25 to 70 µM. Furthermore, the levels of H_2O_2 have been shown to increase up to 20-fold in certain age-related diseases of the eye with marked oxidative stress characteristics *(74)*. Since H_2O_2 is an ultimate byproduct of cellular respiration, it may be present in tear-fluid secretions under normal conditions. In vitro it can also arise spontaneously through a reaction between ascorbic acid, riboflavin, and light (or trace amounts of unbound metals in the absence of light). Given that ascorbate exists in the tear fluid at concentrations of 0.8–0.9 mM *(75)*, and maximal levels of riboflavin have been detected in ocular surface tissues of the rabbit *(76)*, under physiological conditions the former may also assume a pro-oxidant role by generating H_2O_2 as it is consumed. A dynamic equilibrium is essential for H_2O_2 through its continuous production in the ocular surface and tear film, and subsequent elimination *(74)*. It is of interest that we found glutathione secretion by conjunctival epithelial cells may be a critical input for the maintenance of this equilibrium *(77)*. Tear specimens from different human hosts showed a marked inhibition of hydroxy-radical formation, but did not affect superoxide or H_2O_2 levels *(78)*. Minimally stimulated tear samples collected from human subjects do not display a detectable catalase or GSH-peroxidase activity suggesting that the tear film may lack conventional cellular mechanisms that provide significant protection from oxidative properties of H_2O_2 *(79)*. The contribution of vitamin C and its transporters may thus be important in this regard.

Vitamin C given as a dietary supplement has been found to improve the tear-film stability, tear secretion, and health of the ocular surface in general. In normal subjects, improvement was found in tear-film stability with daily administration of vitamin C tablet although the magnitude of improvement was somewhat smaller than a multivitamin/trace element administered group *(80)*. Peponis et al. found that oral vitamin C supplementation along with vitamin E, to non-insulin-dependent diabetic patients, attenuated free radical production, NO levels and improved the tear-function parameters significantly *(81, 82)*. Vitamin C was also found to improve cholesterol-induced microcirculatory changes in rabbits *(83)*.

Animal studies have shown that deficiencies in water- and lipid-soluble vitamins (e.g., C, A, and E) result in the loss of goblet cells in the conjunctiva and abnormal chromatin distribution in the nucleus of epithelial cells *(84)*. Inflammation and loss of vascularity were considered as potential mechanisms for goblet cell loss in ocular surface disorders *(85)*. Given the high vitamin C levels in tears and the antiinflammatory role of vitamin C in the eye *(41, 75, 86)*, further studies will be needed to delineate mechanisms involved in beneficial effects of nutritional supplementation of vitamin C in ocular disorders.

FUTURE DIRECTIONS

Earlier work on vitamin C transport mostly described DHAA transport function via glucose transporters and its relevance in augmenting vitamin C levels in cells and tissues. Following the molecular biological characterization of specific, Na^+-dependent AA transporters, the structural and functional differences between DHAA and AA transporters have been delineated. A knockout mouse model for SVCT2 (Slc23a2(+/−)) has been generated. While the expression pattern of transporters of DHAA and AA in healthy tissues is known, there is a paucity of data on their role and regulation under pathological conditions. Facilitated diffusion of DHAA though glucose-sensitive and glucose-insensitive transporters, facilitated diffusion of AA through channels, exocytosis in secretory vesicles, and secondary active transport through the SVCT family of carrier proteins may play a critical role in prevalent ocular pathologies known to lead to blindness. A number of eye diseases where a decreased vitamin C concentration has been reported in compartments such as those containing aqueous or vitreous fluid, or specific ocular tissue and cell types, cannot be explained entirely in terms of reduction of dietary intake. Evidence gathered regarding independent transport pathways and the regulation of DHAA and AA would, however, facilitate efforts in the optimal design of modalities for elevation of vitamin C level in patients through the local or systemic administration of AA or DHAA. Given the known link between vitamin C and the redox status of the cell, potential use of transgenic mice and gene-therapy approaches to combat ocular diseases presents another area for future research. Overexpression of vitamin C transporters under conditions of deficiency of other antioxidants such as glutathione (GSH) and redoxins can offer rescue mechanisms. Signal transduction mechanisms linked to vitamin C transporter function and the parallel modulation of ion channels offer additional clues for therapy, particularly in corneal wound healing.

REFERENCES

1. Taylor A, Jacques PF, Nadler D, Morrow F, Sulsky SI, Shepard D. Relationship in humans between ascorbic acid consumption and levels of total and reduced ascorbic acid in lens, aqueous humor, and plasma. Curr Eye Res 1991;10(8):751–759.
2. McGahan MC. Ascorbic acid levels in aqueous and vitreous humors of the rabbit: effects of inflammation and ceruloplasmin. Exp Eye Res 1985;41(3):291–298.
3. Delamere NA. Ascorbic acid and the eye. Subcell Biochem 1996;25:313–329.
4. Rose RC, Bode AM. Ocular ascorbate transport and metabolism. Comp Biochem Physiol A 1991;100(2):273–285.
5. Packer L, Fuchs J. Vitamin C in health and disease. New York: M. Dekker; 1997.
6. Levine M, Morita K. Ascorbic acid in endocrine systems. Vitam Horm 1985;42:1–64.
7. Reiss GR, Werness PG, Zollman PE, Brubaker RF. Ascorbic acid levels in the aqueous humor of nocturnal and diurnal mammals. Arch Ophthalmol 1986;104(5):753v755.
8. Reddy VN, Giblin FJ, Lin LR, Chakrapani B. The effect of aqueous humor ascorbate on ultraviolet-B-induced DNA damage in lens epithelium. Invest Ophthalmol Vis Sci 1998;39(2):344–350.
9. Mackic JB, Ross-Cisneros FN, McComb JG et al. Galactose-induced cataract formation in guinea pigs: morphologic changes and accumulation of galactitol. Invest Ophthalmol Vis Sci 1994;35(3):804–810.
10. Wolff SP, Wang GM, Spector A. Pro-oxidant activation of ocular reductants. 1. Copper and riboflavin stimulate ascorbate oxidation causing lens epithelial cytotoxicity in vitro. Exp Eye Res 1987;45(6):777–789.
11. Giblin FJ, McCready JP, Kodama T, Reddy VN. A direct correlation between the levels of ascorbic acid and H_2O_2 in aqueous humor. Exp Eye Res 1984;38(1):87–93.
12. Sasaki K. Cataract classification systems in epidemiological studies. Dev Ophthalmol 1991;21:97–102.
13. Spector A. Oxidative stress-induced cataract: mechanism of action. FASEB J 1995;9(12): 1173–1182.
14. Welch RW, Wang Y, Crossman A, Jr., Park JB, Kirk KL, Levine M. Accumulation of vitamin C (ascorbate) and its oxidized metabolite dehydroascorbic acid occurs by separate mechanisms. J Biol Chem 1995;270(21):12584–12592.

15. Dyer DL, Kanai Y, Hediger MA, Rubin SA, Said HM. Expression of a rabbit renal ascorbic acid transporter in *Xenopus laevis* oocytes. Am J Physiol 1994;267(1 Pt 1):C301–C306.
16. Vera JC, Rivas CI, Fischbarg J, Golde DW. Mammalian facilitative hexose transporters mediate the transport of dehydroascorbic acid. Nature 1993;364(6432):79–82.
17. Corti A, Ferrari SM, Lazzarotti A et al. UV light increases vitamin C uptake by bovine lens epithelial cells. Mol Vis 2004;10:533–536.
18. Hughes RE, Hurley RJ. In vitro uptake of ascorbic acid by the guinea pig eye lens. Exp Eye Res 1970;9(2):175–180.
19. Kern HL, Zolot SL. Transport of vitamin C in the lens. Curr Eye Res 1987;6(7):885–896.
20. DiMattio J. Active transport of ascorbic acid into lens epithelium of the rat. Exp Eye Res 1989;49(5):873–885.
21. DiMattio J. A comparative study of ascorbic acid entry into aqueous and vitreous humors of the rat and guinea pig. Invest Ophthalmol Vis Sci 1989;30(11):2320–2331.
22. Daruwala R, Song J, Koh WS, Rumsey SC, Levine M. Cloning and functional characterization of the human sodium-dependent vitamin C transporters hSVCT1 and hSVCT2. FEBS Lett 1999;460(3):480–484.
23. Rajan DP, Huang W, Dutta B et al. Human placental sodium-dependent vitamin C transporter (SVCT2): molecular cloning and transport function. Biochem Biophys Res Commun 1999;262(3):762–768.
24. Tsukaguchi H, Tokui T, Mackenzie B et al. A family of mammalian Na+-dependent L-ascorbic acid transporters. Nature 1999;399(6731):70–75.
25. Wang H, Dutta B, Huang W et al. Human Na(+)-dependent vitamin C transporter 1 (hSVCT1): primary structure, functional characteristics and evidence for a non-functional splice variant. Biochim Biophys Acta 1999;1461(1):1–9.
26. Wang Y, Mackenzie B, Tsukaguchi H, Weremowicz S, Morton CC, Hediger MA. Human vitamin C (L-ascorbic acid) transporter SVCT1. Biochem Biophys Res Commun 2000;267(2):488–494.
27. Eck P, Erichsen HC, Taylor JG et al. Comparison of the genomic structure and variation in the two human sodium-dependent vitamin C transporters, SLC23A1 and SLC23A2. Hum Genet 2004;115(4):285–294.
28. Wilson JX. Regulation of vitamin C transport. Annu Rev Nutr 2005;25:105–125.
29. Korcok J, Dixon SJ, Lo TC, Wilson JX. Differential effects of glucose on dehydroascorbic acid transport and intracellular ascorbate accumulation in astrocytes and skeletal myocytes. Brain Res 2003;993(1–2):201–207.
30. Sotiriou S, Gispert S, Cheng J et al. Ascorbic-acid transporter Slc23a1 is essential for vitamin C transport into the brain and for perinatal survival. Nat Med 2002;8(5):514–517.
31. Kuo SM, MacLean ME, McCormick K, Wilson JX. Gender and sodium-ascorbate transporter isoforms determine ascorbate concentrations in mice. J Nutr 2004;134(9):2216–2221.
32. Liang WJ, Johnson D, Ma LS, Jarvis SM, Wei-Jun L. Regulation of the human vitamin C transporters expressed in COS-1 cells by protein kinase C [corrected]. Am J Physiol Cell Physiol 2002;283(6):C1696–C1704.
33. Michels AJ, Joisher N, Hagen TM. Age-related decline of sodium-dependent ascorbic acid transport in isolated rat hepatocytes. Arch Biochem Biophys 2003;410(1):112–120.
34. Kannan R, Stolz A, Ji Q, Prasad PD, Ganapathy V. Vitamin C transport in human lens epithelial cells: evidence for the presence of SVCT2. Exp Eye Res 2001;73(2):159–165.
35. Obrenovich ME, Fan X, Satake M et al. Relative suppression of the sodium-dependent Vitamin C transport in mouse versus human lens epithelial cells. Mol Cell Biochem 2006;293:53–62.

36. Mody VC, Jr., Kakar M, Elfving A, Soderberg PG, Lofgren S. Ascorbate in the guinea pig lens: dependence on drinking water supplementation. Acta Ophthalmol Scand 2005;83(2):228–233.
37. Lutsenko EA, Carcamo JM, Golde DW. A human sodium-dependent vitamin C transporter 2 isoform acts as a dominant-negative inhibitor of ascorbic acid transport. Mol Cell Biol 2004;24(8):3150–3156.
38. Pirie A, Wood C. Effects of vitamin A deficiency in the rabbit: 1. On vitamin C metabolism. 2. On power to use preformed vitamin A. Biochem J 1946;40(4):557–560.
39. Reim M, Luthe P. Compartmentation of redox metabolites in the anterior eye segment? Albrecht Von Graefes Arch Klin Exp Ophthalmol 1977;204(2):135–140.
40. Ringvold A, Anderssen E, Kjonniksen I. Distribution of ascorbate in the anterior bovine eye. Invest Ophthalmol Vis Sci 2000;41(1):20–23.
41. Paterson CA, O'Rourke MC. Vitamin C levels in human tears. Arch Ophthalmol 1987;105(3):376–377.
42. Schell DA, Bode AM. Measurement of ascorbic acid and dehydroascorbic acid in mammalian tissue utilizing HPLC and electrochemical detection. Biomed Chromatogr 1993;7(5):267–272.
43. Varma SD, Chand D, Sharma YR, Kuck JF, Jr., Richards RD. Oxidative stress on lens and cataract formation: role of light and oxygen. Curr Eye Res 1984;3(1):35–57.
44. Levene CI, Bates CJ. Ascorbic acid and collagen synthesis in cultured fibroblasts. Ann NY Acad Sci 1975;258:288–306.
45. Murad S, Tajima S, Johnson GR, Sivarajah S, Pinnell SR. Collagen synthesis in cultured human skin fibroblasts: effect of ascorbic acid and its analogs. J Invest Dermatol 1983;81(2):158–162.
46. Saika S, Kanagawa R, Uenoyama K, Hiroi K, Hiraoka J. L-ascorbic acid 2-phosphate, a phosphate derivative of L-ascorbic acid, enhances the growth of cultured rabbit keratocytes. Graefes Arch Clin Exp Ophthalmol 1991;229(1):79–83.
47. Saika S, Uenoyama K, Hiroi K, Ooshima A. L-ascorbic acid 2-phosphate enhances the production of type I and type III collagen peptides in cultured rabbit keratocytes. Ophthalmic Res 1992;24(2):68–72.
48. Pfister RR, Paterson CA. Ascorbic acid in the treatment of alkali burns of the eye. Ophthalmology 1980;87(10):1050–1057.
49. Pfister RR, Hayes SA, Paterson CA. The influence of parenteral ascorbate on the strength of corneal wounds. Invest Ophthalmol Vis Sci 1981;21(1 Pt 1):80–86.
50. Williams RN, Paterson CA. Modulation of corneal lipoxygenase by ascorbic acid. Exp Eye Res 1986;43(1):7–13.
51. Williams RN, Paterson CA. A protective role for ascorbic acid during inflammatory episodes in the eye. Exp Eye Res 1986;42(3):211–218.
52. Candia OA, Askew WA. Active sodium transport in the isolated bullfrog cornea. Biochim Biophys Acta 1968;163(2):262–265.
53. Klyce SD. Transport of Na, Cl, and water by the rabbit corneal epithelium at resting potential. Am J Physiol 1975;228(5):1446–1452.
54. Klyce SD, Wong RK. Site and mode of adrenaline action on chloride transport across the rabbit corneal epithelium. J Physiol 1977;266(3):777–799.
55. Zadunaisky JA. Active transport of chloride across the cornea. Nature 1966;209(5028):1136–1137.
56. McGahan MC, Bentley PJ. Stimulation of transepithelial sodium and chloride transport by ascorbic acid. Induction of Na+ channels is inhibited by amiloride. Biochim Biophys Acta 1982;689(2):385–392.

57. Scott WN, Cooperstein DF. Ascorbic acid stimulates chloride transport in the amphibian cornea. Invest Ophthalmol 1975;14(10):763–766.
58. Reid B, Song B, McCaig CD, Zhao M. Wound healing in rat cornea: the role of electric currents. FASEB J 2005;19(3):379–386.
59. DiMattio J. Ascorbic acid entry into cornea of rat and guinea pig. Cornea 1992;11(1):53–65.
60. DiMattio J. Decreased ascorbic acid entry into cornea of streptozotocin-diabetic rats and guinea-pigs. Exp Eye Res 1992;55(2):337–344.
61. Bode AM, Vanderpool SS, Carlson EC, Meyer DA, Rose RC. Ascorbic acid uptake and metabolism by corneal endothelium. Invest Ophthalmol Vis Sci 1991;32(8):2266–2271.
62. Talluri RS, Katragadda S, Pal D, Mitra AK. Mechanism of L-ascorbic acid uptake by rabbit corneal epithelial cells: evidence for the involvement of sodium-dependent vitamin C transporter 2. Curr Eye Res 2006;31(6):481–489.
63. Dreyer R, Rose RC. Lacrimal gland uptake and metabolism of ascorbic acid. Proc Soc Exp Biol Med 1993;202(2):212–216.
64. Srinivasan BD, Jakobiec FA, Iwamoto T. Conjunctiva. In: Jakobiec F.A., editor. Ocular Anatomy, Embryology, and Teratology. Philadelphia: Harper and Row; 1982. p. 733–760.
65. Shiue MH, Kulkarni AA, Gukasyan HJ, Swisher JB, Kim KJ, Lee VH. Pharmacological modulation of fluid secretion in the pigmented rabbit conjunctiva. Life Sci 2000;66(7):L105–L111.
66. Hosoya K, Kompella UB, Kim KJ, Lee VH. Contribution of Na(+)-glucose cotransport to the short-circuit current in the pigmented rabbit conjunctiva. Curr Eye Res 1996;15(4):447–451.
67. Shi XP, Candia OA. Active sodium and chloride transport across the isolated rabbit conjunctiva. Curr Eye Res 1995;14(10):927–935.
68. Kompella UB, Kim KJ, Shiue MH, Lee VH. Possible existence of Na(+)-coupled amino acid transport in the pigmented rabbit conjunctiva. Life Sci 1995;57(15):1427–1431.
69. Hosoya K, Horibe Y, Kim KJ, Lee VH. Nucleoside transport mechanisms in the pigmented rabbit conjunctiva. Invest Ophthalmol Vis Sci 1998;39(2):372–377.
70. Kompella UB, Kim KJ, Lee VH. Active chloride transport in the pigmented rabbit conjunctiva. Curr Eye Res 1993;12(12):1041–1048.
71. Horibe Y, Hosoya K, Kim KJ, Ogiso T, Lee VH. Polar solute transport across the pigmented rabbit conjunctiva: size dependence and the influence of 8-bromo cyclic adenosine monophosphate. Pharm Res 1997;14(9):1246–1251.
72. Hosoya K, Lee VH. Cidofovir transport in the pigmented rabbit conjunctiva. Curr Eye Res 1997;16(7):693–697.
73. Saha P, Uchiyama T, Kim KJ, Lee VH. Permeability characteristics of primary cultured rabbit conjunctival epithelial cells to low molecular weight drugs. Curr Eye Res 1996;15(12):1170–1174.
74. Rose RC, Richer SP, Bode AM. Ocular oxidants and antioxidant protection. Proc Soc Exp Biol Med 1998;217(4):397–407.
75. Gogia R, Richer SP, Rose RC. Tear fluid content of electrochemically active components including water soluble antioxidants. Curr Eye Res 1998;17(3):257–263.
76. Batey DW, Eckhert CD. Analysis of flavins in ocular tissues of the rabbit. Invest Ophthalmol Vis Sci 1991;32(7):1981–1985.
77. Gukasyan HJ, Lee VH, Kim KJ, Kannan R. Net glutathione secretion across primary cultured rabbit conjunctival epithelial cell layers. Invest Ophthalmol Vis Sci 2002;43(4):1154–1161.
78. Kuizenga A, van Haeringen NJ, Kijlstra A. Inhibition of hydroxyl radical formation by human tears. Invest Ophthalmol Vis Sci 1987;28(2):305–313.

79. Crouch RK, Goletz P, Snyder A, Coles WH. Antioxidant enzymes in human tears. J Ocul Pharmacol 1991;7(3):253–258.
80. Patel S, Plaskow J, Ferrier C. The influence of vitamins and trace element supplements on the stability of the pre-corneal tear film. Acta Ophthalmol (Copenh) 1993;71(6):825–829.
81. Peponis V, Papathanasiou M, Kapranou A et al. Protective role of oral antioxidant supplementation in ocular surface of diabetic patients. Br J Ophthalmol 2002;86(12):1369–1373.
82. Peponis V, Bonovas S, Kapranou A et al. Conjunctival and tear film changes after vitamin C and E administration in non-insulin dependent diabetes mellitus. Med Sci Monit 2004;10(5): CR213–CR217.
83. Freyschuss A, Xiu RJ, Zhang J et al. Vitamin C reduces cholesterol-induced microcirculatory changes in rabbits. Arterioscler Thromb Vasc Biol 1997;17(6):1178–1184.
84. Amemiya T. The eye and nutrition. Jpn J Ophthalmol 2000;44(3):320.
85. Tseng SC, Hirst LW, Maumenee AE, Kenyon KR, Sun TT, Green WR. Possible mechanisms for the loss of goblet cells in mucin-deficient disorders. Ophthalmology 1984;91(6):545–552.
86. Choy CK, Benzie IF, Cho P. Ascorbic acid concentration and total antioxidant activity of human tear fluid measured using the FRASC assay. Invest Ophthalmol Vis Sci 2000;41(11):3293–3298.

II
Transporters of the Ciliary Epithelium

4
Mechanisms of Aqueous Humor Formation
Cellular Model of Aqueous Inflow

Chi-wai Do, Chi-wing Kong, Chu-yan Chan, Mortimer M. Civan, and Chi-ho To

CONTENTS

INTRODUCTION
POSSIBLE MECHANISMS FOR THE SECRETION OF AQUEOUS HUMOR
ION TRANSPORT BY CILIARY EPITHELIUM
Na^+ SECRETION
HCO_3^- SECRETION
Cl^- SECRETION
Cl^- REABSORPTION
REGULATION OF AQUEOUS HUMOR FORMATION
CONCLUSION
REFERENCES

INTRODUCTION

Glaucoma is a leading cause of irreversible blindness in the world *(1)*. Primary open-angle glaucoma (POAG) is the most common form of glaucoma and its onset, as well as progression, is often insidious, leading to significant visual loss before being clinically diagnosed. POAG is frequently associated with elevated intraocular pressure (IOP) of the eye. It is incurable at present, although its progression can be retarded by lowering the IOP by medication and/or surgery. If pharmacological agents fail to attain a targeted hypotensive response within the framework of the clinical characteristics and course, surgery will be indicated.

The IOP is physiologically maintained within a relatively narrow range bracketing a mean of approximately 15 mmHg. The level of IOP reflects a dynamic balance between secretion (inflow) and drainage (outflow) of aqueous humor. The aqueous humor is a transparent fluid that is formed by the ciliary processes (epithelium) of the eye. After its production, aqueous humor flows from the posterior chamber to the anterior chamber of the eye via the pupil (Fig. 1). In the anterior chamber, the temperature difference between the warmer iris and the cooler cornea results in a convectional fluid circulation.

From: *Ophthalmology Research: Ocular Transporters in Ophthalmic Diseases and Drug Delivery*
Edited by: J. Tombran-Tink and C. J. Barnstable © Humana Press, Totowa, NJ

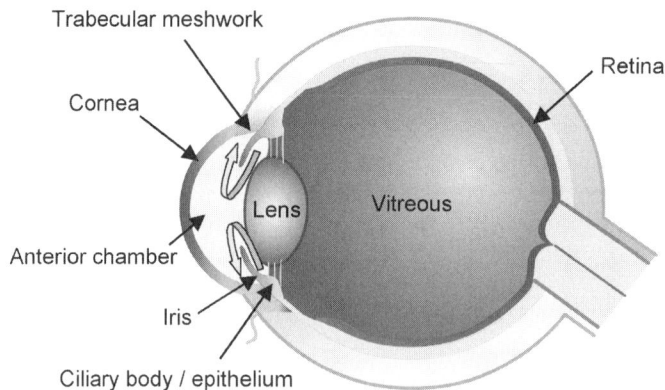

Fig. 1. A schematic diagram showing the anatomical structures of the eye and the route of aqueous flow (indicated by the direction of the arrows).

In principle, the surgical therapy for glaucoma, peripheral iridectomy, would be expected to short-circuit this convective flow, but positional changes of the eyes likely maintain adequate mixing and delivery of fresh aqueous humor to the avascular anterior segment. Eventually, aqueous humor exits the eye through one of the two major routes: the pressure-dependent trabecular pathway and the pressure-independent uveoscleral pathway *(2)*.

The flow of aqueous humor serves many important ocular functions *(2, 3)*: (i) to maintain an optimal level of IOP for structural integrity and normal optical functioning of the eye; (ii) to provide nutrients for, as well as to remove metabolic wastes from, avascular tissues of the anterior segment; (iii) to maintain a high concentration of ascorbate in most species, including human, as ascorbate has been shown to function as an antioxidant to protect the globe from oxidative damages; and (iv) to participate in cellular and humoral immune responses under adverse conditions such as inflammation and infection.

Many anti-glaucoma drugs aim to lower IOP by either reducing the inflow or facilitating the outflow. Currently, lowering of the IOP is the only effective method known to slow the onset and progression of glaucomatous vision loss *(4–6)*. However, many patients develop resistance to current clinical anti-glaucoma drugs and the side-effects of these drugs are frequently significant, eventually reducing patient's compliance. Such limitations point to the need for novel and more-potent medications. In this chapter, we shall focus only on the cellular mechanisms that are responsible for the secretion of aqueous humor, which may aid the design of novel hypotensive drugs to combat the worldwide problem of glaucoma.

POSSIBLE MECHANISMS FOR THE SECRETION OF AQUEOUS HUMOR

The aqueous humor is formed via the ciliary epithelium of the ciliary body. In theory, the formation of aqueous humor can be achieved by three interdependent physiologic processes: diffusion, ultrafiltration, and active secretion. Diffusion is the passive movement of solutes across the cell membrane down a concentration and/or electrical gradient. However, the fact that the concentrations of some solutes (e.g., Cl⁻ and ascorbate) in the aqueous humor are higher than expected from electrochemical equilibrium of the blood plasma suggests that mechanisms other than diffusion are required to produce the aqueous humor *(2, 7, 8)*. Likewise, ultrafiltration is the passive flow of water and water-soluble substances across the cell membrane in response to the hydrostatic pressure difference. It was suggested that ultrafiltration could account for approximately 80% of aqueous humor formation because of the high hydraulic conductivity of the ciliary processes *(9)*. Nevertheless, the hydrostatic pressure in the ciliary stroma was later shown to be smaller than the sum of the IOP and the oncotic pressure, thereby favoring reabsorption rather than ultrafiltration of aqueous humor *(10)*. At present, ultrafiltration is believed to play a minor role (~20%) only in aqueous humor formation *(11, 12)*.

Active secretion has been proposed to play a dominant role and is responsible for about 80% of aqueous humor formation under physiological condition *(11)*. Experimental evidence has shown that the rate of aqueous humor secretion can be suppressed by a variety of metabolic inhibitors *(13–15)*, hypothermia *(16)*, and anoxia *(17, 18)*, supporting the functional significance of active secretion in mediating the aqueous inflow.

ION TRANSPORT BY CILIARY EPITHELIUM

The aqueous humor is secreted by the ciliary epithelium which comprises the pigmented ciliary epithelial (PE) layer facing the ciliary stroma and the nonpigmented ciliary epithelial (NPE) layer facing the aqueous humor (Fig. 2). This bilayered epithelium is unique in orientation, with the apical surfaces of the two cell layers juxtaposed to each other. Neighboring cells between the PE and NPE layers are coupled by intercellular gap junctions, thereby forming a functional syncytium *(19)*.

In essence, the secretion of aqueous humor is driven primarily by active ion secretion across the ciliary epithelium in the direction from stroma to aqueous humor surface, which is followed by osmotic water movement. The transepithelial secretion involves at least three transport steps: (i) uptake of ions from ciliary stroma by the PE cells; (ii) transfer of ions from PE to NPE cells via the gap functions; and (iii) release of ions from the NPE cells into the posterior chamber *(20)*. Since Na⁺, HCO_3^- and Cl⁻ are the major ions in the aqueous humor, they are suggested to participate in the process and will be discussed in detail below.

NA⁺ SECRETION

In the ciliary epithelium, Na⁺ transport is primarily mediated by an enzyme Na⁺, K⁺-adenosine triphosphatase (ATPase) that consists of two α-subunits and two β-subunits. The α-subunit is the catalytic unit which is responsible for ion transport and ATPase activity. The β-subunit facilitates the proper folding of the α-subunit, thereby ensuring the trafficking of the ATPase to the plasma membrane and modifying the corresponding activity *(21)*. Na⁺, K⁺-ATPase acts by extruding three Na⁺, in exchange for the ingress

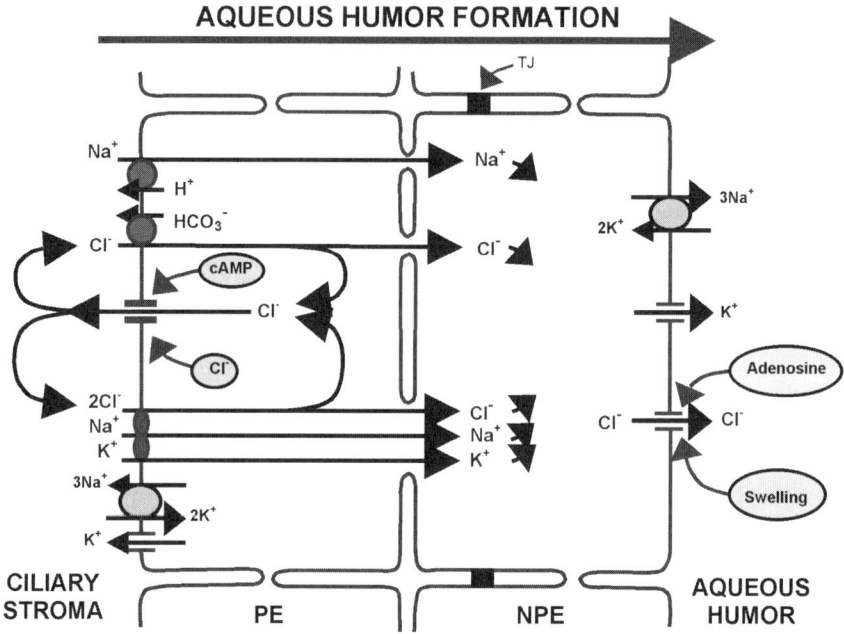

Fig. 2. A simplified model of aqueous humor formation. NaCl is taken up by the pigmented epithelial (PE) cells from the ciliary stroma by the Na$^+$-K$^+$-2Cl$^-$ symport and parallel Na$^+$/H$^+$ and Cl$^-$/HCO$_3^-$ antiports, and diffuses to the nonpigmented epithelial (NPE) cells through gap junctions. Cytosolic carbonic anhydrase (CA II) enhances turnover of the PE antiports, both by enhancing the rate of delivery of H$^+$ and HCO$_3^-$ and by directly activating the antiports. Additionally, Cl$^-$ channels on both PE and NPE cells provide potential pathways for modulating the aqueous humor secretion. TJ=Tight junctions.

of two K$^+$ into the cell, at the expense of one ATP *(22)*. This results in the generation of the transmembrane Na$^+$ and K$^+$ gradients that are of great importance for a variety of cellular functions *(23)*. The activities of the Na$^+$, K$^+$-ATPase can be blocked by cardiac glycosides such as ouabain. The administration of ouabain via different routes has been shown to lower the IOP in experimental animals *(13, 24)*, primarily by reducing the rate of aqueous humor formation *(15, 25, 26)*.

Na$^+$, K$^+$-ATPase has been identified in the ciliary epithelium of rabbit *(27–29)* and ox *(30)* by histochemical methods, and by ^{3}H-ouabain localization in the rabbit *(28)*. Functional expression of this enzyme has also been demonstrated in different species including rabbit *(14, 31)*, cat *(25)*, ox *(31, 32)*, monkey *(14, 33)*, and human *(34)*. Isoform-specific antibodies have been used to map the distribution of the enzyme subunits in rat and mouse ciliary epithelium *(35)*. The Na$^+$, K$^+$-ATPase is predominantly localized at the basolateral infoldings and the interdigitations of both PE and NPE cells *(28, 36)*, but is differentially expressed in the two cell layers. Higher levels of expression are observed in the NPE cells *(28, 31)*, and the isoform composition differs between the PE and NPE cells *(37, 38)*. The distribution of Na$^+$, K$^+$-ATPase isoforms

Model of Aqueous Inflow

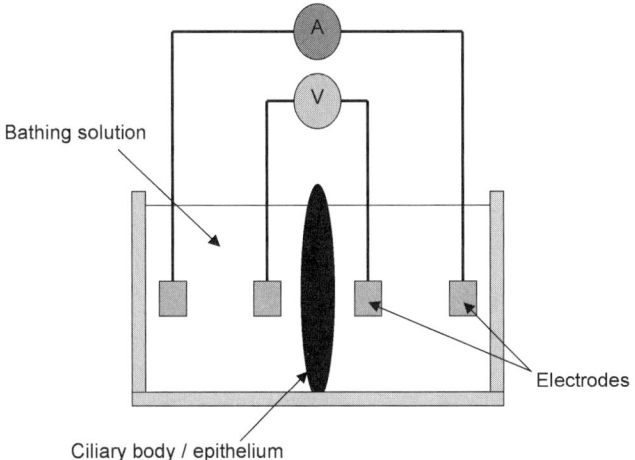

Fig. 3. A simplified diagram of an Ussing–Zerahn-type chamber. The epithelium can be viewed as a partition between two half-chambers, thus forming two compartments, one on each side of the preparation. Two pairs of electrodes are used: one is for the measurement of transepithelial potential difference (PD), whereas the other is for passing electric current through the preparation to measure the tissue resistance.

also varies in different regions of the ciliary epithelium *(30)*. For example, both the α- and β-isoforms have been shown to be more abundant in the pars plicata than in the pars plana *(37)*.

Measurement of the transepithelial potential difference (PD) is a direct approach used to study the ion secretion across the ciliary epithelium (Fig. 3). This was first demonstrated by Cole, who mounted an in vitro ciliary body preparation in the Ussing chamber *(39, 40)*. He demonstrated a PD with aqueous-side positive across the ciliary body of ox and rabbit. Replacing the bathing Na^+ decreased the magnitude of the PD *(39, 40)*. In parallel with this finding, he detected a positive PD by inserting an electrode into the posterior chamber of the rabbit eye *in vivo (41)*. Based on these results, he postulated that the active secretion of Na^+ by the Na^+, K^+-ATPase at the basolateral surface of NPE cells causes a local accumulation of Na^+ at the intercellular clefts and in the invaginations of basal-cell membrane. This generates a local hyperosmotic environment that subsequently drives water into the clefts, leading to the formation of aqueous humor. This hypothesis was later supported by other studies which have investigated the rate of Na^+ accession from the blood plasma to the posterior chamber. It was shown that the rate of Na^+ accession is correlated with the aqueous humor formation rate *(42)*, and intravenous administration of ouabain reduced both Na^+ accession and aqueous humor formation by 30–50% in some species, including cat *(26)*, dog *(42, 43)*, and monkey *(44)*, but not in dogfish *(45)*. He suggested that such discrepancy may be due to the large diffusional Na^+ flux which masks the net Na^+ secretion.

The notion of Na^+ transport was later questioned by several electrophysiologic studies where a negative PD across the ciliary epithelium was observed *(17, 33, 46–49)*. At that time, the negative PD suggested a primacy of an active anion (possibly HCO_3^- and/or Cl^-) over the Na^+ transport. Furthermore, no net Na^+ flux was detected across the ciliary body of rabbit *(50, 51)*, cat *(47)*, toad *(52)*, and ox *(53)*. Interestingly, ouabain did not inhibit, but rather slightly stimulated, the unidirectional Na^+ fluxes when applied to one or both sides of the preparations *(51, 52)*. These results were consistent with the notion that the unidirectional Na^+ fluxes are high, so that relatively small differences in the oppositely directed fluxes may be difficult to detect. This problem was later addressed by measuring the unidirectional Na^+ fluxes under a reduced bathing Na^+ concentration, such that the diffusional fluxes could be minimized *(54)*. At 30 mM bathing Na^+, statistically significant net Na^+ transport in the blood-to-aqueous direction was detected. However, the measured net Na^+ transport was still insufficient to account for the *in vivo* rate of aqueous humor formation. The reason for this discrepancy has been unclear, but might reflect the relatively slow delivery of ions and water to the epithelial surface of the *in vitro* ciliary epithelium. Delivery of stromal solution is expected to be much faster through the microvasculature *in vivo*. This interpretation is consistent with recent findings that the measured rate of aqueous humor formation using the arterially perfused eye *(55, 56)* was much higher than the calculated inflow rate using the excised preparation of ciliary epithelium *(57)* in the same species. In parallel, it has been shown in the living rabbit that aqueous humor secretion is dependent on blood flow when the ciliary blood flow falls below 74% of control values *(58)*.

HCO_3^- SECRETION

It has long been recognized that HCO_3^- transport is involved in the formation of aqueous humor and that carbonic anhydrase (CA) plays a crucial role, either directly or indirectly, in regulating HCO_3^- movement across the ciliary epithelium. CA is an important enzyme that is present in virtually all secretary epithelia. It catalyzes the reversible hydration of CO_2:

$$CO_2 + H_2O \xleftrightarrow{CA} H_2CO_3 \longleftrightarrow H^+ + HCO_3^-$$

At least seven isozymes of mammalian CA (CA I to VII) have been cloned and sequenced. They display different kinetic properties, susceptibility to inhibitors, intracellular location, and tissue distribution *(59, 60)*. Besides, they also participate in a variety of physiological functions such as pH regulation, CO_2 and HCO_3^- transport *(59)*. The presence of CA has been histochemically demonstrated in the ciliary epithelium of rabbit, monkey, and human *(61, 62)*. Also, a number of biochemical studies have demonstrated the CA activity in the ciliary epithelium in different species *(63–67)*. Both the cytosolic CA II *(68)* and membrane-bound CA IV *(67, 69, 70)* have been identified in the ciliary epithelium and their combined effect may give rise to a net transepithelial HCO_3^- transport into the posterior chamber.

The role of CA in secreting aqueous humor was initially proposed by Friedenwald *(71)*. He postulated that CA facilitates the buffering of OH^- by CO_2, leading to the formation of HCO_3^- that is subsequently released into the aqueous humor. The HCO_3^-

secretion is electrically neutralized by Na⁺ entry from the blood plasma *(44)*. Subsequent studies have demonstrated that the rate of HCO_3^- accession to the posterior chamber is very rapid and the administration of acetazolamide, a carbonic anhydrase inhibitor (CAI), decreases the accumulation of HCO_3^- in the aqueous humor *(72–74)*. This finding is consistent with the measurement of the steady-state HCO_3^- concentration in the aqueous humor as compared to its concentration in the blood plasma *(75)*. It has been shown that, in species like rabbit and guinea pig, the HCO_3^- concentration in the aqueous humor is higher than that of the blood plasma, suggesting the presence of an active HCO_3^- transport into the eye. The excess HCO_3^- is necessarily accompanied by a deficit of Cl⁻ in the aqueous humor *(8, 75)*, because of the demands of electroneutrality and isotonicity. Thus, in those species, the aqueous humor is alkaline. In contrast, primates such as the monkey and human secrete acidic aqueous humor with an excess of Cl⁻ and a deficit of HCO_3^- *(74, 75)*. Interestingly, systemic administration of acetazolamide produces opposite effects on these two apparently different species *(74)*. In rabbit, it reduces the aqueous-to-plasma ratio of HCO_3^- but increases the ratio of Cl⁻, whereas in human, it decreases the aqueous-to-plasma ratio of Cl⁻ but increases the ratio of HCO_3^-. Nevertheless, the administration of CAI reduces aqueous humor formation and IOP both in experimental animals *(76–79)* and human *(80, 81)*. These observations are consistent with the concept that the underlying transport mechanisms are common in all these species, regardless of whether the final aqueous humor is slightly acidic or alkaline. Clinically, CAIs have been used in anti-glaucoma treatment to lower IOP for several decades *(82, 83)*. As discussed below, the CAIs are likely to act by inhibiting the first step in secretion of aqueous humor, namely the uptake of NaCl from the stromal fluid.

The importance of anion transport is suggested by the polarity of the PD of the aqueous with respect to the stroma *(17, 33, 46–49, 57, 84)*. When bathed with HCO_3^--rich solution, the rabbit ciliary epithelium displays a negative polarity which is reversible by omitting HCO_3^- from the bathing solution *(17, 46)*. This suggests that HCO_3^- is essential for the maintenance of ionic currents across the ciliary body and is consistent with the notion of predominant anion secretion into the aqueous humor. In studies of intracellular pH (pH_i) regulation and ²²Na uptake, a Cl⁻-dependent, 2,2′-(1,2-ethenediyl)bis (5-isothiocyanatobenzenesulfonic acid) (DIDS)-sensitive Na⁺-HCO_3^- symport was found in cultured bovine PE cells *(85, 86)*. Similarly, a Na⁺-dependent HCO_3^- uptake mechanism has been identified in rabbit PE cells *(87)*, providing a potential source for HCO_3^- accumulation in the ciliary epithelium. In addition, HCO_3^- exit pathways into the posterior chamber could be provided both by the Cl⁻/HCO_3^- exchangers at the basolateral membrane of NPE cells *(88, 89)*, and by the Cl⁻ channels *(90, 91)*. Taken together, these results indicate that transepithelial HCO_3^- secretion can certainly proceed across the ciliary epithelium. However, HCO_3^- constitutes only 20–25% of the total anion composition of the aqueous humor in all mammalian species, so that cations secreted are anticipated to be predominantly accompanied by Cl⁻. Indeed, a recent transport study failed to identify day net HCO_3^- transport across bovine ciliary epithelium *(92)*, underlining the likelihood that HCO_3^- may play an indirect role in aqueous humor formation by modulating the net Cl⁻ secretion *(92)*.

Cl⁻ SECRETION

The importance of Cl⁻ transport had not been fully recognized until recent years. At present, it is generally believed that anion transport is important in the aqueous humor formation and Cl⁻ is likely to be the key player in driving the aqueous inflow *(3)*.

Whether Cl⁻ transport plays a role in mediating aqueous humor secretion had been controversial for some time. Several investigators have demonstrated a clear Cl⁻ dependence of short-circuit current (I_{sc}) across the ciliary body of cat *(47)*, toad *(49)*, and rabbit *(46)*. Moreover, addition of furosemide, a widely used loop diuretic, to the bathing solution reduces the I_{sc} in the ciliary body preparation of different species *(46, 48, 93)*. Nevertheless, neither a Cl⁻-dependent nor furosemide-sensitive I_{sc} was demonstrated in other studies of rabbit *(17, 51)*. Similarly, the data on the net transepithelial Cl⁻ transport across the ciliary epithelium was conflicting. A substantial net Cl⁻ transport in the blood-to-aqueous direction was observed in cat *(47)*, toad *(52)*, and rabbit *(50)*, albeit not observed in a subsequent study of the rabbit eye *(51)*. As mentioned above, Cl⁻ secretion across the ciliary epithelium should proceed through three sequential events and they will be summarized in detail below.

Cl⁻ Uptake by PE Cells

The intracellular Cl⁻ concentration in the ciliary epithelium is significantly higher than that predicted from the electrochemical equilibrium, indicating the presence of an active Cl⁻ uptake pathway in the PE cells *(94–97)*. This Cl⁻ uptake is likely to be mediated through a furosemide-sensitive mechanism because the intracellular Cl⁻ activity can be significantly lowered by applying this transport inhibitor *(94)*.

It has been shown that the uptakes of radiolabeled ^{22}Na and ^{36}Cl in cultured bovine PE cells are interdependent *(66)*. ^{22}Na uptake into the PE cells is stimulated by the presence of Cl⁻ in the bathing media, and vice versa. Based on the measurements of pH_i and radiolabeled tracer uptake, two major pathways have been proposed for Cl⁻ uptake into PE cells *(20, 66, 98–100)*: Na⁺-K⁺-2Cl⁻ symport and parallel Cl⁻/HCO_3^- and Na⁺/H⁺ antiports. Both of the pathways subserve the electroneutral uptake of Na⁺ and Cl⁻ into the ciliary epithelium. Results of subsequent work have indicated that both sets of mechanisms are likely operative *(55, 57, 92, 101, 102)*, although their relative contributions may vary with species and with experimental conditions.

Na⁺-K⁺-2Cl⁻ Symport Versus Parallel Cl⁻/HCO_3^- and Na⁺/H⁺ Antiports

A recent immunocytochemical study of young-calf ciliary epithelium has shown that the majority of Na⁺-K⁺-2Cl⁻ cotransporters are localized along the basolateral border of PE cells in the anterior pars plicata region displaying prominent ciliary processes *(38)*. Blocking the Na⁺-K⁺-2Cl⁻ transport activity with the selective inhibitor bumetanide or nonspecific inhibitor furosemide has been found to reduce Na⁺, Cl⁻ or water uptake by PE cells or ciliary epithelium. For example, furosemide reduces the intracellular Cl⁻ activity of shark ciliary epithelium *(94)*. Bumetanide inhibits the ^{22}Na and ^{36}Cl uptake by cultured bovine PE cells *(66)* and shrinks native bovine PE cells *(103)*. Additionally, the shrinkage of PE cells induced by reducing the bathing Na⁺, K⁺, and Cl⁻ concentrations can be prevented by bumetanide *(103)*. These results clearly demonstrate the presence of a functional, bumetanide-sensitive Na^+-K^+-$2Cl^-$ cotransporter in the PE cells.

Blocking the Na^+-K^+-$2Cl^-$ cotransporter with bumetanide reduces transepithelial Cl^- transport and aqueous humor formation *in vitro*. Stromal bumetanide inhibits both I_{sc} and net Cl^- secretion across rabbit *(101)*, porcine *(102)* and bovine *(57)* ciliary epithelium. The addition of bumetanide to the arterial perfusate also slows aqueous humor secretion by isolated bovine eyes *(55)*. Interestingly, bumetanide inhibits the net Cl^- secretion not only by reducing the blood-to-aqueous Cl^- flux (secretion), but also by increasing the aqueous-to-blood Cl^- transport (reabsorption) *(57, 101, 102)*. The exact reason for the increase in back-flux is not entirely clear, but may reflect two possibilities: (i) the enhancement of NPE-cell reabsorptive pathway *(38, 70, 104–107)* triggered by a reduction of intracellular Cl^- concentration when bumetanide is added to inhibit the Cl^- uptake into the cells; or (ii) bumetanide-altered the permeability of the paracellular pathways.

Parallel Cl^-/HCO_3^- and Na^+/H^+ antiports are likely to be important both in regulating pH_i and Cl^- secretion across ciliary epithelium. Both Cl^-/HCO_3^- and Na^+/H^+ exchangers are involved in the ^{22}Na and ^{36}Cl uptake by cultured bovine PE cells *(20, 66)*. A later study also showed that they play a significant part in the fluid uptake, ^{22}Na entry and pH_i regulation in cultured bovine PE cells *(108)*. In this later study, the pharmacologic profile of Na^+/H^+-exchange inhibition led to the identification of the NHE-1 isoform. Reverse-transcriptase polymerase chain reaction (RT-PCR) amplification of RNA from human ciliary body, and immunostaining of the cultured bovine PE cells, indicated that the Cl^-/HCO_3^- antiport is the AE2 isoform *(108)*. The turnover of these antiports can be accelerated by the enzyme CA that, as noted above, catalyzes the reversible formation of H^+ and HCO_3^- from CO_2 and water *(109)*. The cytosolic CA II enhances the delivery of H^+ and HCO_3^- to the antiports and the membrane-bound CA IV catalyzes the reformation of CO_2 and water from H^+ and HCO_3^- at the stromal surface of the PE cells *(104)*. Carbonic anhydrase II has a more-direct action, as well, indirectly increasing the turnover rates of the NHE-1 *(110)* and AE2 antiports *(111)*. It is likely that CAIs reduce aqueous inflow by interfering with the effects of CA II and CA IV in catalyzing NaCl uptake by the PE cells *(92)*.

Considerable evidence indicates that both sets of electroneutral transporters can subserve NaCl uptake by the PE cells from the stroma, but their relative contributions may vary among different species and experimental conditions. For example, data from the isolated bovine ciliary epithelia and arterially perfused bovine eye preparation suggest approximately equal contributions from the two pathways *(55, 57, 101, 112)*. In addition, bumetanide inhibits the blood-to-aqueous Cl^- transport by half in rabbit, and the other half may be contributed by the Cl^-/HCO_3^- exchanger *(101)*. However, another study has shown that the parallel antiports are the dominant Cl^- uptake pathway in the rabbit ciliary epithelium *(113)*. Electron microprobe analysis has demonstrated that, in the presence of CO_2/HCO_3^-, blocking CA with acetazolamide *(113)* or blocking the Na^+/H^+ exchanger with dimethylamiloride *(114)* reduces the intracellular Cl^- content. Interestingly, blocking the Na^+-K^+-$2Cl^-$ cotransporter with bumetanide lowers the Cl^- content only in HCO_3^--free conditions *(113)*. In the presence of a physiologic concentration of HCO_3^-, bumetanide produces a paradoxical increase of Cl^- content.

This leads to the conclusion that the parallel antiports may have increased the intracellular Cl^- concentration to a sufficient level to reverse the thermodynamic driving force imposed on the Na^+-K^+-$2Cl^-$ symport under HCO_3^--rich conditions *(115)*. In that case, blocking the symport could have inhibited Cl^- release rather than uptake in rabbit. Another possibility is that the increase in Cl^- content with bumetanide might have reflected the upregulation of the Cl^-/HCO_3^- antiport activity, thereby over-compensating for the reduction of Cl^- entry into the PE cells when the Na^+-K^+-$2Cl^-$ symport is blocked.

Irrespective of the exact mechanism underlying the paradoxical effect of bumetanide, subsequent studies of IOP in living mice have shown that topical application of selective inhibitors of Na^+/H^+ exchanger or of acetazolamide lower baseline IOP, whereas bumetanide by itself does not reduce baseline IOP in the living mouse *(116)* or living cynomolgus monkey *(117)*. However, after blocking the Na^+/H^+ antiport, inhibition of the Na^+-K^+-$2Cl^-$ cotransporter with bumetanide significantly reduces mouse IOP. Apparently, the quantitative contribution of these two sets of transporters in mediating Cl^- uptake may be different in different species. In species such as rabbit and mouse, the role of Cl^-/HCO_3^- antiport may be more pronounced as they secrete HCO_3^- to a higher concentration, whereas in other species such as the cow, the role of Na^+-K^+-$2Cl^-$ symport may be more important *(118)*.

Cl^- Transfer from PE to NPE Cells

Gap junctions provide a route for exchange of solutes and water between PE and NPE cells, as well as between adjacent cells within the same cell layer. The intercellular gap junctions have been well documented by structural *(19, 119)*, biochemical *(120–123)*, and functional *(57, 94–97, 102, 103, 112, 124, 125)* studies, suggesting that the ciliary epithelium is a functional syncytium *(2)*. It has been demonstrated that the connexins Cx40 and Cx43 are present in the PE-NPE cells whereas the connexins Cx26 and Cx31 are found in the NPE-NPE cells of rat ciliary epithelium *(121)*. The identity of the connexins between the PE-PE cells is yet unclear. The communication and transport between these layers can be disrupted by common non-selective gap-junction inhibitors such as octanol *(125)* and heptanol *(126)*. For example, heptanol reduces I_{sc} by approximately 80% across isolated ciliary epithelia of different species *(57, 102, 127)* and the inhibition of I_{sc} is primarily associated with a reduction of net Cl^- secretion *(57, 102)*.

In addition to a reduction of net Cl^- secretion, heptanol triggers a slight reduction (around 7%) in tissue resistance *(57, 102, 112)*. The effect is reversible upon removal of heptanol from the bathing solution *(112)*. Since the transmural conductance (i.e. the reciprocal of the tissue resistance) is dominated by the paracellular conductance, it is likely that heptanol exerts at least two separate effects on ciliary epithelium: (i) interrupting the gap junctions between PE and NPE cells, and (ii) increasing the permeability of the paracellular pathway.

Cl^- Release from NPE Cells into the Posterior Chamber

Cl^- efflux from the NPE cells to the aqueous humor is the last step in Cl^- secretion. It has been suggested that the Cl^- release is likely the rate-limiting step in the aqueous humor secretion *(128, 129)* because: (i) the intracellular Cl^- concentration is several-fold higher than that predicted from its electrochemical equilibrium, indicating that the uptake of Cl⁻ from the stroma by the PE cells is not rate-limiting; (ii) the membrane potentials and intracellular ionic content of the PE and NPE cells are similar, suggesting a free exchange of ions between PE and NPE cells through the gap junctions; and (iii) under baseline condition, the activities of Na⁺,K⁺-ATPase and K⁺ channels at the basolateral membrane of NPE cells are high, indicating that they are not rate-limiting. This hypothesis is supported by the fact that a Cl⁻-channel blocker 5-nitro-2-(3-phenylpropylamino) benzoic acid (NPPB, 0.1 mM) significantly inhibits both net Cl⁻ secretion *(57)* and aqueous humor formation *(55) in vitro*.

The baseline activity of NPE Cl⁻ channels can be stimulated by a number of perturbations, including cyclic adenosine monophosphate (cAMP) *(130–132)*, hypotonic cell swelling *(133, 134)*, inhibition of protein kinase C (PKC) activity *(135–138)*, and stimulation of A_3 adenosine receptors (A_3ARs) *(139, 140)*. At present, the molecular identity of the NPE Cl⁻ channels is still unclear. At least two different channels or channel regulators, namely swelling-activated ClC-3 and pI_{Cln}, are likely involved *(134)*. Several lines of evidence have suggested that the swelling-activated ClC-3 may be the dominant Cl⁻ channel, channel component or channel regulator in NPE cells *(141, 142)*: (i) NPE cells express both ClC-3 transcripts and protein *(141)*; (ii) PKC reduces Cl⁻ channel activity in NPE cells *(135, 141, 143)*, a characteristic of ClC-3-associated Cl⁻ currents *(144)*; and (iii) antisense deoxynucleotides downregulate messages for ClC-3 and swelling-activated Cl⁻ channels in NPE cells *(145)*. Recently, a functional blocking antibody specific for ClC-3 *(146)* has been found to reduce Cl⁻ current of bovine *(147)* and cultured rabbit *(148)* NPE cells. Despite these, the role of ClC-3 in Cl⁻ transport across the plasma membranes of NPE and other cells remains controversial *(149, 150)*, particularly because the swelling-activated Cl⁻ channel activity of mouse pancreatic acinar cells and hepatocytes is retained in ClC-3-negative mice *(151)*. One possible explanation of the conflicting results could be that ClC-3 is of importance in only one of several swelling-activated Cl⁻ channels expressed in different cells *(152)* and ClC-3 may constitute only in part of the complex of proteins comprising a swelling-activated Cl⁻ channel. Additional evidence has shown that pI_{Cln} potentially regulates NPE-cell Cl⁻ channels *(153)*, and the first human form of this protein cloned from NPE cells *(135, 154)*, pI_{Cln}, has also been found in native bovine NPE cells *(155)*, and antisense down-regulation of pI_{Cln} reduces the immunological staining and inhibits the swelling-activated Cl⁻ current. However, subsequent work has shown that pI_{Cln} is localized primarily in the cytoplasm of NPE cells, and neither a translocation of pI_{Cln} from the cytoplasm to the plasma membrane nor changes in pI_{Cln} expression are detected upon exposing the cells to hypotonic solution *(156)*. This suggests that pI_{Cln} may not be the key Cl⁻ channel and that its effect on the Cl⁻ channel conduit could be indirect, possibly through cytostructural restructuring.

The putative importance of swelling-activated Cl⁻ channels *(103, 133)* rests upon the tacit assumptions that swelling-activated Cl⁻ currents observed in isolated NPE cells contribute to Cl⁻ secretion across ciliary epithelium. The distribution of these channels is predominantly at the basolateral surface of the NPE cells facing the aqueous humor. Therefore, activation of these Cl⁻ channels is expected to enhance Cl⁻ secretion. Very recently, it has been demonstrated that the stimulation of swelling-activated Cl⁻ current in NPE cells increases the I_{sc} across the intact ciliary epithelium, leading to an increase of Cl⁻ secretion *(157)*. The time course of the hypotonically triggered stimulation of I_{sc} is comparable to that of the regulatory volume decrease in isolated NPE cells, suggesting that the swelling-activated Cl⁻ channels are indeed predominantly localized at the aqueous surface of NPE cells and their net effect is to enhance aqueous humor formation *(157)*.

Cl⁻ REABSORPTION

At the Basolateral Surface of NPE Cells

Studies of NPE cells of different species, including human, have provided considerable evidence for functional activity of parallel Na⁺/H⁺ and Cl⁻/HCO₃⁻ antiport, a Na⁺-Cl⁻ symport and an amiloride-sensitive Na⁺ channel *(67, 70, 89, 107)*. In addition, several investigators have found Na⁺-K⁺-2Cl⁻ symport activity in cultured human *(107, 158)* and rabbit *(106)* NPE cells. Besides, a messenger for the α-subunit of the epithelial ENaC Na⁺ channel has been shown to be expressed in human ciliary body RNA *(129)*. It is anticipated that the stimulation of these ion transporters and channels facilitate NaCl recycling at the basolateral membrane of NPE cells, thereby increasing the Cl⁻ reabsorption. The increased reabsorption may enhance fluid movement through the PE cells back to the ciliary stroma, reducing the net secretion of aqueous humor.

At the Stromal Surface of PE Cells

The symports and antiports in PE cells have been considered to facilitate uptake of ions from the ciliary stroma *(20)*. PE cells also express Na⁺, K⁺-ATPase, and Cl⁻ channels at their basolateral surfaces *(128)*. The Na⁺, K⁺-ATPase, and Cl⁻ channels could subserve Na⁺ extrusion and Cl⁻ release into the stroma, recycling NaCl at the stromal surface of the PE cells and thereby reducing net NaCl secretion into the aqueous humor. This hypothesis is supported by both electrophysiologic and volumetric measurement of immortalized bovine PE cells *(159)*. Addition of ATP to the bath produces activation of Cl⁻ channels, and thereby shrinkage, of PE cells. ATP activates P_2Y_2 ATP receptors *(160)*, triggering sequential increases in free intracellular Ca²⁺ concentration, phospholipase A_2 activity, prostaglandin E_2 (PGE_2) formation and release, and cAMP formation *(159)*. Interestingly, the R_p stereoisomer of 8-bromo-adenosine 3',5'-cyclic monophosphothioate, an inhibitor of cAMP-activated kinase (PKA), mimics the effects of cAMP, indicating that cAMP may act directly on the Cl⁻ channel rather than through PKA. The antiestrogen tamoxifen enhances the ATP-induced activation of PE-cell Cl⁻ channels, but the mechanism is unclear *(161)*.

The results obtained with transformed PE cells have been extended by patch-clamping native bovine PE cells *(162)*. Cyclic AMP activates maxi-Cl⁻ channels in excised inside-out and outside-out patches. These channels have been identified in many cells, including PE cells *(163)*. However, the channel's open probability (P_o) is known to be highest at a nonphysiologic transmembrane potential between +20 and −20 mV, and therefore its physiologic significance has been obscure. It has been shown that the P_o of cAMP-activated maxi-Cl⁻ channels increases progressively with increasing intracellular Cl⁻ concentration, implying that the maxi-Cl⁻ channels might be of particular importance under conditions of rapid NaCl uptake by the PE cells. For example, if the NPE cells could not release NaCl at the same rate into the aqueous humor, the Cl⁻ concentration would increase in PE cells, thereby enhancing Cl⁻ release through the cAMP-activated maxi-Cl⁻ channels by increasing P_o *(162)* (Fig. 2). Although the physiologic trigger for the cAMP production is still not clear, PE cells are known to release ATP upon cell swelling *(164)*, and still possibly trigger the sequential cascade leading to cAMP formation. Therefore, PE cells that have taken up large amounts of NaCl and water might display negative feedback, release ATP, and trigger the events that lead to cAMP-induced Cl⁻ release back into the stroma. This recycle would reduce net Cl⁻ transfer to the NPE cells and might even constitute in part a pathway for the reabsorption of aqueous humor.

REGULATION OF AQUEOUS HUMOR FORMATION

It has been known that the level of IOP is not constant, but displays diurnal variation. The fluctuation of IOP is observed both in normal and glaucomatous patients, although the latter may demonstrate a more significant variation *(165)*. Rates of aqueous humor secretion have been reported to be 2.61 µl min⁻¹ during the day and 1.08 µl min⁻¹ at night in normal human subjects *(166)*. However, the diurnal changes in IOP are not synchronous with those in inflow. The IOP of the normal human peaks in the dark-sleep period *(167)* at a time when aqueous humor secretion is half of that observed later in the day *(168)*. Evidently, the circadian rhythms in IOP and inflow are regulated through independent, as yet unidentified, mechanisms.

The aqueous humor secretion is subjected to modulation by a number of different cascades including catecholamines, antinatriuretic peptide, endothelin, purines, muscarinic receptors, cannabinoids, tumor necrosis factor-alpha, glucocorticoids, prostanoids, cAMP, cyclic guanosine 3′,5′-monophosphate (cGMP), nitric oxide, Ca^{2+}, pH, PKC, PKA, tyrosine kinase, and MAP kinase *(137, 138, 143, 169–179)*. Among these potential cascades, the mechanisms of three common regulatory pathways involving cAMP, A_3 adenosine receptor, and nitric oxide on aqueous inflow will be discussed below.

Cyclic AMP

Cyclic AMP is the most frequently studied intracellular modulator of aqueous humor secretion, however, the exact physiologic significance of cAMP in mediating aqueous humor formation remains controversial. For example, isoproterenol, a β-adrenergic agonist that increases cAMP production, has been shown to activate the Na^+-K^+-$2Cl^-$ symport of PE cells and stimulate net Cl⁻ secretion across the intact rabbit ciliary epithelium *(101)*. In addition, cAMP likely activates NPE-cell Cl⁻ channels *(130)*, so that cAMP might be expected to enhance Cl⁻ secretion by acting on both surfaces of the ciliary epithelium. Nevertheless, cAMP does not markedly increase the rate of aqueous humor formation with forskolin, which stimulates endogenous cAMP production. Instead, it decreases the rate of inflow *(180, 181)*. In parallel with this finding, other studies have shown that cAMP tends to reduce, rather than increase, net Cl⁻ secretion. Activation of cAMP-triggered Cl⁻ channels at the stromal surface of PE cells is thought to inhibit the net Cl⁻ secretion *(162)*. Recent data also suggest that cAMP uncouples the intercellular gap junctions between PE and NPE cells, leading to a reduction of I_{sc} and net Cl⁻ secretion *(182)*. cAMP has been reported to inhibit

Na$^+$, K$^+$-ATPase activity in NPE cells *(183, 184)*, which would reduce Cl$^-$ secretion. These effects may play a role in the observed forskolin-triggered reduction of aqueous humor secretion *(180, 181)*.

On the other hand, β-adrenergic-receptor antagonists such as timolol are known to reduce endogenous cAMP production and are clinically used to reduce aqueous humor formation and thus IOP. Whether their effects on inflow are solely or primarily mediated via cAMP or other messengers remains to be proven *(114, 185)*. Interestingly, timolol is found to reduce the intracellular K$^+$ and Cl$^-$ content in rabbit ciliary epithelium and the fluid uptake by cultured bovine PE cells, but neither effect can be reversed by simultaneous application of cAMP *(114)*. The complexity of the effects produced by forskolin and β-adrenergic-receptor antagonists likely reflects the multiple actions exerted by cAMP on the transporters and channels of ciliary epithelium, which are dependent on the routes and concentrations of the drugs applied, the baseline activities of the target proteins, and also the species being studied. Given this broad spectrum of actions, it seems improbable that the integrated response of cells, tissues, and organs could be determined by the average intracellular cAMP concentration, suggesting the significance of understanding the compartmentalization within the microenvironments of the plasma membrane *(186)*.

A_3 Adenosine Receptors (A_3ARs)

Among the many potential modulators of aqueous humor secretion, adenosine is proposed to be an important regulator for the modulation of IOP *(187, 188)*. Moreover, adenosine receptors are of particular interest because knockout of A_3ARs has recently found to reduce IOP in the living mouse *(189)*. It has been demonstrated that the level of adenosine in the aqueous humor correlates with the level of IOP *(190)*. Recently, adenosine has been shown to increase the rate of aqueous humor formation without affecting the ciliary blood flow *(191)*. Clinically, patients with pseudoexfoliation syndrome (PEX), which is usually associated with elevated IOP and glaucoma, show an elevated level of A_3ARs at the basolateral surface of NPE cells by an order of magnitude *(192)*, suggesting that A_3AR may play a crucial role in regulating the IOP.

Adenosine can be physiologically delivered to the two surfaces of the ciliary epithelium by ATP release from both PE and NPE cells, and subsequent ecto-enzymatic conversion *(164)*. Adenosine has been reported to stimulate whole-cell Cl$^-$ currents of freshly harvested bovine NPE cells *(193)*. A_3AR-selective agonists also activate whole-cell Cl$^-$ currents of cultured human NPE cells, and such activation could be blocked by a selective A_3AR antagonist *(140)*. The effects of purinergic agents on cell volume have also been studied with cultured human NPE cells. A_3AR-selective agonists stimulate shrinkage and this effect was blocked by A_3AR-selective antagonists *(139)*. Additionally, shrinkage triggered by the non-selective P-1 adenosine agonist was blocked by the A_3AR-selective antagonists *(139)*. It has been shown that transcripts for A_3ARs are expressed in cultured human cells and rabbit ciliary processes *(139)*. Electrophysiological study of cultured human NPE cells has shown that the biophysical characteristics of A_3AR-activated Cl$^-$ channels are similar to those of swelling-activated Cl$^-$ channels, suggesting that A_3-selective agonists and hypotonic swelling might activate a common population of Cl$^-$ channels *(140)*. Using intact bovine ciliary epithelium, it has been recently demonstrated that the swelling-activated Cl$^-$ channels in NPE cells are predominantly located at the basolateral surface of the NPE cells which may contribute to Cl⁻ secretion across ciliary epithelium *(157)*. Activation of these swelling-activated, A_3AR-activated Cl⁻ channels is thought to enhance Cl⁻ release at the aqueous surface, the net Cl⁻ secretion, and aqueous humor formation.

These *in vitro* results have been extended to the living mouse. Consistent with the electrophysiologic and volumetric findings, A_3AR-selective agonists increase IOP, and A_3AR-selective antagonists both reduce IOP and markedly inhibit the subsequent response to adenosine *in vivo (189, 194)*. Similarly, A_3-knockout (Adora3$^{-/-}$) mice display a reduced baseline IOP and diminished responses to adenosine, an A_3AR-selective agonist and an A_3AR-selective antagonist *(189)*.

These results suggest that A_3AR-selective antagonists might be of pharmacologic relevance to human *(142)*, but responses of receptors to these antagonists display substantial species variation *(195, 196)*. Therefore, the full implications of these results for human are yet to be elucidated. In contrast, responses to A_3AR agonists are much more conserved across species. Recently, an A_3AR-selective antagonist (MRS 1292) has been generated by structural modification of an A_3AR-selective agonist *(197)*. This putative cross-species A_3AR-selective antagonist has been found effective on a cellular level, to inhibit adenosine-triggered shrinkage of immortalized human NPE cells, and in the living animal, to lower the mouse IOP *(198)*.

Nitric Oxide

Nitric oxide (NO) has been suggested to regulate several physiological functions in the eye including aqueous humor dynamics *(199)*. Due to its potent pharmacological significance in anti-glaucoma therapy, the effect of NO on IOP has been extensively studied. However, the results of these studies have not been conclusive because of the multiple effects of NO on both inflow and outflow pathways *(200)*. NO donors have been demonstrated to reduce IOP in normal and glaucomatous animals *(201–203)*, and in normal human subjects *(204)*. The hypotensive effect can be primarily explained by the effects of NO on vasodilatation and/or relaxation of smooth muscle, including modulation of ocular blood flow *(205)*, trabecular meshwork and ciliary muscle relaxation *(206)*, and aqueous humor outflow resistance *(202)*.

Classically, the direct effect of NO is brought about by binding to the heme complexes in molecules such as guanylate cyclase (GC). The most notable heme protein that forms an iron-NO adduct *in vivo* is soluble GC (sGC) *(207)*. When NO binds to sGC, it alters the protein configuration and activates the sGC. The activated sGC facilitates the conversion of guanosine triphosphate (GTP) to cGMP *(208)*. Subsequently, the second messenger cGMP acts on the effector proteins, including cGMP-dependent protein kinase (PKG) *(209)*. The NO-sGC-cGMP signal transduction pathway is a major NO signaling pathway *(210)*.

Recent demonstration of nitric oxide synthase (NOS) activity in porcine *(211)* and bovine *(212)* ciliary processes, and the localization of NOS in porcine ciliary epithelium *(213)* further support the involvement of NO in regulating the aqueous humor formation. In an arterially perfused bovine eye study, a nitrovasodilator, sodium azide (AZ), was shown to lower the IOP by acting on the ciliary epithelium directly, but not by relaxing the vascular smooth muscles *(214)*. Consistent with this finding, it has been shown that

L-arginine, a precursor of NO, produces a concomitant reduction of aqueous humor formation and IOP in the arterially perfused porcine eye. The inhibition is blocked by L-NAME, an NOS inhibitor *(215)*. Similarly, a number of agents, including 8-pCPT-cGMP (a cGMP analog), AZ, and sodium nitroprusside (SNP, a NO donor), trigger a significant inhibition of aqueous humor formation and IOP. The inhibitory effects elicited by AZ and SNP can be blocked by ODQ, a sGC inhibitor *(215)*. These findings strongly support the inhibitory influence of NO on aqueous humor formation, and thereby IOP. In excised ciliary epithelium, NO donors such as SNP produce a simultaneous steady-state inhibition of I_{sc} and net transepithelial Cl⁻ secretion across the native porcine ciliary epithelium, supporting the functional significance of NO in mediating the Cl⁻ secretion into the posterior chamber *(216)*. Unlike the results of perfused-eye studies *(215)*, 8-pCPT-cGMP produces a paradoxical stimulation of both I_{sc} and net Cl⁻ flux in porcine ciliary epithelium *(216)*. This finding is consistent with another study using the same tissue, where the activation of the NO-sGC-cGMP pathway stimulated a transmembrane Cl⁻ current *(217)*. In parallel with this finding, activation of the sGC-cGMP-PKG pathway caused a transient stimulation of the I_{sc} *(218)*. The exact reason for the discrepancy is not apparent because of the complexity of the system; thus it awaits further investigation. It might suggest that NO has at least two opposing effects on Cl⁻ secretion, which are differentially activated in the perfused-eye and isolated epithelial preparations.

CONCLUSION

As the population continues to age, it is expected that the number of glaucoma patients will increase by 50% by the year 2020 *(219)*. Therefore, more-effective and specific anti-glaucoma therapies are very much needed. This challenge demands a more thorough understanding of the mechanism of aqueous humor formation and its control.

The aqueous humor is secreted by the bilayered ciliary epithelium comprising PE and NPE facing the stroma and aqueous humor, respectively. So far, it is well established that Cl⁻ secretion is the major driving force for aqueous humor formation in human. This net Cl⁻ movement is likely to proceed in three steps: stromal Cl⁻ uptake into PE cells by electroneutral transporters, diffusion through gap junctions to NPE cells, from which Cl⁻ exits into the aqueous humor through Cl⁻ channels. The recycling of Cl⁻ occurs at both surfaces of the ciliary epithelium, facilitates Cl⁻ reabsorption, and reduces the net secretion. At present, many of the transport mechanisms for Cl⁻ secretion have been identified, yet their regulation and integration are still poorly understood. The evidence for signaling cascades that may affect IOP has gradually emerged, and it will certainly enhance our clinical armamentarium for the regulation of aqueous inflow. In particular, the signaling cascades involving cAMP, A_3ARs, and NO represent promising targets for lowering IOP.

REFERENCES

1. Quigley HA. Number of people with glaucoma worldwide. Br J Ophthalmol 1996;80(5): 389–93.
2. Krupin T, Civan MM. Physiologic basis of aqueous humor formation. In: Ritch R, Shields MB, Krupin T, editors. The glaucomas: basic sciences. St. Louis: Mosby; 1995. p. 251–80.

3. Civan MM, Transport components of net secretion of the aqueous humor and their integrated regulation. In: Civan MM, editor. Curr Topics Membranes 1998. p. 1–24.
4. Collaborative Normal-Tension Glaucoma Study Group. The effectiveness of intraocular pressure reduction in the treatment of normal-tension glaucoma. Am J Ophthalmol 1998;126(4):498–505.
5. Collaborative Normal-Tension Glaucoma Study Group. Comparison of glaucomatous progression between untreated patients with normal-tension glaucoma and patients with therapeutically reduced intraocular pressures. Am J Ophthalmol 1998; 126(4):487–97.
6. The AGIS investigators. The advanced glaucoma intervention study (AGIS): 7. The relationship between control of intraocular pressure and visual field deterioration. Am J Ophthalmol 2000;130(4):429–40.
7. Davson H. The aqueous humour and the intraocular pressure. In: H. Davson H, editor. Physiology of the eye. London: Macmillan; 1990. p. 3–95.
8. Kinsey VE. Comparative chemistry of aqueous humor in posterior and anterior chambers of rabbit eye. Arch Ophthalmol 1953;50:401–17.
9. Green K, Pederson JE. Contribution of secretion and filtration to aqueous humor formation. Am J Physiol 1972;222(5):1218–26.
10. Bill A. The role of ciliary blood flow and ultrafiltration in aqueous humor formation. Exp Eye Res 1973;16(4):287–98.
11. Bill A. Blood circulation and fluid dynamics in the eye. Physiol Rev 1975;55(3):383–417.
12. Cole DF. Secretion of the aqueous humour. Exp Eye Res 1977;25(Suppl):161–76.
13. Becker B. Ouabain and aqueous humor dynamics in the rabbit eye. Invest Ophthalmol 1963;2(4):325–31.
14. Becker B. Vanadate and aqueous humor dynamics. Proctor Lecture. Invest Ophthalmol Vis Sci 1980;19(10):1156–65.
15. Kodama T, Reddy VN, Macri FJ. Pharmacological study on the effects of some ocular hypotensive drugs on aqueous humor formation in the arterially perfused enucleated rabbit eye. Ophthalmic Res 1985;17(2):120–4.
16. Cole DF. Evidence for active transport of chloride in ciliary epithelium of the rabbit. Exp Eye Res 1969;8(1):5–15.
17. Krupin T, et al. Transepithelial electrical measurements on the isolated rabbit iris-ciliary body. Exp Eye Res 1984;38(2):115–23.
18. Chu TC, Candia OA. Active transport of ascorbate across the isolated rabbit ciliary epithelium. Invest Ophthalmol Vis Sci 1988;29(4):594–9.
19. Raviola G, Raviola E. Intercellular junctions in the ciliary epithelium. Invest Ophthalmol Vis Sci 1978;17(10):958–81.
20. Wiederholt M, Helbig H, Korbmacher C. Ion transport across the ciliary epithelium: Lessons from cultured cells and proposed role of the carbonic anhydrase. In Botré F, Gross G, Storey BT, editors. Carbonic Anhydrase. New York: VCH; 1991. p. 232–244.
21. Geering K. The functional role of beta subunits in oligomeric P-type ATPases. J Bioenerg Biomembr 2001;33(5):425–38.
22. Glynn IM. Annual review prize lecture. 'All hands to the sodium pump'. J Physiol 1993;462:1–30.
23. Mobasheri A, et al. Na+,K+-ATPase isozyme diversity; comparative biochemistry and physiological implications of novel functional interactions. Biosci Rep 2000;20(2):51–91.
24. Waitzman MB, Jackson RT. Effects of topically administered ouabain on aqueous humor dynamics. Exp Eye Res 1965;4(3):135–45.
25. Oppelt WW, White ED. Effect of ouabain on aqueous humor formation rate in cats. Invest Ophthalmol 1968;7(3):328–33.

26. Garg LC, Oppelt WW. The effect of ouabain and acetazolamide on transport of sodium and chloride from plasma to aqueous humor. J Pharmacol Exp Ther 1970;175(2):237–47.
27. Cole DF. Location of ouabain-sensitive adenosine triphosphatase in ciliary epithelium. Exp Eye Res 1964;3:72–75.
28. Usukura J, Fain GL, Bok D. Ouabain localization of Na-K ATPase in the epithelium of rabbit ciliary body pars plicata. Invest Ophthalmol Vis Sci 1988;29(4):606–14.
29. Flügel C, Lütjen-Drecoll E. Presence and distribution of Na+/K+-ATPase in the ciliary epithelium of the rabbit. Histochemistry 1988;88(3–6):613–21.
30. Ghosh S., et al. Cellular distribution and differential gene expression of the three alpha subunit isoforms of the Na,K-ATPase in the ocular ciliary epithelium. J Biol Chem 1990;265(5):2935–40.
31. Riley MV, Kishida K. ATPases of ciliary epithelium: cellular and subcellular distribution and probable role in secretion of aqueous humor. Exp Eye Res 1986;42(6):559–68.
32. Helbig H, Korbmacher C, Wiederholt M. K+-conductance and electrogenic Na+/K+ transport of cultured bovine pigmented ciliary epithelium. J Membr Biol 1987;99(3):173–86.
33. Chu TC, Candia OA, Podos SM. Electrical parameters of the isolated monkey ciliary epithelium and effects of pharmacological agents. Invest Ophthalmol Vis Sci 1987;28(10): 1644–8.
34. Helbig H. et al. Electrical membrane properties of a cell clone derived from human nonpigmented ciliary epithelium. Invest Ophthalmol Vis Sci 1989;30(5):882–9.
35. Wetzel RK, Eldred ED. Immunocytochemical localization of NaK-ATPase isoforms in the rat and mouse ocular ciliary epithelium. Invest Ophthalmol Vis Sci 2001;42(3):763–9.
36. Mori N, Yamada E, Sears ML. Immunocytochemical localization of Na/K-ATPase in the isolated ciliary epithelial bilayer of the rabbit. Arch Histol Cytol 1991;54(3):259–65.
37. Ghosh S, et al. Expression of multiple Na+,K(+)-ATPase genes reveals a gradient of isoforms along the nonpigmented ciliary epithelium: functional implications in aqueous humor secretion. J Cell Physiol 1991;149(2):184–94.
38. Dunn JJ, Lytle C, Crook RB. Immunolocalization of the Na-K-Cl cotransporter in bovine ciliary epithelium. Invest Ophthalmol Vis Sci 2001;42(2):343–53.
39. Cole DF. Electrical potential across the isolated ciliary body observed {in vitro}. Br J Ophthalmol 1961;45:641–53.
40. Cole DF. Transport across the isolated ciliary body of ox and rabbit. Br J Ophthalmol 1962;46:577–91.
41. Cole DF. Electrochemical changes associated with the formation of the aqueous humour. Br J Ophthalmol 1961;45:202–17.
42. Maren TH. The rates of movement of Na+, Cl-, and HCO3- from plasma to posterior chamber: effect of acetazolamide and relation to the treatment of glaucoma. Invest Ophthalmol 1976;15(5):356–64.
43. Zimmerman TJ, et al. The effect of acetazolamide on the movement of sodium into the posterior chamber of the dog eye. J Pharmacol Exp Ther 1976;199(3):510–7.
44. Maren TH. Ion secretion into the posterior aqueous humor of dogs and monkeys. Exp Eye Res 1977;25(Suppl):245–7.
45. Maren TH, et al. The rates of ion movement from plasma to aqueous humor in the dogfish, Squalus acanthias. Invest Ophthalmol 1975;14(9):662–73.
46. Kishida K, et al. Electric characteristics of the isolated rabbit ciliary body. Jpn J Ophthalmol 1981;25:407–416.
47. Holland MG, Gipson CC, Chloride ion transport in the isolated ciliary body. Invest Ophthalmol 1970;9(1):20–9.

48. Iizuka S, et al. Electrical characteristics of the isolated dog ciliary body. Curr Eye Res 1984;3(3):417–21.
49. Watanabe T, Saito Y. Characteristics of ion transport across the isolated ciliary epithelium of the toad as studied by electrical measurements. Exp Eye Res 1978;27(2):215–26.
50. Kishida K, et al. Sodium and chloride transport across the isolated rabbit ciliary body. Curr Eye Res 1982;2(3):149–52.
51. Pesin SR, Candia OA. Na+ and Cl- fluxes, and effects of pharmacological agents on the short-circuit current of the isolated rabbit iris-ciliary body. Curr Eye Res 1982;2(12):815–27.
52. Saito Y, Watanabe T. Relationship between short-circuit current and unidirectional fluxes of Na and Cl across the ciliary epithelium of the toad: demonstration of active Cl transport. Exp Eye Res 1979;28(1):71–9.
53. To CH, et al. Chloride and sodium transport across bovine ciliary body/epithelium (CBE). Curr Eye Res 1998;17(9):896–902.
54. Candia OA, Shi XP, Chu TC. Ascorbate-stimulated active Na+ transport in rabbit ciliary epithelium. Curr Eye Res 1991;10(3):197–203.
55. Shahidullah M, et al. Effects of ion transport and channel-blocking drugs on aqueous humor formation in isolated bovine eye. Invest Ophthalmol Vis Sci 2003;44(3):1185–91.
56. Wilson WS, Shahidullah M, Millar C. The bovine arterially-perfused eye: an in vitro method for the study of drug mechanisms on IOP, aqueous humour formation and uveal vasculature. Curr Eye Res 1993;12(7):609–20.
57. Do CW, To CH. Chloride secretion by bovine ciliary epithelium: a model of aqueous humor formation. Invest Ophthalmol Vis Sci 2000;41(7):1853–60.
58. Reitsamer HA, Kiel JW. Relationship between ciliary blood flow and aqueous production in rabbits. Invest Ophthalmol Vis Sci 2003;44(9):3967–71.
59. Sly WS, Hu PY. Human carbonic anhydrases and carbonic anhydrase deficiencies. Annu Rev Biochem 1995;64:375–401.
60. Lindskog S. Structure and mechanism of carbonic anhydrase. Pharmacol Ther 1997;74(1):1–20.
61. Lütjen-Drecoll E, Lonnerholm G. Carbonic anhydrase distribution in the rabbit eye by light and electron microscopy. Invest Ophthalmol Vis Sci 1981;21(6):782–97.
62. Lütjen-Drecoll E, Lonnerholm G, Eichhorn M. Carbonic anhydrase distribution in the human and monkey eye by light and electron microscopy. Graefes Arch Clin Exp Ophthalmol 1983;220(6):285–91.
63. Dobbs PC, Epstein DL, Anderson PJ. Identification of isoenzyme C as the principal carbonic anhydrase in human ciliary processes. Invest Ophthalmol Vis Sci, 1979;18(8):867–70.
64. Wistrand PJ, Garg LC. Evidence of a high-activity C type of carbonic anhydrase in human ciliary processes. Invest Ophthalmol Vis Sci 1979;18(8):802–6.
65. Muther TF, Friedland BR. Autoradiographic localization of carbonic anhydrase in the rabbit ciliary body. J Histochem Cytochem 1980;28(10):1119–24.
66. Helbig H, et al. Coupling of 22Na and 36Cl uptake in cultured pigmented ciliary epithelial cells: a proposed role for the isoenzymes of carbonic anhydrase. Curr Eye Res 1989;8(11):1111–9.
67. Wu Q, Delamere NA, Pierce WM. Membrane-associated carbonic anhydrase in cultured rabbit nonpigmented ciliary epithelium. Invest Ophthalmol Vis Sci 1997;38(10):2093–102.
68. Wistrand PJ, Schenholm M, Lonnerholm G. Carbonic anhydrase isoenzymes CA I and CA II in the human eye. Invest Ophthalmol Vis Sci 1986;27(3):419–28.
69. Ridderstrale Y, Wistrand PJ, Brechue WF. Membrane-associated CA activity in the eye of the CA II-deficient mouse. Invest Ophthalmol Vis Sci 1994;35(5):2577–84.

70. Matsui H, et al. Membrane carbonic anhydrase (IV) and ciliary epithelium. Carbonic anhydrase activity is present in the basolateral membranes of the non-pigmented ciliary epithelium of rabbit eyes. Exp Eye Res 1996;62(4):409–17.
71. Friedenwald JS. The formation of the intraocular fluid. Proctor Lecture. Am J Ophthalmol 1949;32:9–27.
72. Kinsey VE, Reddy DVN. Turnover of total carbon dioxide in the aqueous humors and the effect thereon of acetazolamide. Arch Ophthalmol 1959;62:78–83.
73. Becker B. The effects of carbonic anhydrase inhibitor, acetazolamide, on the composition of the aqueous humor. Am J Ophthalmol 1955;40:129–36.
74. Becker B. Carbonic anhydrase and the formation of aqueous humor. The Friedenwald Lecture. Am J Ophthalmol 1959;47:342–61.
75. Davson H. Physiology of the ocular and cerebrospinal fluids. London: Churchill; 1956.
76. Stein A, et al. The effect of topically administered carbonic anhydrase inhibitors on aqueous humor dynamics in rabbits. Am J Ophthalmol 1983;95(2):222–8.
77. Wang RF, et al. MK-507 (L-671,152), a topically active carbonic anhydrase inhibitor, reduces aqueous humor production in monkeys. Arch Ophthalmol 1991;109(9):1297–9.
78. Bar-Ilan A, Pessah NI, Maren TH. The effects of carbonic anhydrase inhibitors on aqueous humor chemistry and dynamics. Invest Ophthalmol Vis Sci 1984;25(10):1198–205.
79. Kishida K, Miwa Y, Iwata C. 2-Substituted 1, 3, 4-thiadiazole-5-sulfonamides as carbonic anhydrase inhibitors: their effects on the transepithelial potential difference of the isolated rabbit ciliary body and on the intraocular pressure of the living rabbit eye. Exp Eye Res 1986;43(6):981–95.
80. Dailey RA, Brubaker RF, Bourne WM. The effects of timolol maleate and acetazolamide on the rate of aqueous formation in normal human subjects. Am J Ophthalmol 1982;93(2):232–7.
81. Rosenberg LF, et al. Combination of systemic acetazolamide and topical dorzolamide in reducing intraocular pressure and aqueous humor formation. Ophthalmology, 1998;105(1):88–92.
82. McLaughlin MA, Chiou GC. A synopsis of recent developments in antiglaucoma drugs. J Ocul Pharmacol 1985;1:101–21.
83. Hoyng PF, van Beek LM. Pharmacological therapy for glaucoma: a review. Drugs 2000;59(3):411–34.
84. Sears ML, et al. The isolated ciliary bilayer is useful for studies of aqueous humor formation. Trans Am Ophthalmol Soc 1991;89:131–54.
85. Helbig H, et al. Role of HCO_3^- in regulation of cytoplasmic pH in ciliary epithelial cells. Am J Physiol 1989;257(4 Pt 1):C696–705.
86. Helbig H, et al. Sodium bicarbonate cotransport in cultured pigmented ciliary epithelial cells. Curr Eye Res 1989;8(6):595–8.
87. Butler GA, et al. Na^+- Cl^- and HCO_3^--dependent base uptake in the ciliary body pigment epithelium. Exp Eye Res 1994;59(3):343–9.
88. Wolosin JM, et al. Bicarbonate transport mechanisms in rabbit ciliary body epithelium. Exp Eye Res 1991;52(4):397–407.
89. Wolosin JM, et al. Separation of the rabbit ciliary body epithelial layers in viable form: identification of differences in bicarbonate transport. Exp Eye Res 1993;56(4):401–9.
90. Tabcharani JA, et al. Bicarbonate permeability of the outwardly rectifying anion channel. J Membr Biol 1989;112(2):109–22.
91. Nicholl AJ, et al. The role of bicarbonate in regulatory volume decrease (RVD) in the epithelial-derived human breast cancer cell line ZR-75-1. Pflugers Arch 2002;443(5-6):875–81.
92. To CH, et al. Model of ionic transport for bovine ciliary epithelium: effects of acetazolamide and HCO_3^-. Am. J. Physiol. Cell Physiol 2001;280(6):C1521–30.

93. Saito Y, et al. Mode of action of furosemide on the chloride-dependent short-circuit current across the ciliary body epithelium of toad eyes. J Membr Biol 1980;53(2):85–93.
94. Wiederholt M, Zadunaisky JA. Membrane potentials and intracellular chloride activity in the ciliary body of the shark. Pflügers Arch 1986;407(Suppl 2):S112–5.
95. Green K, et al. An electrophysiologic study of rabbit ciliary epithelium. Invest Ophthalmol Vis Sci 1985;26(3):371–81.
96. Carré DA, et al. Effect of bicarbonate on intracellular potential of rabbit ciliary epithelium. Curr Eye Res 1992;11(7):609–24.
97. Bowler JM, et al. Electron probe X-ray microanalysis of rabbit ciliary epithelium. Exp Eye Res 1996;62(2):131–9.
98. Helbig H, et al. Kinetic properties of Na+/H+ exchange in cultured bovine pigmented ciliary epithelial cells. Pflugers Arch 1988;412(1–2):80–5.
99. Helbig H, et al. Na+/H+ exchange regulates intracellular pH in a cell clone derived from bovine pigmented ciliary epithelium. J Cell Physiol 1988;137(2):384–9.
100. Helbig H, et al. Characterization of Cl-/HCO3- exchange in cultured bovine pigmented ciliary epithelium. Exp Eye Res 1988;47(4):515–23.
101. Crook RB, et al. The role of NaKCl cotransport in blood-to-aqueous chloride fluxes across rabbit ciliary epithelium. Invest Ophthalmol Vis Sci 2000;41(9):2574–83.
102. Kong CW, Li KK, To CH. Chloride secretion by porcine ciliary epithelium: New insight into species similarities and differences in aqueous humor formation. Invest Ophthalmol Vis Sci 2006;47(12):5428–36.
103. Edelman JL, Sachs G, Adorante JS. Ion transport asymmetry and functional coupling in bovine pigmented and nonpigmented ciliary epithelial cells. Am J Physiol 1994;266(5 Pt 1):C1210–21.
104. Wu Q, Pierce WR, Delamere NA. Cytoplasmic pH responses to carbonic anhydrase inhibitors in cultured rabbit nonpigmented ciliary epithelium. J Membr Biol 1998;162(1):31–8.
105. Crook RB, von Brauchitsch DK, Polansky JR. Potassium transport in nonpigmented epithelial cells of ocular ciliary body: inhibition of a Na+, K+, Cl- cotransporter by protein kinase C. J Cell Physiol 1992;153(1):214–20.
106. Dong J, Delamere NA. Protein kinase C inhibits Na(+)-K(+)-2Cl- cotransporter activity in cultured rabbit nonpigmented ciliary epithelium. Am J Physiol 1994;267(6 Pt 1):C1553–60.
107. Civan MM, Coca-Prados M, Peterson-Yantorno K. Regulatory volume increase of human non-pigmented ciliary epithelial cells. Exp Eye Res 1996;62(6):627–40.
108. Counillon L, et al. Na+/H+ and Cl-/HCO3-antiporters of bovine pigmented ciliary epithelial cells. Pflügers Arch 2000;440(5):667–78.
109. Meldrun NU, Roughton RJW. Carbonic anhydrase. Its preparation and properties. J Physiol 1933;80:113–142.
110. Li X, et al. Carbonic anhydrase II binds to and enhances activity of the Na+/H+ exchanger. J Biol Chem 2002;277(39):36085–91.
111. Sterling D, Reithmeier RA, Casey JR. A transport metabolon. Functional interaction of carbonic anhydrase II and chloride/bicarbonate exchangers. J Biol Chem 2001;276(51):47886–94.
112. Do CW. Characterization of chloride and bicarbonate transport across the isolated bovine ciliary body/epithelium (CBE) PhD Thesis, Department of Optometry and Radiography. Hong Kong: The Hong Kong Polytechnic University; 2002.
113. McLaughlin CW, et al. Effects of HCO_3^- on cell composition of rabbit ciliary epithelium: a new model for aqueous humor secretion. Invest Ophthalmol Vis Sci 1998;39(9):1631–41.
114. McLaughlin CW, et al. Timolol may inhibit aqueous humor secretion by cAMP-independent action on ciliary epithelial cells. Am J Physiol Cell Physiol 2001;281(3):C865–75.

115. McLaughlin CW, et al. Regional differences in ciliary epithelial cell transport properties. J Membr Biol 2001;182(3):213–22.
116. Avila MY, et al. Inhibitors of NHE-1 Na+/H+ exchange reduce mouse intraocular pressure. Invest Ophthalmol Vis Sci 2002;43(6):1897–902.
117. Gabelt BT, et al. Anterior segment physiology after bumetanide inhibition of Na-K-Cl cotransport. Invest Ophthalmol Vis Sci 1997;38(9):1700–7.
118. Gerometta RM, et al. Cl- concentrations of bovine, porcine and ovine aqueous humor are higher than in plasma. Exp Eye Res 2005;80(3):307–12.
119. Reale E. Freeze-fracture analysis of junctional complexes in human ciliary epithelia. Albrecht Von Graefes Arch Klin Exp Ophthalmol 1975;195(1):1–16.
120. Coca-Prados M, et al. Expression and cellular distribution of the alpha 1 gap junction gene product in the ocular pigmented ciliary epithelium. Curr Eye Res 1992;11(2):113–22.
121. Coffey KL, et al. Molecular profiling and cellular localization of connexin isoforms in the rat ciliary epithelium. Exp Eye Res 2002;75(1):9–21.
122. Sears J, Nakano T, Sears M. Adrenergic-mediated connexin43 phosphorylation in the ocular ciliary epithelium. Curr Eye Res 1998;17(1):104–7.
123. Wolosin JM, Schütte M, Chen S. Connexin distribution in the rabbit and rat ciliary body. A case for heterotypic epithelial gap junctions. Invest Ophthalmol Vis Sci 1997;38(2):341–8.
124. Oh J, et al. Dye coupling of rabbit ciliary epithelial cells in vitro. Invest Ophthalmol Vis Sci 1994;35(5):2509–14.
125. Stelling JW, Jacob TJ. Functional coupling in bovine ciliary epithelial cells is modulated by carbachol. Am J Physiol 1997;273(6Pt1):C1876–81.
126. Mitchell CH, Civan MM. Effects of uncoupling gap junctions between pairs of bovine NPE-PE ciliary epithelial cells of the eye. FASEB J 1997?
127. Wolosin JM, et al. Effect of heptanol on the short circuit currents of cornea and ciliary body demonstrates rate limiting role of heterocellular gap junctions in active ciliary body transport. Exp Eye Res 1997;64(6):945–52.
128. Jacob TJ, Civan MM. Role of ion channels in aqueous humor formation. Am J Physiol 1996;271(3Pt1):C703–20.
129. Civan MM, et al. Potential contribution of epithelial Na+ channel to net secretion of aqueous humor. J Exp Zool 1997;279(5):498–503.
130. Chen S, et al. Role of cyclic AMP-induced Cl conductance in aqueous humour formation by the dog ciliary epithelium. Br J Pharmacol 1994;112(4):1137–45.
131. Edelman JL, Loo DD, Sachs G. Characterization of potassium and chloride channels in the basolateral membrane of bovine nonpigmented ciliary epithelial cells. Invest Ophthalmol Vis Sci 1995;36(13):2706–16.
132. Chen S, Sears M. A low conductance chloride channel in the basolateral membranes of the non-pigmented ciliary epithelium of the rabbit eye. Curr Eye Res 1997;16(7):710–8.
133. Yantorno RE, et al. Whole cell patch clamping of ciliary epithelial cells during anisosmotic swelling. Am J Physiol 1992;262(2Pt1):C501–9.
134. Zhang JJ, Jacob TJ. Three different Cl- channels in the bovine ciliary epithelium activated by hypotonic stress. J Physiol 1997;499(Pt2):379–89.
135. Coca-Prados M, et al. PKC-sensitive Cl- channels associated with ciliary epithelial homologue of pICln. Am J Physiol 1995;268(3Pt1):C572–9.
136. Civan MM, et al. Prolonged incubation with elevated glucose inhibits the regulatory response to shrinkage of cultured human retinal pigment epithelial cells. J Membr Biol 1994;139(1):1–13.

137. Shi C, et al. Protein tyrosine kinase and protein phosphatase signaling pathways regulate volume-sensitive chloride currents in a nonpigmented ciliary epithelial cell line. Invest Ophthalmol Vis Sci 2002;43(5):1525–32.
138. Shi C, et al. A3 adenosine and CB1 receptors activate a PKC-sensitive Cl(-) current in human nonpigmented ciliary epithelial cells via a Gbetagamma-coupled MAPK signaling pathway. Br J Pharmacol 2003;139(3):475–86.
139. Mitchell CH, et al. A_3 adenosine receptors regulate Cl⁻ channels of nonpigmented ciliary epithelial cells. Am J Physiol 1999;276(3Pt1):C659–66.
140. Carré DA, et al. Similarity of A(3)-adenosine and swelling-activated Cl(-) channels in nonpigmented ciliary epithelial cells. Am J Physiol Cell Physiol 2000;279(2):C440–51.
141. Coca-Prados M, et al. Association of ClC-3 channel with Cl- transport by human nonpigmented ciliary epithelial cells. J Membr Biol 1996;150(2):197–208.
142. Civan MM. The fall and rise of active chloride transport: implications for regulation of intraocular pressure. J Exp Zoolog A Comp Exp Biol 2003;300(1):5–13.
143. Civan MM, Coca-Prados M, Peterson-Yantorno K. Pathways signaling the regulatory volume decrease of cultured nonpigmented ciliary epithelial cells. Invest Ophthalmol Vis Sci 1994;35(6):2876–86.
144. Kawasaki M, et al. Cloning and expression of a protein kinase C-regulated chloride channel abundantly expressed in rat brain neuronal cells. Neuron 1994;12(3):597–604.
145. Wang L, Chen L, Jacob TJ. The role of ClC-3 in volume-activated chloride currents and volume regulation in bovine epithelial cells demonstrated by antisense inhibition. J Physiol 2000;524(1):63–75.
146. Wang GX, et al. Functional effects of novel anti-ClC-3 antibodies on native volume-sensitive osmolyte and anion channels in cardiac and smooth muscle cells. Am J Physiol Heart Circ Physiol 2003;285(4):H1453–63.
147. Do CW, et al. Inhibition of swelling-activated Cl- currents by functional anti-ClC-3 antibody in native bovine non-pigmented ciliary epithelial cells. Invest Ophthalmol Vis Sci 2005;46(3):948–55.
148. Vessey JP, et al. Hyposmotic activation of ICl,swell in rabbit nonpigmented ciliary epithelial cells involves increased ClC-3 trafficking to the plasma membrane. Biochem Cell Biol 2004;82(6):708–18.
149. Jentsch TJ, et al. Molecular structure and physiological function of chloride channels. Physiol Rev 2002;82(2):503–68.
150. Hermoso M, et al. ClC-3 is a fundamental molecular component of volume-sensitive outwardly rectifying Cl- channels and volume regulation in HeLa cells and Xenopus laevis oocytes. J Biol Chem 2002;277(42):40066–74.
151. Stobrawa SM, et al. Disruption of ClC-3, a chloride channel expressed on synaptic vesicles, leads to a loss of the hippocampus. Neuron 2001;29(1):185–96.
152. Yamamoto-Mizuma S, et al. Altered properties of volume-sensitive osmolyte and anion channels (VSOACs) and membrane protein expression in cardiac and smooth muscle myocytes from Clcn3-/- mice. J Physiol 2004;557(2):439–56.
153. Paulmichl M, et al. New mammalian chloride channel identified by expression cloning. Nature 1992;356(6366):238–41.
154. Anguíta J, et al. Molecular cloning of the human volume-sensitive chloride conductance regulatory protein, pICln, from ocular ciliary epithelium. Biochem Biophys Res Commun 1995;208(1):89–95.
155. Chen L, Wang L, Jacob TJ. Association of intrinsic pICln with volume-activated Cl- current and volume regulation in a native epithelial cell. Am J Physiol 1999;276(1Pt1):C182–92.

156. Sanchez-Torres J, et al. Effects of hypotonic swelling on the cellular distribution and expression of pI(Cln) in human nonpigmented ciliary epithelial cells. Curr Eye Res 1999;18(6):408–16.
157. Do CW, Peterson-Yantorno K, Civan MM. Swelling-activated Cl- channels support Cl- secretion by bovine ciliary epithelium. Invest Ophthalmol Vis Sci 2006;47(6):2576–82.
158. Crook RB, Polansky JR. Stimulation of Na+,K+,Cl- cotransport by forskolin-activated adenylyl cyclase in fetal human nonpigmented epithelial cells. Invest Ophthalmol Vis Sci 1994;35(9):3374–83.
159. Fleischhauer JC, et al. PGE_2, Ca^{2+}, and cAMP mediate ATP activation of Cl^- channels in pigmented ciliary epithelial cells. Am J Physiol Cell Physiol 2001;281(5):C1614–23.
160. Shahidullah M, Wilson WS. Mobilisation of intracellular calcium by P2Y2 receptors in cultured, non-transformed bovine ciliary epithelial cells. Curr Eye Res 1997;16(10):1006–16.
161. Mitchell CH, et al. Tamoxifen and ATP synergistically activate Cl- release by cultured bovine pigmented ciliary epithelial cells. J Physiol 2000;525(1):183–93.
162. Do CW, et al. cAMP-activated maxi-Cl(–) channels in native bovine pigmented ciliary epithelial cells. Am J Physiol Cell Physiol 2004;287(4):C1003–11.
163. Mitchell CH, Wang L, Jacob TJ. A large-conductance chloride channel in pigmented ciliary epithelial cells activated by GTPgammaS. J Membr Biol 1997;158(2):167–75.
164. Mitchell CH, et al. A release mechanism for stored ATP in ocular ciliary epithelial cells. Proc Natl Acad Sci USA 1998;95(12):7174–8.
165. Brubaker RF. Flow of aqueous humor in humans [The Friedenwald Lecture]. Invest Ophthalmol Vis Sci 1991;32(13):3145–66.
166. McCannel CA, Heinrich SR, Brubaker RF. Acetazolamide but not timolol lowers aqueous humor flow in sleeping humans. Graefes Arch Clin Exp Ophthalmol 1992;230(6):518–20.
167. Liu JH, et al. Nocturnal elevation of intraocular pressure in young adults. Invest Ophthalmol Vis Sci 1998;39(13):2707–12.
168. Reiss GR, et al. Aqueous humor flow during sleep. Invest Ophthalmol Vis Sci 1984;25(6): 776–8.
169. Brubaker RF. Clinical measurement of aqueous dynamics: Implications for addressing glaucoma. In: Civan MM, editor. Eye's Aqueous Humor: From Secretion to Glaucoma. San Diego: Academic Press; 1998. p. 234–284.
170. Carré DA, Civan MM. cGmp modulates transport across the ciliary epithelium. J Membr Biol 1995;146(3):293–305.
171. Chu TC, Candia OA. Effects of adrenergic agonists and cyclic AMP on the short-circuit current across the isolated rabbit iris-ciliary body. Curr Eye Res 1985;4(4):523–9.
172. Chu TC, Candia OA, Iizuka S. Effects of forskolin, prostaglandin F{2 alpha}, and Ba2+ on the short-circuit current of the isolated rabbit iris-ciliary body. Curr Eye Res 1986;5(7):511–6.
173. Horio B, et al. Regulation and bioelectrical effects of cyclic adenosine monophosphate production in the ciliary epithelial bilayer. Invest Ophthalmol Vis Sci 1996;37(4):607–12.
174. Liu R, Flammer J, Haefliger IO. Forskolin upregulation of NOS I protein expression in porcine ciliary processes: a new aspect of aqueous humor regulation. Klin Monatsbl Augenheilkd 2002;219(4):281–3.
175. Shahidullah M, Wilson WS. Atriopeptin, sodium azide and cyclic GMP reduce secretion of aqueous humour and inhibit intracellular calcium release in bovine cultured ciliary epithelium. Br J Pharmacol 1999;127(6):1438–46.
176. Crook RB, Chang AT. Differential regulation of natriuretic peptide receptors on ciliary body epithelial cells. Biochem J 1997;324(1):49–55.

177. Wu R, et al. Reduction of nitrite production by endothelin-1 in isolated porcine ciliary processes. Exp Eye Res 2003;77(2):189–93.
178. Ellis DZ, et al. Carbachol and nitric oxide inhibition of Na,K-ATPase activity in bovine ciliary processes. Invest Ophthalmol Vis Sci 2001;42(11):2625–31.
179. Zhang X, et al. Dexamethasone regulates endothelin-1 and endothelin receptors in human non-pigmented ciliary epithelial (HNPE) cells. Exp Eye Res 2003;76(3):261–72.
180. Caprioli J, et al. Forskolin lowers intraocular pressure by reducing aqueous inflow. Invest Ophthalmol Vis Sci 1984;25(3):268–77.
181. Lee PY, et al. Effect of topically applied forskolin on aqueous humor dynamics in cynomolgus monkey. Invest Ophthalmol Vis Sci 1984;25(10):1206–9.
182. Do CW, Kong CW, To CH. cAMP inhibits transepithelial chloride secretion across bovine ciliary body/epithelium. Invest Ophthalmol Vis Sci 2004;45(10):3638–43.
183. Nakai Y, et al. Genistein inhibits the regulation of active sodium-potassium transport by dopaminergic agonists in nonpigmented ciliary epithelium. Invest Ophthalmol Vis Sci 1999;40(7):1460–6.
184. Delamere NA, King KL. The influence of cyclic AMP upon Na,K-ATPase activity in rabbit ciliary epithelium. Invest Ophthalmol Vis Sci 1992;33(2):430–5.
185. Yorio T. Cellular mechanisms in the actions of antiglaucoma drugs. J Ocul Pharmacol 1985;1(4):397–422.
186. Huang P, et al. Compartmentalized autocrine signaling to cystic fibrosis transmembrane conductance regulator at the apical membrane of airway epithelial cells. Proc Natl Acad Sci USA 2001;98(24):14120–5.
187. Okamura T, et al. Structure-activity relationships of adenosine A3 receptor ligands: new potential therapy for the treatment of glaucoma. Bioorg Med Chem Lett 2004;14(14):3775–9.
188. Crosson CE, Gray T. Characterization of ocular hypertension induced by adenosine agonists. Invest Ophthalmol Vis Sci 1996;37(9):1833–9.
189. Avila MY, Stone RA, Civan MM. Knockout of A(3) adenosine receptors reduces mouse intraocular pressure. Invest Ophthalmol Vis Sci 2002;43(9):3021–3026.
190. Daines BS, et al. Intraocular adenosine levels in normal and ocular-hypertensive patients. J Ocul Pharmacol Ther 2003;19(2):113–9.
191. Kiel JW. Effect of intravenous adenosine infusion on ciliary blood flow and aqueous production in rabbit. In ARVO 2006. Fort Lauderdale.
192. Schlotzer-Schrehardt U, et al. Selective upregulation of the A3 adenosine receptor in eyes with pseudoexfoliation syndrome and glaucoma. Invest Ophthalmol Vis Sci 2005;46(6):2023–34.
193. Carré DA, et al. Adenosine stimulates Cl⁻ channels of nonpigmented ciliary epithelial cells. Am J Physiol 1997;273(4Pt1):C1354–61.
194. Avila MY, Stone RA, Civan MM. A(1)-, A(2A)- and A(3)-subtype adenosine receptors modulate intraocular pressure in the mouse. Br J Pharmacol 2001;134(2):241–5.
195. Jacobson KA, et al. Pharmacological characterization of novel A_3 adenosine receptor-selective antagonists. Neuropharmacology 1997;36(9):1157–65.
196. Linden J. Molecular approach to adenosine receptors: receptor-mediated mechanisms of tissue protection. Annu Rev Pharmacol Toxicol 2001;41:775–87.
197. Gao ZG, et al. Structural determinants of A(3) adenosine receptor activation: nucleoside ligands at the agonist/antagonist boundary. J Med Chem 2002;45(20):4471–84.
198. Yang H, et al. The cross-species A3 adenosine-receptor antagonist MRS 1292 inhibits adenosine-triggered human nonpigmented ciliary epithelial cell fluid release and reduces mouse intraocular pressure. Curr Eye Res 2005;30(9):747–54.

199. Nathanson JA, McKee M. Identification of an extensive system of nitric oxide-producing cells in the ciliary muscle and outflow pathway of the human eye. Invest Ophthalmol Vis Sci 1995;36(9):1765–73.
200. Becquet F, Courtois Y, Goureau O. Nitric oxide in the eye: multifaceted roles and diverse outcomes. Surv Ophthalmol 1997;42(1):71–82.
201. Nathanson JA. Nitrovasodilators as a new class of ocular hypotensive agents. J Pharmacol Exp Ther 1992;260(3):956–65.
202. Schuman JS, Erickson K, Nathanson JA. Nitrovasodilator effects on intraocular pressure and outflow facility in monkeys. Exp Eye Res 1994;58(1):99–105.
203. Wang RF, Podos SM. Effect of the topical application of nitroglycerin on intraocular pressure in normal and glaucomatous monkeys. Exp Eye Res 1995;60(3):337–9.
204. Chuman H, et al. The effect of L-arginine on intraocular pressure in the human eye. Curr Eye Res 2000;20(6):511–6.
205. Schmetterer L, Polak K. Role of nitric oxide in the control of ocular blood flow. Prog Retin Eye Res 2001:20(6):823–47.
206. Wiederholt M, Sturm A, Lepple-Wienhues A. Relaxation of trabecular meshwork and ciliary muscle by release of nitric oxide. Invest Ophthalmol Vis Sci 1994;35(5):2515–20.
207. Murad F. The nitric oxide-cyclic GMP signal transduction system for intracellular and intercellular communication. Recent Prog Horm Res 1994;49:239–48.
208. Miranda KM, et al. The chemical biology of nitric oxide. In: Ignarro LJ, editor. Nitric oxide: biology and pathobiology. San Diego: Academic Press: 2000. p. 41–56.
209. Ignarro LJ. Nitric oxide: biology and pathobiology. San Diego: Academic Press: 2000. p. 3–19.
210. Ignarro LJ, et al. Activation of purified guanylate cyclase by nitric oxide requires heme. Comparison of heme-deficient, heme-reconstituted and heme-containing forms of soluble enzyme from bovine lung. Biochim Biophys Acta 1982;718(1):49–59.
211. Haufschild T, et al. Spontaneous calcium-independent nitric oxide synthase activity in porcine ciliary processes. Biochem Biophys Res Commun 1996;222(3):786–9.
212. Geyer O, Podos SM, Mittag T. Nitric oxide synthase activity in tissues of the bovine eye. Graefes Arch Clin Exp Ophthalmol 1997;235(12):786–93.
213. Meyer P, et al. Localization of nitric oxide synthase isoforms in porcine ocular tissues. Curr Eye Res 1999;18(5):375–80.
214. Millar JC, Shahidullah M, Wilson WS. Intraocular pressure and vascular effects of sodium azide in bovine perfused eye. J Ocul Pharmacol Ther 2001;17(3):225–34.
215. Shahidullah M, Yap M, To CH. Cyclic GMP, sodium nitroprusside and sodium azide reduce aqueous humour formation in the isolated arterially perfused pig eye. Br J Pharmacol 2005;145(1):84–92.
216. Kong CW. Chloride (Cl-) transport and its regulation by nitric oxide (NO) in porcine ciliary body / epithelium (CBE). PhD Thesis, School of Optometry, the Hong Kong Polytechnic University, Hong Kong; 2005.
217. Fleischhauer JC, et al. NO/cGMP pathway activation and membrane potential depolarization in pig ciliary epithelium. Invest Ophthalmol Vis Sci 2000;41(7):1759–63.
218. Wu R, et al. Role of anions in nitric oxide-induced short-circuit current increase in isolated porcine ciliary processes. Invest Ophthalmol Vis Sci 2004;45(9):3213–22.
219. Friedman DS, et al. Prevalence of open-angle glaucoma among adults in the United States. Arch Ophthalmol 2004;122(4):532–8.

III
Lens Transporters

5
Membrane Transporters
New Roles in Lens Cataract

Paul J. Donaldson and Julie Lim

CONTENTS

INTRODUCTION
LENS STRUCTURE AND FUNCTION
DIFFERENTIAL EXPRESSION OF MEMBRANE TRANSPORTERS IN THE LENS
ROLE OF MEMBRANE TRANSPORTERS IN THE NORMAL AND DIABETIC LENS
NUTRIENT TRANSPORTERS IN THE NORMAL AND CATARACTIC LENS
ACKNOWLEDGEMENTS
REFERENCES

INTRODUCTION

The function of the lens is to focus light onto the retina. To achieve this function the lens is required to maintain its transparency over many decades. A failure of lens transparency results in cataract, the leading cause of blindness in the world today *(1)*. Clinically, four main forms of lens cataract are recognized: sub-capsular, cortical, nuclear and mixed (nuclear and cortical). Of these classes, age-related nuclear (ARN) cataract is the most common, followed by cortical cataracts such as those seen in diabetics. Cataract is routinely treated by surgical replacement of the opaque lens with a plastic intraocular implant. However, because of escalating levels of diabetes and the ageing of the world's population, it is expected that the prevalence of cataract will reach epidemic proportions placing severe pressure on health resources *(2)*. An alternative to surgical intervention is the development of medical treatments to delay the onset of cataract. The rational design of anti-cataract therapies requires an understanding of the cellular mechanisms involved in the maintenance of lens transparency and how these mechanisms are impaired in the different forms of lens cataract. Here we review lens structure and function with a specific focus on the involvement of transporters in the maintenance of lens transparency and how their dysfunction may contribute to the initiation of diabetic and ARN cataract.

From: *Ophthalmology Research: Ocular Transporters in Ophthalmic Diseases and Drug Delivery*
Edited by: J. Tombran-Tink and C. J. Barnstable © Humana Press, Totowa, NJ

LENS STRUCTURE AND FUNCTION

The transparency of the lens is closely linked to the unique structure and function of its fibre cells. These highly differentiated cells are derived from equatorial epithelial cells that exit the cell cycle and embark upon a differentiation process that produces extensive cellular elongation, the loss of cellular organelles and nuclei, and the expression of fibre-specific proteins *(3, 4)*. At the lens poles, elongating fibres from opposite hemispheres meet and interdigitate to form the lens sutures *(5)*. Since this process continues throughout life, a gradient of fibre cells at different stages of differentiation is established around an internalized nucleus of mature, anucleate fibre cells (Fig. 1A). While the transparent properties of the lens are a direct result of its highly ordered tissue architecture, the lens however, should not be considered a purely passive optical element. Because of its size, the avascular lens cannot rely on passive diffusion alone to transport nutrients to deeper-lying cells, or to transport waste products back to the surface *(6, 7)*. Thus, maintenance of this architecture requires special mechanisms not only to supply the deeper-lying fibre cells with nutrients, but also to control the volume of these cells.

A common feature of all vertebrate lenses studied to date *(8)*, is the existence of a standing flow of ionic current that is directed inward at the poles and outward at the

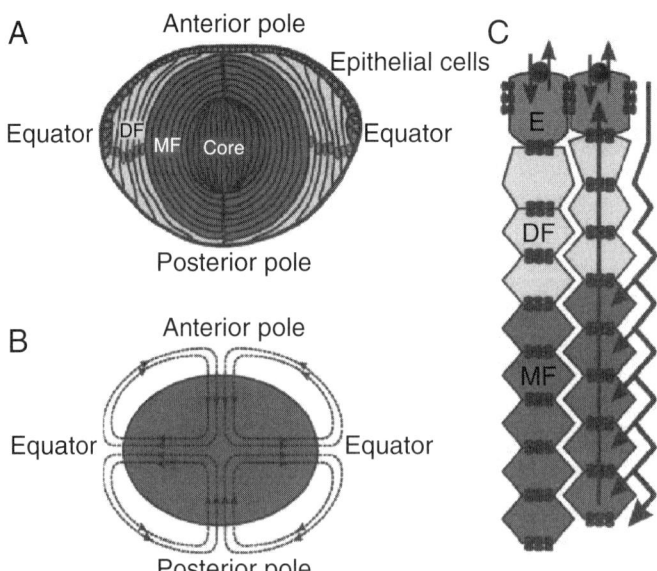

Fig. 1. Structure and function of the lens. (**A**) Architecture of the lens showing anterior epithelial monolayer, nucleated differentiating fibre cells (DF) in the outer cortex and non-nucleated mature fibre cells (MF) in the inner cortex and core. (**B**) Current flow through the lens that underpins the internal circulation system. (**C**) Representative cross-section through the equator. Current and solute are proposed to flow into the lens via the extracellular space, to cross fibre cell membranes, and to flow outward towards the surface epithelium (E) via an intracellular pathway mediated by gap junction channels. Reproduced with permission from Donaldson et al. 2001 *(6)*.

equator (Fig. 1B). Mathias and colleagues have suggested that this standing current generates a unique internal microcirculatory system that is responsible for maintaining lens transparency *(7, 9)*. This working model states that the current, which is carried primarily by Na^+, enters via locations all around the lens along the extracellular clefts between fibre cells (Fig. 1B). It eventually crosses the fibre-cell membranes and then flows from cell to cell towards the surface via an intracellular pathway mediated by gap junction channels. Data reviewed in Mathias et al. suggest that the surface cells, including epithelial cells and newly differentiating fibre cells contain Na^+/K^+ pumps and K^+ channels, which together generate a negative intracellular potential *(7)*. Fibre cells deeper in the lens lack functional Na^+/K^+ pumps and K^+ channels. Instead, their permeability is dominated by Na^+ and Cl^- leak conductances. In these inner cells, a negative membrane potential is maintained by connections to the surface cells via gap junctions. This electrical connection, together with the different membrane properties of the surface and inner cells, causes a standing current to flow (Fig. 1C). In this model, the circulating current creates a net flux of solute that generates fluid flow. The extracellular flow of fluid causes convection of nutrients towards the deeper-lying fibre cells, while the intracellular flow removes wastes and creates a well-stirred intracellular compartment. Thus transport by surface cells is able to regulate the ion composition of inner fibre cells and allows them to maintain the constant volume necessary for the maintenance of overall lens transparency.

While not accepted by all, recent electrical *(10)*, morphological *(11–14)* and molecular *(11, 15–17)* evidence in favor of the model is accumulating. Candia and co-workers *(10)*, using modified Ussing chambers have independently confirmed the existence of outwardly and inwardly directed currents at the lens equator and poles, respectively. In a series of morphological studies using confocal microscopy to monitor changes in fibre-cell volume and molecular approaches to localize chloride cotransporters (Table 1), our laboratory has identified spatially distinct zones of ion influx and efflux that interact to maintain overall lens volume. A number of investigators have also shown that fiber cells express a variety of transporters (Table 1) that enable deeper fiber cells to directly uptake the nutrients convected to them via the circulation system. In the remainder of this chapter, we will initially discuss some of the general mechanisms responsible for establishing the observed spatial changes in transporter distribution before focusing on how our recently accumulated knowledge of the transporters involved in volume regulation and nutrient uptake in the lens affords us novel insights into the initiation of diabetic and ARN cataract.

DIFFERENTIAL EXPRESSION OF MEMBRANE TRANSPORTERS IN THE LENS

While a central tenet of the lens circulation system is the existence of spatial differences in membrane transport processes, most researchers have tended to focus solely on characterizing transporters in the lens epithelium and often only characterize transporters in cultured epithelial cell lines. In contrast, our laboratory has made a concerted effort to identify and localize key membrane transport proteins in different regions of the lens *(9)*. Utilizing a combination of reverse-transcriptase polymerase chain reaction (RT-PCR) and Western blotting, we have identified the specific isoforms of key membrane transporter

Table 1 Molecular inventory of transporter protein expression in different regions of the lens

Transporters	Species	CE	NE	OC	IC	C	Mode of action
A: Ion homeostasis and cell volume regulation							
Na$^+$/K$^+$ ATPase *(18–26)*							
α 1	Rabbit	√^{A}	√				Primary active transport
	Rat	√^{A}	√				
	Porcine	√^{A,E}	√				
	Bovine	√^{A,E}					
	Frog	√^{A,E}					
α 2	Rabbit	√^{A}					
	Rat	√^{A}					
	Bovine	√^{E}	√				
	Human		√				
	Frog	√^{A,E}					
α 3	Rabbit	√^{A}	√	√	√		
	Rat	√^{A}					
	Bovine	√^{A,E}	√				
	Human		√				
β1	Porcine	√^{A,E}					
	Rabbit	√	√				
Plasma membrane Ca^{2+} ATPase *(25, 27)*							
PMCA1	Human	√	√				Primary active transport
PMCA2	Human	√	√				
PMCA4	Human	√	√				
PMCA2	Porcine	√^{A,E}					
PMCA4	Porcine	√^{A,E}					
Sarco/endoplasmic reticulum Ca^{2+} ATPase *(28)*							
SERCA2b	Human	√					
SERCA3	Human	√					
Na$^+$/K$^+$/2Cl$^-$ cotransporter *(29)*							
NKCC1	Human	√					Secondary active transport
K$^+$/Cl$^-$ cotransporter *(11, 30)*							
KCC1	Rat			√	√	√	Secondary active transport
KCC1	Human	√					
KCC3	Rat			√	√	√	
KCC3	Human	√					
KCC4	Rat		√	√	√	√	
KCC4	Human	√					
B: Nutrient uptake							
Glucose transporters *(17, 31, 32)*							
GLUT1	Rat	√					Facilitative diffusion
GLUT3	Rat			√	√		
SGLUT2	Rat			√	√	√	Secondary active transport

(continued)

Table 1 (continued)

Transporters	Species	Localization CE NE OC IC C					Mode of action
Cystine/glutamate exchanger (X_c^-) *(15)*							
xCT	Rat			√	√	√	Sodium-independent antiporter
Glutamate transporters *(15, 16)*							
EAAT1	Rat			√	√		Secondary active transport
EAAT2	Rat		√	√			
EAAT3	Rat		√	√	√		
EAAT4	Rat			√	√		
EAAT5	Rat			√			
ASCT2	Rat		√	√	√	√	
Glycine transporters *(16)*							
GLYT1	Rat		√	√	√		Secondary active transport
GLYT2	Rat		√	√	√	√	
Vitamin C transporters *(33)*							
SVCT2	Human	√					Secondary active transport

CE = cultured epithelial cells; NE = native epithelial cells; OC = outer cortex; IC = inner cortex; C = core; A = anterior surface; E = equatorial surface.

families that are expressed in the lens, and have used immunocytochemistry to localize these isoforms to specific areas of the lens at subcellular resolution *(15, 17, 32, 34)*. The results we have obtained for transporters involved in ion homeostasis and cell volume regulation, and in nutrient uptake are summarized in Table 1, as are other transporter isoforms identified in the lens by other researchers. From inspection of Table 1, it is clear that spatial differences in the expression of membrane transporters exist in the lens.

This raises the question of how these spatial differences are established and maintained? Since the lens is continually adding new fibre cells at its equator, it would appear that establishing and maintaining the spatial differences in membrane transport proteins necessary to generate the lens circulation system is an integral part of fibre-cell differentiation. Furthermore, anucleate cells in the center of the lens have different energy requirements and occupy a different physical environment (low pH and oxygen tension) than nucleated peripheral fibre cells. In this regard, how do mature anucleate fibre cells that have lost their ability to synthesize new proteins alter their complement of membrane transport proteins to reflect their changing position in the lens, and to compensate for changes in their local external environment?

To address some of these questions, we have developed procedures to map membrane protein distributions which utilize high quality cryosections that are systematically imaged to produce high-resolution data sets *(35)*. Such an image-based data set contains information not only on how the spatial distribution of a labeled membrane protein changes as a function of fibre-cell differentiation, but also information on the subcellular distribution of the protein. This is important because differentiating fibre cells are essentially elongated epithelial cells, which retain distinct apical, basal and lateral membrane domains (36, 37). In these cells, the lateral membranes are further divided into the broad and narrow sides, which contribute to the distinctive hexagonal profile of the fibre cells. Thus, in addition to radial differences in membrane protein distribution that can occur as a consequence of fibre-cell differentiation, other changes may be evident in an axial direction (pole–equator–pole) along the length of a fibre cell, or within the lateral membrane domains (broad versus narrow side).

The cumulative evidence (9) obtained by adopting these imaging procedures to systematically localize membrane transporters in the lens indicates that differentiation-dependent changes in the molecular composition, subcellular distribution and age-dependent processing of a variety of transport proteins all contribute to the establishment and maintenance of the observed spatial gradients in transporter localization (Fig. 2). In the epithelium and nucleated fibre cells of the outer cortex, it is appar-

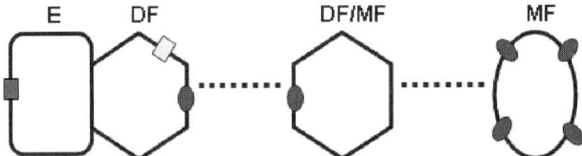

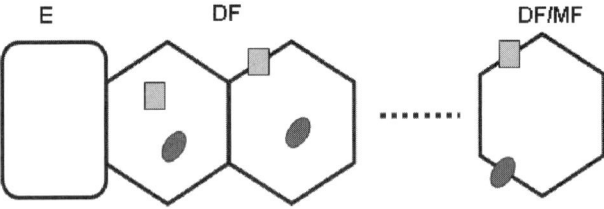

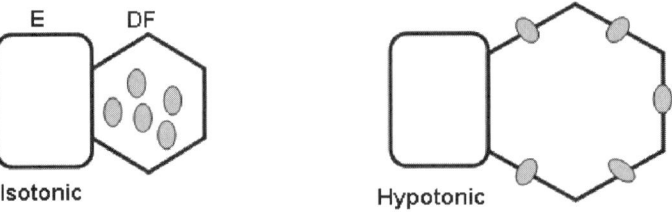

Fig. 2. Differentiation-dependent changes in the spatial location of membrane proteins in the lens. Schematic diagram depicting some of the mechanisms responsible for the establishment of the spatial differences in membrane transport processes that contribute to the lens circulation system. (**A**) Differential expression of transporter genes. (**B**) Differentiation-dependent membrane insertion of transport proteins from a cytoplasmic pool. (**C**) Dynamic membrane insertion from a cytoplasmic pool in response to a change in cell volume. See text for further details. E = epithelium, DF = differentiating fibre cells, MF = mature fibre cells.

ent that simple differential expression of transporter genes can account for some of the observed differences (Fig. 2A). For example, the facilitative glucose transporters GLUT1 and GLUT3 are differentially expressed in the epithelium and fibre cells, respectively *(17, 32)*, while in the differentiating fibre cells there appears to be differential expression of excitatory amino acid transporter (EAAT) isoforms, EAAT1-5, involved in the uptake of glutamate *(15)*.

Interestingly, many of the membrane transporter proteins we have localized were initially found to be more abundant in the cytoplasm than in the membrane of differentiating fibre cells. However, with distance into the lens they became incorporated into the membrane, where they presumably became functional. For transporters such as the cystine/glutamate exchanger, X_c^- *(15, 16)*, and the glucose transporter, GLUT3, this shift to the membrane occurred early in fibre-cell differentiation. For other transporters, such as the potassium chloride cotransporter 4 (KCC4) *(11)*, the sodium-dependent glucose cotransporter 2 (SGLT2) *(31)* and the glutamine/glutamate transporter 2 (ASCT2) *(16)*, membrane insertion occurred at a later stage of fibre-cell differentiation that coincided with abrupt loss of cell nuclei that marks the transition between the outer and inner cortex of the lens (Fig. 2B). Based on these observations, we have hypothesized that the cytoplasmic pool of transporters represents a store of proteins that are inserted into membranes at discrete stages of fibre-cell differentiation to compensate for the inability of mature anucleate fibre cells to perform *de novo* protein synthesis. In addition to this differentiation-dependent programmed membrane insertion, we have also shown that transporter insertion can be stimulated by changes in environmental conditions (Fig. 2C). Chee and colleagues showed that exposure of the lens to hypotonic conditions induced fibre-cell swelling and the recruitment to the membrane of the volume-sensitive KCC isoforms 1 and 4, but not the volume-insensitive isoform, KCC3 *(11)*. Since the lens is capable of volume regulation this indicates, at least for KCC1 and 4, that the observed pool of cytoplasmic transporters represents a functional pool of transporters that can be recruited to the membrane in response to volume stress.

Finally, it appears that membrane resident transporters can undergo changes that alter their subcellular distribution in the membrane. In the outer cortex, some transporters are preferential expressed on either the narrow (GLUT3) *(17)* or broad sides (X_c^-) *(15)* of fibre cells (Figs. 2A and 2B), and are presumably tethered in position by interactions with cytoskeletal elements, and/or via adhesion junctions which define the boundaries of the membrane domains *(38)*. However, during the course of fibre-cell differentiation, the factors that restrict membrane transporters to distinct membrane domains are lost *(38, 39)*, and the transporters become more evenly distributed around the perimeter of fibre cells.

Thus it appears that the lens has a number of ways of establishing and maintaining the spatial differences in membrane transporters required to generate the lens's internal microcirculation system. In the following sections, we will detail how these spatial differences in membrane transporters contribute to the regulation of lens volume and the delivery of nutrients to the lens nucleus and how impairment of these processes can lead to diabetic cataract and ARN cataract, respectively.

ROLE OF MEMBRANE TRANSPORTERS IN THE NORMAL AND DIABETIC LENS

Diabetic Cataract: a Problem with Lens-Volume Regulation?

Diabetics have a high incidence of cortical cataract and develop cataracts several years earlier than the general population *(40)*. The morphological and biochemical changes associated with diabetic cataract have been extensively studied. Morphologically, diabetic cataract manifests itself as a discrete zone of tissue liquefaction in the outer cortex of the lens, surrounded by relatively normal fibre cells (Fig. 3A). This is an intriguing damage phenotype, when one considers that lens fibre cells are extensively coupled by gap junctions *(41, 42)*. Using a diabetic rat model, Bond et al. *(43)*, subsequently showed that this tissue liquefaction was initiated by the swelling of individual fibre cells within this

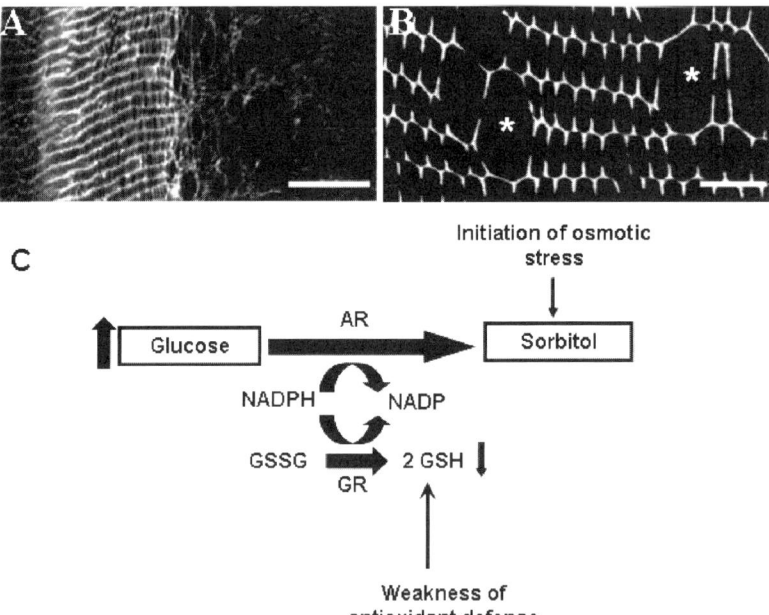

Fig. 3. Dysfunction of volume regulation is an initiating factor in diabetic cataract. Confocal images of equatorial sections taken from rat lenses made diabetic by injection with streptozotocin. **(A)** One month post-injection, a distinct zone of tissue liquefaction in the outer cortex of the lens, surrounded by relatively normal fibre cells is evident. **(B)** At one week post-injection, a higher-power image of cells located in this zone indicates that tissue liquefaction is initiated by the swelling of individual fibre cells. **(C)** A schematic of some of the biochemical mechanisms that may contribute to diabetic cataract. In hyperglycemia an increase in the conversion of glucose to the impermeable osmolyte sorbitol by aldose reductase (AR), induces osmotic stress. Since this increase in AR activity consumes reducing equivalents (NADPH), the regeneration of oxidized glutathione (GSSH) by glutathione reductase (GR) to its reduced form (GSH) is decreased. The fall is GSH compromises the antioxidant defense system of the fibre cells, thereby exposing them to the damaging effects of oxidative stress.

discrete damage zone (Fig. 3B), suggesting that dysfunction of volume regulation was an initiating factor in diabetic cataract *(44)*. This view that osmotic stress contributes to the initiation of diabetic cataract was supported by early biochemical studies in rat lenses *(45)*, which showed that hyperglycaemia causes an overactivation of the pentose phosphate pathway resulting in the conversion of glucose to the inorganic osmolyte sorbitol by the enzyme aldose reductase (Fig. 3C). These observations led to the promotion and development of aldose reductase inhibitors as potential anti-cataract therapies *(46)*. However, while they where extremely effective in rat lenses, they had minimal effect in mice and human lenses, which were later shown to have minimal aldose reductase activity *(47)*. To account for this discrepancy, it was proposed that the main biochemical effect of hyperglycaemia was the induction of oxidative stress, the depletion of the key lens antioxidant glutathione (GSH), and oxidative damage to the transport processes involved in regulating the volume of lens fibre cells *(46, 47)*.

Regardless of the actual biochemical mechanism involved, the morphological consequences of hyperglycaemia appear to indicate that fibre cells in a discrete area of the lens lose their ability to regulate their volume. Earlier studies have shown that lenses placed in either hyposmotic or hyperosmotic medium are capable of regulating their volume, and recover from the volume changes induced by the osmotic stress via the loss [regulatory volume decrease (RVD)], or gain [regulatory volume increase (RVI)] of osmolytes and obligatory water movement *(48)*. In other cell types, RVD is associated with a loss of KCl that is mediated by the activation of K^+ and Cl^- channels *(49, 50)*, and/or potassium chloride cotransporters (KCCs) *(51)*. In contrast, RVI is often effected by uptake of K^+ and Cl^- mediated by transporters such as the sodium-potassium-chloride cotransporter (NKCC) *(52)*. In the lens, the relative contributions of Cl^- channels and transporters to volume regulation have been addressed by a series of experiments that examined the effects on lens volume, transparency, and fibre-cell morphology. Rat lenses were cultured under isosmotic conditions in the presence of reagents that modulated the activity of Cl^- channels and transporters *(11–14, 53)*. In all cases, the presence of reagents in the culture media that inhibited either Cl^- channels, KCC or NKCC produced an increase in lens volume and light scattering. This indicates that under isosmotic conditions, a variety of channels and transporters contribute to a constitutively active flux of Cl^- ions that regulates fibre-cell volume and maintains lens transparency.

Interestingly, subsequent histological analysis of treated lenses revealed that blocking components of this Cl^- flux with a specific transport inhibitor induced either one of two spatially distinct tissue-damage phenotypes, or on occasion a combination of the two phenotypes. The NKCC and Cl^- channel inhibitors, bumetanide and 5-nitro-2-(3-phenylpropylamino) benzoic acid (NPPB) caused a localized band of extracellular fluid accumulation between fibre cells located around 150 μm in from the capsule. In contrast, the predominant effect of the KCC inhibitor [(dihydronindenyl)oxy] alkanoic acid (DIOA), was the swelling of fibre cells located at the lens periphery, although some deeper extracellular-space dilations were evident. The two distinct damage phenotypes, generated by the inhibitors can be explained with reference to the circulation model (Fig. 1C) and experimental measurements of lens membrane potential *(7)*. By measuring radial differences in transmembrane potential and the concentration of Cl^- in the whole lens, the electrochemical gradient for Cl^- ion movement (E_{Cl}) can be calculated at different depths into the lens *(54, 55)*. This analysis predicts that Cl⁻ would move from the extracellular space into fibre cells in the inner lens, but will move from the cytoplasm of fibre cells to the extracellular space in the lens periphery (Fig. 4A). Therefore, one would expect that an inhibition of Cl⁻ fluxes in the inner lens would block the uptake of Cl⁻ from the extracellular space by fibre cells. This would cause an accumulation of Cl⁻ ions and water in the tortuous extracellular space and the subsequent formation of extracellular-space dilations (Fig. 4B). In the lens periphery, the efflux of Cl⁻ ions from fibre cells would be blocked thereby causing an intracellular accumulation of osmolytes and resultant fibre-cell swelling (Fig. 4C). These two spatially segregated zones of cell swelling and extracellular-space dilations were deemed to be due to the inhibition of Cl⁻ influx and efflux in deeper and peripheral fibre cells, respectively *(14)*. Since these pathways are coupled together by gap junctions, they generate a circulating flux of Cl⁻ ions which contributes to the maintenance of steady-state lens volume (Fig. 5D). Thus the pharmacological incubation experiments indicate that NKCC and Cl⁻ channels both mediate ion uptake in the deeper cells while KCC mediates ion efflux in peripheral fibre cells.

This relative simple view of circulating Cl⁻ fluxes has been recently confirmed and extended by a series of electrophysiological, molecular localization, and additional morphological studies. Firstly, patch clamp experiments on isolated fibre cells have allowed the channel(s) that mediates Cl⁻ influx to be functionally characterized in the influx and efflux zones *(53)*. Long fibre cells (>120 µm in length) isolated from the influx zone exhibited a constitutively active outwardly rectifying chloride conductance that exhibited a lyotrophic anion selectivity sequence (I⁻>Cl⁻), reminiscent of volume-sensitive Cl⁻ conductances seen in many cell types *(56)*. In contrast, shorter fibre cells isolated from the efflux zone initially lacked a constitutively active Cl⁻ conductance, however, a Cl⁻ conductance could be activated by isosmotic cell swelling induced by inhibition of KCC-mediated ion efflux that appears to dominate in these cells*(57)*. Thus, while KCC transporters normally mediate Cl⁻ efflux and maintain cell volume in peripheral fibre cells, under conditions of sustained or serve osmotic stress these cells can activate a normally quiescent Cl⁻ conductance to mediate Cl⁻ efflux and relieve osmotic stress.

Secondly, molecular experiments have shown that NKCC1*(29)* and three KCC isoforms (KCC1, KCC3 and KCC4) *(11)* are differentially expressed in the rat lens, supporting their initial pharmacological identification. In the case of KCC, isoform-specific changes in the subcellular distribution of the KCC transporters in response to changes in fibre-cell volume were observed. Fibre-cell swelling evoked either isosmotically by the DIOA-induced blockage of Cl⁻ efflux in peripheral fibre cells, or anisosmotically via hyposmotic challenge, both caused a noticeable increase in membrane insertion of KCC1 and KCC4, but not KCC3. This observed recruitment of KCCs 1 and 4 from a cytoplasmic pool to the membrane (Fig. 2C) indicates that the contribution of KCC to the regulation of fibre-cell volume is a dynamic process that can be up regulated in response to osmotic stress.

Finally, if our interpretation of the morphological data is correct, then we would expect stimulating Cl⁻ fluxes in the lens to have opposite effects to those induced by pharmacologically blocking Cl⁻ efflux and influx. To test this, we utilized the ability of N-ethylmaleimide (NEM) to stimulate KCC activity *(58, 59)* (Fig. 5A). Exposure of

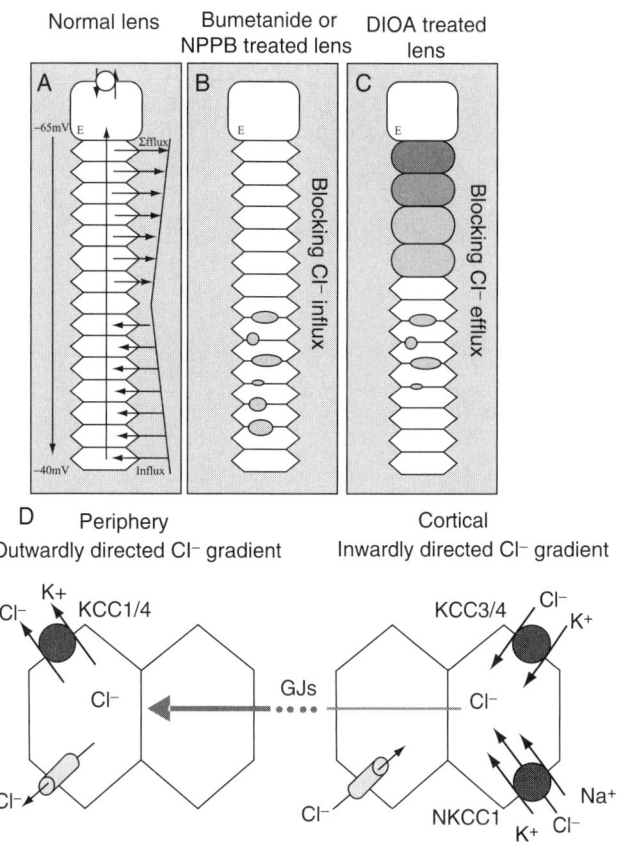

Fig. 4. A circulating flux of Cl⁻ ions contributes to steady state volume regulation in the rat lens. (**A–C**) Schematic summarizing the effects inhibitors of Cl⁻ transport have on the morphology of organ-cultured lens. (**A**) In the absence of inhibitors measurements of membrane potential and E_{Cl^-} predict that in deeper fibre cells the electrochemical gradient favors Cl⁻ influx while in the periphery it promotes Cl⁻ efflux. (**B**) Blocking Cl⁻ influx mediated by either Cl⁻ channels or sodium-potassium-chloride cotransporters (NKCC) with 5-nitro-2-(3-phenylpropylamino) benzoic acid (NPPB) and bumetanide, respectively, results in the accumulation of Cl⁻ ions and water between deeper fibre cells and the formation of extracellular space dilations. (**C**) Blocking Cl⁻ channel efflux mediated by the potassium chloride cotransporter (KCC) with [(dihydroindenyl)oxy] alkanoic acid (DIOA) results in the intracellular accumulation of osmolytes and the swelling of peripheral fibre cells. Some deeper extracellular space dilations were also evident. (**D**) Emerging molecular model of transporters and channels that contribute to volume regulation by mediating ion influx and efflux in different regions of the lens. Since the zones of ion influx and efflux are connected by gap junction channels (GJs) a circulating flux is established which helps to maintain cell volume. Patch clamp studies indicate that in peripheral fibre cells a volume-sensitive Cl⁻ channel becomes activated following cell swelling induced by either blockage of KCC (isosmotic cell swelling) or exposure to hyposmotic solutions.

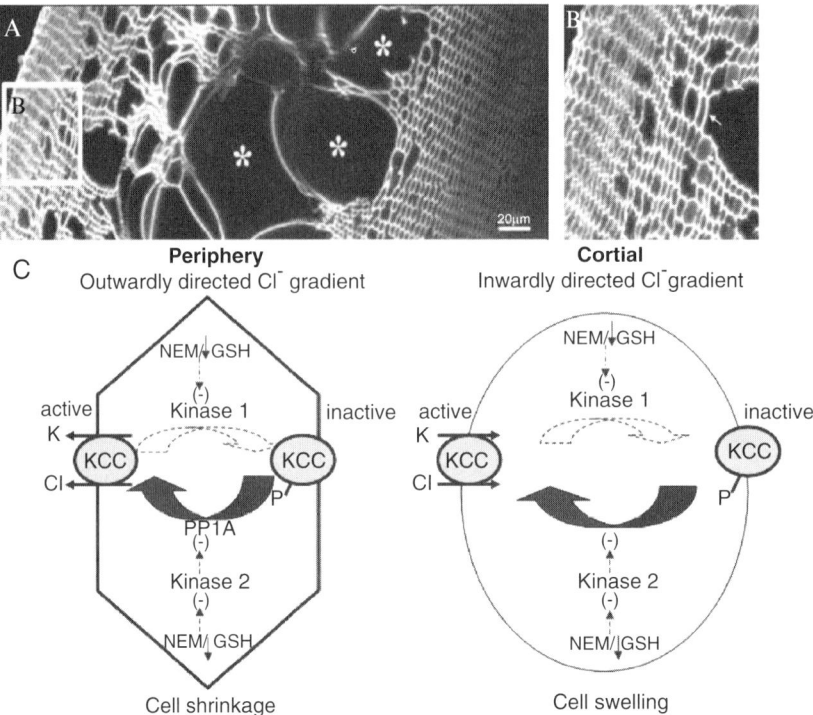

Fig. 5. Mimicking diabetic cataract by activation of the potassium chloride cotransporter (KCC). (A) Equatorial section from a lens organ cultured in the KCC activator N-ethylmaleimide (NEM) for 18 hours results in shrinkage of peripheral fibre cells and the swelling of deeper lying fibre cells producing a band of localized tissue liquefaction that mimics that seen in diabetic cataract. (B) A close-up view of a region from (A) showing the two distinct phenotypes induced by activation of KCC. (C) Schematic diagrams outlining how activation of KCC via either NEM or depletion of glutathione (GSH) affects the phosphorylation status and activity of KCC in the lens periphery (left panel) and in deeper zones of tissue liquefaction (right panel), NEM or GSH depletion inhibits the activity of two kinase that either directly (Kinase 1), or indirectly (Kinase 2) via inhibition of a phosphatase (PP1A), phosphorylates KCC and renders it inactive. The resultant dephosphorylation and activation of KCC promotes cell shrinkage in the periphery and cell swelling in deeper fibre cells.

organ-cultured lenses to NEM, induces the overactivation of KCC and resulted in the shrinkage of peripheral fibre cells (Fig. 5B) in the efflux zone and the swelling of deeper fibre cells (Fig. 5A) in the influx zone that ultimately caused the formation of a localized band of tissue liquefaction *(11)*. This indicates that the normally minimal KCC activity is massively upregulated in the influx zone by NEM activation. The activity of the KCC transporter is normally modulated by its phosphorylation status *(51)*. Two separate kinases are involved in the phosphorylation-dependent inactivation of KCC, while dephosphorylation via a PP1A-type phosphatase causes an increase in KCC activity. NEM stimulates KCC transporter activity via thiol inactivation of the two kinases that control transporter activity *(60)* (Fig. 5C).

The similarity between the liquefaction zone seen in NEM-treated (Fig. 5A) and diabetic lenses (Fig. 3A) is so striking that it prompts speculation that overactivation of KCC transport is occurring in the diabetic cataract. In red blood cells it has been shown that depletion of glutathione (GSH) levels by oxidative stress, causes activation of KCC presumably via oxidation of the critical thiol groups in the two regulatory kinases *(61)*. In the diabetic lens, hyperglycaemia is known to deplete GSH levels *(46)*, thus it is possible that excess glucose compromises the ability of fibre cells to regulate their volume via two pathways: the accumulation of the impermeable osmolyte, sorbitol; and the stimulation of ion influx mediated by the activation of KCC. These osmotic stresses are exacerbated by the activation of volume-sensitive chloride and cation channels, which because of the direction of the ion gradients in this region of the lens, import more osmolytes and water, causing uncontrolled cell swelling *(53)*. Finally, membrane leakage of calcium causes the activation of calcium-dependent proteases, leading to fibre cell vesiculation *(62–64)* and localized tissue liquefaction that is characteristic of diabetic cataract*(43)*. The potential involvement of KCC in the initiation of diabetic cataract opens up new avenues for the development of novel anti-cataract therapies.

NUTRIENT TRANSPORTERS IN THE NORMAL AND CATARACTIC LENS

Age-Related Nuclear (ARN) Cataract - a Transport Problem?

As the world's population ages, ARN cataract has become the leading cause of blindness. It is characterized by a drastic decline in GSH concentration in the nucleus of the lens, but not the cortex, exposing the centre of the lens to the damaging effects of oxygen radicals, causing protein aggregation, increased light scattering and ultimately nuclear cataract *(65)*. However, unlike diabetic cataract, in ARN cataract there appears to no significant morphological changes to the cellular architecture of the lens. Biochemically, however, extensive modifications to proteins in the lens nucleus are observed that include oxidation of methionine residues and sulfhydryl groups, insolubilization of crystallins, and protein cross-linking to form mixed disulphides, all of which contribute to light scattering *(66)*.

Oxidative damage is a key feature of ARN cataracts. In the transparent lens, a robust oxygen-radical scavenger system which utilizes glutathione (GSH) as its principal antioxidant, guards against oxidative stress *(67)*. In the lens, GSH exists in unusually high concentrations (around 10 mM). These high levels are established via the direct uptake of GSH from the aqueous humor*(68–70)* and/or the endogenous synthesis of GSH from its precursor amino acids cysteine, glutamate, and glycine *(71)* by the sequential actions of the enzymes γ-glutamylcysteine synthetase (γGCS) and glutathione synthetase (GS) in the lens cortex. GSH levels are then maintained by the regeneration of GSH from the oxidized form of GSH (GSSG) by the enzyme glutathione reductase (GR), and the consumption of reducing equivalents such as NADPH *(72)*.

In ARN cataract, the levels of GSH are abruptly reduced in the nucleus relative to the cortex, making the centre of the lens especially susceptible to oxidative damage *(65, 73)*. Since the levels of GSH and the activities of its associated enzymes have been shown to progressively decline as a function of age, it has been assumed that ARN cataract is the result of a failure of enzymatic activity *(74)*. However, while the specific activities of enzymes were reduced with increasing age, these reductions were not only deemed to be insufficient to account for the decrease in GSH levels observed in the nucleus, but they do not explain the abrupt fall in GSH levels seen in ARN cataract *(75)*. Rather, nutrient transport to the lens nucleus may be an underlying factor in the development of ARN cataract *(76)*. This raises the question of whether the fall in GSH in the lens nucleus observed during ARN cataract progression is in fact a transport problem.

Differential Expression of Nutrient Transporters in the Lens

While mature fibre cells in the lens nucleus are devoid of cellular organelles, they still require a supply of nutrients for anaerobic metabolism, the replenishment of antioxidants and the maintenance of transparency. The traditional view is that the lens nucleus receives its nutrients from the cortex via an intercellular pathway mediated by gap junctions. Indeed, Sweeney and Truscott have proposed that with advancing age, a barrier develops that restricts the intercellular diffusion of GSH from the cortex to the lens nucleus *(76)*. The alternative view is that the circulation system causes convection of nutrients and antioxidants into the lens nucleus via an extracellular route faster than would be achieved by passive diffusion alone *(6, 7)*. If this notion is correct, we would expect mature fibre cells to express transporters to accumulate the molecules delivered to them by the circulation system. Consistent with this view, molecular studies (Table 1) have shown that fibre cells express a full repertoire of transporters that mediate the uptake of glucose *(17, 32)* and of the amino acids *(15, 16, 77)* involved in the synthesis of the GSH.

Furthermore, immunocytochemical localization of these transporters has shown that differences exist in the complement of transporters expressed in the cortex and nucleus of the lens, suggesting that regional differences in nutrient uptake may exist. Merriman-Smith et al. *(17, 31, 32)* have shown that while lens epithelial cells express the facilitative glucose transporter GLUT1, fibre cells express both the higher affinity GLUT3 isoform, and the sodium-dependent glucose transporter SGLT2 (Fig. 6A). Both the GLUT3 and SGLT2 transporters appear to be present predominantly in the cytoplasm of peripheral fibre cells and become inserted into the plasma membrane at different stages of fibre-cell differentiation. The GLUT3 transporter is initially inserted into the narrow sides of the differentiating fibre cells, while SGLT2 is abruptly inserted into the membranes of mature fibre cells at the transition zone between differentiating and mature fibre cells that coincides with the loss of cell nuclei. We have proposed that the observed differential expression of glucose transporters establishes an affinity gradient which increases the ability of deeper fibre cells to extract a diminishing supply of glucose from the extracellular space. In this regard, the expression of SGLT2 in the lens nucleus means that mature fibre cells are able to utilize the energy stored in the sodium gradient to accumulate glucose above its concentration gradient, where it can be used for anaerobic metabolism.

A similar change in the complement of transporters that mediate cystine uptake in the cortex and nucleus has also been observed in the rat lens *(15, 16)*. Cystine, the dimeric oxidized form of cysteine, is more stable than cysteine, and is more abundant in the aqueous humor *(69)*. Upon intracellular accumulation cystine is rapidly reduced

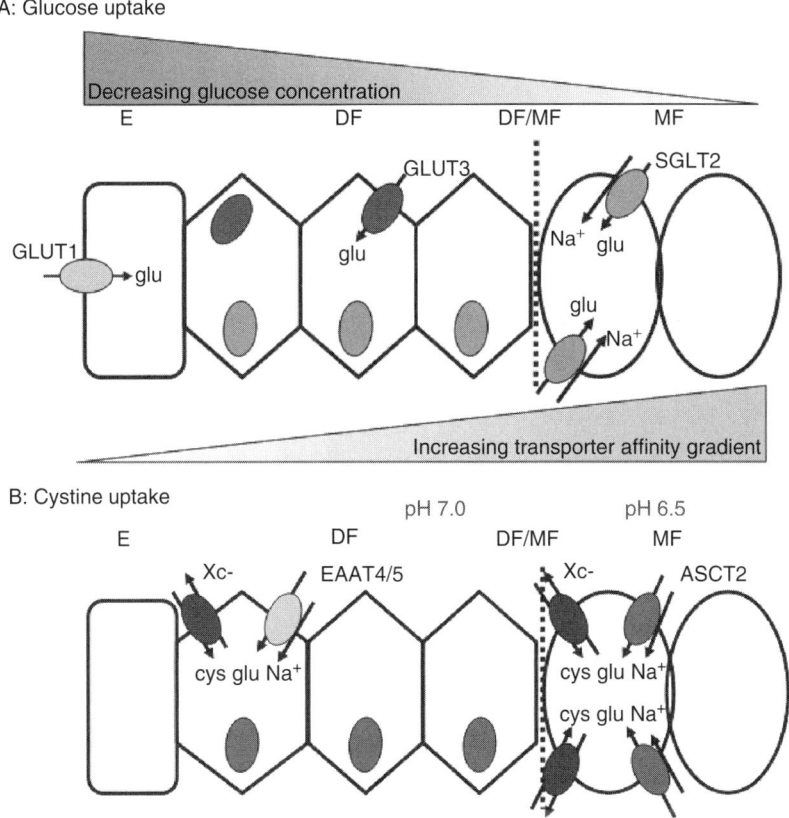

Fig. 6. Regional differences in nutrient uptake in the lens. (**A**) Glucose uptake in the lens is mediated by at least three different transporters. In the epithelium GLUT1 is expressed, while in fibre cells GLUT3 and SGLT2 are expressed. During the course of fibre cell differentiation GLUT3 first becomes inserted into the narrow sides of differentiating fibre cells (DF) in the outer cortex. In contrast SGLT2 is inserted into membranes of mature fibre cells (MF) from a cytoplasmic pool at the DF/MF transition. It is proposed that this differential expression of glucose transporters establishes an affinity gradient, which increases the ability of deeper fibre cells to extract a diminishing supply of glucose from the extracellular space. (**B**) The uptake of cystine from the extracellular space occurs in exchange for intracellular glutamate and is mediated by X_{c^-}. Glutamate is then recovered from the extracellular space by a number of different Na⁺-dependent transporters that belong to the X_{AG^-} family. In the lens cortex X_{c^-} works with EAAT4/5, while in the acidic nucleus of the lens its partner is ASCT2, a transporter that at low pH specifically accumulates glutamate. This change in the transporters responsible for the uptake of glutamate ensures that cystine is accumulated even under the low pH conditions experienced in the lens nucleus.

to cysteine*(78)*, which is the rate-limiting substrate for GSH synthesis. Cystine uptake in a variety of tissues is mediated by the amino acid transport system X_{c^-} *(15, 79)*, a heterodimer composed of a heavy chain (4F2hc) and a light chain (xCT) *(80)*. Cystine uptake is mediated by xCT and involves the exchange of extracellular cystine for intracellular glutamate. Previous biochemical studies indicate that the primary source of glutamate in the lens is via the uptake of glutamine, which is twice as abundant as glutamate in the aqueous humor *(81)*, and its subsequent deamidation to glutamate by the enzyme glutaminase *(82)*. The high intracellular glutamate concentration established can then be utilized by X_{c^-} to drive cystine uptake. To maintain the glutamate gradient, glutamate needs to be actively removed from the extracellular space. In other tissues this glutamate recycling is mediated by members of the X_{AG^-} amino acid transport family *(79, 83)*, a multigene family of Na^+-dependent amino acid transporters, which include the excitatory amino acid transporters (EAAT1-5) and the alanine-serine-cysteine transporters (ASCT1-2) *(84)*. In the lens, we have showed that xCT works in combination with different members of the X_{AG^-} family (Fig. 6B). In the outer cortex, xCT expression was shown to overlap with EAAT4/5 expression *(15)*, while in the nucleus, xCT expression colocalized with ASCT2 expression *(16)*. We hypothesized that the observed switch in glutamate uptake mechanisms from EAAT4/5 to ASCT2 reflects the ability of ASCT2 to preferentially accumulate glutamate at low intracellular pH *(85)*, a finding consistent with the known acidic environment in the lens nucleus *(86)*. Thus, all fibre cells appear to contain the appropriate transport mechanisms for cystine/glutamate exchange and the recycling of glutamate to maintain the glutamate concentration gradient.

Nutrient Delivery to the Lens Nucleus via the Sutures

The molecular identification and localization of transporters to different regions of the lens does not mean they are functionally active. Indeed, the functionality of many of these transporters, especially those in the lens nucleus, has yet to be proven and will be technically difficult to achieve using conventional assays. Recently, Li et al. developed an immunocytochemical assay that utilizes antibodies designed to detect free amino-acid levels *(87)* to map free amino-acid levels quantitatively throughout the rat lens with high spatial resolution *(77)*. Since this assay also utilizes essentially the same equatorial and axial lens cryosections used to localize amino acid transporters *(15, 16)*, it allows the expression of a specific transporter to be correlated to the distribution of its substrate at subcellular resolution. In this assay, active amino acid uptake is deemed to be occurring in a specific region of the lens, if cells contain both membrane labelling for the transporters, and cytoplasmic accumulation of the appropriate amino acid.

By adopting such an immunocytochemical approach, Li et al. were able to directly correlate cystine distribution to lens morphology and membrane labelling for the cystine uptake system, X_{c^-}, previously characterized in our laboratory*(15)*. In both equatorial sections and axial sections, a similar bimodal distribution of cystine was observed (Fig. 7A). Cystine was initially high in the outer cortex, a result consistent with X_{c^-} membrane labelling in this region. Cystine levels then declined to reach a minimum, indicating that the delivery of cystine to deeper-lying fibre cells in the outer cortex is diffusion limited. Thus while cells deeper in the outer cortex express X_{c^-}, insufficient cystine is delivered to them, making cystine accumulation difficult to detect. In the inner cortex, the fall-off in intracellular cystine suddenly plateaus at the differentiating fibre-cell (DF) to mature fibre-cell (MF) transition, before eventually rising again in the nucleus. Since membrane associated X_{c^-} transporters are found in both the inner cortex and nucleus, the observed cystine labelling pattern suggests that the extracellular delivery of cystine to X_{c^-} in the inner cortex is impaired and that a secondary delivery route for X_{c^-} to the lens

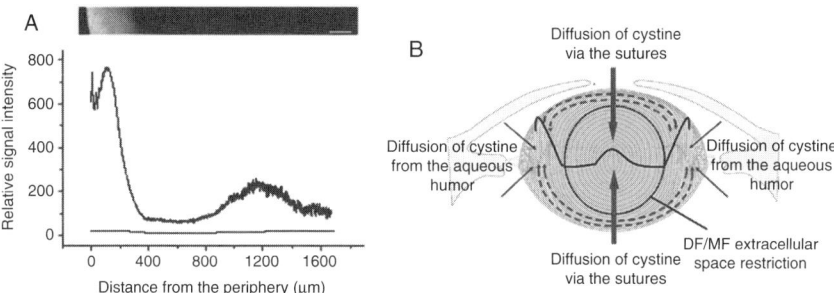

Fig. 7. Cystine delivery pathways in the rat lens. **(A)** A seamless image montage of cystine labelling at the equator of the rat lens (top panel) and an intensity profile extracted from this image montage (lower panel) shows cystine levels to be high in the outer cortex and low in the inner cortex, before increasing again in the nucleus. This bimodal intensity profile can not be explained by passive diffusion alone. **(B)** The bimodal cystine intensity profile has been overlaid on a schematic of lens structure to illustrate the alternative delivery pathways for cystine. In the outer cortex, cystine delivery occurs either via direct diffusion from the surrounding aqueous humor (blue arrows) or via axial diffusion from the sutures (dashed red arrows). In the inner cortex, extracellular delivery of cystine is limited by a restriction of the extracellular space at the differentiating fibre cell/mature fibre cell (DF/MF) region causing cystine levels to reach a minimum in this region. In the nucleus, cystine delivery is mediated by the sutures (red arrows).

nucleus exists (Fig. 7B). These conclusions are supported by two observations. Firstly, it has been recently shown that at the DF to MF transition the membrane insertion of the adhesion protein MP20 is associated with an abrupt restriction of the extracellular diffusion of small molecules *(88)*, which could potentially limit the delivery of cystine to X_{c^-} transporters located in this area. Secondly, careful analysis of cystine uptake in axial lens sections revealed the morphological location of the delivery pathway to the lens nucleus *(77)*. In this orientation, the lens sutures, formed when elongating fibre cells from opposing hemispheres meet at the poles, are observed. In these sections, cystine labelling was strongly co-localized with the sutures, indicating that cystine is preferentially delivered to the very centre of the lens nucleus via the sutures and diffuses out from the nucleus towards the inner cortex (Fig. 7B). These results are extremely exciting since they indicate that it may be possible to use the sutures to deliver small therapeutic agents directly to the nucleus of the lens to delay or prevent the oxidative damage thought to initiate ARN cataract in this area of the lens.

A Model for Age-Related Nuclear Cataract

As well as identifying the sutures as a delivery route to the nucleus of the lens, our work on the localization of membrane transporters in the lens also suggest a mechanism for the age-dependent onset of ARN cataract. Inspection of Table 1 shows that fibre cells express a variety of transporters that either directly or indirectly utilize energy stored in the form of the Na^+ gradient to drive the accumulation of nutrients against their concentration gradients. The circulation model shows that the Na^+ gradient is ultimately maintained by the active removal of Na^+ at the lens surface. Thus, in addition to generating the circulating ion fluxes that underpin the convective delivery of the nutrients to deeper fibre cells, Na^+ efflux at the lens surface also serves to maintain the Na^+ gradient utilized by secondary active transporters to accumulate nutrients in the lens nucleus. It can be predicted therefore that any dissipation of the Na^+ gradient in the nucleus of the lens will selectively reduce nutrient uptake in this region of the lens. Since ARN cataract is known to be associated with a progressive fall in cysteine and GSH levels in the nucleus, but not the cortex *(75, 89)*, this suggests that with advancing age the ability of the circulation system to maintain appropriate nutrient and antioxidant levels in the lens nucleus declines, exposing crystallin proteins in this region to the oxidative damage that precipitates cataract formation.

While pathological dysfunction of any component of the circulation system (extracellular delivery, intracellular uptake, or intercellular efflux pathways) could result in cataract formation, ARN cataract may simply be due to an inability of the circulation system to compensate for the decades of continuing lens growth. It should be remembered that the lens grows throughout life by laying down new fibre cells that internalize the older cells. Since each additional fibre cell layer will contribute to the overall Na^+ leak, an age-dependent increase in passive Na^+ influx occurs, which necessitates an increase in active Na^+ efflux at the lens surface if ionic homeostasis is to be maintained. Unfortunately, the surface area of the lens available for mediating Na^+ efflux does not increase at the same rate. This line of argument is supported by measurements in human lenses that showed lens potential depolarized and the intracellular Na^+ concentration increased with advancing age *(90)*. This age-dependent decline in the Na^+ gradient continues throughout life and eventually leads to initiation of ARN cataract.

If this interpretation is correct, therapeutic interventions to prevent the oxidative damage that initiates ARN cataract should focus on enhancing the activity of the circulation system. Stimulation of the circulation system in the ageing lens will not only increase the delivery of nutrients and antioxidants to the lens nucleus via the sutures, but will also enhance their uptake by restoring the Na^+ gradient utilized by the transporters expressed in this region of the lens. This mechanistic explanation for the onset of ARN cataract illustrates potential benefits of pursuing basic research into the cellular and molecular mechanisms responsible for lens transparency. Furthermore, by continually testing, challenging, and refining the lens internal microcirculation model, we predict novel targets for the development of anti-cataract therapies will be generated.

ACKNOWLEDGEMENTS

The authors wish to acknowledge the Health Research Council of New Zealand, the Marsden Fund (NZ), the Lotteries Grant Board (NZ), the Maurice and Phyllis Paykel Trust and the University of Auckland Research Committee.

REFERENCES

1. Brain, G. & Taylor, H. (2001). Cataract Blindness-challenges for the 21st century. *Bull World Health Organ* **79**, 249–256.
2. Foster, A. (2001). Cataract and "Vision 2020-the right to sight" initiative. *Br J Ophthalmol* **85**, 635–637.
3. Bassnett, S. (2002). Lens organelle degradation. *Exp Eye Res* **74**, 1–6.

4. Menko, A. S. (2002). Lens epithelial cell differentiation. *Exp Eye Res* **75**, 485–490.
5. Kuszak, J. R., Zoltoski, R. K. & C.E., T. (2004). Development of lens sutures. *Int J Dev Biol* **48**, 889–902.
6. Donaldson, P., Kistler, J. & Mathias, R. T. (2001). Molecular solutions to mammalian lens transparency. *News Physiol Sci* **16**, 118–123.
7. Mathias, R. T., Rae, J. L. & Baldo, G. J. (1997). Physiological properties of the normal lens. *Physiol Rev* **77**, 21–50.
8. Robinson, K. R. & Patterson, J. W. (1983). Localization of steady currents in the lens. *Curr Eye Res* **2**, 843–7.
9. Donaldson, P. J., Grey, A. C., Merriman-Smith, B. R., Sisley, A. M., Soeller, C., Cannell, M., B. & Jacobs, M. D. (2004). Functional imaging: new views on lens structure and function. *Clin Exp Pharmacol Physiol* **31**, 890–5.
10. Candia, O. A. & Zamudio, A. C. (2002). Regional distribution of the Na+ and K+ currents around the crystalline lens of rabbit. *Am J Physiol Cell Physiol* **282**, C252–C262.
11. Chee, K.-S. N., Kistler, J. & Donaldson, P. J. (2006). Roles for KCC Transporters in the Maintenance of Lens Transparency *Invest Ophthalmol Vis Sci* **47**, 673–682.
12. Merriman-Smith, B. R., Young, M. A., Jacobs, M. D., Kistler, J. & Donaldson, P. J. (2002). Molecular identification of P-glycoprotein: A role in lens circulation? *Invest Ophthalmol Vis Sci* **43**, 3008–3015.
13. Tunstall, M. J., Eckert, R., Donaldson, P. & Kistler, J. (1999). Localised fibre cell swelling characteristic of diabetic cataract can be induced in normal rat lens using the chloride channel blocker 5-nitro-2-(3-phenylpropylamino) benzoic acid. *Ophthalmic Res* **31**, 317–320.
14. Young, M. A., Tunstall, M. J., Kistler, J. & Donaldson, P. J. (2000). Blocking chloride channels in the rat lens: Localized changes in tissue hydration support the existence of a circulating chloride flux. *Invest Ophthalmol Vis Sci* **41**, 3049–3055.
15. Lim, J., Lam, Y. C., Kistler, J. & Donaldson, P. J. (2005). Molecular characterization of the cystine/glutamate exchanger and the excitatory amino acid transporters in the rat lens. *Invest Ophthalmol Vis Sci* **46**, 2869–2877.
16. Lim, J., Lorentzen, K. A., Kistler, J. & Donaldson, P. J. (2006). Molecular identification and characterisation of the glycine transporter (GLYT1) and the glutamine/glutamate transporter (ASCT2) in the rat lens. *Exp Eye Res* **83**, 447–455.
17. Merriman-Smith, B. R., Krushinsky, A., Kistler, J. & Donaldson, P. J. (2003). Expression patterns for glucose transporters GLUT1 and GLUT3 in the normal rat lens and in models of diabetic cataract. *Invest Ophthalmol Vis Sci* **44**, 3458–3466.
18. Merriman-Smith, R., Donaldson, P. & Kistler, J. (1999). Differential expression of facilitative glucose transporters GLUT1 and GLUT3 in the lens. *Invest Ophthalmol Vis Sci* **40**, 3224–3230.
19. Li, L., Lim, J., Jacobs, M. D. & Donaldson, P. J. (2006). Regional differences in cystine accumulation point to a sutural delivery pathway in the lens core. *Invest Ophthalmol Vis Sci* **148**, 1253–1260.
20. Jacobs, M. D., Donaldson, P. J., Cannell, M. B. & Soeller, C. (2003). Resolving morphology and antibody labeling over large distances in tissue sections. *Microsc Res Tech* **62**, 83–91.
21. Bassnett, S., Missey, H. & Vucemilo, I. (1999). Molecular architecture of the lens fiber cell basal membrane complex. *J Cell Sci* **112**, 2155–2165.
22. Zampighi, G. A., Eskandari, S. & Kreman, M. (2000). Epithelial organization of the mammalian lens. *Exp Eye Res* **71**, 415–435.
23. Merriman-Smith, B. R., Grey, A. C., Varadaraj, R., Kistler, J., Mathias, R. T. & Donaldson, P. J. (2004). The functional implications of the differentiation-dependent expression of glucose transporters in the rat lens. . *Exp Eye Res* **79S**, 105.

24. Beebe, D. C., Vasiliev, O., Guo, J. L., Shui, Y. B. & Bassnett, S. (2001). Changes in adhesion complexes define stages in the differentiation of lens fiber cells. *Invest Ophthalmol Vis Sci* **42**, 727–734.
25. Lee, A., Fischer, R. S. & Fowler, V. M. (2000). Stabilization and remodeling of the membrane skeleton during lens fiber cell differentiation and maturation. *Devel Dyn* **217**, 257–270.
26. Janghorbani, M. B., Jones, R. B. & Allison, S. P. (2000). Incidence of and risk factors for cataract among diabetes clinic attenders. *Ophthalmic Epidemiol* **7**, 13–25.
27. Goodenough, D. A. (1992). The crystalline lens. A system networked by gap junctional intercellular communication. *Semin Cell Biol* **3**, 49–58.
28. Kistler, J., Ling, J. S., Bond, J., Green, C., Eckert, R., Merriman, R., Tunstall, M. & Donaldson, P. (1999). Connexins in the lens: are they to blame in diabetic cataractogenesis? *Novartis Found Symp* **219**, 97–108.
29. Bond, J., Green, C., Donaldson, P. & Kistler, J. (1996). Liquefaction of cortical tissue in diabetic and galactosemic rat lenses defined by confocal laser scanning microscopy. *Invest Ophthalmol Vis Sci* **37**, 1557–1565.
30. Jacob, T. J. C. (1999). The relationship between cataract, cell swelling and volume regulation. *Prog Retinal Eye Res* **18**, 223–233.
31. Kinoshita, J. H., Merola, L. O., Satoh, K. & Dikmak, E. (1962). Osmotic changes caused by the accumulation of dulcitol in the lenses of rats fed with galactose. *Nature* **194**, 1085–7.
32. Kyselova, Z., Stefek, M. & Bauer, V. (2004). Pharmacological prevention of diabetic cataract. *J Diabetes Complications* **18**, 129–140.
33. Crabbe, M. J. C. & Goode, D. (1998). Aldose Reductase: a Window to the Treatment of Diabetic Complications? *Prog Retinal Eye Res* **17**, 313–383.
34. Patterson, J. W. (2006). Volume regulation in rat lens. In *Red Blood Cell and Lens Metabolism*. Elsevier, Amsterdam. **148**, 1253–1250.
35. Niemeyer, M. I., Cid, L. P. & Sepulveda, F. V. (2001). K^+ conductance activated during regulatory volume decrease. The channels in Ehrlich cells and their possible molecular counterpart. *Comp Biochem Physiol A Mol Integr Physiol* **130**, 565–575.
36. Sardini, A., Amey, J. S., Weylandt, K. H., Nobles, M., Valverde, M. A. & Higgins, C. F. (2003). Cell volume regulation and swelling-activated chloride channels. *Biochimica Et Biophysica Acta-Biomembranes* **1618**, 153–162.
37. Lauf, P. K. & Adragna, N. C. (2000). K-Cl cotransport: properties and molecular mechanism. *Cell Physiol Biochem* **10**, 341–54.
38. Russell, J. M. (2000). Sodium-potassium-chloride cotransport. *Physiol Rev* **80**, 211–276.
39. Webb, K. F., Merriman-Smith, B. R., Stobie, J. K., Kistler, J. & Donaldson, P. J. (2004). Cl- Influx into Rat Cortical Lens Fiber Cells Is Mediated by a Cl- Conductance That Is Not ClC-2 or -3. *Invest Ophthalmol Vis Sci* **45**, 4400–4408.
40. Mathias, R. T. (1985). Steady-state voltages, ion fluxes, and volume regulation in syncytial tissues. *Biophys J* **48**, 435–448.
41. Mathias, R. T. & Rae, J. L. (1985). Transport properties of the lens. *Am J Physiol* **249**, C181–C190.
42. Strange, K., Emma, F. & Jackson, P. S. (1996). Cellular and molecular physiology of volume-sensitive anion channels. *Am J Physiol Cell Physiol* **270**, C711–C730.
43. Donaldson, P. J., Chee, K. N., Webb, K. F. & Kistler, J. (2005). Spatially Distinct Cl- Influx and Efflux Pathways Interact to Maintain Lens Volume and Transparency *Invest Ophthalmol Vis Sci* **46**, E-abstract:1129.
44. Alvarez, L. J., Candia, O. A., Turner, H. C. & Polikoff, L. A. (2001). Localization of a Na^+–K^+–$2Cl(-)$ cotransporter in the rabbit lens. *Exp Eye Res* **73**, 669–680.

45. Lauf, P. K. & Theg, B. E. (1980). A chloride dependent K$^+$ flux induced by N-ethylmaleimide in genetically low K$^+$ sheep and goat erythrocytes. *Biochem Biophys Res Commun* **92**, 1422–1428.
46. Logue, P., Anderson, C., Kanik, C., Farquharson, B. & Dunham, P. (1983). Passive potassium transport in LK sheep red cells. Modification with N-Ethyl-Maleimide. *J Gen Physiol* **81**, 861–885.
47. Lauf, P. K. (1985). K:Cl cotransport: sulfhydryls, divalent cations, and the mechanism of volume activation in a red cell. *J Memb Biol* **88**, 1–13.
48. Lauf, P. K., Adragna, N. C. & Agar, N. S. (1995). Glutathione removal reveals kinases as common targets for K-Cl cotransport stimulation in sheep erythrocytes. *Am J Physiol Cell Physiol* **269**, C234–241.
49. Bhatnagar, A., Ansari, N. H., Wang, L. F., Khanna, P., Wang, C. S. & Srivastava, S. K. (1995). Calcium-mediated disintegrative globulization of isolated ocular lens fibers mimics cataractogenesis. *Exp Eye Res* **61**, 303–310.
50. Wang, L., Christensen, B. N., Bhatnagar, A. & Srivastava, S. K. (2001). Role of calcium-dependent protease(s) in globulization of isolated rat lens cortical fiber cells. *Invest Ophthalmol Vis Sci* **42**, 194–199.
51. Wang, L. F., Dhir, P., Bhatnagar, A. & Srivastava, S. K. (1997). Contribution of osmotic changes to disintegrative globulization of single cortical fibers isolated from rat lens. *Exp Eye Res* **65**, 267–275.
52. Lou, M. F. (2003). Redox regulation in the lens. *Prog Retin Eye Res* **22**, 657–682.
53. Truscott, R. J. W. (2005). Age-related nuclear cataract - oxidation is the key. *Exp Eye Res* **80** 709–725.
54. Reddy, V. N. (1990). Glutathione and its function in the lens- an overview. *Exp Eye Res* **50**, 771–778.
55. Kannan, R., Yi, J. R., Zlokovic, B. V. & Kaplowitz, N. (1995). Molecular characterization of a reduced glutathione transporter in the lens. *Invest Ophthalmol Visual Sci* **36**, 1785–1792.
56. Mackic, J. B., Kannan, R., Kaplowitz, N. & Zlokovic, B. V. (1997). Low de novo glutathione synthesis from circulating sulfur amino acids in the lens epithelium. *Exp Eye Res* **64**, 615–626.
57. Stewart-DeHaan PJ, D. T., Trevithick JR. (1999). Modelling cortical cataractogenesis XXIV: uptake by the lens of glutathione injected into the rat. *Mol Vis* **22**, 37.
58. Ganea, E. & Harding, J. (2006). Glutathione related enymes and the eye. *Curr Eye Res* **31**, 1–11.
59. Rathbun, W. B. & Bovis, M. G. (1986). Activity of glutathione peroxidase and glutathione reductase in the human lens related to age. *Curr Eye Res* **5**, 381–5.
60. Rathbun, W. B. & Murray, D. L. (1991). Age-related cysteine uptake as rate-limiting in glutathione synthesis and glutathione half-life in the cultured human lens. *Exp Eye Res* **53**, 205–12.
61. Sweeney, M. H. & Truscott, R. J. (1998). An impediment to glutathione diffusion in older normal human lenses: a possible precondition for nuclear cataract. *Exp Eye Res* **67**, 587–95.
62. Wang, X. F. & Cynader, M. S. (2000). Astrocytes provide cysteine to neurons by releasing glutathione. *J Neurochem* **74**, 1434–1442.
63. McBean, G. J. & Flynn, J. (2001). Molecular mechanisms of cystine transport. *Biochem Soc Trans* **29**, 717–722.
64. Sato, H., Tamba, M., Ishii, T. & Bannai, S. (1999). Cloning and expression of a plasma membrane cystine/glutamate exchange transporter composed of two distinct proteins. *J Biol Chem* **274**, 11455–11458.
65. Kern, H. L. & Ho, C. K. (1973). Transport of L-glutamic acid and L-glutamine and their incorporation into lenticular glutathione. *Exp Eye Res* **17**, 455–462.

66. Jernigan, H. M. J. & Zigler, J. S. J. (1987). Metabolism of glutamine and glutamate in monkey lens. *Exp Eye Res* **44**, 871–876.
67. McBean, G. J. (2002). Cerebral cystine uptake: a tale of two transporters. *Trends Pharmacol Sci* **23**, 299–302.
68. Gegelashvili, G. & Schousboe, A. (1997). High affinity glutamate transporters: regulation of expression and activity. *Mol Pharmacol* **52**, 6–15.
69. Utsunomiya-Tate, N., Endou, H. & Kanai, Y. (1996). Cloning and functional characterization of a system ASC-like Na$^+$ dependent neutral amino acid transporter. *J Biol Chem* **271**, 14883–14890.
70. Baldo, G. J. & Mathias, R. T. (1992). Spatial variations in membrane properties in the intact rat lens. *Biophys J* **63**, 518–29.
71. Marc, R., Murry, R. & Basinger, S. (1995). Pattern recognition of amino acid signatures in retinal neurons. *J Neurosci* **15**, 5106–5129.
72. Grey, A. C., Jacobs, M. D., Gonen, T., Kistler, J. & Donaldson, P. J. (2003). Insertion of MP20 into lens fibre cell plasma membranes correlates with the formation of an extracellular diffusion barrier. *Exp Eye Res* **77**, 567–574.
73. Duncan, G., Hightower, K. R., Gandolfi, S. A., Tomlinson, J. & G., M. (1989). Human lens membrane cation permeability increases with age. *Invest Ophthalmol Vis Sci* **30**, 1855–9.
74. Dean, W. L., Delamere, N. A., Borchman, D., Moseley, A. E. & Ahuja, R. P. (1996). Studies on lipid and the activity of Na,K-ATPase in lens fibre cells. *Biochem J* **314**, 961–967.
75. Delamere, N. A., Dean, W. L., Stidam, J. M. & Moseley, A. E. (1996). Differential expression of sodium pump catalytic subunits in the lens epithelium and fibers. *Ophthalmic Res* **28** (Suppl 1), 73–6.
76. Delamere, N. A. & Tamiya, S. (2004). Expression, regulation and function of Na,K-ATPase in the lens. *Prog Retin Eye Res* **23**, 593–615.
77. Gao, J., Sun, X., Yatsula, V., Wymore, R. S. & Mathias, R. T. (2000). Isoform-specific function and distribution of Na/K pumps in the frog lens epithelium. *J Memb Biol* **178**, 89–101.
78. Garner, M. H. & Kong, Y. L. (1999). Lens epithelium and fiber na,K-ATPases: Distribution and localization by immunocytochemistry. *Invest Ophthalmol Vis Sci* **40**, 2291–2298.
79. Moseley, A. E., Dean, W. L. & Delamere, N. A. (1996). Isoforms of Na,K-ATPase in rat lens epithelium and fiber cells. *Invest Ophthalmol Vis Sci* **37**, 1502–1508.
80. Ong, M. D., Payne, M. D. & Garner, M. H. (2003). Differential protein expression in lens epithelial whole mounts and lens epithelial cell cultures. *Exp Eye Res* **77**, 35–49.
81. Tamiya, S., Dean, W. L., Paterson, C. A. & Delamere, N. A. (2003). Regional distribution of Na,K-ATPase activity in porcine lens epithelium. *Invest Ophthalmol Vis Sci* **44**, 4395–4399.
82. Tao, Q. F., Hollenberg, N. K. & Graves, S. W. (1999). Sodium pump inhibition and regional expression of sodium pump alpha-isoforms in lens. *Hypertension* **34**, 1168–74.
83. Marian, M. J., Li, H. & Borchman, D. (2005). Plasma membrane Ca^{2+}ATPase expression in the human lens *Exp Eye Res* **81**, 57–64.
84. Liu, L., Bian, L., Borchman, D. & Paterson, C. A. (1999). Expression of sarco/endoplasmic reticular Ca^{2+}ATPase in human lens epithelial cells and cultured human lens epithelial B-3 cells. *Curr Eye Res* **19**, 389–94.
85. Misri, S., Chimote, A. A., Adragna, N. C., Warwar, R., Brown, T. L. & Lauf, P. K. (2006). KCC isoforms in a human lens epithelial cell line (B3) and lens tissue extracts. *Exp Eye Res* **83**, 1287–94.
86. Kannan, R., Stolz, A., Ji, Q., Prasad, P. & Gannapathy, V. (2001). Vitamin C transport in human lens epithelial cells:evidence for the presence of SVCT2. *Exp Eye Res* **73**, 159–165.

6
Lens Na⁺, K⁺-ATPase

Nicholas A. Delamere and Shigeo Tamiya

CONTENTS

SUMMARY
NA, K-ATPASE DISTRIBUTION IN THE LENS
LENS NA, K-ATPASE TURNOVER
REGULATION OF LENS NA, K-ATPASE ACTIVITY
LENS OSMOREGULATION AND OPACIFICATION
ACKNOWLEDGMENTS
REFERENCES

SUMMARY

Mammalian cells are required to export sodium and import potassium in an active manner. At the resting state, a normal cell has a cytoplasmic sodium concentration $[Na^+]_i$ of approximately 15 mM, while the concentration of extracellular sodium is around 140 mM. Extracellular potassium is generally around 5 mM while cytoplasmic potassium is around 120 mM. Although cells generally have a fairly high permeability to potassium, the tendency for outward potassium diffusion, driven by the concentration gradient across the plasma membrane, is more or less balanced by the tendency for inward potassium movement due to the electrical voltage (cytoplasm negative) across the plasma membrane. In the case of sodium, both the ion concentration gradient and the negative voltage across the plasma membrane favor the movement of extracellular sodium toward the cytoplasm. Sodium permeability of the plasma membrane is considerably lower than potassium permeability but nevertheless sodium leaks into the cell. More sodium enters via sodium-coupled cotransporters and counter-transporters that shift other solutes using the electrochemical sodium gradient as a driving force. To prevent the cytoplasmic sodium concentration from rising, sodium entry must be balanced by continuous outwardly directed active transport and this is carried out by the Na,K-adenosine triphosphatase (-ATPase), a P-type ATPase that comprises of two membrane-spanning proteins, the α and β subunits. Na,K-ATPase catalyzes the hydrolysis of adenosine triphosphate (ATP) to adenosine diphosphate (ADP) and in so doing shifts three sodium ions outward and two potassium ions inward. Na,K-ATPase structure and function has been reviewed elsewhere *(1)*.

From: *Ophthalmology Research: Ocular Transporters in Ophthalmic Diseases and Drug Delivery*
Edited by: J. Tombran-Tink and C. J. Barnstable © Humana Press, Totowa, NJ

NA, K-ATPASE DISTRIBUTION IN THE LENS

In many respects, the lens is no different from any other tissue in that Na,K-ATPase is essential for cell survival. On the other hand, specializations that give rise to lens transparency make the task of ion regulation rather unusual in the lens. For example, the lens is made up of millions very tightly packed cells, lens fibers, most of which have no nucleus or mitochondria. There is little or no turnover of cells; each fiber cell is retained for the lifetime of the individual. Moreover, the lens is avascular, so fiber cells in the center of the lens are situated far distant from the nearest blood supply.

The lens relies to a large extent on the epithelium for Na,K-ATPase-mediated ion transport. The lens epithelium is a monolayer of cells that covers the anterior, but not the posterior, surface (Fig. 1). Thus the epithelium represents a tiny fraction of lens cells. A very small population of epithelial cells located at the lens equator are the only lens cells that undergo mitosis. The bulk of the lens is made up of millions of tightly packed fiber cells. Fibers are formed by a complex process of differentiation following epithelial mitosis. Each fiber elongates until it stretches from the anterior to posterior lens pole. New fibers are formed continuously. In the humans lens it is estimated 40 new fibers are made each day *(2)*. Newly made fibers overlay existing cells, compressing older cells towards the center (nucleus) of the lens. The completion of the fiber differentiation process involves loss of mitochondria and the nucleus. Consequently, most fibers are incapable of mitochondrial ATP production.

Na,K-ATPase activity varies considerably between different regions of the lens. The lens epithelium has a generally higher Na,K-ATPase-specific activity than the fibers *(3–5)*. Within the epithelial monolayer, Na,K-ATPase-specific activity is highest at the equator and much lower in the central zone *(6–8)* (Fig. 2). Thus, active Na-K transport capacity is concentrated at the periphery of the epithelial sheet. Although Na,K-ATPase-specific activity in cortex is low, the sheer bulk of the cortex compared to the epithelium monolayer means that cortical fibers account for most of the lens Na,K-ATPase activity *(3)*. In the fiber mass, Na,K-ATPase activity is undetectable in the lens nucleus *(4)*. In the outer cortex, the highest Na,K-ATPase activity is found at the equator *(3)*.

Enrichment of Na,K-ATPase activity at the surface region of the lens, particularly at the equator, is one of several factors thought to give rise to circulating electrical currents

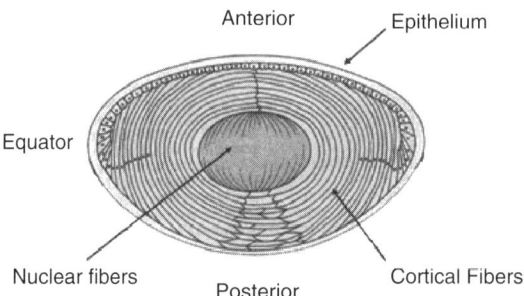

Fig. 1. Diagrammatic cross-section of the lens (adapted from *44*).

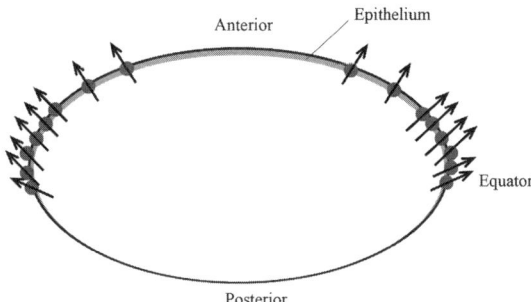

Fig. 2. Na,K-ATPase-mediated outward sodium transport (arrows) is more active at the lens equator. Na,K-ATPase activity in the epithelium is highest at the equator. Fiber cells at the equator have an Na,K-ATPase activity that is higher than fibers elsewhere in the lens.

that flow around the lens *(9)*. Detected first with a unique vibrating electrode probe, currents flow outward at the equator and inward at the anterior and posterior poles of the lens *(10,11)*. It has been proposed that such circulating currents might support electro-osmotic flow of solutes that speeds penetration of dissolved substances to fibers deep in the lens. This could explain why labeled glucose penetrates the lens fiber mass at a faster rate than the rate predicted by simple diffusion *(12)*.

To some extent, the close proximity of a mitochondrial ATP source may dictate the observed restriction of high Na,K-ATPase activity to cells in the anterior lens surface and equatorial fibers. Regions of the lens with high Na,K-ATPase activity are also the regions where mitochondria are found. This makes sense because Na,K-ATPase-mediated ion transport consumes ATP at a high rate. In some epithelial tissues, it has been estimated that Na,K-ATPase utilizes more than 50% of the ATP supply. In the lens, mitochondria are restricted to the epithelial monolayer and to the younger, therefore superficially located, fibers at the lens equator. According to this distribution, there are a few mitochondria in the central light path through the lens. It is thought that mitochondrial distribution is an adaptation that minimizes the light scatter in the optical path. Could it be that the restriction of mitochondria to the equatorial region of the lens influences Na,K-ATPase activity distribution?

Fibers, which make up almost the entire bulk of the lens, are differentiated epithelial cells. The lack of emphasis on Na,K-ATPase activity is just one of the many characteristics of the differentiated fiber. Fiber shape and size is different from the undifferentiated lens epithelium. As mentioned above, the fully differentiated fiber loses it nucleus, endoplasmic reticulum and mitochondria. Not surprisingly, fibers express some proteins not expressed in the epithelium and vice versa. Connexins (gap junction proteins) and aquaporins in fibers and epithelium are distinct. In some species, lens epithelium and fibers have a different pattern of Na,K-ATPase expression. In the rat lens, the epithelium expresses three different isoforms of Na,K-ATPase catalytic subunit, $\alpha 1$, $\alpha 2$ and $\alpha 3$, while in fibers only $\alpha 1$ is detected *(13)*. The α subunit is the ~100 kDa membrane-spanning protein responsible for sodium and potassium ion transport translocation and ATP hydrolsis (see *1* for review). Less is known about the lens distribution of the β subunit, the membrane glycoprotein that is the second component of the α/β Na,K-ATPase pump unit.

Expression of multiple Na,K-ATPase isoforms in the rat lens epithelium fits with the notion that the epithelium maintains a higher Na,K-ATPase-specific activity than fibers. However, the story is more complex. Lenses from other species have a different pattern of Na,K-ATPase isoform expression. In porcine lens, for example, the Na,K-ATPase α1 isoform is the principle isoform in epithelium and fibers and the α2 and α3 isoforms are either absent or expressed in small amounts (8,14). What makes the situation interesting is the fact that the Na,K-ATPase expression pattern can change. In porcine lenses exposed for 12 hours or more to amphotericin B, an agent that increases the apparent permeability to sodium and other ions, Na,K-ATPase α2 isoform protein becomes detectable in the lens epithelium (14). Boosting the expression of the α2 Na,K-ATPase isoform could be an adaptation that permits the porcine lens to cope with situations where additional Na,K-ATPase transport capacity is required. A similar pattern of Na,K-ATPase α2 upregulation was observed in porcine lenses where Na,K-ATPase was partially inhibited by dihydro-ouabain (15). The ability of the lens epithelial cells to respond to a challenge by synthesizing additional Na,K-ATPase protein suggests that Na,K-ATPase activity is linked to the amount of Na,K-ATPase protein. This is logical, but suprisingly, the situation is not so simple. While lens epithelium and fibers have very different Na,K-ATPase activity, both cell types have been found to contain relatively similar amount of Na,K-ATPase polypeptide (4,16); this points to a situation where Na,K-ATPase protein in cells from different regions of the lens is not necessarily active to the same degree.

At this point we do not know precisely what determines the activity of Na,K-ATPase in lens cells. In all likelihood, several factors may be involved. Recent studies suggest that turnover of Na,K-ATPase protein could be important and that protein phosphorylation might play a role. These issues are discussed below.

LENS NA, K-ATPASE TURNOVER

Measuring the abundance of Na,K-ATPase protein, or any other protein, does not give a sense of the age of the protein. Some lens cells are very old yet the protein constituents found in the cell at any point in time may be much younger, reflecting a continuous cycle of protein synthesis and degradation. Na,K-ATPase turnover, defined here as the replacement of old Na,K-ATPase polypeptide with newly-synthesized polypeptide, has been studied in the lens using methionine labeled with ^{35}S (17). Rat lenses incubated for four hours in ^{35}S-methionine-containing culture medium showed distinct labeling of many proteins including Na,K-ATPase α1 polypeptide. This indicates incorporation of the radioactive amino acid into newly synthesized Na,K-ATPase protein. In studies with a human lens epithelium cell line, HLE-B3, a period of ^{35}S-methionine incorporation was followed by exposure of the cells to non-radioactive medium (17). In this phase of the experiment, newly synthesized proteins do not contain the ^{35}S-labeled amino acid, so radioactive protein synthesized during the ^{35}S-methionine incorporation phase is gradually replaced with non-labeled protein. To home in specifically on Na,K-ATPase, Cui and his coworkers (17) used an immunoprecipitation technique to isolate Na,K-ATPase α1 subunit. In HLE-B3 cells, these investigators found an apparent half-time of approximately 30 min for synthesis of Na,K-ATPase α1 protein (Fig. 3). Although the net amount of Na,K-ATPase α1 protein assessed by Western blot remains unchanged, part of the Na,K-ATPase protein pool is continuously being replaced with newly synthesized pro-

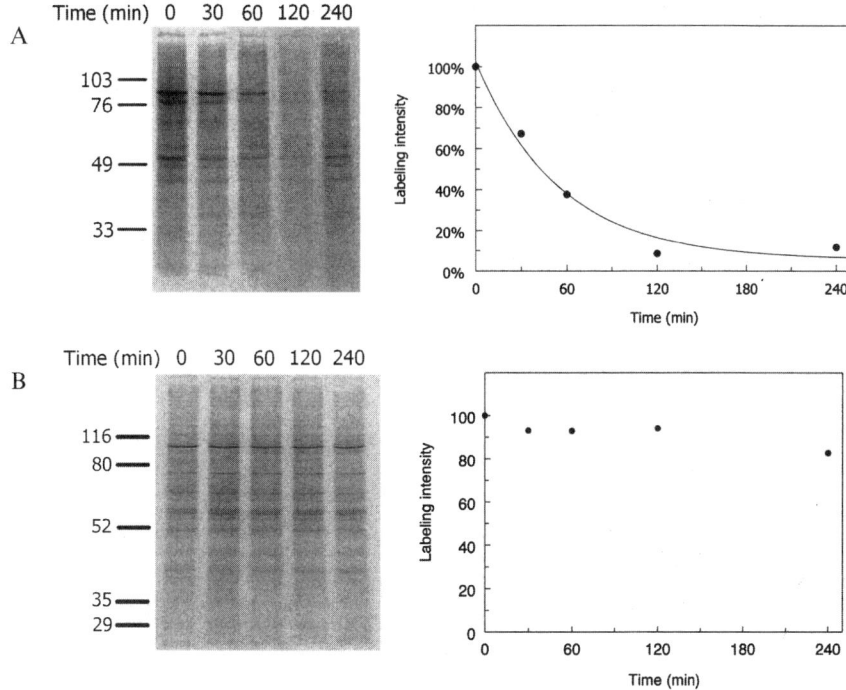

Fig. 3. Turnover of Na, K-ATPase α 1 protein in a human lens epithelial cell line, HLE-B3. Following a two-hour ^{35}S-methionine incorporation period, HLE-B3 cells were incubated in normal media containing nonradioactive methionine, with or without 100 μM cycloheximide, for up to 240 minutes. Left panels show autoradiographs of immunoprecipitated Na,K-ATPase α 1 protein (102-kDa band). Quantified values of the α 1 band labeling intensity (normalized to the $t = 0$ control value) are plotted against time in the right panels. (A) In control cells, radioactive label intensity of the Na,K-ATPase α 1 protein decreases over time as it is replaced by newly synthesized nonlabeled Na,K-ATPase α 1 protein. (B) Label intensity does not change since protein synthesis is inhibited by cycloheximide. (From 17; used with permission.)

tein. When turnover was inhibited by cycloheximide, an agent that suppresses protein synthesis, Na,K-ATPase activity was found to diminish although the overall abundance of Na,K-ATPase was unchanged. One interpretation of the cycloheximide response is that Na,K-ATPase protein loses enzyme activity over time. On this basis, cells that do not continually synthesize Na,K-ATPase protein may be expected to have low Na,K-ATPase activity. This fits with the low Na,K-ATPase activity found in fully differentiated fibers cells in the lens nucleus. These fibers are incapable of protein synthesis. It remains to be determined whether Na,K-ATPase protein synthesis occurs in younger fiber cells that still possess organelles and a nucleus. There are also important questions to be answered regarding the stimuli that influence the rate of Na,K-ATPase turnover in the lens epithelium. Na,K-ATPase activity is highest at the equatorial region of the epithelium where the lens is closest to the ciliary body. According to studies by Coca Prados and colleagues, the ciliary body functions as a neuroendocrine tissue capable of secreting neuropeptides and other substances that potentially can alter cell function in other eye tissues *(18)*. Substances emanating from the ciliary body may act as chemical signals which instruct the equatorial lens epithelium to maintain high Na,K-ATPase activity. This idea remains to be tested.

REGULATION OF LENS NA, K-ATPASE ACTIVITY

Protein synthesis maybe one of several factors that influence Na,K-ATPase activity. Short-term changes of Na,K-ATPase activity, elicited by agonists such as endothelin-1 and thrombin, seem to occur without an apparent change in Na,K-ATPase protein expression *(19,20)*. The responses to these agents are quick, occurring within minutes. In lenses exposed to thrombin, the epithelium was subsequently removed and assayed under control conditions in which no thrombin was present, but even so, the treated epithelial cell material was found to have significantly diminished Na,K-ATPase activity compared to epithelium samples isolated from control lenses *(19)*. The results suggest V_{max} of Na,K-ATPase is decreased in the epithelium of thrombin-treated lenses. This idea is reinforced by the observation that thrombin has a pronounced inhibitory effect on Na,K-ATPase in lenses that are exposed to amphotericin B. Amphotericin B stimulates Na,K-ATPase by permitting cytoplasmic sodium to rise, thereby making Na,K-ATPase function at or close to the V_{max} activity and eliminating the possibility that the Na-K transport is rate-limited by restriction of sodium entry. It is a point of interest that responses to thrombin and endothelin-1 occur in parallel with pronounced changes in tyrosine phosphorylation and there is evidence that implicates tyrosine phosphorylation as a mechanism that regulates Na,K-ATPase activity. Importantly, the Na,K-ATPase response to thrombin is eliminated if tyrosine kinase inhibitors are applied at the same time as thrombin. Tyrosine kinase inhibitors also prevent the Na,K-ATPase response of the lens to endothelin-1 *(20)*.

It should be noted that many proteins become tyrosine phosphorylated in thrombin-treated or endothelin-1-treated lenses, and the observed inhibition of Na,K-ATPase activity could potentially be an indirect result of altered function of another protein. Nevertheless, immunoprecipitation studies points to tyrosine phosphorylation of the Na,K-ATPase α1 catalytic subunit itself *(21)*. Feraille and coworkers have presented evidence implicating a specific tyrosine residue, Tyr-10, in the Na,K-ATPase response of a kidney tubule cell line to insulin *(22)*. The phosphorylation of Na,K-ATPase was not observed in cells expressing a mutant form of Na,K-ATPase α1 that lacked this tyrosine residue.

Various tyrosine kinases exist within a cell but a particular family of tyrosine kinases, the Src family of tyrosine kinases (SFKs), has been implicated in the regulation of ion transporters and channels, including the Na,K-ATPase, in several tissues and cell types. In terms of the lens, Bozulic and colleagues have shown that addition of the SFK member Lyn to a lens membrane preparation results in phosphorylation of multiple membrane proteins and inhibition of Na,K-ATPase activity *(21)*. The same authors followed up with a interesting study that showed different SFK members caused different patterns of membrane protein phosphorylation and different Na,K-ATPase activity responses *(23)*. While Na,K-ATPase activity was decreased by Lyn and Fyn SFK treatment, Src increased Na,K-ATPase activity and another SFK member, Lck, had no effect on the Na,K-ATPase activity. Although the complex mechanism underlying this finding still needs to be resolved, it clearly demonstrates that fine tuning of the Na,K-ATPase activity is possible by protein phosphorylation. It should be noted that SFK protein abundance,

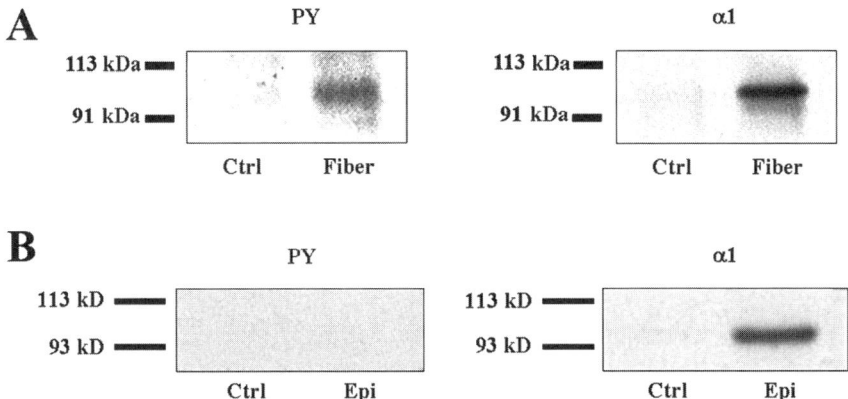

Fig. 4. Phosphorylation status of immunoprecipitated Na,K-ATPase α 1 protein from (**A**) porcine lens fiber and (**B**) epithelium membrane material. Following immunoprecipitation of Na, K-ATPase α 1 protein using a polyclonal antibody, immunoprecipitates were probed by Western blot for phosphotyrosine proteins (PY) (left). The same blot was stripped and re-probed with a monoclonal anti-Na,K-ATPase α 1 antibody (right). In the control, the immunoprecipitating antibody was used in the absence of membrane material. (From *16*; used with permission.)

as well as SFK activity, is higher at the equator of the lens epithelium compared to the central anterior region *(24)*. Na,K-ATPase activity is also higher at the equator *(8)*.

Although the two lens cell types have similar Na,K-ATPase protein abundance, the specific Na,K-ATPase activity in lens fibers is much lower compared to the lens epithelium *(4)*. This difference in specific activity between the two cell types can be explained, at least in small part, by protein tyrosine phosphorylation. By immunoprecipitation of the Na,K-ATPase and using a phosphotyrosine-specific antibody, Bozulic and colleagues have shown that Na,K-ATPase in lens fiber cells has detectable tyrosine phosphorylation *(16)* (Fig. 4). In contrast, tyrosine phosphorylation of the Na,K-ATPase in the lens epithelial cells was undetectable. Importantly, treatment of a lens fiber membrane preparation with protein tyrosine phosphatase-1B (PTP-1B) reduced the extent of tyrosine phosphorylation and caused Na,K-ATPase activity to increase. Protein tyrosine phosphatase-1B had no effect on Na,K-ATPase activity of a membrane preparation obtained from the epithelium (Fig. 5). Tyrosine phosphorylation, together with low turnover of Na,K-ATPase protein, may contribute to the low Na,K-ATPase-specific activity in the lens cortical fibers.

LENS OSMOREGULATION AND OPACIFICATION

As discussed above, the Na,K-ATPase directly regulates lens sodium and potassium levels. This is crucial for osmoregulation. If Na,K-ATPase activity ceases, the subsequent rise of cytoplasmic sodium and loss of potassium is accompanied by a increase in lenticular water. The lens swells, cell architecture is disrupted and transparency deteriorates. Furthermore, the sodium gradient created by the Na,K-ATPase plays an important role in various cotransport and counter-transport mechanisms that shift other solutes.

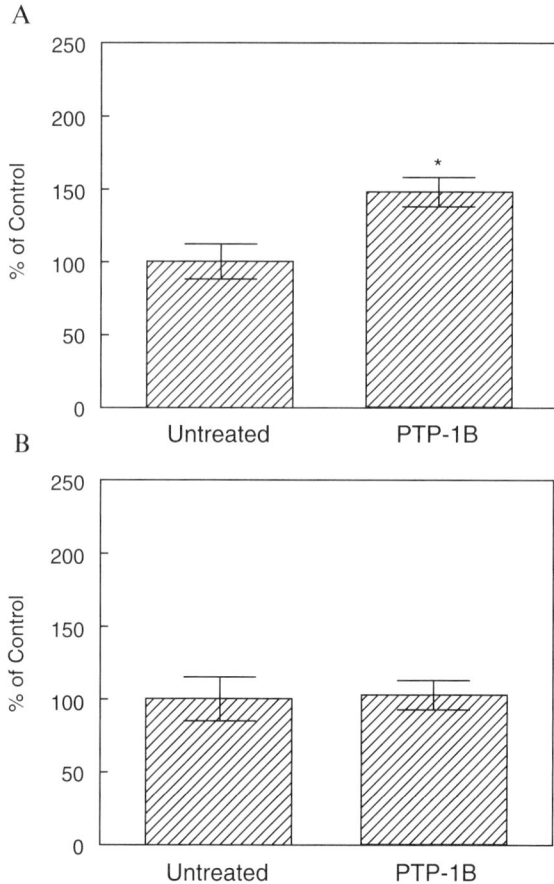

Fig. 5. The effect of PTP-1B treatment (dephosphorylation) of Na, K-ATPase activity of a membrane preparation obtained from (**A**) porcine lens fibers and (**B**) epithelium. Membrane material was incubated in the presence or absence (untreated) of recombinant PTP-1B. After removal of the PTP-1B, Na, K-ATPase activity (ouabain-sensitive ATP hydrolysis) was measured. The data are normalized to the activity value determined in the untreated control for each preparation ($n = 40$ from eight different batches of fiber membrane material and $n = 8$ from two different batches of epithelium membrane material). * Indicates a significant difference ($p < 0.01$) from untreated control. (From *16*; used with permission.)

These transport mechanisms seem to play a vital role in maintenance of lens transparency. Two articles, by Tomlinson et al., using rat lenses, and by Tamiya and Delamere using rabbit lenses, have shown the deleterious effect of a diminished sodium gradient on the lens by culturing lenses in sodium-free or low-sodium medium *(25,26)*. The effect on lens transparency was swift with opacification starting to occur within hours of the sodium gradient collapse. In the study by Tamiya and Delamere, low sodium (21 mM, NaCl substituted with either N-methyl D-glucamine or choline chloride) resulted in a severe opacification of the lens which was accompanied by a significant increase in lens calcium content (Fig. 6). In both studies, omission of calcium from the external

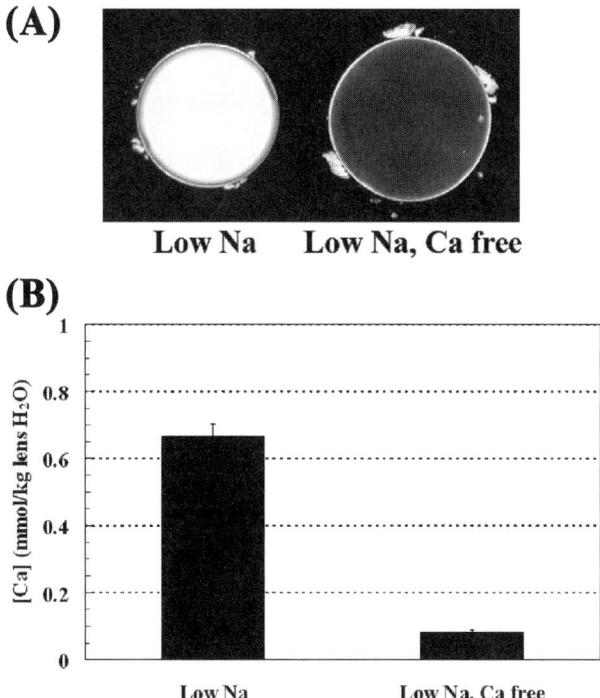

Fig. 6. The effect of diminished sodium gradient on intact rabbit lenses cultured in the presence or absence of external calcium. (**A**) The dark field images illustrates the severe opacity that develops in rabbit lenses cultured for 48 hours in modified M199 medium with reduced external sodium (21mM) (low Na). Transparency in low sodium medium is mostly preserved in the absence (low Na, Ca free) of external calcium. N-methyl D-glucamine was added to substitute for sodium chloride in order to maintain osmolarity. (**B**) Calcium content measured by atomic absorption in lenses incubated under the same conditions (From 25; used with permission).

medium significantly reduced the opacification, suggesting the important role of the sodium gradient-driven calcium export in the maintenance of lens transparency. This points to the importance of the sodium calcium exchanger (NCX), which utilizes the sodium gradient created by the Na,K-ATPase to export intracellular calcium (Fig. 7A). The sodium calcium exchanger can work in the forward mode (calcium exporting) or the reverse mode (calcium importing) depending on the sodium gradient. In the aforementioned studies, the sodium-free or low-sodium conditions used will not only inhibit the NCX working in the forward mode, but potentially reverse the NCX resulting in calcium import (Fig. 7B). Deterioration of transparency is well recognized in lenses that gain calcium.

Cardiac glycosides are highly specific inhibitors of Na,K-ATPase that bind to a clearly defined site on the α subunit. In the laboratory, the most commonly used cardiac glycoside is ouabain. Cardiac glycosides are used therapeutically, the most widely used being digoxin, which is derived from the leaves of *digitalis lanata*. Several investigators have identified endogenous cardiac glycoside-like substances (for review see Jortani and Valdes 1997). Digitalis-like compounds have been detected in bovine, rat, cat, rabbit, and

(A) Normal External (Na+)

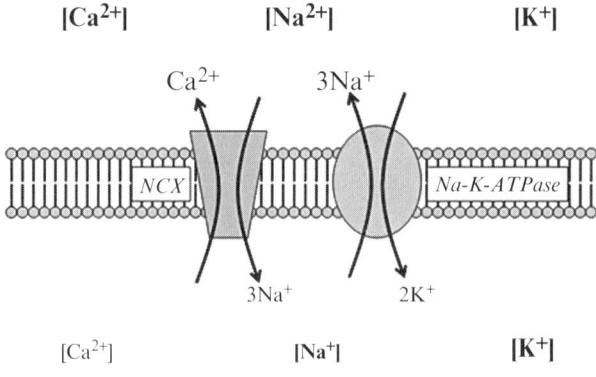

(B) Low External (Na+)

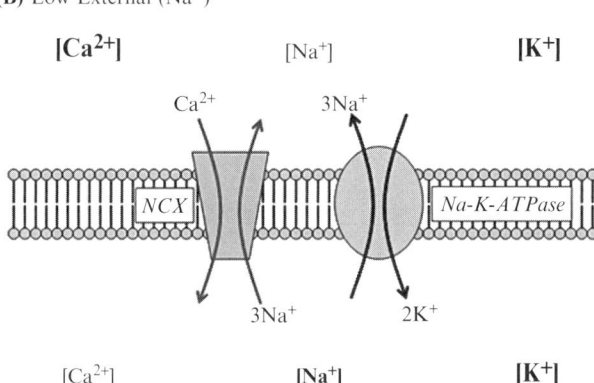

Fig. 7. Na,K-ATPase-mediated transport establishes a sodium gradient that is the driving force for outwardly directed calcium transport by the sodium-calcium exchanger (NCX) (**A**). Under experimental conditions where the sodium concentration in the external medium is reduced, outwardly directed NCX-mediated calcium transport is inhibited. The external sodium concentration can be reduced to a point where NCX reverses and shifts calcium in an inward direction (**B**).

human lenses *(27)*. It has been suggested that endogenous digitalis-like factors could be associated with the development of cataract *(28)* and in this respect it is interesting that Lichtstein and his coworkers detected a higher amount of digitalis-like factor in human cataractous lenses than in transparent human lenses *(27)*. In the Nakano mouse, a strain that results from an autosomal recessive single gene mutation *(29)*, there is evidence that associates a different type of endogenous Na,K-ATPase inhibitor with cataract formation *(30)*. Nakano cataract develops in parallel with an increase of lens sodium and loss of potassium, together with an increase of lens calcium and water *(31)*. The Nakano lens contains an Na,K-ATPase inhibitor with the characteristics of a small polypeptide that is capable of inhibiting not only lens Na,K-ATPase, but also Na,K-ATPase in brain and retina *(32)*. An endogenous Na,K-ATPase inhibitor peptide has also been reported in the lens of the ICR/f rat, a recessive mutant strain that develops cataract *(33)*.

While it is evident that endogenous Na,K-ATPase inhibition has the potential to bring about changes that lead to lens opacification, the development of human nuclear cataract proceeds without detectable alteration of lens sodium or potassium levels, and there is no reason to suspect a defect in Na,K-ATPase or any other ion transport mechanism. On the other hand, age-related cortical cataract in the human lens is very clearly associated with an increase in sodium and decrease in lens potassium, together with an increase of calcium and water *(34,35)*. The ion changes are more pronounced when opacification is severe. Although such a pattern of changes could be explained by insufficient Na,K-ATPase activity in the lens, analysis of Na,K-ATPase activity in cataractous human lenses has produced mixed results. Some studies indicate a decrease of Na,K-ATPase activity in cataractous lenses *(5,36–38)*, while other investigators report Na,K-ATPase activity that is apparently normal *(39,40)* or increased *(41)*. It seems quite likely that alterations of membrane permeability could contribute to the abnormal ion concentrations detected in cataractous lenses *(39,42)*. It is interesting that an age-related increase of membrane permeability appears to occur in the transparent human lens *(43)*. The fact that lens transparency is preserved in the face of increased ion permeability suggest the rate of Na,K-ATPase-mediated ion transport is able to increase sufficiently to balance the additional ion leak. It has yet to be determined whether the abundance or pattern of Na,K-ATPase protein expression or the phosphorylation status of the Na,K-ATPase α subunit is different in transparent human lenses that have an abnormally high ion permeability.

ACKNOWLEDGMENTS

Supported by NIH grant EY09532, an unrestricted grant from Research to Prevent Blindness Inc. and the KY Lions Eye Foundation.

REFERENCES

1. Kaplan JH. Biochemistry of Na,K-ATPase. Annu Rev Biochem 2002;71:511–35.
2. Oyster CW. The lens and vitreous. In: Oyster CW, editor. The human eye : structure and function. Sunderland, Mass: Sinauer; 1999. p. 491–544
3. Alvarez LJ, Candia OA, Grillone LR. Na+-K+ ATPase distribution in frog and bovine lenses. Curr Eye Res 1985;4:143–52.
4. Delamere NA and Dean WL. Distribution of lens sodium-potassium-adenosine triphosphatase. Invest Ophthalmol Vis Sci 1993;34:2159–63.
5. Kobatashi S, Roy D, Spector A. Sodium/potassium ATPase in normal and cataractous human lenses. Curr Eye Res 1982;2:327–34.
6. Candia OA and Zamudio AC. Regional distribution of the Na(+) and K(+) currents around the crystalline lens of rabbit. Am J Physiol Cell Physiol 2002;282:C252–62.
7. Gao J, Sun X, Yatsula V, Wymore RS, Mathias RT. Isoform-specific function and distribution of Na/K pumps in the frog lens epithelium. J Membr Biol 2000;178:89–101.
8. Tamiya S, Dean WL, Paterson CA, Delamere NA. Regional distribution of Na,K-ATPase activity in porcine lens epithelium. Invest Ophthalmol Vis Sci 2003;44:4395–9.
9. Mathias RT, Rae JL, Baldo GJ. Physiological properties of the normal lens. Physiol Rev 1997;77:21–50.

10. Patterson JW. Characterization of the equatorial current of the lens. Ophthalmic Res 1988;20:139–42.
11. Robinson KR and Patterson JW. Localization of steady currents in the lens. Curr Eye Res 1982;2:843–7.
12. Donaldson P, Kistler J, Mathias RT. Molecular solutions to mammalian lens transparency. News Physiol Sci 2001;16:118–23.
13. Moseley AE, Dean WL, Delamere NA. Isoforms of Na,K-ATPase in rat lens epithelium and fiber cells. Invest Ophthalmol Vis Sci 1996;37:1502–8.
14. Delamere NA, Dean WL, Stidam JM, Moseley AE. Influence of amphotericin B on the sodium pump of porcine lens epithelium. Am J Physiol 1996;270:C465–73.
15. Delamere NA, Manning RE Jr, Liu L, Moseley AE, Dean WL. Na,K-ATPase polypeptide upregulation responses in lens epithelium. Invest Ophthalmol Vis Sci 1998;39:763–8.
16. Bozulic LD, Dean WL, Delamere NA. The influence of protein tyrosine phosphatase-1B on Na,K-ATPase activity in lens. J Cell Physiol 2004;200:370–6.
17. Cui G, Dean WL, Delamere NA. The influence of cycloheximide on Na,K-ATPase activity in cultured human lens epithelial cells. Invest Ophthalmol Vis Sci 2002;43:2714–20.
18. Coca-Prados M, Escribano J, Ortego J. Differential gene expression in the human ciliary epithelium. Prog Retin Eye Res 1999;18:403–29.
19. Okafor MC, Dean WL, Delamere NA. Thrombin inhibits active sodium-potassium transport in porcine lens. Invest Ophthalmol Vis Sci 1999;40:2033–8.
20. Okafor MC and Delamere NA. The inhibitory influence of endothelin on active sodium-potassium transport in porcine lens. Invest Ophthalmol Vis Sci 2001;42:1018–23.
21. Bozulic LD, Dean WL, Delamere NA. The influence of Lyn kinase on Na,K-ATPase in porcine lens epithelium. Am J Physiol Cell Physiol 2004;286:C90–6.
22. Feraille E, Carranza ML, Gonin S, Beguin P, Pedemonte C, Rousselot M, Caverzasio J, Geering K, Martin PY, Favre H. Insulin-induced stimulation of Na+,K(+)-ATPase activity in kidney proximal tubule cells depends on phosphorylation of the alpha-subunit at Tyr-10. Mol Biol Cell 1999;10:2847–59.
23. Bozulic LD, Dean WL, Delamere NA. The influence of SRC-family tyrosine kinases on Na,K-ATPase activity in lens epithelium. Invest Ophthalmol Vis Sci 2005;46:618–22.
24. Tamiya S and Delamere NA. Studies of tyrosine phosphorylation and Src family tyrosine kinases in the lens epithelium. Invest Ophthalmol Vis Sci 2005;46:2076–81.
25. Tamiya S and Delamere NA. The influence of sodium-calcium exchange inhibitors on rabbit lens ion balance and transparency. Exp Eye Res 2006;83:1089–95.
26. Tomlinson J, Bannister SC, Croghan PC, Duncan G. Analysis of rat lens 45Ca2+ fluxes: evidence for Na(+)-Ca2+ exchange. Exp Eye Res 1991;52:619–27.
27. Lichtstein D, Gati I, Samuelov S, Berson D, Rozenman Y, Landau L, Deutsch J. Identification of digitalis-like compounds in human cataractous lenses. Eur J Biochem 1993;216:261–8.
28. Tao QF, Hollenberg NK, Graves SW. Sodium pump inhibition and regional expression of sodium pump alpha-isoforms in lens. Hypertension 1999;34:1168–74.
29. Narita M, Wang Y, Kita A, Omi N, Yamada Y, Hiai H. Genetic analysis of Nakano Cataract and its modifier genes in mice. Exp Eye Res 2002;75:745–51.
30. Nakano K, Yamanoto S, Kutsukake G, Ogawa H, Nakajima A, Takano E. Hereditary cataract in mice. Jpn J Clin Ophthalmol 1960;14:1772–1776.
31. Takehana M. Hereditary cataract of the Nakano mouse. Exp Eye Res 1990;50:671–6.
32. Fukui HN, Merola LO, Kinoshita JH. A possible cataractogenic factor in the Nakano mouse lens. Exp Eye Res 1978;26:477–85.

33. Kamei A and Sakai H. Characterization of peptide inducing cataractogenesis in lens of hereditary cataractous rat (ICR/f RAT). Jpn J Ophthalmol 1989;33:348–57.
34. Davies PD, Duncan G, Pynsent PB, Arber DL, Lucas VA. Aqueous humour glucose concentration in cataract patients and its effect on the lens. Exp Eye Res 1984;39:605–9.
35. Duncan G and Bushell AR. Ion analyses of human cataractous lenses. Exp Eye Res 1975;20:223–30.
36. Auricchio G, Rinaldi E, Savastano S, Albini L, Curto A, Landolfo V. The Na-K-ATPase in relation to the Na, K and taurine levels in the senile cataract. Metab Pediatr Ophthalmol 1980;4:15–7.
37. Gupta JD and Harley JD. Decreased adenosine triphosphatase activity in human senile cataractous lenses. Exp Eye Res 1975;20:207–9.
38. Nordmann J and Klethi J. Na-K-ATPase activity in the normal aging crystalline lens and in senile cataract. Arch Ophtalmol (Paris) 1976;36:523–8.
39. Pasino M and Maraini G. Cation pump activity and membrane permeability in human senile cataractous lenses. Exp Eye Res 1982;34:887–93.
40. Paterson CA, Delamere NA, Mawhorter L, Cuizon JV. Na,K-ATPase in simulated eye bank and cryoextracted rabbit lenses, and human eye bank lenses and cataracts. Invest Ophthalmol Vis Sci 1983;24:1534–8.
41. Friedburg D. Enzyme activity patterns in clear human lenses and in different types of human senile cataracts. In: Elliott K, Fitzsimmons DW, editors. The Human Lens in Relation to Cataract, CIBA Foundation Symposium Vol. 19. Amsterdam: Elsevier/North Holland; 1973. p. 117–133.
42. Gandolfi SA, Tomba MC, Maraini G. 86-Rb efflux in normal and cataractous human lenses. Curr Eye Res 1985;4:753–8.
43. Duncan G, Hightower KR, Gandolfi SA, Tomlinson J, Maraini G. Human lens membrane cation permeability increases with age. Invest Ophthalmol Vis Sci 1989;30:1855–9.
44. Jaffe NS and Horwitz J. Lens and cataract. In: Podos SM, Yanoff M, editors.Textbook of ophthalmology. New York, USA: Gower Medical (J.B. Lippincott); 1991. p. 1–19.

IV
Transport Across the Blood–Retinal Barrier

7
Pathophysiology of Pericyte-containing Retinal Microvessels
Roles of Ion Channels and Transporters

Donald G. Puro

CONTENTS

PERICYTES AND CAPILLARY FUNCTION
FUNCTIONAL SPECIALIZATIONS OF THE RETINAL MICROVASCULATURE
DIABETES AND THE PERICYTE-CONTAINING RETINAL MICROVASCULATURE
EFFECTS OF DIABETES ON MICROVASCULAR K_{IR} CHANNELS
THE EFFECT OF DIABETES ON $P2X_7$ PURINOCEPTORS
HYPERGLYCEMIA AND PERICYTE CHANNELS
BREAKDOWN OF THE BLOOD–RETINAL BARRIER AND ION CHANNEL FUNCTION
METABOLIC MODULATION OF MICROVASCULAR FUNCTION
CONCLUSIONS
ACKNOWLEDGMENTS
REFERENCES

PERICYTES AND CAPILLARY FUNCTION

Evidence is accumulating *(1–7)* that capillary perfusion is not only regulated by the smooth-muscle cells that encircle upstream arterioles, but also by the contractile pericytes that are located on the abluminal wall of microvessels (Figs. 1 and 2). Consistent with pericytes regulating microvascular function, it has been known for 20 years that pericytes isolated from the retina and maintained in culture can contract or relax during exposure to a variety of vasoactive molecules *(8–11)*. More recently, time-lapse photography of freshly isolated retinal microvessels successfully demonstrated for the first time that putative vasoactive signals can induce contractile responses in pericytes that are located on the abluminal vascular wall *(3–6)*. [Time-lapse movies of retinal microvessels responding to vasoactive signals can be viewed in the supplemental material of recent publications *(3–6)*.] The demonstration that pericytes can be induced to contract and relax vigorously enough to cause potent constrictions and effective dilations of isolated microvessels complements the observation that the diameter of retinal capillaries in situ change in response to vasoactive signals *(12)*. Although in vivo confirmation of these observations remains elusive due to the inability to simultaneously visualize pericytes,

From: *Ophthalmology Research: Ocular Transporters in Ophthalmic Diseases and Drug Delivery*
Edited by: J. Tombran-Tink and C. J. Barnstable © Humana Press, Totowa, NJ

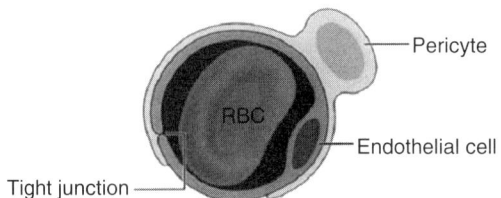

Fig. 1. Schematic showing a cross-section of an endothelial tube and an abluminal pericyte. Endothelial tight junctions form the retinal component of the blood-retinal barrier. Although not illustrated, pericytes and endothelial cells are interconnected by gap junctions. RBC = red blood cell.

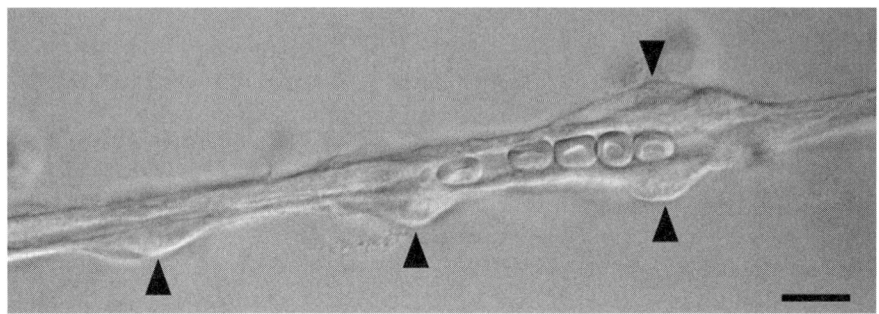

Fig. 2. Differential interference contrast photomicrograph of a segment of a pericyte-containing retinal microvessel freshly isolated from an adult rat. Arrowheads point to pericytes. Five erythrocytes are within the lumen. The cale bar shows 10 µm. From Kawamura et al. *(4)*, with permission.

endothelial cells and intraluminal erythrocytes of capillaries located within the retina, consensus is growing that regulation of blood of flow occurs, not only at pre-capillary sites, but also within pericyte-containing microvessels.

FUNCTIONAL SPECIALIZATIONS OF THE RETINAL MICROVASCULATURE

While nearly all tissues have pericyte-containing microvessels, the function of pericytes may be particularly important in the retina due to the special requirements of this translucent tissue. In the retina, a tight functional link between local perfusion and metabolic demand is especially crucial due to the low density of capillaries, which minimizes interference with the passage of light, but leaves little functional reserve *(12)*. In order to maintain tight coupling between metabolic supply and demand, the retinal vasculature is adapted to enhance the local control of blood flow with a minimum of extrinsic oversight, i.e., perfusion is largely autoregulated.

The autoregulation of retinal blood flow is facilitated by the vasculature's lack of autonomic innervation *(13)*, which in other tissues conveys central nervous system (CNS) input. Also, the presence of tight junctions between vascular endothelial cells in the retinal vasculature prevents circulating vasoactive molecules from directly affecting the contractile tone of mural cells. In addition, the retina's particularly high density of pericytes *(14, 15)* is thought to reflect the importance of these mural cells in the decentralized regulation of capillary perfusion.

Under physiological conditions, autoregulation of retinal vascular function allows blood flow to meet local metabolic demand despite the relative paucity of capillaries. However, due to the limited functional reserve of the retina's circulatory system, even modest vascular dysfunction may result in complications that compromise vision. Furthermore, the unique physiological requirements of the retinal vasculature may contribute to its particularly high vulnerability to the microangiopathic effects of diseases such as diabetes *(16)*.

DIABETES AND THE PERICYTE-CONTAINING RETINAL MICROVASCULATURE

Diabetes is the most important disease that affects the function of the retinal microvasculature. With the accelerating incidence of diabetes, the number of people at risk for the sight-threatening complications of diabetic retinopathy is increasing at an alarming rate *(17)*. For this reason, there is keen interest in elucidating the mechanisms by with diabetes disrupts vascular function in the retina. Recent experimental studies indicate that ion channels and transporters play important roles in the responses of the retinal microvasculature to diabetes.

For more than four decades, attention has focused on the effects of diabetes on the pericyte-containing retinal microvasculature whose loss of mural cells is one of the earliest histopathological signs of diabetic retinopathy *(18)*. More recently, investigators observed that apoptotic death, not only of pericytes, but also of endothelial cells within microvessels of the diabetic retina *(19)*. Unsurprisingly, functional changes in the retinal vasculature are detected well before the onset of microvascular cell death. Early functional changes include a loss of autoregulatory mechanisms for adjusting capillary perfusion of meet local metabolic demand *(20)*, a breakdown of the blood–retinal barrier *(21)*, and a decrease in retinal oxygen partial pressure (PO_2) *(12, 22)*. At present, the mechanisms by which diabetes affects the function of the retinal vasculature are incompletely understood. However, evidence is accumulating that microvascular ion channels and transporters are significantly involved (Fig. 3).

EFFECTS OF DIABETES ON MICROVASCULAR K_{IR} CHANNELS

The recent development of a technique to isolate viable pericyte-containing microvessels from the retinas of normal and diabetic rats has facilitated the quest to elucidate the mechanisms by which diabetes alters microvascular function. With freshly isolated microvessels, it is feasible to monitor pericyte currents via patch-pipettes, measure pericyte calcium levels with fura-2, and visualize pericyte contractions and lumen constrictions using time-lapse photography. Thus, retinal microvessels can be studied from the level of ion channels to pericyte function. Recent analyses of freshly isolated retinal microvessels reveal that insulin-deficient diabetes significantly affects the function of inwardly-rectifying potassium (K_{IR}) channels and $P2X_7$ purinoceptors *(23, 24)*.

Although knowledge of the function of K_{IR} channels in pericyte-containing microvessels is limited, there is substantial evidence from studies of smooth muscle-encircled

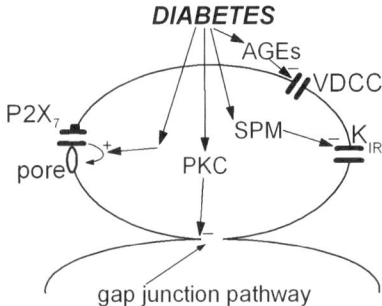

Fig. 3. Schematic of the effects of diabetes on the function of $P2X_7$ purinoceptors, voltage-dependent calcium channels (VDCC), inwardly rectifying potassium (K_{IR}) channels and gap junctions in a pericyte-containing retinal microvessel. AGEs, advanced glycation endproducts; SPM, spermine; PKC, protein kinase C. Experimental evidence indicates that insulin-deficient diabetes: (i) inhibits K_{IR} channels by a mechanism involving an increase in spermine, (ii) increases the transition from activated $P2X_7$ purinoceptors to opened transmembrane pores, (iii) inhibits VDCC activity secondary to the formation of advanced glycated endproducts, and (iv) inhibits microvascular gap junctions by a PKC-dependent mechanism *(28)*.

vessels to suggest that these channels are important in the circulatory system. For example, K_{IR} channels regulate blood flow via their effects on the membrane potential of mural cells and thereby, the activity of voltage-dependent calcium channels (VDCCs), the influx of calcium, and the contractile tone of these myocytes *(25)*. In addition, the phenomenon of metabolic vasodilation is likely to be mediated by the unique ability of K_{IR} channels to generate a hyperpolarizing efflux of K^+ when the concentration of this cation becomes elevated at sites where there is a mismatch between local perfusion and metabolic demand *(26, 27)*.

We recently reported that in the retinal microvasculature there is a topographical heterogeneity of the K_{IR} conductance *(23)*, namely, the K_{IR} current detected at proximal microvascular sites shows weak inward rectification, i.e., K^+ efflux via these channels is relatively large; in contrast, the K_{IR} current in the distal portion of this microvasculature is strongly rectifying. Because of extensive cell-to-cell communication via gap junction pathways within pericyte-containing retinal microvessels *(28, 29)*, the hyperpolarization generated by the efflux of K^+ via proximal K_{IR} channels spreads throughout the microvascular complex. Thus, the activity of proximal K_{IR} channels plays a major role in setting the membrane potential of pericytes and thereby, the basal contractile tone of the retinal microvasculature.

After two months of streptozotocin-induced diabetes, the heterogeneity of the microvascular K_{IR} conductance is minimal, as the efflux of K^+ via proximal K_{IR} channels decreases markedly *(23)*. The diabetes-induced inhibition of these channels is mimicked by the polyamine, spermine, whose concentration is elevated in the diabetic eye *(30)*. Also suggestive of a role for spermine is the fact that exposure of diabetic microvessels to a pharmacological inhibitor of polyamine synthesis reverses the diabetes-induced inhibition of the proximal K_{IR} channels *(23)*.

What are the consequences of diabetes causing the proximal K_{IR} current to become smaller? In agreement with diabetes reducing the efflux of K^+ via K_{IR} channels, the resting membrane potential of pericytes is significantly depolarized in diabetic retinal microvessels of diabetic rats *(23)*. As a consequence of this depolarization, the activity of VDCCs increases, cytoplasmic calcium rises, and pericytes contract. The resulting vasoconstriction may account, at least in part, for the decrease in retinal blood flow observed early in the course of diabetic retinopathy *(31)*.

THE EFFECT OF DIABETES ON $P2X_7$ PURINOCEPTORS

K_{IR} channels are not the only types of microvascular channels affected by insulin-deficient diabetes. We recently found that the function of the ligand-gated $P2X_7$ purinoceptors is also altered in microvessels of the diabetic retina *(24)*. These purinoceptors are involved in the vasoconstrictive response of retinal microvessels to extracellular adenosine triphosphate (ATP) *(3)*, which is likely to be a glial-to-vascular signal in the retina. However, in addition to this physiological role, activation of $P2X_7$ purinoceptors can have a pathological effect due to their peculiar ability to cause large transmembrane pores to form *(32, 33)*. These pores, which are permeable to molecules of less than 900 Da, can cause microvascular cell death by disrupting ionic gradients and/or by providing pathways for an efflux of vital intracellular molecules *(32)*.

Of potential relevance to diabetes, the concentration of $P2X_7$ ligand needed to open pores and to trigger apoptosis is decreased approximately 80-fold in pericyte-containing microvessels isolated from retinas of insulin-deficient rats *(24, 34)*. The increased formation of pores in microvessels of the diabetic retina is not associated with an upregulation of functional $P2X_7$ purinoceptors, but rather appears to be due to a diabetes-induced enhancement of the transition from activated $P2X_7$ receptor/channels to opened pores *(24)*. In this way, normally nonlethal concentrations of $P2X_7$ ligands may trigger microvascular cell death in the diabetic retina.

To characterize the microvascular cell death triggered by $P2X_7$ activation succinctly, we coined the term, 'purinergic vasotoxicity.' Although the idea that signaling molecules can cause cell death in the circulatory system appears to be new, the concept of extracellular signals having both physiological and pathological effects is not novel. For example, it is well known that the excitatory neurotransmitter glutamate can be neurotoxic. Analogous to excitotoxicity, which plays a role in neuronal pathobiology, purinergic vasotoxicity may be a cause for microvascular dysfunction. Our experiments indicate that a diabetes-induced increase in the vulnerability of retinal microvessels to purinergic vasotoxicity may be a previously unrecognized mechanism by which diabetic retinopathy progresses.

HYPERGLYCEMIA AND PERICYTE CHANNELS

With epidemiological studies indicating that hyperglycemia is a critical factor in the development of diabetic retinopathy *(35)*, investigators have assessed the effects of glucose on the function of ion channels in pericytes that were isolated from normal animals, maintained in culture and exposed to high concentrations of glucose. In these studies, the activities of certain calcium-activated potassium channels *(36, 37)* and VDCCs *(38)* are altered directly by glucose and/or secondary to the formation of advanced glycated endproducts *(39, 40)*. Changes in these potassium and calcium channels may also contribute to the dysfunction of microvessels in the diabetic retina.

BREAKDOWN OF THE BLOOD–RETINAL BARRIER AND ION CHANNEL FUNCTION

Although diabetes is the most common cause of breakdown in the endothelial barrier of the retinal vasculature, many other conditions are also associated with this sight-threatening phenomenon. In the case of diabetes, it appears that the upregulation of vascular endothelial growth factor (VEGF) results in the disruption of the endothelial tight junctions *(41, 42)* that limit the access of serum-derived molecules to the extracellular space of the retina. Because pericytes are positioned on the abluminal wall of microvascular endothelial tubes, these mural cells are among the first cells to be exposed to the molecules leaking at sites where the blood-retinal barrier is compromised. Consequently, the effects of blood-derived molecules on pericyte function are likely to influence how the retina responds to defects in the vascular endothelial barrier.

Based on the premise that ion channels play a vital role in the function of the pericyte-containing microvasculature, we examined the effects of serum on the ionic currents of retinal pericytes located on freshly isolated retinal microvessels *(43)*. These studies revealed that exposure to serum is associated with the activation of calcium-permeable nonspecific cation (NSC) channels and calcium-activated chloride (Cl_{Ca}) channels. Because these effects are mimicked by insulin-like growth factor-1 (IGF-1), which is a normal component of the blood *(44)*, IGF-1 may be one of the serum-derived molecules that regulates the physiology of retinal pericytes when there is a breakdown in the blood–retinal barrier.

Calcium influx is an important consequence of NSC and Cl_{Ca} channels being activated during the exposure of retinal microvessels to serum-derived molecules, such as IGF-1 *(43)*. Calcium-permeable NSC channels provide one pathway for the influx of calcium. In addition, the opening of the NSC and Cl_{Ca} channels induces pericyte depolarization and thereby VDCC activation *(43, 45)*. As a consequence of the serum-induced influx of calcium, retinal pericytes contract, and microvascular lumens constrict *(3, 5)*.

The serum-induced contraction of pericytes may be a successful adaptive response to a focal breakdown of the blood–retinal barrier. With vasoconstriction occurring at these sites, blood would be shunted away from leaky microvessels and sight-threatening retinal edema would be minimized. On the other hand, prolonged shunting of blood may result in ischemic damage at leaky sites. Furthermore, sustained Ca^{2+} influx due to persistent serum-induced depolarization would require a substantial of expenditure energy in order to prevent cellular damage secondary to an excessive concentration of cytoplasmic calcium. Perhaps, this metabolic stress contributes to the cell death observed at sites of a defect in the blood–retinal barrier. With the death of microvascular cells, the structural integrity of the vascular wall is further compromised, the leakage of serum and blood increased and the function of the retina further compromised.

METABOLIC MODULATION OF MICROVASCULAR FUNCTION

Recent studies have revealed dynamic roles for ion channels and transporters in the responses of the retinal microvasculature to hypoxia, which occurs in a variety of disorders, including diabetic retinopathy. Due to its low density of capillaries and the relatively low volume of retinal blood flow, the retina is particularly vulnerable to conditions that lead to decreased tissue oxygenation *(12)*. Thus, it is crucial that the vasculature of the retina respond effectively to hypoxia. We recently proposed that the response of the retinal microvasculature to a decrease in PO_2 is facilitated by the ability of extracellular signals, such as platelet-derived growth factor-BB (PDGF-BB) and extracellular lactate, to switch from being vasoconstrictors when energy supplies are ample and to being vasodilators with the onset of hypoxia *(6, 46)*. This bifunctional capability involves microvascular ion channels, transporters and gap junctions.

In the case of lactate, which is present in the extracellular space of the retina under both physiological and pathophysiological conditions, there is an elevation of intracellular calcium, contraction of pericytes and constriction of lumens when retinal microvessels are exposed to this metabolic product under normoxic conditions *(6)*. This lactate-induced vasoconstriction is mediated by a cascade of events resulting in the inhibition of Na^+/Ca^{2+} exchangers (NCXs). The decrease in Ca^{2+} extrusion by NCXs results in the accumulation of intracellular calcium and thereby, the contraction of pericytes. Key events in this cascade include the importation via monocarboxylate transporters of protons along with lactate, and the subsequent efflux of these protons via Na^+/H^+ exchangers (NHEs). Due to the NHE-mediated influx of Na^+, the ability of NCXs to exchange this cation for intracellular Ca^{2+} is impaired. As a consequence, cytoplasmic Ca^{2+} rises, and pericytes contract (Fig. 4). In contrast to the rise in calcium in pericytes observed during exposure to lactate under nomoxic conditions, this metabolic product fails to evoke a calcium increase in hypoxic retinal microvessels *(28)*. In fact, during hypoxia, lactate causes vasodilation, rather than vasoconstriction; most likely this lactate-induced relaxa-

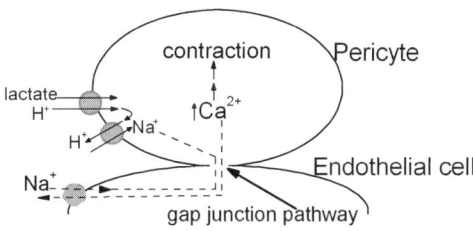

Fig. 4. Schematic of the effects of extracellular lactate on pericyte-containing microvessels under conditions in which pericyte-endothelial gap junctions are opened. As discussed in the text, the transport of lactate into microvascular cells triggers a cascade of events that result in the inhibition of Na^+/Ca^{2+} exchangers (NCXs), which are located in the endothelial cells. Dashed lines show that the rise in intracellular Na^+ diminishes the NCX-mediated importation of Na^+ and thereby, the extrusion of Ca^{2+} via this exchanger. The resulting rise in intracellular Ca^{2+} causes pericyte contraction by a mechanism that is thought to involve the inhibition of myosin light-chain kinase.

tion of pericytes is due to cotransported H⁺ causing intracellular acidity, which inhibits the contractile apparatus.

By what mechanism does hypoxia switch the lactate from being a vasoconstrictor to being a vasodilator? Based on a variety of experimental observations *(6)*, it appears that this switch occurs due to the closure of pericyte/endothelial gap junctions and the resulting inability of NCXs located in the endothelium to regulate the concentration of calcium in abluminal pericytes. Thus, the functional status of the gap junctions interconnecting pericytes and endothelial cells appears to determine whether pericytes contract or relax during exposure of retinal microvessels to lactate. An implication of our studies is that due to a hypoxia-induced closure of gap junctions, extracellular lactate enhances retinal blood flow. In contrast, the lactate-induced vasoconstriction at well-oxygenated sites would reduce local perfusion and thereby shunt blood flow to less adequately oxygenated regions of the retina.

Our experiments also indicate that there is metabolic modulation of the response of the retinal microvasculature to PDGF-BB *(46)*, which is expressed by vascular endothelial cells and whose receptors are on pericytes *(47)*. Under normal conditions, exposure of retinal microvessels to this growth factor results in the activation of depolarizing NSC and Cl_{Ca} currents, the elevation of intracellular calcium, the contraction of pericytes and the constriction of microvascular lumens. On the other hand, the NSC and Cl_{Ca} channels do not open during exposure of ischemic microvessels to PDGF-BB. Rather this growth factor enhances the activation of K_{ATP} channels whose hyperpolarizing effect results in pericyte relaxation and vasodilation *(46)*. Based on these observations, we have proposed that PDGF-BB causes blood flow to increase at hypoxic retinal sites, but to decrease when oxygenation is adequate. It seems likely that the dual vasoactive effects of extracellular molecules, such as PDGF-BB and lactate, provide a mechanism to efficiently adjust capillary perfusion to match local metabolic need.

CONCLUSIONS

Ion channels and transporters are involved in playing an important role in the response of pericyte-containing retinal microvessels to disorders such as diabetes. Due to the limited functional reserve of the retina's circulatory system, it seems likely that even modest dysfunction of the retinal microvasculature can lead to sight-threatening complications. As progress continues in elucidating the role of microvascular channels and transporters, these membrane proteins are likely to become fruitful targets for pharmacological intervention to prevent or minimize the detrimental effects of diseases such as diabetes on the function of the retinal microvasculature.

ACKNOWLEDGMENTS

National Insititue of Health grants EY12505, EY07003 and a senior investigator award from Research to Prevent Blindness, Inc. provided support to the author.

REFERENCES

1. Tilton RG. Capillary pericytes: perspectives and future trends. J Electron Microsc Tech 1991;19:327–344.
2. Hirschi KK and D'Amore PA. Pericytes in the microvasculature. Cardiovasc Res 1996;32:687–698.

3. Kawamura H, Sugiyama T, Wu DM, Kobayashi M, Yamanishi S, Katsumura K, Puro DG. ATP: a vasoactive signal in the pericyte-containing microvasculature of the rat retina. J Physiol 2003;551:787–799.
4. Kawamura H, Kobayashi M, Li Q, Yamanishi S, Katsumura K, Minami M, Wu DM, Puro DG. Effects of angiotensin II on the pericyte-containing microvasculature of the rat retina. J Physiol 2004;561:671–683.
5. Wu DM, Kawamura H, Sakagami K, Kobayashi M, Puro DG. Cholinergic regulation of pericyte-containing retinal microvessels. Am J Physiol 2003;284:H2083–H2090.
6. Yamanishi S, Katsumura K, Kobayashi T, Puro DG. Extracellular lactate as a dynamic vasoactive signal in the rat retinal microvasculature. Am J Physiol 2006;290:H925–H934.
7. Puro DG. Vasoactive signals and pericyte function in the retina. In: Shepro D and D'Amore PA, editors. Microvascular Research: Biology and Pathology. USA: Elsevier; 2006. p. 265–269.
8. Dodge AB, Hechtman HB, Shepro D. Microvascular endothelial-derived autacoids regulate pericyte contractility. Cell Motil Cytoskeleton 1991;18:180–188.
9. Haefliger IO, Zschauer A, Anderson DR. Relaxation of retinal pericyte contractile tone through the nitric oxide-cyclic guanosine monophosphate pathway. Invest Ophthalmol Vis Sci 1994;35:991–997.
10. Kelley C, D'Amore P, Hechtman HB, Shepro D. Microvascular pericyte contractility in vitro: comparison with other cells of the vascular wall. J Cell Biol 1987;104:483–490.
11. Kelley C, D'Amore P, Hechtman HB, Shepro D. Vasoactive hormones and cAMP affect pericyte contraction and stress fibres in vitro. J Muscle Res Cell Motil 1988;9:184–194.
12. Funk RH. Blood supply of the retina. Ophthalmic Res 1997;29:320–325.
13. Ye XD, Laties AM, Stone RA. Peptidergic innervation of the retinal vasculature and optic nerve head. Invest Ophthalmol Vis Sci 1990;31:1731–1737.
14. Shepro D and Morel NM. Pericyte physiology. FASEB J 1993;7:1031–1038.
15. Frank RN, Turczyn TJ, Das A. Pericyte coverage of retinal and cerebral capillaries. Invest Ophthalmol Vis Sci 1990;31:999–1007.
16. Puro DG. Physiology and pathobiology of the pericyte-containing retinal microvasculature: new developments. Microcirculation 2007;14:1–10.
17. Zimmet P, Alberti KG, Shaw J. Global and societal implications of the diabetes epidemic. Nature 2001;414:782–787.
18. Cogan DG, Toussaint D, Kuwabara T. Retinal vascular patterns. IV. Diabetic retinopathy. Arch Ophthalmol 1961;166:366–378.
19. Mizutani M, Kern TS, Lorenzi M. Accelerated death of retinal microvascular cells in human and experimental diabetic retinopathy. J Clin Invest 1996;97:2883–2890.
20. Kohner EM, Patel V, Rassam SM. Role of blood flow and impaired autoregulation in the pathogenesis of diabetic retinopathy. Diabetes 1995;44:603–607.
21. Cunha-Vaz J, Faria de Abreu JR, Campos AJ. Early breakdown of the blood-retinal barrier in diabetes. Br J Ophthalmol 1975;59:649–656.
22. Trick GL and Berkowitz BA. Retinal oxygenation response and retinopathy. Prog Retin Eye Res 2005;24:259–274.
23. Matsushita K and Puro DG. Topographical heterogeneity of K_{IR} currents in pericyte-containing microvessels of the rat retina: effect of diabetes. J Physiol 2006;573:483–495.
24. Sugiyama T, Kobayashi M, Kawamura H, Li Q, Puro DG. Enhancement of $P2X_7$-induced pore formation and apoptosis: an early effect of diabetes on the retinal microvasculature. Invest Ophthalmol Vis Sci 2004;45:1026–1032.
25. Chrissobolis S, and Sobey CG. Inwardly rectifying potassium channels in the regulation of vascular tone. Curr Drug Targets 2003;4:281–289.

26. Zaritsky JJ, Eckman DM, Wellman GC, Nelson MT, Schwarz TL. Targeted disruption of Kir2.1 and Kir2.2 genes reveals the essential role of the inwardly rectifying K^+ current in K^+-mediated vasodilation. Circ Res 2000:87:160–166.
27. Edwards FR, Hirst GD, Silverberg GD. Inward rectification in rat cerebral arterioles; involvement of potassium ions in autoregulation. J Physiol 1988;404:455–466.
28. Oku H, Kodama T, Sakagami K, Puro DG. Diabetes-induced disruption of gap junction pathways within the retinal microvasculature. Invest Ophthalmol Vis Sci 2001;42: 1915–1920.
29. Wu DM, Miniami M, Kawamura H, Puro DG. Electrotonic transmission within pericyte-containing retinal microvessels. Microcirculation 2006;13:353–363.
30. Nicoletti R, Venza I, Ceci G, Visalli M, Teti D, Reibaldi A. Vitreous polyamines spermidine, putrescine, and spermine in human proliferative disorders of the retina. Br J Ophthalmol 2003;87:1038–1042.
31. Grunwald J and Bursell S-E. Hemodyanmic changes as early markers of diabetic retinopathy. Curr Opin Endocrin Diab 1996;3:298–306.
32. North RA. Molecular physiology of P2X receptors. Physiol Rev 2002;82:1013–1067.
33. Sugiyama T, Kawamura H, Yamanishi S, Kobayashi M, Katsumura K, Puro DG. Regulation of $P2X_7$-induced pore formation and cell death in pericyte-containing retinal microvessels. Am J Physiol 2005;288:C568–576.
34. Liao SD and Puro DG. NAD^+-induced vasotoxicity in the pericyte-containing microvasculature of the rat retina: effect of diabetes. Invest Ophthalmol Vis Sci 2006; in press.
35. The Diabetes Control and Complications Trial Research Group. The effect of intensive treatment of diabetes on the development and progression of long-term complications in insulin-dependent diabetes mellitus. N Engl J Med 1993;329:977–986.
36. Wiederholt M, Berwick S, Helbig H. Electrophysiological properties of cultured retinal capillary pericytes. Prog Retina Eye Res 1995;14:437–451.
37. Quignard JF, Harley EA, Duhault J, Vanhoutte PM, Feletou M. K^+ channels in cultured bovine retinal pericytes: effects of beta-adrenergic stimulation. J Cardiovasc Pharmacol 2003;42:379–388.
38. McGinty A, Scholfield CN, Liu WH, Anderson P, Hoey DE, Trimble ER. Effect of glucose on endothelin-1-induced calcium transients in cultured bovine retinal pericytes. J Biol Chem 1999;274:25250–25253.
39. Stitt, A.W., and Curtis, T.M. (2005) Advanced glycation and retinal pathology during diabetes. Pharmacol Rep 2005;57(Suppl):156–168.
40. Hughes SJ, Wall N, Scholfield CN, McGeown JG, Gardiner TA, Stitt AW, Curtis TM. Advanced glycation endproduct modified basement membrane attenuates endothelin 1 induced $[Ca^{2+}]_i$ signalling and contraction in retinal microvascular pericytes. Mol Vis 2004;10:996–1004.
41. Antonetti DA, Barber AJ, Khin S, Lieth E, Tarbell JM, Gardner TW. Vascular permeability in experimental diabetes is associated with reduced endothelial occludin content: vascular endothelial growth factor decreases occludin in retinal endothelial cells. Diabetes 1998;47:1953–1959.
42. Antonetti DA, Lieth E, Barber AJ, Gardner TW. Molecular mechanisms of vascular permeability in diabetic retinopathy. Semin Ophthalmol 1999;14:240–248.
43. Sakagami K, Wu DM, Puro DG. Physiology of rat retinal pericytes: modulation of ion channel activity by serum-derived molecules. J Physiol 1999;521:637–650.
44. Le Roith D. Seminars in medicine of the Beth Israel Deaconess Medical Center. Insulin-like growth factors. N Engl J Med 1997;336:633–640.

45. Sakagami K, Kawamura H, Wu DM, Puro DG. Nitric oxide/cGMP-induced inhibition of calcium and chloride currents in retinal pericytes. Microvasc Res 2001;62:196–203.
46. Sakagami K, Kodama T, and Puro DG. PDGF-induced coupling of function with metabolism in microvascular pericytes of the retina. Invest Ophthalmol Vis Sci 2001;42:1939–1944.
47. Lindahl P, Johansson BR, Leveen P, Betsholtz C. Pericyte loss and microaneurysm formation in PDGF-B-deficient mice. Science 1997;277:242–245.

8
Molecular Mechanisms of the Inner Blood-Retinal Barrier Transporters

Masatoshi Tomi and Ken-ichi Hosoya

CONTENTS

SUMMARY
INTRODUCTION
ENERGY SUPPLY AND STORAGE
ANTIOXIDANT SUPPLY
AMINO ACID SUPPLY
ADENOSINE TRANSPORT
ORGANIC ANION TRANSPORT
ATP-BINDING CASSETTE (ABC) TRANSPORTERS
CONCLUSIONS
ACKNOWLEDGEMENTS
REFERENCES

SUMMARY

The inner blood–retinal barrier (inner BRB) forms complex tight junctions of retinal capillary endothelial cells to prevent the free diffusion of substances between the circulating blood and the neural retina. Thus, understanding of the inner BRB transport mechanisms could provide a basis for strategies of drug delivery to the retina. Recent development of several analytical methods has succeeded in showing that the inner BRB is equipped with several membrane transporters such as GLUT1, monocarboxylate transporter 1 (MCT1), creatine transporter (CRT), xCT, L-type amino acid transporter 1 (LAT1), taurine transporter (TauT), equilibrative nucleoside transporter 2 (ENT2), organic anion transporter polypeptide 1a4 (Oatp1a4), multidrug resistance 1a (mdr1a), and ATP-binding cassette transporter G2 (Abcg2). These transporters play essential roles in supplying nutrients to the retina and are responsible for the efflux of neurotransmitter metabolites, toxins, and xenobiotics.

INTRODUCTION

The blood–retinal barrier (BRB), which forms complex tight junctions of retinal capillary endothelial cells (inner BRB) and retinal pigment epithelial cells (RPE, outer BRB),

From: Ophthalmology Research: Ocular Transporters in Ophthalmic Diseases and Drug Delivery
Edited by: J. Tombran-Tink and C. J. Barnstable © Humana Press, Totowa, NJ

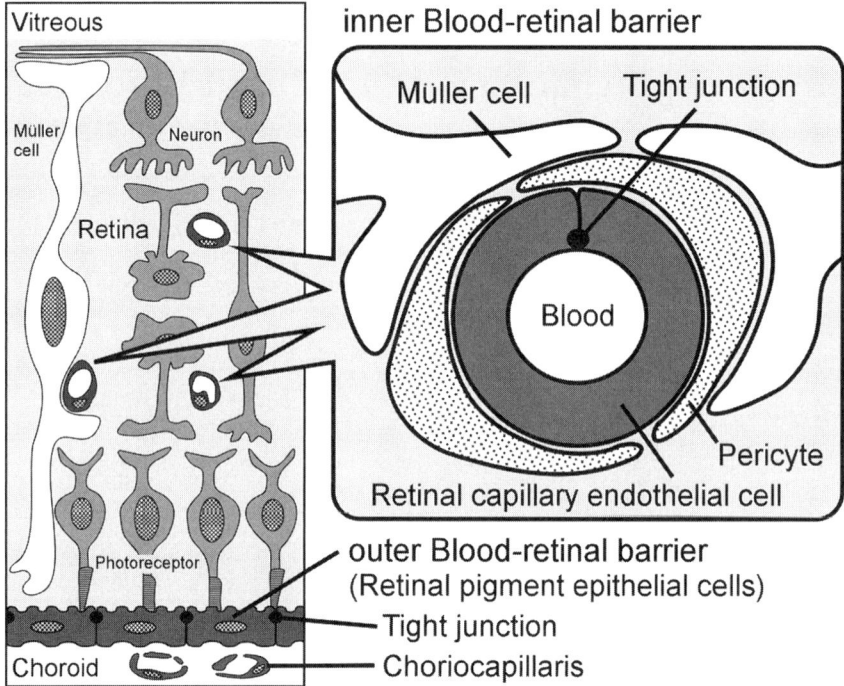

Fig. 1. Schematic diagram of the blood–retinal barrier (BRB). The BRB is composed of retinal capillary endothelial cells (inner BRB) and retinal pigment epithelial cells (RPE, outer BRB). Retinal capillary endothelial cells are surrounded by the pericyte and Müller cell foot process.

restricts nonspecific transport between the neural retina and the circulating blood and maintains a constant milieu in the neural retina (Fig. 1) *(1, 2)*. The inner two thirds of the human retina is nourished by the inner BRB and the remainder is covered by choriocapillaris via the outer BRB. Compared with peripheral capillary endothelial cells, the inner BRB expresses a variety of unique transporters which play essential roles in supplying nutrients to the retina and are responsible for the efflux of neurotransmitter metabolites, toxins, and xenobiotics. A better understanding of the transport mechanisms at the inner BRB will provide important information about the effective delivery of drugs to the retina as well as the physiological role of the inner BRB.

The concept of the BRB was first proposed by Schnaudigel in 1913 *(3)*, who found that dyes injected into veins did not stain the retina. Although information about transporter expressions and functions at the inner BRB had been very limited for a long time, the development of several in vivo, ex vivo, and in vitro methods in recent years has significantly increased progress in inner BRB transport research. In vivo transport studies using the retinal uptake index (RUI) method have been performed to investigate influx transport into the retina *(4–7)*. Moreover, the in vivo microdialysis method can be used for evaluating the efflux transport from the vitreous humor/retina by monitoring the vitreous humor concentration of a test substrate and bulk-flow marker *(8)*. These methods have the advantage of being able to estimate the transport under physiological conditions; however, it is difficult to distinguish between substrates that are transported across the inner BRB and the outer BRB. Isolation techniques applied to retinal capillaries or endothelial cells are very useful for investigating the expression and function of transporters at the inner BRB ex vivo *(9–11)*, although only a few capillaries or cells can be obtained. Conditionally immortalized rat retinal capillary endothelial cell lines (i.e., TR-iBRB cells) have been successfully developed and used for the acquisition of information regarding transport mechanisms and transporter expression levels at the inner BRB in vitro *(12)*. TR-iBRB cells maintain certain in vivo functions as described in another review *(2)*, however, there is no way of knowing whether the expression and functions of transporters are changed by culture passages and conditions. Therefore, when we investigate the mechanism(s) governing inner BRB transport, it is essential to understand the limitations of each method and select the combination of methods that are most suited to particular studies.

In this chapter, we present an overview of the latest biological research focusing on the transport mechanism at the inner BRB. We will also discuss hypothetical roles of the inner BRB transport systems from physiological, pathophysiological, and pharmacological points of view.

ENERGY SUPPLY AND STORAGE

In the retina, neuronal cells, including photoreceptor cells, require a large amount of metabolic energy for phototransduction and neurotransduction, which is maintained by ionic gradients across the plasma membrane. Therefore, one of the most important roles of transporters at the inner BRB is to meet the high energy demand in the retina by supplying a source of energy from the circulating blood.

Glucose

Adequate D-glucose transport from the circulating blood to the retina is essential for energy metabolism in the retina. Betz and Goldstein *(10)* were the first to report a carrier-mediated transport process for D-glucose at the inner BRB; using isolated bovine retinal capillaries and they showed that [^{14}C]3-O-methyl-D-glucose (3-OMG, an analog of D-glucose that cannot be metabolized) uptake was saturable and could be inhibited by substrates and inhibitors of facilitative D-glucose transporters (i.e., GLUT). Ennis et al. *(13)* obtained a Michaelis constant (K_m) for the rat blood-to-retina transport of D-glucose of 7.81 mM using a modification of the RUI method. In 1992, Takata et al. *(14)* demonstrated the immunolocalization of GLUT1 (Slc2a1) in the retina and showed that GLUT1 is localized at both the luminal (blood) and abluminal (retinal) sides of the inner BRB, and at both the apical and basolateral sides of the outer BRB. The expression of GLUT1 at the abluminal membrane of the inner BRB is approximately two- and threefold greater than that at the luminal membrane in humans and rats, respectively *(15, 16)*. This asymmetrical distribution of GLUT1 at the inner BRB suggests that D-glucose transport is limited at the blood-to-luminal rather than the abluminal-to-interstitial interface. There are some arguments in favor of GLUT3 (Slc2a3) being expressed and playing a role in transporting D-glucose at the inner BRB *(17)*. Although GLUT3 mRNA was at the lower limit of detection, GLUT1 mRNA was well expressed in isolated rat retinal vascular endothelial cells (Fig. 2A) *(11)*. Therefore, GLUT1 is believed to be responsible for the transport of D-glucose at the rat inner BRB.

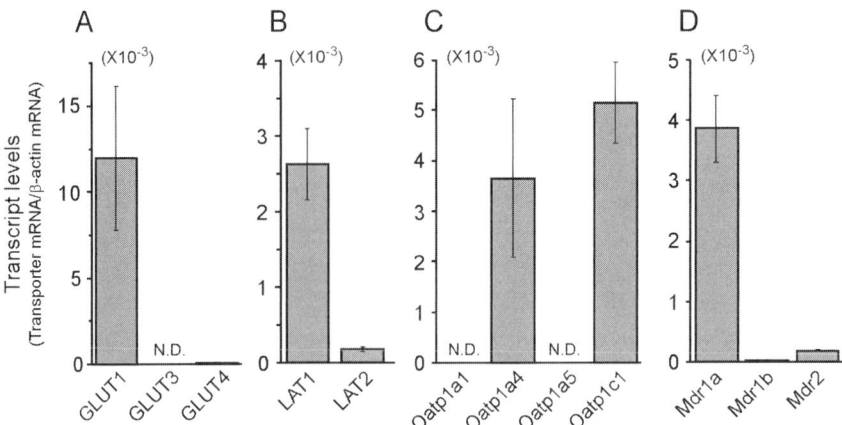

Fig. 2. Transcript levels of GLUT **(A)**, L-type aminio acid transporter (LAT) **(B)**, organic anion transporter polypeptide (Oatp) **(C)**, and multidrug resistance (mdr) **(D)** transporters in isolated retinal vascular endothelial cells. The transcript level of transporters in rat retinal vascular endothelial cells isolated using magnetic beads coated with anti-rat CD31 antibody was determined by quantitative real-time polymerase chain reaction (PCR) analysis. Each column represents the mean ± the standard error of the mean (S.E.M.) of at least three different samples. N.D., not detected. Data taken from Journal of Neurochemistry, 94, Tomi and Hosoya, Application of magnetically isolated rat retinal vascular endothelial cells for the determination of transporter gene expression levels at the inner blood-retinal barrier. 1244–1248, 2004, with permission from Blackwell Publishing.

Diabetic retinopathy is generally regarded to primarily affect the retinal microvasculature. The amount of D-glucose transport into the retina would be a factor in the development diabetic retinopathy, since this is the major determinant of the retinal D-glucose concentration. Thus, the regulation of GLUT1 levels in retinal capillary endothelial cells subjected to hyperglycemia has been of special interest. The amount of GLUT1 protein was reduced in the retina and vascular endothelium of streptozotocin-induced diabetic rats *(18,19)* and in the retinas of diabetic Goto Kakizaki (GK) rats and alloxan-induced diabetic rabbits *(20)*. On the other hand, ultrastructural localization of GLUT1 protein shows the increased expression of GLUT1 at the plasma membrane in retinal capillaries of diabetic patients *(16)*, or no change in GLUT1 levels at the plasma membrane in retinal capillaries of diabetic GK rats *(15)*. Taken together, these observations suggest that the reduction in the total pool of GLUT1 caused by diabetes may not limit GLUT1 targeting to plasma membranes but may result in a reduction in the available pool of intracellular GLUT1. In hyperglycemia, the absence of compensatory down-regulation of GLUT1 expression at the inner BRB may expose the retina to elevated free D-glucose concentrations and, therefore, accelerate the development of diabetic retinopathy.

Lactate

In the retina, L-lactate is a major nutrient for photoreceptors and it reaches a concentration 10-fold greater than that in the blood. Müller glia produce L-lactate from D-glucose and release it to photoreceptors *(21)*. Gerhart et al. *(22)* provided immunohistochemical evidence to show that H^+-dependent monocarboxylate transporter (MCT)

1 (Slc16a1) is localized at both the luminal and abluminal sides of the inner BRB. In an investigation of the transport characteristics of L-lactate, RUI studies have provided the evidence that the rat blood-to-retina transport of L-lactate is saturable and pH-dependent *(5)*, suggesting the involvement of MCT in L-lactate transport. MCT1 mRNA is expressed in TR-iBRB cells, and [^{14}C]L-lactate uptake by TR-iBRB cells was shown to be a temperature-, H$^+$-, and concentration-dependent process with a K_m of 1.7 mM *(23)*. Knockout mice for CD147/basigin, which is responsible for targeting of MCT to the cell membrane, exhibit physiological and morphological alterations of the retina *(24)*, implying the physiological importance of MCT in the retina. MCT1 at the inner BRB plays a role in maintaining an adequate L-lactate concentration in the retina by regulating L-lactate transport between the circulating blood and the neural retina.

Creatine

Creatine plays a vital role in the storage and transmission of adenosine triphosphate (ATP) due to the conversion of creatine to phosphocreatine. High levels of creatine in the retina (3 mM) and in photoreceptor cells (10 to 15 mM) suggest that the creatine/phosphocreatine system plays a significant role in the retina. Indeed, gyrate atrophy of the choroid and retina (GA) is characterized by hyperornithinemia and hypocreatinemia, implying that excessive ornithine leads to chorioretinal degeneration through suspension of creatine synthesis. It has been shown that S-adenosyl-L-methionine:N-guanidinoacetate methyltransferase, a key enzyme for creatine synthesis, is preferentially expressed and creatine is biosynthesized in Müller glia of the retina *(25)*. Moreover, an in vivo intravenous administration study demonstrated that [^{14}C]-labeled creatine is transported from the blood to the retina. [^{14}C]Creatine uptake by TR-iBRB cells took place in an Na$^+$-, Cl$^-$-, and concentration-dependent manner with a K_m of 15 μM and was inhibited by creatine transporter (CRT/Slc6a8) inhibitors. CRT mRNA and protein were expressed in the retina and TR-iBRB cells, and immunohistological electron microscopy investigations revealed the localization of CRT immunoreactivity at both the luminal and abluminal sides of the inner BRB *(26)*. In light of these findings, the high level of creatine in the retina appears to be maintained by local biosynthesis in the Müller glia and supplied by the circulating blood through the BRB. Since photoreceptors require enormous quantities of metabolic energy for phototransduction and, therefore, accumulate creatine, the retinal creatine is most likely taken up into photoreceptors and plays a pivotal role in energy storage and regeneration of ATP, which is generated from L-lactate in photoreceptors.

ANTIOXIDANT SUPPLY

The retina is an ideal environment for the generation of reactive oxygen species for several reasons *(27, 28)*. Firstly, the retina is subject to high levels of cumulative irradiation, which causes the production of reactive oxygen species, and it contains an abundance of chromophores such as rhodopsin, melanin, and lipofuscin. Secondly, oxygen consumption by the retina is much greater than that by any other tissue, since the retina requires a large amount of metabolic energy for neurotransduction, as mentioned above. Thirdly, photoreceptor outer-segment membranes are rich in polyunsaturated fatty acids, which are readily oxidized and can initiate a cytotoxic chain reaction. Finally, the process of phagocytosis by the RPE is itself an oxidative stress, and results in the generation of reactive oxygen species. Oxidative damage in the retina is thought to be involved in retinal diseases, such as diabetic retinopathy *(29)* and age-related macular degeneration *(28)*. It has been shown that transporters at the inner BRB protect the retina by supplying antioxidants.

Vitamin C

Vitamin C acts as a free-radical scavenger as well as a cofactor in the enzymatic biosynthesis of collagen, catecholamine, and peptide neurohormones. The concentration of vitamin C in the retina is more than 10-fold greater than that in the plasma, suggesting that vitamin C in the retina is supplied from the circulating blood through a specific transport process at the BRB. The reduced form vitamin C, ascorbic acid (AA), and the oxidized form of vitamin C, dehydroascorbic acid (DHA), are known to be substrates of Na$^+$-dependent L-ascorbic acid transporter (SVCT/Slc23a) and GLUT, respectively. An in vivo intravenous administration study demonstrated that the apparent influx permeability clearance per gram retina ($K_{in,retina}$) of [^{14}C]DHA was about 38-fold greater than that of [^{14}C]AA (Fig. 3A) *(30)*. Moreover, most of the [^{14}C]DHA injected into the femoral vein is converted to [^{14}C]AA in the retina, suggesting that vitamin C is mainly transported as the DHA form across the BRB and accumulates as the AA form in the rat retina. The initial uptake rate of [^{14}C]DHA in TR-iBRB cells was also found to be 37-fold greater than that of [^{14}C]AA. [^{14}C]DHA uptake by TR-iBRB cells took place in an Na$^+$-independent and concentration-dependent manner with a K_m of 93.4 µM and was inhibited by substrates and inhibitors of GLUT *(30)*. The expression of GLUT1 at both the inner and outer BRB was mentioned above, and TR-iBRB cells also express GLUT1 protein at 55 kDa and exhibit Na$^+$-independent 3-OMG uptake with a K_m of 5.6 mM *(12)*. In the light of these findings, GLUT1 at the inner BRB plays an important role in supplying D-glucose, as well as vitamin C, to the neural retina.

In diabetic mellitus, it is expected that hyperglycemia restricts the supply of vitamin C to the retina due to inhibition of GLUT1. Indeed, [^{14}C]DHA uptake by TR-iBRB cells was inhibited by D-glucose and 3-OMG in a concentration-dependent manner with an IC$_{50}$ of 5.56 and 16.9 mM, respectively (Fig. 3B) *(30)*. The $K_{in,retina}$ of [^{14}C]DHA was reduced by 66% in streptozocin-induced diabetic rats compared with normal rats (Fig. 3A) *(31)*. Accordingly, diabetic patients may experience enhanced oxidative stress and metabolic perturbations in the retina following a reduction in the influx transport of DHA, leading to the hypothesis that diabetic retinopathy involves a dysfunction of DHA influx transport at the BRB.

Cystine

Glutathione, which is a tripeptide composed of L-glutamic acid, L-cysteine, and glycine, plays an important role in protecting cells against reactive oxygen species and other toxic agents *(32)*. Therefore, the glutathione concentration in the retina is relatively high (1.2 µmol/g wet weight in freshly excised rat retina) *(33)*. To maintain the intracellular glutathione concentration at an appropriate level, transport of glutathione and/or its thiol source, L-cyst(e)ine, into the retina is critical for the health of the retina. Wang and Cynader *(34)* have reported that the plasma of the carotid artery and internal jugular vein contain high concentrations of L-cystine (35 and 26 µM, respectively) and there is

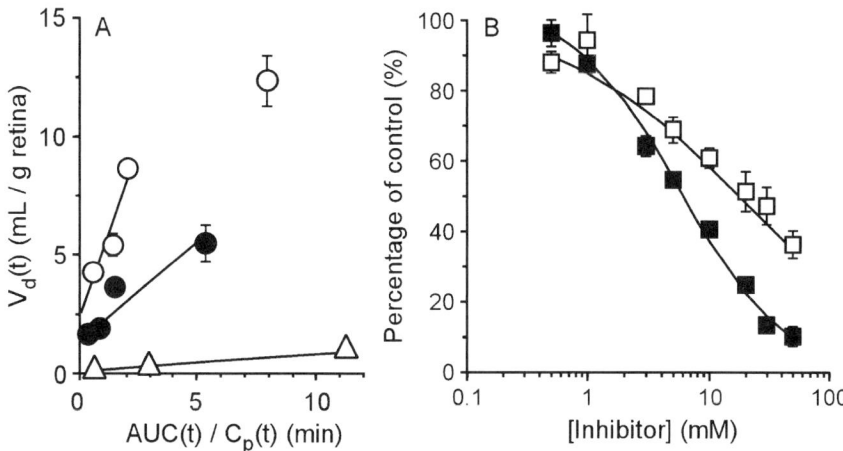

Fig. 3. Characteristics of vitamin C transport at the inner blood-retinal barrier. (**A**) The initial uptake of [^{14}C]dehydroascorbic acid (DHA) by the retina of normal (open circle) and streptozotocin-induced diabetic (closed circle) rats and [^{14}C]ascorbic acid (AA) (open triangle) by the normal rat retina after intravenous administration. The apparent influx permeability clearance of [^{14}C]substrate per gram retina ($K_{in,retina}$) can be obtained from the initial slope of a plot of the apparent retina-to-plasma concentration ratio (V_d) over the time-period of the experiment (t) versus $AUC(t)/C_p(t)$ where $AUC(t)$ and $C_p(t)$, respectively, represent the area under the plasma concentration-time curve of [^{14}C]substrate from time 0 to t and the plasma [^{14}C]substrate concentration at time t. The $K_{in,retina}$ of [^{14}C]DHA in normal rats, [^{14}C]DHA in streptozotocin-induced diabetic rats, and [^{14}C]AA in normal rats are respectively 2.44×10^3 μL/(min·g retina), 841 μL/(min·g retina), and 65.4 μL/(min·g retina). (**B**) Concentration-dependent inhibition of D-glucose (closed square) and 3-O-methyl-D-glucose (open square) involving the uptake of [^{14}C]DHA by TR-iBRB cells. [^{14}C]DHA uptake was performed in the presence or absence of inhibitors at two minutes and 37°C. The IC_{50} values for D-glucose and 3-O-methyl-D-glucose are 5.56 and 16.9 mM, respectively. Each point represents the mean ± the standard error of the mean (SEM) ($n = 3-5$). Data taken from Investigative Ophthalmology & Visual Science, 45, Hosoya et al., Vitamin C transport in oxidized from across the rat blood-retinal barrier. 1232–1239, 2004, with permission from Association for Research in Vision and Ophthalmology.

an obvious positive arteriovenous concentration difference, implying that L-cystine is transported from the circulating blood to the brain and retina in substantial amounts. On the other hand, the plasma concentrations of glutathione and L-cysteine are lower than that of L-cystine and exhibit no arteriovenous differences.

An in vivo intravenous administration study has shown that L-cystine uptake by the eye is activated by pretreatment with diethyl maleate (DEM), a reagent used to deplete intracellular glutathione in order to induce oxidative stress. This uptake is inhibited in the presence of L-glutamic acid and L-α-aminoadipic acid, selective substrates of the system x_c^- (*35*), which is composed of xCT (Slc7a11) and the heavy chain of 4F2 cell surface antigen (4F2hc/CD98/Slc3a2). TR-iBRB cells express xCT and 4F2hc mRNA and L-cystine uptake by TR-iBRB cells takes place in a Na$^+$-independent and concentration-dependent manner with a K_m of 9.2 μM, and is inhibited by system x_c^- substrates and inhibitors. DEM treatment causes significant induction of xCT mRNA, L-cystine uptake, and an increase in the glutathione concentration in TR-iBRB cells *(36)*. These results suggest that L-cystine influx transport at the inner BRB is mediated by system x_c^- and induced under oxidative stress by enhanced transcription of the xCT gene.

Retinal glutathione is preferentially distributed in Müller cells *(37, 38)* and, therefore, L-cystine in retinal interstitial fluid needs to be taken up into Müller cells. It has been shown that L-α-aminoadipic acid is preferentially taken up into Müller cells in the retina, suggesting that system x_c^- is present in these cells *(39)*. TR-MUL cells, a type of conditionally immortalized retinal Müller cell lines, express xCT and 4F2hc and exhibit system x_c^--mediated L-cystine uptake *(40)*. Taking all these findings into consideration, it appears that system x_c^- mediates L-cystine influx transport not only from the circulating blood to the retinal interstitial fluid but also from the interstitial fluid to Müller cells, and the L-cystine in Müller cells is used for glutathione synthesis. This supply pathway of L-cystine into the retina would be of great value in protecting the retina from oxidative stress.

AMINO ACID SUPPLY

Leucine

L-Glutamate and its metabolic product, γ-aminobutyric acid (GABA), are the main neurotransmitters in the retina. The retinal L-glutamate pool is exclusively derived from de novo L-glutamate synthesis in the retina, since there is virtually no uptake of plasma L-glutamate into the retina *(6)*. Therefore, the retina requires branched-chain amino acids (L-leucine, L-isoleucine, and L-valine), particularly L-leucine, and D-glucose as, respectively, the main nitrogen and carbon precursors of retinal L-glutamate synthesis *(41, 42)*. Branched-chain amino acids also serve as a source of carbon skeletons for the tricarboxylic acid cycle and as a substrate for protein synthesis. Branched-chain amino acids need to be supplied from the circulating blood to the retina since they are essential amino acids.

In 1978, L-leucine uptake by isolated retinal capillaries was demonstrated to take place in a concentration-dependent and Na$^+$-independent manner, and was inhibited by L-valine and L-dihydroxyphenylalanine *(9)*, suggesting that system L, which is composed of L-type amino acid transporters (LATs) and 4F2hc, mediates L-leucine transport at the inner BRB. L-Leucine uptake into the retina has also been observed in vivo using the RUI method *(6)*. [^{3}H] L-Leucine uptake by TR-iBRB cells was an Na$^+$-independent and concentration-dependent process with a K_m of 14.1 µM. This process was more potently cis-inhibited by substrates of LAT1 (Slc7a5), D-leucine, D-phenylalanine, and D-methionine, than those of LAT2 (Slc7a8), L-alanine and L-glutamine (Fig. 4A). [^{3}H]L-Leucine efflux from TR-iBRB cells was trans-stimulated by substrates of LAT1 (Fig. 4B). The expression of LAT1 mRNA was 100- and 15-fold greater than that of LAT2 in TR-iBRB and isolated rat retinal vascular endothelial cells, respectively (Fig. 2B). The expression of LAT1 protein was observed in TR-iBRB and primary-cultured human retinal endothelial cells and immunostaining of LAT1 was observed along the rat retinal capillaries *(43)*. Accordingly, LAT1 is predominantly involved in blood-to-retina transport of L-leucine at the inner BRB and seems to be closely involved in visual functions by supplying neurotransmitter precursors.

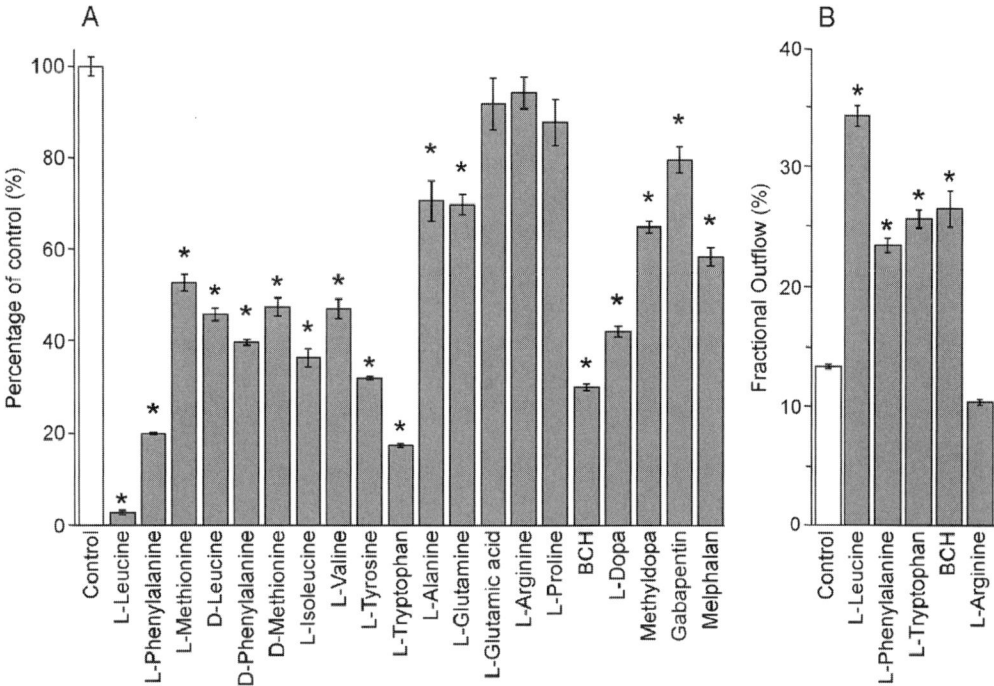

Fig. 4. Effect of several amino acids and drugs on the uptake (**A**) and efflux (**B**) of [^{3}H]L-leucine in TR-iBRB cells. [^{3}H]L-Leucine uptake was measured in the absence (control) or presence of inhibitors at five minutes and 37°C. The efflux of preloaded [^{3}H]L-leucine was performed in the absence (control) or presence of extracellularly applied amino acids at four minutes and 37°C. Concentrations of amino acids and drugs were 2 mM except for melphalan (100 µM). Each column represents the mean ± the standard error of the mean (SEM). (n = 4). $^*p$ < 0.01, significantly different from the control; BCH, 2-aminobicyclo-(2,2,1)-heptane-2-carboylic acid. Data taken from Investigative Ophthalmology & Visual Science, 46, Tomi et al., L-type amino acid transporter 1-mediated L-leucine transport at the inner blood-retinal barrier. 2522–2530, 2005, with permission from Association for Research in Vision and Ophthalmology.

L-Dopa is the most widely used drug for Parkinson's disease, since L-dopa and its metabolite, 3-O-methyldopa, are able to be transported via LAT1 at the blood-brain barrier *(44)*. [^{3}H]L-Leucine uptake by TR-iBRB cells is inhibited by L-dopa and methyldopa (Fig. 4A), suggesting that these compounds are also transported at the inner BRB via LAT1 *(43)*. Many patients with Parkinson's disease have blurred vision or other visual disturbances, which are reflected in the reduced retinal dopamine concentration and the delayed visual-evoked potentials *(45)*. L-Dopa administration has been reported to reduce these delayed visual-evoked potentials in Parkinson's disease *(46)*. Moreover, neutral amino acid-mimetic drugs such as gabapentin and melphalan significantly inhibit [^{3}H]L-leucine uptake by TR-iBRB cells (Fig. 4A) *(43)*. Gabapentin has been shown to be an effective treatment for some patients with acquired pendular nystagmus *(47)*. Chemotherapy with melphalan is used in patients with retinoblastoma *(48)*. Therefore, it appears that LAT1 at the inner BRB plays a key role in transporting amino acid mimetic drugs from the circulating blood to the retina and contributes to their pharmacological actions in the retina.

Taurine

Taurine is a nonessential amino acid in humans, but is considered to be an essential amino acid during fetal growth and lactation. In the retina, taurine exerts a number of neuroprotective functions, acting as an osmolyte and antioxidant, and accounts for more than 50% of the free amino acid content *(49)*. It is physiologically important to elucidate the transport mechanism of taurine in the retina, since a loss of vision due to severe retinal degeneration is reported to be exhibited in taurine transporter-(TauT/Slc6a6) knockout mice *(50)*. The activity of cysteine sulfinic acid decarboxylase, a rate-limiting enzyme for taurine biosynthesis from L-cysteine, is low in rat retina in comparison with the abundance of retinal taurine *(51)*, leading the hypothesis that the blood-to-retina transport system(s) of taurine play a key role in maintaining the taurine concentration in the retina. It has been shown that [^{3}H]taurine is transported from the circulating blood to the retina across the BRB and this is inhibited by a substrate of TauT *(6)*. [^{3}H]Taurine uptake by TR-iBRB cells is Na$^+$-, Cl$^-$-, and concentration-dependent with a K_m of 22.2 µM and inhibited by TauT substrates. TauT is expressed in TR-iBRB and primary cultured human retinal endothelial cells *(52)*. Consequently, TauT is most likely involved in taurine transport at the inner BRB.

The uptake of [^{3}H]taurine and the expression of TauT mRNA in TR-iBRB cells is increased under hypertonic conditions *(52)*. Some retinal diseases, such as those relating to ischemia and reperfusion, diabetic retinopathy, macular edema, and neurodegeneration, are associated with fluctuations in cell volume *(53)*. Retinal taurine is an ideal organic osmolyte because of its high concentration, so that this upregulation of TauT may affect the achievement of osmotic equilibrium by accumulation of the osmolyte taurine in the endothelial cells.

ADENOSINE TRANSPORT

Adenosine is an important intercellular signaling molecule and it plays a number of roles in retinal neurotransmission, blood flow, vascular development, and response to ischemia, through cell-surface adenosine receptors *(54, 55)*. Two classes of nucleoside transporters have been identified: Na$^+$-dependent concentrative nucleoside transporters (CNTs) and Na$^+$-independent equilibrative nucleoside transporters (ENTs). Equilibrative nucleoside transporters can be further subdivided into the nitrobenzylmercaptopurine riboside (NBMPR)-sensitive ENT1 (Slc29a1) and the NBMPR-insensitive ENT2 (Slc29a2).

TR-iBRB cells exhibit an Na$^+$-independent, NBMPR-insensitive, and concentration-dependent [^{3}H]adenosine uptake with a K_m of 28.5 µM, suggesting that ENT2 is involved in [^{3}H]adenosine uptake by TR-iBRB cells. Adenosine, inosine, uridine, and thymidine inhibit this process by more than 60%, while cytidine inhibits it by 30%. Quantitative real-time polymerase chain reaction (PCR) techniques have shown that ENT2 mRNA is predominantly expressed in TR-iBRB cells. An RUI study has demonstrated that [^{3}H]adenosine uptake from the circulating blood to the retina is inhibited by adenosine and thymidine, but unaffected by cytidine, which is similar to the results obtained in TR-iBRB cells *(7)*. These results suggest that ENT2 most likely mediates adenosine transport at the inner BRB and it is expected to have the ability to modulate retinal functions by regulating the adenosine concentration in retinal interstitial fluid.

From a pharmacological viewpoint, ENT2 has greater binding affinity for some antiviral or anticancer nucleoside drugs, like 3′-azido-3′-deoxythymidine (AZT), 2′ 3′-dideoxycytidine (ddC), 2′ 3′-dideoxyinosine (ddI), cladribine, cytarabine, fludarabine, gemcitabine, and capecitabine as substrates (56, 57). This has led to the hypothesis that ENT2 at the inner BRB could be a potential route for delivering nucleoside drugs from the circulating blood to the retina.

ORGANIC ANION TRANSPORT

Understanding the transport mechanisms of organic anions at the inner BRB will provide pharmacologically important information about the effective delivery of many anionic drugs to the retina. Moreover, the transport processes for organic anions at the inner BRB are physiologically important since hormones and neurotransmitters are mostly metabolized in the form of organic anions in the retina, and need to undergo efflux transport from the retina to the circulating blood. Vitreous fluorophotometry has demonstrated that the transport of fluorescein, an organic anion, in the vitreous humor-to-blood direction is more than 100-fold greater than that in the opposite direction and is inhibited in the presence of an organic anion, probenecid (58). The efflux transport of [^{3}H]estradiol 17-β glucuronide (E17βG) and [^{14}C]D-mannitol, which were used as a model compound for amphipathic organic anions and as a bulk flow marker, respectively, from rat vitreous humor/retina has been determined by the use of a microdialysis probe placed in the vitreous humor. [^{3}H]E17βG efflux transport from rat vitreous humor/retina was approximately twofold greater than that of [^{14}C]D-mannitol and was significantly inhibited by organic anions including digoxin, a specific substrate for organic anion transporter polypeptide (Oatp) 1a4 (Slco1a4; Oatp2) (8). Gao et al. have provided immunohistochemical evidence that Oatp1a4 is present at the rat inner and outer BRB (59). Moreover, Oatp1a4 and 1c1 (Slco1c1; Oatp14) mRNA are predominantly expressed in isolated rat retinal vascular endothelial cells (Fig. 2C) (11). Consequently, Oatp1a4 is thought to be involved in the efflux transport of E17βG at the BRB.

Betz and Goldstein (10) demonstrated that p-aminohippuric acid (PAH) uptake by isolated retinal capillaries is slightly greater than that of the extracellular marker sucrose, and is inhibited by fluorescein and penicillin. Organic anion transporter polypeptide does not transport PAH, though organic anion transporters (OATs/Slc22a) prefer PAH as a substrate. Therefore, in addition to Oatp, OAT is also involved in the organic anion transport at the inner BRB. Nevertheless, further investigations regarding organic anion transporters at the inner BRB are needed, since their expression and function at the inner BRB remain largely unknown, including the function of Oatp1c1 and the expression of OAT.

ATP-BINDING CASSETTE (ABC) TRANSPORTERS

P-glycoprotein, which is encoded by multidrug resistance (mdr) 1a (Abcb1a), mdr1b (Abcb1b), or mdr2 (Abcb4), exhibits a protective role by restricting the entry of a wide variety of xenobiotics (e.g., cyclosporine A, rhodamine 123, mitoxantrone, and doxorubicin). P-glycoprotein is present at the luminal membrane of the inner BRB (2, 60), and the expression of mdr1a mRNA is 200- and 24-fold greater than that of mdr1b and mdr2, respectively, in isolated rat retinal vascular endothelial cells (Fig. 2D) *(11)*. Moreover, no cyclosporine A is detected in the intraocular tissues of cyclosporine A-treated rabbits, although the blood level of cyclosporine A was within the therapeutic window *(61)*. TR-iBRB cells express P-glycoprotein *(12)*, and rhodamine 123 accumulation in TR-iBRB cells is enhanced in the presence of inhibitors of P-glycoprotein *(62)*.

ATP-binding cassette transporter G2 (Abcg2) is also known to act as an efflux transporter and is reported to be expressed at the luminal membrane of the inner BRB *(63)*. Abcg2 shows great affinity for not only drugs (e.g., mitoxantrone and doxorubicin), but also photosensitive toxins including pheophorbide-a, a chlorophyll-derived dietary phototoxin related to porphyrin. The retina is subject to high levels of cumulative irradiation and, therefore, is vulnerable to light-induced damage caused by a variety of phototoxic compounds including porphyrins *(64)*. TR-iBRB cells express Abcg2 protein and Ko143, an Abcg2 inhibitor, inhibits the excretion of pheophorbide-a from TR-iBRB cells *(63)*. Taking these findings into consideration, mdr1a-encoding P-glycoptorein, as

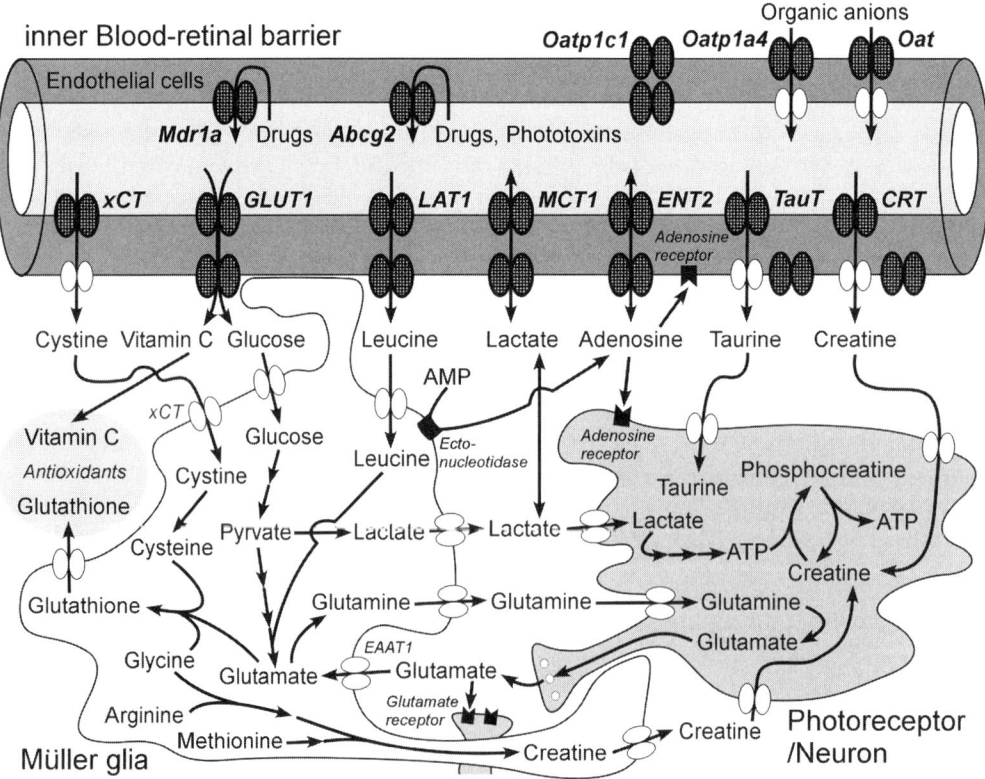

Fig. 5. Hypothetical localization and physiological function of the inner blood-retinal barrier transporters. Inner blood-retinal barrier transporters play an essential role in supplying nutrients to the retina and carrying out the efflux transport of neurotransmitter metabolites, toxins, and xenobiotics to maintain a constant milieu in the neural retina.

well as Abcg2 at the luminal membrane of the inner BRB, could act by restricting the distribution of xenobiotics, including drugs and phototoxins, in the retina. However, the contribution of each ABC transporter subtype to the inner BRB efflux transport is an important issue that remains to be resolved.

CONCLUSIONS

Recent molecular identification and functional analyses of carrier-mediated transport systems for endogenous compounds as well as xenobiotics show that the inner BRB is equipped with a variety of membrane transport mechanisms, for both influx and efflux, to maintain homeostasis in the neural retina (Fig. 5). Retinal diseases, such as age-related macular degeneration and diabetic retinopathy, involve severe vision loss and are a serious social problem especially in developed countries. Topical instillation of drugs does not result in sufficient quantities reaching the retina and so, in order to achieve the maximum efficacy of drugs acting at the retina given by systemic administration, it is important to understand the transport mechanisms at the inner BBB. In addition to transport processes covered in the present chapter, there must be other influx and efflux transport processes at the inner BRB. Identification and characterization of all the transporters at the inner BRB will provide a basis for more-successful strategies for drug delivery to the retina.

ACKNOWLEDGEMENTS

This work was supported, in part, by a grant-in-aid for scientific research from the Japan Society for the Promotion of Science and a grant for research on sensory and communicative disorders by the Ministry of Health, Labor, and Welfare, Japan. The authors thank Drs. T. Terasaki, S. Ohtsuki, K. Katayama, T. Kondo, and M. Tachikawa and Messrs. T. Funaki, T. Isobe, H. Abukawa, A. Minamizono, M. Mori, Y. Ohshima, T. Terayama, and K. Nagase for valuable suggestions and technical assistance.

REFERENCES

1. Cunha-Vaz JG. The blood–retinal barriers system. Basic concepts and clinical evaluation. Exp Eye Res 2004;78:715–21.
2. Hosoya K and Tomi M. Advances in the cell biology of transport via the inner blood–retinal barrier: establishment of cell lines and transport functions. Biol Pharm Bull 2005;28:1–8.
3. Schnaudigel O. Die vitale farbung mit trypanblau an auge. Graefe Arch Ophthal 1913;86: 93–105.
4. Alm A and Törnquist P. The uptake index method applied to studies on the blood–retinal barrier. I. A methodological study. Acta Physiol Scand 1981;113:73–9.
5. Alm A and Törnquist P. Lactate transport through the blood-retinal and the blood–brain barrier in rats. Ophthalmic Res 1985;17:181–4.
6. Törnquist P and Alm A. Carrier-mediated transport of amino acids through the blood-retinal and the blood-brain barriers. Graefes Arch Clin Exp Ophthalmol 1986;224:21–5.
7. Nagase K, Tomi M, Tachikawa M, Hosoya K. Functional and molecular characterization of adenosine transport at the rat inner blood-retinal barrier. Biochim Biophys Acta 2006;1758:13–9.
8. Katayama K, Ohshima Y, Tomi M, Hosoya K. Application of microdialysis to evaluate the efflux transport of estradiol 17-beta glucuronide across the rat blood-retinal barrier. J Neurosci Methods 2006;156:249–56.

9. Hjelle JT, Baird-Lambert J, Cardinale G, Specor S, Udenfriend S. Isolated microvessels: the blood-brain barrier in vitro. Proc Natl Acad Sci U S A 1978;75:4544–8.
10. Betz AL, Goldstein GW. Transport of hexoses, potassium and neutral amino acids into capillaries isolated from bovine retina. Exp Eye Res 1980;30:593–605.
11. Tomi M, Hosoya K. Application of magnetically isolated rat retinal vascular endothelial cells for the determination of transporter gene expression levels at the inner blood–retinal barrier. J Neurochem 2004;91:1244–8.
12. Hosoya K, Tomi M, Ohtsuki S, Takanaga H, Ueda M, Yanai N, Obinata M, Terasaki T. Conditionally immortalized retinal capillary endothelial cell lines (TR-iBRB) expressing differentiated endothelial cell functions derived from a transgenic rat. Exp Eye Res 2001;72:163–72.
13. Ennis SR, Johnson JE, Pautler EL. In situ kinetics of glucose transport across the blood-retinal barrier in normal rats and rats with streptozocin-induced diabetes. Invest Ophthalmol Vis Sci 1982;23:447–56.
14. Takata K, Kasahara T, Kasahara M, Ezaki O, Hirano H. Ultracytochemical localization of the erythrocyte/HepG2-type glucose transporter (GLUT1) in cells of the blood-retinal barrier in the rat. Invest Ophthalmol Vis Sci 1992;33:377–83.
15. Fernandes R, Suzuki K, Kumagai AK. Inner blood-retinal barrier GLUT1 in long-term diabetic rats: an immunogold electron microscopic study. Invest Ophthalmol Vis Sci 2003;44:3150–4.
16. Kumagai AK, Vinores SA, Pardridge WM. Pathological upregulation of inner blood-retinal barrier Glut1 glucose transporter expression in diabetes mellitus. Brain Res 1996;706:313–7.
17. Knott RM, Robertson M, Muckersie E, Forrester JV. Regulation of glucose transporters (GLUT-1 and GLUT-3) in human retinal endothelial cells. Biochem J 1996;318:313–7.
18. Tang J, Zhu XW, Lust WD, Kern TS. Retina accumulates more glucose than does the embryologically similar cerebral cortex in diabetic rats. Diabetologia 2000;43:1417–23.
19. Badr GA, Tang J, Ismail-Beigi F, Kern TS. Diabetes downregulates GLUT1 expression in the retina and its microvessels but not in the cerebral cortex or its microvessels. Diabetes 2000;49:1016–21.
20. Fernandes R, Carvalho AL, Kumagai A, Seica R, Hosoya K, Terasaki T, Murta J, Pereira P, Faro C. Downregulation of retinal GLUT1 in diabetes by ubiquitinylation. Mol Vis 2004;10:618–28.
21. Poitry-Yamate CL, Poitry S, Tsacopoulos M. Lactate released by Müller glial cells is metabolized by photoreceptors from mammalian retina. J Neurosci 1995;15:5179–91.
22. Gerhart DZ, Leino RL, Drewes LR. Distribution of monocarboxylate transporters MCT1 and MCT2 in rat retina. Neuroscience 1999;92:367–75.
23. Hosoya K, Kondo T, Tomi M, Takanaga H, Ohtsuki S, Terasaki T. MCT1-mediated transport of L-lactic acid at the inner blood-retinal barrier: a possible route for delivery of monocarboxylic acid drugs to the retina. Pharm Res 2001;18:1669–76.
24. Hori K, Katayama N, Kachi S, Kondo M, Kadomatsu K, Usukura J, Muramatsu T, Mori S, Miyake Y. Retinal dysfunction in basigin deficiency. Invest Ophthalmol Vis Sci 2000;41:3128–33.
25. Nakashima T, Tomi M, Tachikawa M, Watanabe M, Terasaki T, Hosoya K. Evidence for creatine biosynthesis in Müller glia. Glia 2005;52:47–52.
26. Nakashima T, Tomi M, Katayama K, Tachikawa M, Watanabe M, Terasaki T, Hosoya K. Blood-to-retina transport of creatine via creatine transporter (CRT) at the rat inner blood-retinal barrier. J Neurochem 2004;89:1454–61.
27. Ohia SE, Opere CA, Leday AM. Pharmacological consequences of oxidative stress in ocular tissues. Mutat Res 2005;579:22–36.
28. Beatty S, Koh H, Phil M, Henson D, Boulton M. The role of oxidative stress in the pathogenesis of age-related macular degeneration. Surv Ophthalmol 2000;45:115–34.

29. Baynes JW and Thorpe SR. Role of oxidative stress in diabetic complications: a new perspective on an old paradigm. Diabetes 1999;48:1–9.
30. Hosoya K, Minamizono A, Katayama K, Terasaki T, Tomi M. Vitamin C transport in oxidized form across the rat blood-retinal barrier. Invest Ophthalmol Vis Sci 2004;45:1232–9.
31. Minamizono A, Tomi M, Hosoya K. Inhibition of dehydroascorbic acid transport across the rat blood-retinal and -brain barriers in experimental diabetes. Biol Pharm Bull 2006;29:2148–50.
32. Shan XQ, Aw TY, Jones DP. Glutathione-dependent protection against oxidative injury. Pharmacol Ther 1990;47:61–71.
33. Winkler BS and Giblin FJ. Glutathione oxidation in retina: effects on biochemical and electrical activities. Exp Eye Res 1983;36:287–97.
34. Wang XF and Cynader MS. Astrocytes provide cysteine to neurons by releasing glutathione. J Neurochem 2000;74:1434–42.
35. Hosoya K, Saeki S, Terasaki T. Activation of carrier-mediated transport of L-cystine at the blood-brain and blood-retinal barriers in vivo. Microvasc Res 2001;62:136–42.
36. Tomi M, Hosoya K, Takanaga H, Ohtsuki S, Terasaki T. Induction of xCT gene expression and L-cystine transport activity by diethyl maleate at the inner blood-retinal barrier. Invest Ophthalmol Vis Sci 2002;43:774–9.
37. Schutte M and Werner P. Redistribution of glutathione in the ischemic rat retina. Neurosci Lett 1998;246:53–6.
38. Pow DV and Crook DK. Immunocytochemical evidence for the presence of high levels of reduced glutathione in radial glial cells and horizontal cells in the rabbit retina. Neurosci Lett 1995;193:25–8.
39. Pow DV. Visualising the activity of the cystine-glutamate antiporter in glial cells using antibodies to aminoadipic acid, a selectively transported substrate. Glia 2001;34:27–38.
40. Tomi M, Funaki T, Abukawa H, Katayama K, Kondo T, Ohtsuki S, Ueda M, Obinata M, Terasaki T, Hosoya K. Expression and regulation of L-cystine transporter, system x_c^-, in the newly developed rat retinal Müller cell line (TR-MUL). Glia 2003;43:208–17.
41. LaNoue KF, Berkich DA, Conway M, Barber AJ, Hu LY, Taylor C, Hutson S. Role of specific aminotransferases in de novo glutamate synthesis and redox shuttling in the retina. J Neurosci Res 2001;66:914–22.
42. Lieth E, LaNoue KF, Berkich DA, Xu B, Ratz M, Taylor C, Hutson SM. Nitrogen shuttling between neurons and glial cells during glutamate synthesis. J Neurochem 2001;76:1712–23.
43. Tomi M, Mori M, Tachikawa M, Katayama K, Terasaki T, Hosoya K. L-type amino acid transporter 1-mediated L-leucine transport at the inner blood-retinal barrier. Invest Ophthalmol Vis Sci 2005;46:2522–30.
44. Kageyama T, Nakamura M, Matsuo A, Yamasaki Y, Takakura Y, Hashida M, Kanai Y, Naito M, Tsuruo T, Minato N, Shimohama S. The 4F2hc/LAT1 complex transports L-DOPA across the blood-brain barrier. Brain Res 2000;879:115–21.
45. Bodis-Wollner I. Visual electrophysiology in Parkinson's disease: PERG, VEP and visual P300. Clin Electroencephalogr 1997;28:143–7.
46. Bhaskar PA, Vanchilingam S, Bhaskar EA, Devaprabhu A, Ganesan RA. Effect of L-dopa on visual evoked potential in patients with Parkinson's disease. Neurology 1986;36:1119–21.
47. Averbuch-Heller L, Tusa RJ, Fuhry L, Rottach KG, Ganser GL, Heide W, Buttner U, Leigh RJ. A double-blind controlled study of gabapentin and baclofen as treatment for acquired nystagmus. Ann Neurol 1997;41:818–25.
48. Kaneko A and Suzuki S. Eye-preservation treatment of retinoblastoma with vitreous seeding. Jpn J Clin Oncol 2003;33:601–7.
49. Pasantes-Morales H, Klethi J, Ledig M, Mandel P. Free amino acids of chicken and rat retina. Brain Res 1972;41:494–7.

50. Heller-Stilb B, van Roeyen C, Rascher K, Hartwig HG, Huth A, Seeliger MW, Warskulat U, Haussinger D. Disruption of the taurine transporter gene (taut) leads to retinal degeneration in mice. FASEB J 2002;16:231–3.
51. Heinamaki AA. Endogenous synthesis of taurine and GABA in rat ocular tissues. Acta Chem Scand B 1988;42:39–42.
52. Tomi M, Terayama T, Isobe T, Egami F, Morito A, Kurachi M, Ohtsuki S, Kang YS, Terasaki T, Hosoya K. Function and regulation of taurine transport at the inner blood-retinal barrier. Microvasc Res 2007; 73:100–106.
53. Pasantes-Morales H, Ochoa de la Paz LD, Sepulveda J, Quesada O. Amino acids as osmolytes in the retina. Neurochem Res 1999;24:1339–46.
54. Ghiardi GJ, Gidday JM, Roth S. The purine nucleoside adenosine in retinal ischemia-reperfusion injury. Vision Res 1999;39:2519–35.
55. Lutty GA and McLeod DS. Retinal vascular development and oxygen-induced retinopathy: a role for adenosine. Prog Retin Eye Res 2003;22:95–111.
56. Baldwin SA, Beal PR, Yao SY, King AE, Cass CE, Young JD. The equilibrative nucleoside transporter family, SLC29. Pflugers Arch 2004;447:735–43.
57. Yao SY, Ng AM, Sundaram M, Cass CE, Baldwin SA, Young JD. Transport of antiviral 3′-deoxy-nucleoside drugs by recombinant human and rat equilibrative, nitrobenzylthioinosine (NBMPR)-insensitive (ENT2) nucleoside transporter proteins produced in Xenopus oocytes. Mol Membr Biol 2001;18:161–7.
58. Engler CB, Sander B, Larsen M, Koefoed P, Parving HH, Lund-Andersen H. Probenecid inhibition of the outward transport of fluorescein across the human blood-retina barrier. Acta Ophthalmol (Copenh) 1994;72:663–7.
59. Gao B, Wenzel A, Grimm C, Vavricka SR, Benke D, Meier PJ, Reme CE. Localization of organic anion transport protein 2 in the apical region of rat retinal pigment epithelium. Invest Ophthalmol Vis Sci 2002;43:510–4.
60. Holash JA and Stewart PA. The relationship of astrocyte-like cells to the vessels that contribute to the blood-ocular barriers. Brain Res 1993;629:218–24.
61. BenEzra D and Maftzir G. Ocular penetration of cyclosporin A. The rabbit eye. Invest Ophthalmol Vis Sci 1990;31:1362–6.
62. Shen J, Cross ST, Tang-Liu DD, Welty DF. Evaluation of an immortalized retinal endothelial cell line as an in vitro model for drug transport studies across the blood-retinal barrier. Pharm Res 2003;20:1357–63.
63. Asashima T, Hori S, Ohtsuki S, Tachikawa M, Watanabe M, Mukai C, Kitagaki S, Miyakoshi N, Terasaki T. ATP-binding cassette transporter G2 mediates the efflux of phototoxins on the luminal membrane of retinal capillary endothelial cells. Pharm Res 2006;23:1235–42.
64. Boulton M, Rozanowska M, Rozanowski B. Retinal photodamage. J Photochem Photobiol B 2001;64:144–61.

V
Transport Across the Retinal Pigment Epithelium

9
Regulation of Transport in the RPE

Adnan Dibas and Thomas Yorio

CONTENTS

INTRODUCTION
TRANSPORT OF NUTRIENTS AND IONS FROM THE SUBRETINAL SPACE
  TO THE CHORIOCAPILLARIS
MECHANISM OF TRANSPORT OF IONS IN RETINAL PIGMENT EPITHELIUM
THE Na^+,K^+-ATPase PUMP
K^+ AND Na^+ CHANNELS
Cl^- CHANNELS
CALCIUM-ACTIVATED Cl^- CHANNELS
AGONISTS ENHANCING FUNCTION OF CALCIUM-ACTIVATED Cl^- CHANNELS
Na^+/Ca^{2+} EXCHANGER FUNCTIONS IN RPE TO REDUCE $[Ca^{2+}]_i$ AND ANTAGONIZES
  CALCIUM-ACTIVATED K^+ AND Cl^- CHANNELS
ELECTRONEUTRAL CATION-Cl^- COTRANSPORTERS (SLC12)
REGULATION OF $[Cl^-]_i$ AND THE RELATIONSHIP WITH BICARBONATE TRANSPORTERS
BICARBONATE TRANSPORTERS IN RETINAL PIGMENT EPITHELIUM AND FUNCTIONAL
  COUPLING TO OTHER RETINAL CELL LAYERS
BICARBONATE TRANSPORTERS CANNOT FUNCTION PROPERLY WITHOUT FUNCTIONAL
  CARBONIC ANHYDRASES (CAS)
CARBONIC ANHYDRASES
CARBONIC ANHYDRASE INHIBITORS OF THERAPEUTIC VALUE IN RETINAL
  PIGMENT EPITHELIUM-RELATED DISEASE
Na^+/H^+ (NH) ANTIPORTERS
MONOCARBOXYLATE TRANSPORTERS (MCT)
AQP1, AQP9 AND EDEMA
BUFFERING OF IONS IN THE SUBRETINAL SPACE IN THE LIGHT-DARK CYCLE
  AND EXPLANATION OF ELECTRORETINOGRAM (ERG) C-WAVE AND DELAYED
  HYPERPOLARIZATION
OTHER IMPORTANT RETINAL PIGMENT EPITHELIUM TRANSPORT MECHANISMS
REGULATION OF RETINAL PIGMENT EPITHELIUM TRANSPORT BY HORMONES
  AND AGONISTS
UNANSWERED QUESTIONS REGARDING RETINAL PIGMENT EPITHELIUM TRANSPORT
ACKNOWLEDGEMENTS
REFERENCES

From: *Ophthalmology Research: Ocular Transporters in Ophthalmic Diseases and Drug Delivery*
Edited by: J. Tombran-Tink and C. J. Barnstable © Humana Press, Totowa, NJ

INTRODUCTION

The retinal pigment epithelium (RPE) is a monolayer of post-mitotic pigmented epithelial cells juxtaposed between the neural retina and the choroid. It separates the outer surface of the neural retina from the choriocapillaris and functions as a part of the blood–retina barrier. The RPE's long apical microvilli surround the light-sensitive outer segments establishing a small extracellular domain known as the subretinal space, whereas the basolateral membrane faces Bruch's membrane, which separates the RPE from the choriocapillaris. The volume and chemical composition of the subretinal space fluctuates with the light and dark cycle. Such changes are attributed to the many volume and transport systems associated with RPE.

A healthy RPE is essential for photoreceptor survival, and maintenance of photoreceptor viability depends on viable RPE. The RPE functions both as a protective barrier and a nutrient controller for the photoreceptors. In fact, RPE regulates the transport of ions, nutrients (glucose, retinol and fatty acids) and waste products (metabolic endproducts) to and from the retina *(1)*. In addition, RPE acts as a waste/recycling facility by ingesting and degrading the shed photoreceptor discs, enabling outer segment renewal *(2)*. Furthermore, RPE absorbs the light energy focused by the lens onto the retina, and therefore neutralizes excessive high-energy light *(3)*. More importantly, RPE cells secrete valuable factors necessary for growth of neighboring cells *(4)*. Known growth factors released by RPE cells include: vascular endothelial growth factor (VEGF), lens epithelium-derived growth factor (LEDGF), ciliary neurotrophic factor (CNTF), fibroblast growth factor (FGF), transforming growth factor-β (TGF-β), insulin-like growth factor-I (IGF-I), platelet-derived growth factor (PDGF), and pigment epithelium-derived growth factor (PEDF).

Equally important is the exchange between photoreceptors and the RPE. Photon absorption converts 11-*cis*-retinal into all-*trans*-retinal that cannot be changed into 11-*cis*-retinal by photoreceptors. The RPE carries out the recycling of all-*trans*-retinal and delivery services of 11-*cis*-retinal for the photoreceptors, a process known as the visual cycle of retinal. Vitamin A (all-*trans*-retinal) uptake from the bloodstream represents a minor additional supply to photoreceptors *(5)*. The uptake of vitamin A occurs in a receptor-mediated process, with recognition by a serum retinol-binding protein/transthyretin (RBP/TTR) complex *(6)*. Equally important, is the delivery of docosahexaenoic acid by RPE to photoreceptors [docosahexaenoic acid (22:6ω3 fatty acid) cannot be synthesized by photoreceptors] where it is needed to build phospholipids *(7)*.

TRANSPORT OF NUTRIENTS AND IONS FROM THE SUBRETINAL SPACE TO THE CHORIOCAPILLARIS

Retinal pigment epithelium transepithelial transport has two unique characteristics: firstly, RPE contains very adhesive apical tight junctions that block diffusion through the paracellular spaces; secondly, there is a novel polarized distribution of transporters carrying out vectorial transport across the monolayer. High metabolic turnover in the retina produces a large amount of water that must be eliminated from the subretinal space, and that may explain the presence of a unique adhesive force between RPE and the retina. Published studies have estimated the paracellular resistance to be more than

10 times higher than the transcellular resistance *(8)*. Retinal pigment epithelium cells need such a resistant barrier because they are subjected to varying osmotic pressures due to light-dependent changes in the ionic composition of the subretinal space, and because RPE cells may undergo large volume changes during phagocytosis of shed photoreceptor discs. Photoreceptors shed around 10% of the length of their outer segments daily, generating a large amount of debris that must be cleared by RPE *(2)*. In addition, the transition from light to dark increases the K^+ concentration from approximately 2 to 5 mM in the subretinal space *(9)*. Therefore, RPE must be able to control the volume and chemical composition of the subretinal and choroidal extracellular spaces, allowing for a stabilization of ions that is essential for viable photoreceptor function. However, this tight barrier may be compromised due to increased expression of metalloproteases (MMPs) *(10)*. It has been shown that the retinas of diabetic animals have elevated levels of MMP-2, MMP-9 and MMP-14 messenger RNA (mRNA), and that high glucose concentrations induce a significant increase in the production of MMP-9. This can be detrimental since MMPs are known to degrade the tight junction protein occludin, and therefore disrupt the overall tight junction complex. Compromise of transport pathways of ion and fluid movement across the RPE is likely to cause abnormal accumulation of fluid and ions in the subretinal space leading to pathological disorders. A key factor in age-related macular degeneration is the development of drusen deposits. Drusen are deposits of backed-up waste products from various layers of the retina, which are deposited in Bruch's membrane. They act as a barrier, depriving photoreceptors of oxygen and nutrients and therefore cause retinal degeneration *(11)*.

MECHANISM OF TRANSPORT OF IONS IN RETINAL PIGMENT EPITHELIUM

The transport of ions can be summarized in the following steps: firstly, a gradient is established by the apically located Na^+, K^+-adenosine triphosphatase (-ATPase), which causes sodium efflux and potassium influx, creating a driving force for K^+ efflux; secondly, K^+ and Cl^- uptake is mediated by apical Na^+-K^+-$2Cl^-$ and K^+-Cl^- contransporters. Accumulation of Cl^- results in intracellular chloride concentration ($[Cl^-]_i$) of around 20 to 60 mM and this gradient drives uptake of Na^+ and HCO_3^- through the Na^+-HCO_3^- cotransporter, with intracellular Cl^- being extruded at the basal membranes through at least four distinct Cl^- channels and K^+/Cl^- cotransporters. Voltage-activated Cl^- channels (ClC), calcium-activated Cl^- channels, bestrophin (also activated by Ca^{2+}) and cyclic adenosine monophosphate (cAMP)activated cysticfibrosis regulator conductance (CFTR) channel, pump Cl^- extracellularly. Chloride is predominantly transported from the subretinal space to the choriocapillaris and such activity results in an apical-positive transepithelial potential of approximately 5 to 15 mV. Potassium ions are also extruded by at least two distinct K^+ channels at the basal membranes (voltage-activated K^+ channels and Ca^{2+}-activated K^+ channels) as well as by K^+/Cl^- cotransporters. In fact, the basolateral K^+ conductance is higher than the apical K^+ conductance, leading to a net transepithelial K^+ transport from the subretinal space to the choroidal site. Depending on changes in the K^+ concentration in the subretinal space, this transport direction can be reversed. Overall, this transepithelial transport drives water through the cell's water channels, aquaporins,

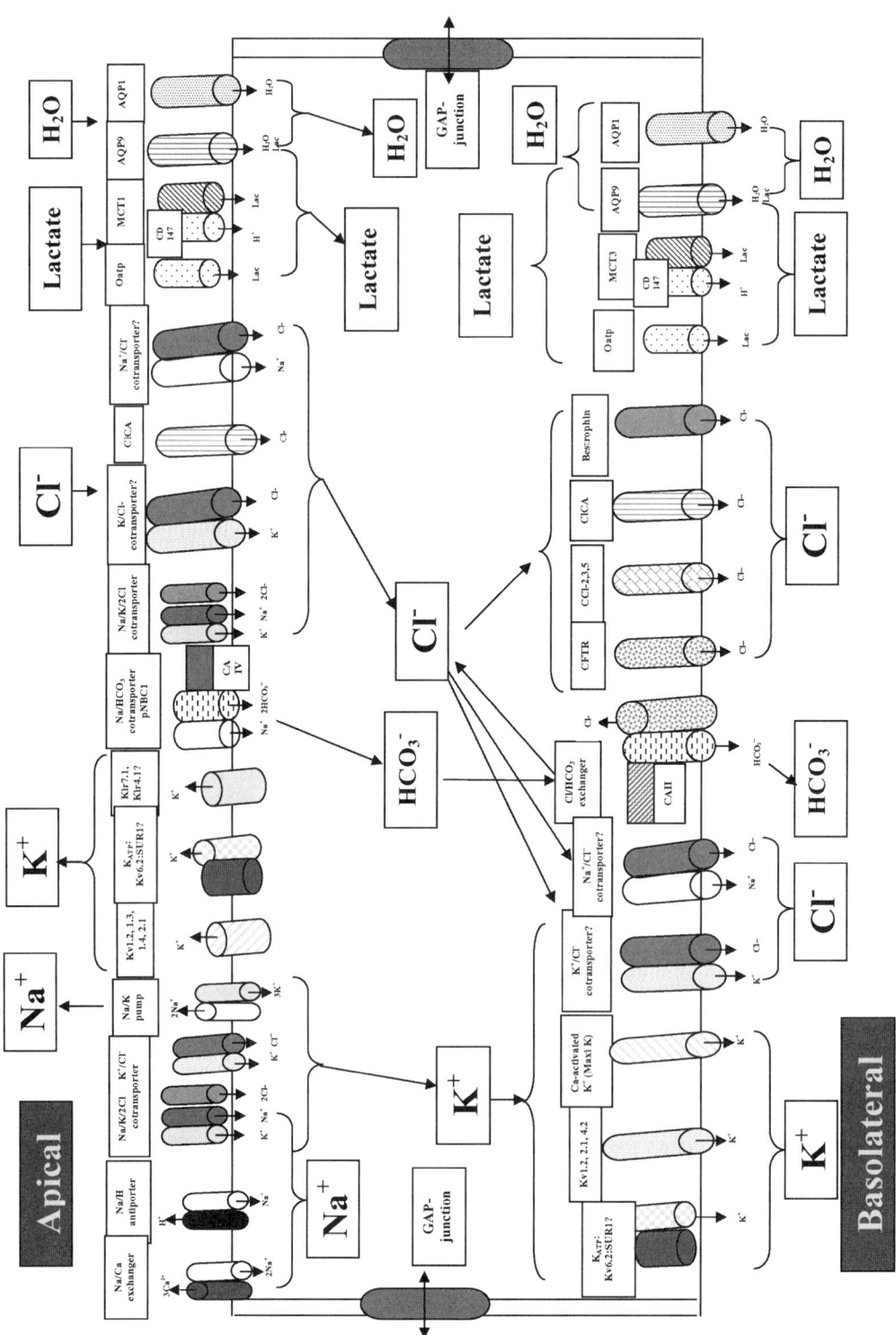

Fig. 1. Ion transporters involved in retinal pigment epithelium (RPE) transepithelial transport. Channels and transporters involved in Na$^+$ homeostasis include the apically located Na$^+$, K$^+$-ATPase. There are no studies on the expression of a neutral Na$^+$/Cl$^-$ cotransporter in RPE yet. Also, voltage Na$^+$ channels may not be expressed in vivo in RPE. The Na$^+$,K$^+$-ATPase pump helps create a gradient by mediating Na$^+$ efflux and K$^+$ influx, constituting a driving force for K$^+$ efflux. K$^+$ and Cl$^-$ uptake is mediated by apical Na$^+$-K$^+$-2Cl$^-$ and K$^+$-Cl$^-$ cotransporters (the identity and membrane localization are yet to be determined). Accumulation of Cl$^-$ results in an intracellular chloride concentration ([Cl$^-$]$_i$) of around 20 to 60 mM. Also, this gradient drives uptake of Na$^+$ and HCO$_3^-$ through the Na$^+$-HCO$_3^-$ cotransporter, with intracellular Cl$^-$ extruded at the basal membranes through at least four distinct Cl$^-$ channels and K$^+$/Cl$^-$ cotransporters. Voltage-activated Cl$^-$ channels (ClC), Calcium-activated Cl$^-$ channels (CACL), bestrophin and cAMP-activated cytic fibrosis regulator conductance (CFTR) channel, function to pump Cl$^-$ extracellularly. Cl$^-$ is predominantly transported from the subretinal space to the choriocapillaris resulting in an apical-positive transepithelial potential of approximately 5 to 15 mV. K$^+$ is also extruded by at least two distinct K$^+$ channels at the basal membranes (voltage-activated K$^+$ channels and Ca^{2+}-activated K$^+$ channels) as well as K$^+$/Cl$^-$ cotransporters. The basolateral K$^+$ conductance is higher than the apical K$^+$ conductance leading to a net transepithelial K$^+$ transport from the subretinal space to the choroidal site. Depending on changes in the K$^+$ concentration in the subretinal space, this transport direction can be reversed. Overall, this transepithelial transport drives water through aquaporins from the subretinal space to the choriocapillaris. There are least two distinct aquaporin channels (AQP1 and AQP9).

As a consequence of the above steps, intracellular Ca^{2+} is decreased by Ca^{2+}-pumps and the Na$^+$/Ca^{2+} exchanger. This will deactivate Ca^{2+}-dependent Cl$^-$ and K$^+$ channels. In addition, apical K$_{ir}$ channels and voltage-activated K$^+$ channels (Kv1.2, Kv1.4 and Kv4.2) mediate efflux of K$^+$. Furthermore, a basolateral Cl$^-$/HCO$_3^-$ exchanger extrudes HCO$_3^-$ from cells and increases intracellular Cl$^-$. Lactate transport can be mediated by at least three different proteins; aquaporin-9 (AQP9), monocarboxylate transporters (MCT) and the Na$^+$-dependent organic-anion transporting protein (Oatp). Finally, all of cellular signaling and ion fluxes are synchronized between RPE cells through gap junction proteins (connexin 43).

from the subretinal space to the choriocapillaris. As a consequence of these processes, intracellular Ca^{2+} is decreased by Ca^{2+}-pumps and the Na^+/Ca^{2+} exchanger. This will deactivate Ca^{2+}-dependent Cl^- and K^+ channels. In addition, apical inwardly rectifying K^+ channels (K_{ir} channels) and voltage-activated K^+ channels (Kv1.2, Kv1.4, and Kv4.2) mediate efflux of K^+. Furthermore, a basolateral Cl^-/HCO_3^- exchanger extrudes HCO_3^- from cells and increases intracellular Cl^-. Finally, all cellular signaling processes and ion fluxes are synchronized between RPE cells through gap junction proteins (e.g., connexin 43) that allow synchronization in response. Unfortunately, gap junctions are reduced in diabetic animals, leading to the loss of such uniform responses by RPE cells (12). Figure 1 lists the transporters and ion channels involved in RPE transport, while Table 1 details the diseases associated with impaired transporter function in retina.

Table 1 Diseases associated with impaired transporter function in the retina

Transporter affected	Pathology	Reference
ClC2 (voltage-gated Cl channel)	Disrupted gene in mice produces a phenotype similar to retinitis pigmentosa.	Bosl et al. 2001 *(43)*
cAMP-activated cystic-fibrosis transmembrane regulator (CFTR) Cl channel	Patients with cystic fibrosis do not develop retinal degeneration, but have reduced amplitudes of the fast oscillation in the EOG.	Miller et al. 1992 *(50)*
Bestrophin Cl^- channel	Mutation in human bestrophin-1 are responsible for an early type of macular degeneration called Best vitelliform macular dystrophy.	O'Gorman et al. 1998, *(52)*, Sun et al. 2002 *(53)*,
Na^+/Ca^{2+} exchanger	Abnormal expression occurred in retinal pigment epithelium that varied with different ocular complications.	Loeffler et al. 1998 *(67)*
Sodium bicarbonate cotransporter-3 (NBC3)	Although retinal pigment epithelium remains normal, mice showed complete loss of photoreceptors.	Bok et al. 2003 *(82)*
Carbonic anhydrase-IV (CA-IV)	Human with missense mutations in CA IV are suffering from an autosomal dominant rod-cone dystrophy due to a dysfunctional NBC1.	Yang et al. 2005 *(89)*

Table 1 (continued)

Transporter affected	Pathology	Reference
Cl⁻/HCO₃ anion exchanger (AE)	Royal College of Surgeons rats characterized by defective phagocytosis showed increased and redistribution of CA activities indicating altered AE function, since both proteins colocalized together.	Eichhorn et al. 1996 *(96)*
Monocarboxylate cotransporters (MCT)	MCT form a dimmer with CD147 protein. CD147 knockout mice showed loss of membrane expression of MCT1, MCT3, & MCT4 which results in reduction of lactate transport.	Philp et al. 2003, Ochrietor et al 2004, Clamp et al 2004 *(111–113)*

THE Na⁺,K⁺-ATPase PUMP

Unlike most epithelial cells where the Na⁺, K⁺-ATPase is concentrated at the basolateral membranes, this Na⁺/K⁺-pump is primarily localized at the apical membranes in RPE *(13–15)*. The pump establishes a gradient for sodium from the extracellular space to the intracellular compartment. This Na⁺ gradient serves to drive the uptake of K⁺ and Cl⁻ on the apical side resulting in a high intracellular Cl⁻ concentration. Interestingly, removal of K⁺ reduced the apical sodium pump levels *(13)*. Ouabain, a potent inhibitor of the sodium pump, decreased transepithelial electrical resistance (TER) in human RPE cells *(13)* and inhibited elimination of water from the subretinal space *(16)*. More importantly, elevated glucose levels were associated with impaired sodium-pump activity in bovine RPE; this action may explain complications related to diabetic retinopathy *(17)*.

K⁺ AND Na⁺ CHANNELS

Potassium channels serve a variety of functions in epithelial cells, but in general their activation is associated with the dampening of membrane excitability. There are at least four main K⁺ channel families: (i) voltage-activated K⁺ channels (at least eight subtypes: Kv1–Kv8, where each has different isoforms [Kv1.1, etc.]) *(18,19)*; (ii) Ca²⁺-activated K⁺ channels (classified according to their conductance into three subtypes; large conductance [BK_{Ca}, also known as MaxiK], intermediate conductance [IK_{Ca}], and small conductance [SK_{Ca}]) *(20)*; (iii) inwardly-rectifying K⁺ channels (at least seven members: $K_{ir}1$–$K_{ir}7$ with each subtype has different isoforms, i.e. $K_{ir}1.1$ or $K_{ir}1.2$ etc.) *(21,22)*; and (iv) adenosine triphosphate (ATP)-sensitive K⁺ channels (unique channel formed by a combination of inwardly rectifying Kir6.2 and sulfonylurea SUR subunit [member of the sulphonylurea receptor family, with two known members SUR1 and SUR2]) *(23)*. Interestingly, all four K⁺ channel families are expressed in RPE and play a number of important roles including the establishment of a membrane potential, transcellular transport and balancing of K$^+$ between the subretinal space and choroid, as well as the generation of the c-wave of the electroretinogram.

Pinto et al. *(24)* have reported the expression of at least five voltage-activated K$^+$ channels (Kv) in mouse RPE. While Kv1.3 and Kv1.4 are apical, Kv4.2 is localized at the basolateral membranes. However, Kv1.2 and Kv2.1 are expressed at both membrane surfaces. Similarly, Wollmann et al. *(25)* reported a delayed-rectifier K$^+$ channel (Kv1.3) expressed in mouse RPE and Takhira et al. *(26)* have shown the existence of an outwardly rectifying K$^+$ channel in bovine RPE that was sensitive to tetraethylammonium (TEA), while Kv1.2 and Kv1.4 are inhibited by acidosis *(27)*. Another K$^+$ channel that may play an essential role in the regulation of membrane potential and ion transport is the calcium-activated K$^+$ channel. Tao et al. *(28)* have provided evidence for the presence of a Ca^{2+}-activated K$^+$ channel in rabbit RPE. The channel is inhibited by iberiotoxin and charybdotoxin, known inhibitors of Ca^{2+}-activated K$^+$ channels. Sheu et al. *(29)* also reported the presence of Ca^{2+}-activated K$^+$ channels in human RPE. Known oxidizing agents (e.g., t-butyl hydroperoxide, thimerosal, and 4,4'-dithiodipyridine) appeared to inhibit this channel, however, the membrane localization of such a channel has yet to be determined (i.e., apical vs. basolateral).

The inwardly rectifying K$^+$ channels (K$_{ir}$) are also expressed in human, rabbit and mouse RPE *(25,30–32)*. It appears there are least three subtypes of K$_{ir}$ in RPE. While Yang et al. *(33)* have demonstrated the presence of inwardly rectifying K$_{ir}$7.1, but not K$_{ir}$4.1, in native bovine RPE, Kusaka et al. *(34)* have detected K$_{ir}$4.1 at rat apical RPE cells (expression was confirmed by RT-PCR analysis, in situ hybridization and immunohistochemistry) suggesting a species difference. In bovine RPE, K$_{ir}$7.1 was localized at the apical membranes and colocalized with the Na$^+$, K$^+$-pump suggesting synergistic function of both channel and pump. While K$_{ir}$7.1 promotes efflux of K$^+$ through the apical membrane, the Na$^+$, K$^+$-pump allows influx of K$^+$. Inwardly rectifying K$^+$ channels appear to be activated by acidification *(35)* and require MgATP for sustaining their activity *(36)*. In rats, the expression of K$_{ir}$4.1 appears at 10 days after birth in RPE cells, which correlates with the maturation of retinal neuronal activity as represented by the a- and b-waves of the electroretinogram *(34)*. K$_{ir}$6.2 is also expressed in rat RPE as part of the K$_{ATP}$ channels *(37)*.

Ettaiche et al. *(37)* have also reported the expression of ATP-sensitive channels in rat RPE (a combination of K$_{ir}$6.2:SUR1). These K$_{ATP}$-sensitive channels are inhibited by ATP (by direct interaction with K$_{ir}$) as well as sulfonylurea drugs (via interaction with the SUR subunit [e.g., glibenclamide, glipizide, tolbutamide]) *(23)*. K$_{ATP}$-sensitive channels help stabilize the membrane potential and act as a brake on excitability. K$_{ATP}$ channels also act as an electrical transducer of the metabolic state of the cell and are inhibited when cellular phosphorylation potential is high, whereas they are activated and open when metabolism decreases. However, the contribution of K$_{ATP}$ channels in RPE ion transport has yet to be studied in detail.

Other voltage-activated channels that are of significance in RPE functions are the Ca^{2+} and Na$^+$ channels. Since Ca^{2+} channels are covered in another chapter in this book, they will not be addressed here. Botchkin et al. *(38)* have reported the presence of at least two distinct Na$^+$ voltage channels in rat RPE, with one being blocked by tetrodotoxin while the other is insensitive to tetrodotoxin (up to 10 µM). However, such expression may be due to the differentiation of RPE and may not actually occur in vivo.

Cl⁻ CHANNELS

Increasing evidence indicates that intracellular chloride concentration ($[Cl^-]_i$) is an important regulatory signal in epithelial ion transport and cell-volume homeostasis. As K^+ continuously exits the cell, leading to a build-up of positive charges extracellularly; the only way to neutralize the charge build-up is through the intracellular chloride pool that can supply the neutralizing Cl^- to the other side of the epithelial membrane. At the same time, in order to sustain secretion, the loss from intracellular chloride pool must be compensated by influx from the basolateral membranes. Apical Cl^- transporters and channels are distinct from basolateral Cl^- transporters and channels.

$[Cl^-]_i$ can be modulated in RPE by the uptake of at least two different solute carriers (SLC), a family that includes electroneutral cation-Cl^--coupled cotransporters (also known as the SLC12 family) and a bicarbonate transporter family (also known as the SLC4 family). $[Cl^-]_i$ can be reduced by at least four channel families that include voltage-gated Cl^- channels (ClC), cAMP-activated Cl^- channels [cystic-fibrosis transmembrane regulator (CFTR)], swelling-activated Cl^- channels, and Ca^{2+}-activated Cl^- channels. This last family has members encoded by distinct genes that include the betrophin and CLCA families. There is still a controversy surrounding the CLCA family, as to whether it contains the sole Ca^{2+}-activated Cl^- channels or whether there are other unidentified members that constitute the real Ca^{2+}-activated Cl^- channels that are occasionally termed CaCC (see review in *39*). In addition, the identity of the swelling-activated Cl^- channels is not known and therefore will not be discussed.

The ClC Family

The ClC channel is a voltage-gated Cl^- channel associated with at least nine different mammalian genes *(40)*. ClC (around 100 KDa) appears to function as a dimer with 13 potential transmembrane domains. Weng et al. *(41)* have detected at least three different ClC channels in human adult RPE. ClC-2, ClC-3, and ClC-5, were detected by RT-PCR, Western blot, and immunofluorescence microscopy. Interestingly, the outwardly rectifying chloride current was inhibited by the oxidant H_2O_2. Similar findings were reported in the human fetal-cell line RPE, 28 SV4 *(42)*, which expresses both ClC-3 and ClC-5. In a transgenic mouse model with a disrupted gene for ClC-2 channels, the resulting phenotype has retinal degeneration similar to retinitis pigmentosa. The disease is characterized by a lack of both epithelial transport of Cl^-, and transepithelial potential, in RPE *(43)*. However, the exact mechanism for the retinal degeneration is yet to be determined. Both ClC-1 and ClC-2 are inhibited by micromolar concentrations of Zn^{2+} *(44)*. Myopic patients with retinal detachment have also elevated zinc and copper concentrations *(45)*.

Cyclic AMP-activated Cystic-Fibrosis Transmembrane Regulator (CFTR) Channels

Another Cl^- channel that may regulate transport in RPE is the cystic-fibrosis transmembrane regulator (CFTR). Unlike the ClC family, CFTR is voltage insensitive but requires ATP for maximal activation and is inhibited by adenosine diphosphate (ADP) *(46)*. Amino-acid sequence predicted up to 15 potential protein kinase A (PKA) phosphorylation sites in the CFTR. In unstimulated cells, CFTR is kept dephosphorylated *(47)* and it is subsequently activated through increases in cAMP. The source for production of cAMP is adenylyl cyclase (AC) type 7, which appears as the only isoform expressed in bovine RPE *(48)*. While CFTR channels are expressed in apical membranes in different epithelial cells, it appears that the majority is localized at the basolateral membrane in RPE *(49)*. Cystic-fibrosis transmembrane regulator channel (CTFR) expression has been reported in human fetal and adult RPE by RT-PCR, immunolocalization, and electrophysiological techniques *(41,42)* and it has been suggested that the fast oscillation (FO) component of the electro-oculogram (EOG) is mediated directly or indirectly by CFTR. As expected, Cl^- currents were stimulated by cAMP, but inhibited by 1 mM 2,2'-(1,2-ethenediyl)bis(5-isothiocyanatobenzenesulfonic acid) (DIDS) and the oxidative agent, hydrogen peroxide. However, CFTR may not play an essential role in fluid transport in RPE, and only part of the transepithelial transport of Cl^- is dependent on cAMP-activated Cl^- conductance. Patients with cystic fibrosis do not develop retinal degeneration, but have reduced amplitude of the FO on EOG. This could be due to compensatory activation of other Cl^- channels, such as the Ca^{2+}-dependent Cl^- channel or bestrophin, in the RPE *(50)*.

CALCIUM-ACTIVATED Cl^- CHANNELS

The Bestrophin Family

Bestrophins are a newly identified family of calcium-activated and volume-sensitive Cl^- channels with an approximate molecular weight of 68 KDa. At least four human bestrophin genes have been cloned. Human bestrophins 1, 2 and 4 are commonly expressed in the eye *(51)*. Bestrophins are localized at the basolateral membranes of RPE and are involved in the efflux of Cl^- *(49)*. Mutations in human bestrophin 1 (hBest1) are responsible for a type of early-onset macular degeneration called Best's vitelliform macular dystrophy, characterized by the accumulation of a yellowish fluid between the retina and RPE, suggesting impaired fluid transport *(52,53)*. Patients with Best's vitelliform macular degeneration have a reduced light-peak-to-dark ratio in their EOG suggesting that the light peak results from the activation of basolateral Cl^- conductance. However, some patients with bestrophin mutations have normal light peaks or an onset of light-peak reduction that occurs after the onset of macular dystrophy *(54–56)*. In addition, overexpression of wild-type bestrophin did not change light-peak amplitude, but desensitized the luminance response in rats *(57)*. Therefore, dysfunctional Cl^- bestorphin channels cannot simply explain all the complications observed in Best's dystrophy patients.

As expected, bestrophin currents were activated by ionomycin (a calcium ionophore), but, in contrast, they were inhibited by glibenclamide and 5-nitro-2-(3-phenylpropylamino)benzoate in basal dog RPE *(58)*. Qu et al. *(59)* have shown that mouse bestrophin-2 is also permeable to SCN^-, I^-, Br^-, Cl^-, and F^-. Bestrophins are also deactivated by hypertonicity and activated by hypotonicity *(59–61)*. The activation by hypotonicity raises the following question: are bestrophins the 'missing' swelling-activated Cl^- channels?

Calcium-Activated Cl Channels (CACL or CaCC)

At least four human Ca^{2+}-activated Cl^- channels have been cloned, with an estimated molecular weight of 100 KDa *(62)*. Sequence analysis of human CACL predicts at least four potential transmembrane domains with multiple potential phosphorylation sites (16 sites) for protein kinase C (PKC) and Ca^{2+}-calmodulin-dependent kinase *(63)*. Mouse CACL5 is intensely expressed in eyes, but the ocular distribution is yet to be studied *(64)*. Antibodies against porcine CLCA1 labeled the apical membranes of canine RPE and its presence was confirmed by RT-PCR. Overexpression of CLCA1 increased cAMP-dependent Cl⁻ transport, suggesting that CLCA1 plays more of a regulatory role for other Cl⁻ channels *(58)*.

AGONISTS ENHANCING FUNCTION OF CALCIUM-ACTIVATED Cl⁻ CHANNELS

Many agonists have been shown to increase intracellular Ca^{2+} concentration $[Ca^{2+}]i$ including; ATP, uridine triphosphate (UTP), P2Y2 agonists, fibroblast growth factor-β (βFGF), neuropeptide Y, atrial natriuretic peptide (ANP), as well as hypotonicity, and therefore these factors are likely to activate bestrophin or other Ca^{2+}-activated Cl⁻ channels (see section on regulation of RPE transport by hormones and agonists).

Na^+/Ca^{2+} EXCHANGER FUNCTIONS IN RPE TO REDUCE $[Ca^{2+}]_I$ AND ANTAGONIZES CALCIUM-ACTIVATED K⁺ AND Cl⁻ CHANNELS

Among the known families of transport proteins that can catalyze net Ca^{2+} efflux across the plasma membrane, is the Na^+/Ca^{2+} exchanger (also known as the SL8 family). This antiporter was first discovered in heart (termed the cardiac antiporter) and exchanges three Na^+ ions for one Ca^{2+} ion (i.e., it is electrogeneic). It is different from the retinal rod exchanger, which belongs to a different solute-carrier family known as SLC24 and works via a different stoichiometry (four Na^+ ions in exchange for one Ca^{2+} ion and one K^+ ion). In the forward mode, it operates as a calcium pump mediating Ca^{2+} efflux and Na^+ influx. However, it can operate in the reverse mode and mediate Ca^{2+} influx and Na^+ efflux *(55,56)*. The cardiac Na^+/Ca^{2+} exchanger has been detected in human RPE *(67,68)*. Loeffler et al. *(67)* reported variable staining intensities in RPE among the different specimens with reactive RPE cells revealing the most intense labeling, which leads to the assumption that there is a link between some RPE-related diseases and abnormal expression of the Na^+/Ca^{2+} exchanger.

ELECTRONEUTRAL CATION-Cl⁻ COTRANSPORTERS (SLC12)

This family of solute carriers transport a cation (Na^+ or K^+) and a Cl⁻ ion in a one-to-one stoichiometry and therefore, ion translocation produces no change in transmembrane potential (that observation explains their being named electroneutral cation-Cl⁻ coupled cotransporters) (for an excellent review see 69; for the latest classification and nomenclature of different solute carriers, see the review*108)*. There are at least seven members including one gene encoding the thiazide-sensitive Na^+-Cl⁻ cotransporter, two genes encoding bumetanide-sensitive Na^+-K^+-$2Cl^-$ cotransporters, and four genes encoding dihydroindenyloxyalkanoic acid (DIOA)-sensitive K^+-Cl⁻ cotransporters. Higher concentrations of DIOA also inhibit the Na^+-K^+-$2Cl^-$. The K^+-Cl⁻ cotransporters, while inhibited by DIOA, are activated by swelling and *N*-ethylmaleimide (NEM) *(70)*. Although this family of cotransporters are sometimes referred to as secondary transporters, because ion translocation is not dependent on ATP hydrolysis but rather on gradients generated by the Na^+, K^+-ATPase pump, they play important roles in regulating ion absorption, secretion, cellular volume regulation, and adjusting $[Cl^-]_i$. While changes in Na^+ and K^+ by SLC12 cotransporters are quickly corrected by Na^+, K^+-ATPase, that is not the case with Cl^- and therefore their activation means the net accumulation of Cl^- intracellularly or extracellularly.

Hu et al. *(71)* have shown that bumetanide hyperpolarized and reduced net Cl^- flux by 83% by reducing the unidirectional apical-to-basal Cl^- flux. While some epithelial cells possess such cotransporter activity at both apical and basal membranes, basal bumetanide had no effect on electrical parameters in human RPE, suggesting lack of basal expression in human RPE *(71)*. Epinephrine activation of apical membrane α1-adrenergic receptors increased $[Ca^{2+}]_i$ and activated the bumetanide-sensitive Na^+-K^+-$2Cl^-$ uptake at the apical membrane and K^+/Cl^- efflux at the basolateral membrane resulting in an increased fluid absorption across bovine RPE *(72)*. Evidence for the presence of a K^+/Cl^+ contransporter in human RPE was provided by Kennedy et al. *(73)*. A bumetanide-insensitive Rb^+ influx was activated by hypotonic challenge, as well as by treatment with NEM; the presence of known activators of the K^+/Cl^- cotransporter were reported in human RPE. Unfortunately, there are no available studies into thiazide-sensitive Na^+/Cl^- cotransporter presence in RPE.

REGULATION OF $[Cl^-]_I$ AND THE RELATIONSHIP WITH BICARBONATE TRANSPORTERS

An understanding of the regulation of Cl^- concentration cannot be complete without understanding the bicarbonate buffering system that plays a crucial role in maintaining intracellular and extracellular pH, and cellular volume, in RPE. Bicarbonate concentration in the blood is kept at around 25 mM by the kidney *(74)*. However, bicarbonate cannot move across the plasma membrane without selective ubiquitous proteins known as bicarbonate transporters. Bicarbonate transporter proteins are encoded by at least two unrelated gene families; SLC4 and SLC26 (sulfate permease). One key difference between the SLC26 and SLC4 families is the transport of Na^+ by most of the SLC4 transporters. All the SLC26 subtypes (at least 10) transport a variety of monovalent anions (e.g., Cl^-, I, OH^-, HCO_3^-, and $HCOO^-$) and divalent anions (e.g., sulfate and oxalate). The SLC26 family subtypes have yet to be characterized in RPE and there are no available studies on their potential significance (see reviews *75–80*).

The SLC4 family shares the common property of transporting bases (HCO_3^-/CO_3^{2-}), but members differ in their ability to mediate the concomitant transport of Na^+ and or Cl^-. The SLC4 family is divided into three classes: (i) Na^+-independent Cl^-/HCO_3^- anion exchangers (AEs) mediating electroneutral exchanges (examples are: AE1, 2, 3, 4); (ii) Na^+-driven Cl^-/HCO_3^- anion exchangers mediating exchange of Cl^- for Na^+ and a base (HCO_3^-, CO_3^{2-}), and (iii) Na^+-HCO_3^- cotransporters (NBCs) mediating cotransport of Na^+ and a base *(75)*. Na^+-HCO_3^- cotransporters are inhibited by stilbenes [4,4′-diisothiocyanostilbene-2,2-disulfonic acid (DIDS)]. The NBC is a versatile transporter that can operate in a forward mode (secretion of Na^+ and HCO_3^-) or reverse mode (reabsorption of Na^+ and HCO_3^-). The NBC can be either electroneutral, mediating cotransport of an equal ratio of anions (HCO_3^-/CO_3^{2-}) to positive cations (Na^+) (e.g., NBC2 and NBC3), or electrogenic, where it mediates cotransport of more negative anions (HCO_3^-/CO_3^{2-}) than positive cations (Na^+) (e.g., NBC1 and NBC4) *(74,75,78,79)*. There are conflicting reports regarding the stoichiometry of the Na^+ and HCO_3^- being transported (1:3 vs. 1:2). Such controversy can be due to species differences. More importantly, phosphorylation by PKA has been shown to shift the stoichiometry from 3:1 to 2:1 *(80)*. Therefore, the phosphorylation status of the bicarbonate transporter determines the stoichiometry of its transport. Bok et al. *(81)* have reported the expression of pancreatic-type NBC (pNBC1) at the apical membranes in rat RPE. Interestingly, mutations in NBC3 cause Usher syndrome in mice and loss of photoreceptors, but not RPE cells *(82)*.

BICARBONATE TRANSPORTERS IN RETINAL PIGMENT EPITHELIUM AND FUNCTIONAL COUPLING TO OTHER RETINAL CELL LAYERS

The subretinal, as well as intracellular, pH of RPE is linked to the transepithelial transport of HCO_3^-. Directional HCO_3^- transport is dependent on the apical transmembrane potential and intracellular HCO_3^- concentration, allowing the HCO_3^- transport system to regulate the transport direction in response to pH changes. Under higher intracellular and subretinal pH, HCO_3^- is taken into the RPE cells by the Na^+/HCO_3^- cotransporter in the apical membranes, and leaves the cell through the basolateral membrane in exchange for Cl^- in a manner mediated by the Cl^-/HCO_3^- exchanger (subretinal to choroid directed HCO_3^- transport). At low intracellular and subretinal pH, HCO_3^- is taken up by the Cl^-/HCO_3^- exchanger at the basolateral membrane and leaves the cell through the Na^+/HCO_3^- cotransporter in the apical membrane (HCO_3^- transport from the choroid to the subretinal space) *(84)*. In addition, Segawa et al. *(85)* have shown that elevated bicarbonate concentrations, applied basally, depolarized the apical membrane of cat retinal pigment epithelium-choroid tissue and decreased the potential across the RPE. In contrast, similar doses applied apically hyperpolarized the apical membrane of the RPE and resulted in an increase in the potential across RPE, suggesting the importance of bicarbonate fluxes across retinal pigment epithelium *(85)*. Furthermore, increasing the apical extracellular K^+ to levels mimicking fluctuations during the light-dark cycle have resulted in alkalinization of RPE, in a response that was attenuated by either DIDS or removal of bicarbonate, which further supports a close relationship between photoreceptor function and changes in intracellular pH in RPE *(86)*.

RPE has an electrogeneic Na^+/HCO_3^- cotransporter that operates by secretion of three Na^+ ions and one HCO_3^- ion *(87,88)*, and NBC1 is specifically expressed in the human choriocapillaris *(89)*. Bok et al. *(81)* have reported the expression of pNBC1 at the apical membranes in rat RPE. Lin et al. *(89)* have also shown that apical membranes of frog RPE contain a DIDS-sensitive electrogenic Na^+/HCO_3^- cotransporter and suggested that such a mechanism accounts for around 80% of acid removal in frog RPE. Similarly, Kenyon et al. *(88)* have shown that explants of bovine retinal pigment epithelium-choroid possess DIDS-sensitive electrogenic Na^+/HCO_3^- cotransporters at both apical and basal membranes. Both acid and alkali recovery were HCO_3^- dependent and can be blocked by DIDS applied on either apical or basal membranes. Basal Cl^- removal, or addition of basal HCO_3^-, caused HCO_3^- and Cl-dependent alkalinizations, respectively.

BICARBONATE TRANSPORTERS CANNOT FUNCTION PROPERLY WITHOUT FUNCTIONAL CARBONIC ANHYDRASES (CAS)

While bicarbonate transporters can be regulated through phosphorylation by protein kinases and changes in ion and pH concentrations, they lose their transport function if deprived of functional carbonic anhydrases. At least two reports have provided evidence for a direct association between bicarbonate transporters and isoforms of CA. Firstly, AE1 has been shown to bind carbonic anhydrase CAII, and the inhibition of CAII inhibited AE1 transport *(90,91)*. The same acidic cytosolic domain in AE1 is present in AE2, AE3, and AE4, suggesting a similar potential interaction *(90–93)*. CAII has been detected in the cytoplasm of rabbit RPE *(94)* and may associate similarly with AE. Furthermore, Alvarez et al. *(95)* have shown that CA IV forms a complex with NBC1. Humans with miss-sense mutations in CA IV suffer from an autosomal dominant rod-cone dystrophy that is characterized by a dysfunctional NBC1 cotransporter *(89)*. In fact, the inability of mutant CA IV to form a complex with NBC1 results in defective function. Therefore, impaired CA II or CA IV can lead to impaired bicarbonate flux. It appears that the activity of CA II changes with the metabolic status of cells. Immuno-histochemical studies have shown that healthy RPE from rabbits stains intensely with a peroxidase-linked antibody specific for human CAII, whereas injured RPE stained less intensely *(94)*. Additional evidence for irregular CA activities is documented in the Royal College of Surgeons' (RCS) rats. These RCS rats, characterized by defective phagocytosis of rod outer segments by the retinal pigmented epithelium (RPE), also show increased and redistributed CA activity. While in the RPE of control rats, the apical membranes were most intensely stained, RCS rats showed intense labeling of the basolateral membrane, and surprisingly the adjacent endothelial cells of the choriocapillaris and of retinal capillaries developed staining for CA activity *(96)*.

CARBONIC ANHYDRASES

Carbonic anhydrases are zinc-containing, ubiquitous enzymes that catalyze the reversible hydration/dehydartion of CO_2 in the following manner.

$$CO_2 + H_2O \leftrightarrow HCO_3^- + H^+$$

Such a reaction can either generate HCO_3^- for transport or may consume HCO_3^- that has already been transported: in other words, CAs are suppliers or consumers of HCO_3^-.

There are at least 16 different mammalian isoenzymes to-date. CAs can be classified into cytosolic and plasma membrane isoforms; CA I, II, III, V, VII and XIII are cytosolic isoforms, whereas CA IV, IX, XII, XIV, and XV are membrane bound. Carbonic anhydrases IX, XII and XIV have transmembrane domains whereas CA IV and XV are anchored to membranes via glycosylphosphatidylinositol (GPI). Although CA VIII, X, and XI are termed CA-related enzymes, their exact function is yet to be determined as they lack enzymatic activities and cannot carry out enzymatic hydration of CO_2. Finally, CA V is a unique mitochondrial isoform, whereas CA VI is secreted in the saliva and milk *(97–99)*. The levels of CA can be regulated at the point of synthesis, as well as activity.

For example, synthesis of CAII is activated by PKC *(100)*, whereas PKA enhances the activity of CA II by phopshorylation *(101)*. Other known natural activators of CAs include amines and amino-acids, e.g., dopamine, noradrenaline, adrelanine, histamine, histidine, imidazoles, and phenyalanine *(98)*.

CARBONIC ANHYDRASE INHIBITORS OF THERAPEUTIC VALUE IN RETINAL PIGMENT EPITHELIUM-RELATED DISEASE

Macular edema caused by damage to the blood-retina barrier has been successfully treated by administration of inhibitors of carbonic anhydrase. Inhibition of cytosolic CA by acetazolamide induces acidification of the subretinal space, leading to an increase in epithelial Cl⁻ transport into the choroid, which drives the elimination of water from the retina and therefore, increases retinal adhesiveness of RPE. Such treatment has been suggested to reduce macular edema, especially in patients suffering from cystoid macular edema due to either retinitis pigmentosa or uveitis *(102,103)*. However, it appears that the inhibition of membrane-bound CA is also sufficient to enhance subretinal fluid absorption and retinal adhesiveness. Wolfenberger et al. *(103)* have shown the presence of a CA activity on the apical and basolateral cell membranes of human RPE as determined by Hansson's technique. Immunohistochemical studies have detected the presence of membrane-bound CA IV at apical RPE membrane. Another membrane-bound CA-XIV isoenzyme was detected on both the apical and basal membranes of mouse RPE *(105,106)*. Benzolamide, a membrane-impermeable hydrophilic CA inhibitor, induced acidification in perfused chick retina RPE-choroid preparation. Its effect mimicked the membrane-permeable acetazolamide CA inhibitor and suggests that inhibition of membrane-bound CA is sufficient to decrease subretinal pH as well as volume *(107)*. A similar therapeutic finding was reproduced by the same group in rabbits *(102)*.

NA⁺/H⁺ (NH) ANTIPORTERS

A key transporter that regulates intracellular pH is the Na⁺/H⁺ (NH) antiporter. Unlike the Na⁺/Ca²⁺ exchanger, the NH antiporter operates with one-to-one stoichiometry and therefore is electroneutral (it is also known as the SL9 family of solute carriers). There are at least eight different isoforms of Na/H exchanger *(108)*. The NH antiporter is localized at the apical membrane of dogfish, frog and bovine RPE *(86,88,109)*. Similar to all Na⁺/H⁺ antiporters, it was blocked by amiloride, a potent inhibitor of Na⁺/H⁺ antiporters. It normally operates to extrude H⁺ while permeates Na⁺. The identity of isoforms of the Na⁺/H⁺ antiporter subtypes in RPE are yet to be determined.

MONOCARBOXYLATE TRANSPORTERS (MCT)

Photoreceptor activity is associated with high production of lactic acid. Adler et al. *(110)* report that lactate concentrations can reach up to 13 mM in bovine extracellular interphotoreceptor matrix (IPM, a proteoglycan-rich extracellular matrix that lies in the subretinal space and plays a role in nutrition of the photoreceptors), around

4 mM near the apical RPE membranes, and up to 19 mM in the RPE. These results are similar to other mammalian retinas; 18 mM in rabbit retina and 22 to 33 mM in rat and cat retinas, respectively. Therefore, such high lactate concentrations require efficient removal of lactate to the blood. Interestingly, lactate transport from photoreceptors has been shown to be stereospecific. Lactate can be transported by special proteins known as monocarboxylate transporters (MCTs, 14 subtypes have been cloned and are known as the SL16 family of solute carriers). Monocarboxylate transporters are a family of highly homologous membrane proteins that mediate the one-to-one transport of a proton (supplied to the subretinal space by the apically located Na^+/H^+ exchanger) and a lactate ion. Interestingly, MCT1, MCT3, and MCT4 form heterodimeric complexes with the cell-surface glycoprotein CD147 and exhibit tissue-specific polarized distributions. This is evident based on studies in CD147 knock-out mice where MCT1, MCT3 and MCT4 are no longer expressed in RPE *(111–113)*. In the RPE, MCT1/CD147 is polarized to the apical membrane and MCT3/CD147 to the basolateral membrane in humans *(111,114,115)*. While lactate in the apical bath caused intracellular acidification, basal lactate caused intracellular alkalinization that was dependent on the presence of Na^+ *(116)*. ARPE19 cells have at least three MCTs (MCT1, MCT4, and MCT8, with MCT1 only present at the apical membranes) *(117)*. In addition, lactic acid can be removed from the subretinal space by the Na^+-dependent organic-anion transporting protein (Oatp, which belongs to solute carrier family SLC21). Oatp2 and 3 were confirmed in rat RPE by Northern blot, RT-PCR and Western blot, but no Oatp1 was not found. Immunohistochemistry revealed that Oatp2 was predominantly expressed at the apical surface of the RPE *(118,119)*. Known inhibitors of Oatps are sulfobromophthalein, probenecid and sulfinpyrazone *(118–120)*. Interestingly, Oatps can allow impermeable cAMP in RPE to mediate pigment-granule aggregation *(121)*. Finally, lactate can be also transported via aquaporin-9 (AQP9), which has been discovered by our laboratory in ARPE-19 cells.

AQP1, AQP9 AND EDEMA

Development of edema in the subretinal space is detrimental to retina and often associated with retinal degeneration. Retinal edema results from swelling caused by fluid leaking from damaged blood-retinal barrier. Therefore, the activation of AQPs in RPE is needed to eliminate water that cannot go through paracellular spaces, but only via intracellular routes. Such a function is crucial for the survival of the retina *(1)*. Although AQP1 has been cloned from a cDNA library of human RPE, three separate studies failed, using various immunochemical techniques, to detect AQP1 protein in adult human or rat RPE preparations *(122–124)*. However, Stamer et al. *(125)* reported finding AQP1 in human RPE after enrichment of membrane proteins and it was shown that the transepithelial transport of water was facilitated by the functional presence of AQP1. However, our laboratory also failed to detect AQP1 in ARPE-19 cells (unpublished observations). AQP1 expression may be there, but at too low level to be detected, which raises doubts on whether AQP1 is the sole contributor to water elimination or whether there are other aquaporins yet to be discovered in RPE. While our laboratory failed to detect AQP1 in ARPE-19 cells, we discovered the expression of AQP9, a novel aquaporin, in ARPE-19 cells (Fig. 2); this

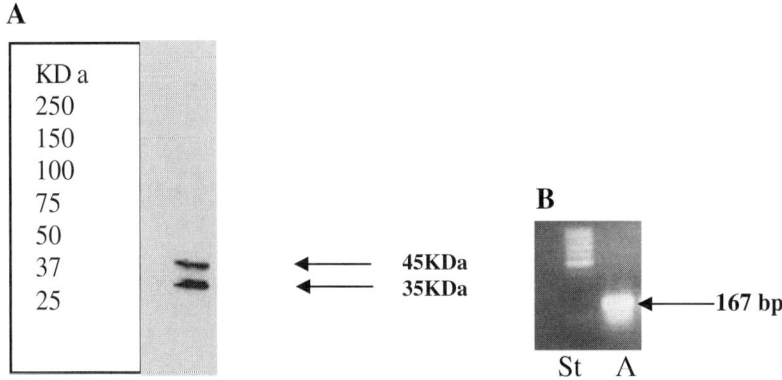

Fig. 2. Evidence for the expression of aquaporin-9 (AQP9) in ARPE-19 cells by Western blotting and reverse transcriptase polymerase chain reaction (RT-PCR). Plasma membranes were purified and processed for Western blotting as described by Dibas et al. 2005 *(126)*. **(A)**, a doublet of 37 and 47 KDa is seen in ARPE-19 cells. Total RNA was extracted with Trizol (Life Technology) as described by the manufacturer. In this study, 5 μg of total RNA was reverse transcribed using the iscript kit (Bio-Rad) according to the manufacturer's instructions. Control RT-PCR reactions were performed in the absence of cDNA templates. Polymerase chain reaction (PCR) products were run on agarose gels and bands were cut and sequenced to verify identity. See Tables 2 and 3 for primer sequences and conditions. **(B)** AQP9 mRNA is expressed in ARPE-19 cells (St: DNA ladder, A: ARPE-19).

Table 2 Gene-specific forward (sense) and reverse (anti-sense) primers

Primer name	Forward primer (5→3′)	Reverse primer (5→3′)	Product size (bp)
Aquaporin 9 (AQP9)	AGCCACCTCT-GGTCTTGCTA	ATGTAGAGCAT-CCCCTGGTG	167
β-actin	TGTGATGGTGG-GAATGGGTCAG	TTTGATGTCACG-CACGATTTCC	514

Table 3 Primer optimization

Primer name	Denaturation	Annealing	Extension	Number of cycles
AQP9	96°C 15 sec	60°C 60 sec	72°C 90 sec	40
β-actin	95°C 60 sec	60°C 60 sec	72°C 120 sec	40

observation was confirmed by Western blotting and PCR. Interestingly, AQP9 expression is up-regulated under hypoxic and hypotonic insults (Fig. 3). AQP9 possesses general features of a water channel, but in addition it is permeable to a wide variety of uncharged solutes such as lactate, β-hydroxybutyrate, glycerol, purines, pyrimidines, urea, mannitol, and sorbitol. Therefore, AQP9 may play a significant role in water and cellular transport in RPE cells. The membrane distribution of AQP9 is currently under investigation by our

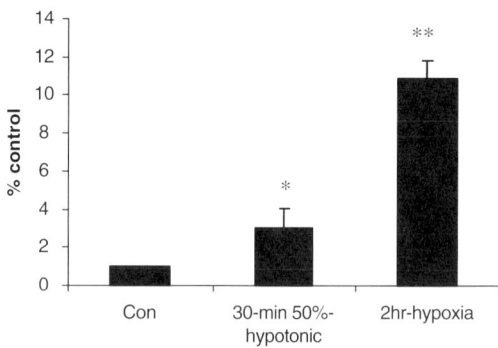

Fig. 3. Effect of hypoxia and hypotonic shock on aquaporin 9 (AQP9) expression in ARPE-19 cells. This test was performed as described by Yamamoto et al 2001 *(127)*. After exposure to hypoxia for the indicated periods (from one to three hours), cells were processed using reverse-transcriptase polymerase chain reaction (RT-PCR). To evaluate the effect of hypotonic shock on AQP9 mRNA, cells were exposed to media that was diluted 1:1 with sterile water. β-actin was used as a housekeeping gene. See Tables 2 and 3 for primer sequences and conditions. Hypotonicity for 30 minutes increased AQP9 by 3.1 ± 1 fold ($^*p < 0.005$ vs. control). Longer times were also tested and cells showed continued up-regulation for up to one to two hours before the mRNA appeared down-regulated (data not shown). Unsurprisingly, hypoxia increased AQP9 expression in ARPE-19 cells by 10.9 ± 0.9 fold ($^{**}p < 0.005$ vs. control).

laboratory. More interestingly, in addition to AQP9 expression at the plasma membrane, AQP9 is enriched in the mitochondria of ARPE-19 cells, human nonpigmented epithelial (HNPE) cells, human optic nerve head astrocytes and retinal ganglion cells (manuscript in preparation). AQP9 transport of lactate is likely to be key in the elimination of the high concentrations that accumulate in RPE cells (around 19 mM).

BUFFERING OF IONS IN THE SUBRETINAL SPACE IN THE LIGHT-DARK CYCLE AND EXPLANATION OF ELECTRORETINOGRAM (ERG) C-WAVE AND DELAYED HYPERPOLARIZATION

Stimulation of photoreceptors by light reduces the dark current. The dark current is the influx of Na^+ via open cyclic guanidine monophosphate (cGMP)-gated cation channels in the outer segments in parallel to an efflux of K^+, via K^+ channels in the inner segment. This leads to a reduction in the subretinal K^+ concentration from around 5 to 2 mM, causing hyperpolarization of the apical membranes of RPE. Such an effect activates the inwardly rectifying K^+ channels and causes a change in the ratio of apical and basolateral K^+ conductance, leading to secretion of K^+ in the subretinal space to replenish the light-induced decrease of subretinal K^+. Furthermore, the $Na^+/K^+/2Cl^-$ cotransporter works in the reverse mode supplying additional K^+ to the subretinal space. Thus, the direction of epithelial transport

Regulation of Transport in the RPE

Fig. 4. Buffering of ions in the subretinal space in the light–dark cycle and explanation of electroretinogram (ERG) c-wave and delayed hyperpolarization.

of K^+ from the subretinal space to the choroidal site is now reversed. Reducing intracellular Cl^- also inhibits the Cl^-/HCO_3^- basolateral exchanger that normally drives Cl^- outside the RPE, which results in the subsequent hyperpolarization of the basolateral membranes of RPE. Such ion fluctuations can be monitored in the electroretinogram (ERG) as the c-wave and delayed hyperpolarization, where the c-wave results from the hyperpolarization of the apical RPE cell membrane and the delayed hyperpolarization results from hyperpolarization of the basolateral membrane (Fig. 4).

OTHER IMPORTANT RETINAL PIGMENT EPITHELIUM TRANSPORT MECHANISMS

Among key nutrients needed for the RPE cells and photoreceptors is vitamin C (ascorbate), which plays a key role in detoxifying free radicals in the retina. Vitamin C concentrations are more than 10 times that of plasma. Khatami et al. *(128)* have shown that the transport of ascorbate by primary cultures of cat RPE cells was Na^+ dependent, and the protein responsible appeared to belong to the SLC23 solute carrier family. In addition, human RPE cells have at least two Na^+-myoinositol distinct cotransporters *(129)*. As important, is the transport and storage of iron in RPE, which shows strong immunoreactivity for ferritin and transferrin and appears to be the main synthesis site of transferrin receptors in rodents *(130,131)*. Furthermore, human RPE plays an important role in providing a sufficient supply of riboflavin (vitamin B2, which is essential for many cellular pathways) to the retina. Vitamin B2 uptake is energy-, temperature- and pH- dependent, but not Na^+-dependent. Also, vitamin B2 uptake is inhibited by DIDS, amiloride, and the sulfhydryl group inhibitor p-chloromercuriphenyl-sulphonate (p-CMPS) *(132)*. Finally, one of the most important cofactors needed by retinal cells, and all cells, is folate, which cannot be synthesized by cells; folate is essential for DNA, RNA, and protein synthesis. The RPE has two unique, polarized folate transporters. The folate receptor α (FRα) is located at the basal membrane, whereas the reduced-folate transporter (RFT-1) is located at the apical membrane *(133,134)*. Therefore, folate is taken from the blood by the basolateral FRα and secreted to retinal cells by αRFT-1 at the apical membrane *(135,136)*. More importantly, this vectorial transport appears to be compromised in diabetic animals. Cultured RPE in the presence of high glucose shows a reduction in FRT-1 mRNA and protein levels *(137)*.

REGULATION OF RETINAL PIGMENT EPITHELIUM TRANSPORT BY HORMONES AND AGONISTS

RPE is unique in its regulation of ion transport by autocrine and exocrine factors. For example, growth factors, pyrimidines and changes in tonicity could trigger ATP release into the subretinal space. Apical ATP (or UTP) increases $[Ca^{2+}]_i$ and basolateral membrane Cl^- conductance, while decreased apical membrane K^+ conductance, a response that is blocked by the P2-purinoceptor blockers suramin and DIDS *(138)*. Activation of P2Y(2) receptors in vivo directly stimulates rabbit RPE active transport as evidenced using the P2Y(2) receptor agonist, INS542 *(139)*. Maminishkis et al. *(140)* have shown that P2Y(2) agonists (e.g., INS37217) increase $[Ca^{2+}]_i$ and stimulate RPE fluid transport in vitro, and retinal reattachment in rats, and therefore may be therapeutically useful for treating retinal disorders associated with fluid accumulation in the subretinal space. While RPE cells can release apical ATP, they also release ATP-hydrolyzing enzymes, and ADP has been detected in ARPE-19 media *(141)*. Another key regulator of RPE transport is neuropeptide Y (NPY). Apical application of NPY increases the transepithelial potential in RPE-choroid preparations *(143)*. Similarly, serotonin increases Cl^- transport and transepithelial transport *(144)*. Furthermore, Kawahara et al. *(145)* have shown that the adenosine agonist, 2–5 -N-ethylcarboxamidoadenosine (NECA), at high doses, enhances the reabsorption of subretinal fluid compared with phosphate-buffered saline (PBS). Finally, in cultured chick and human RPE cells, vasoactive intestinal peptide (VIP) is the most effective stimulator of the cAMP signaling *(146)*. Fujiseki et al. *(147)* have shown that the natriuretic peptide (NP) receptor (NPR-A and NPR-B) mRNAs were present in rabbit RPE cells with NPR-B mRNA more than 10-fold higher than that of NPR-A mRNA. While the expression of NPR-A mRNA was not affected by treatments that may change subretinal fluid transport, NPR-B mRNA was inhibited by transmitters involved in light- and dark-adaptation (e.g., dopamine and melatonin). Expression of NPR-B mRNA was also suppressed by platelet-derived growth factor and transforming growth factor-β. Mikami et al. *(148)* have shown that ANP increases $[Cl^-]_i$ in rabbit RPE cells via activation of cGMP-dependent kinase of the $Na^+/K^+/2Cl^-$ cotransporter.

UNANSWERED QUESTIONS REGARDING RETINAL PIGMENT EPITHELIUM TRANSPORT

While RPE transporters have received great attention, there are still many unanswered questions. The identity of the Na^+/H^+ antiporter subtype has yet to be determined, since different isoforms exist with distinct regulation of function. Does RPE possess Na^+/Cl^- cotransporter activity and where it is localized? What isoform of the K^+/Cl^- cotransporter exists in RPE and where? SLC26 cotransporter family members have yet to be studied in RPE and there is doubt as to the identity of calcium-activated Cl (CACL) channels. The role of AQP9 in RPE transport is yet to be characterized. Perhaps the greatest challenge in linking such transporters to RPE function is the lack of selective inhibitors against RPE transporters and channels.

ACKNOWLEDGEMENTS

This work was supported in part by a grant from Texas Higher Education Coordinating Board for T. Yorio, and a grant from the National Glaucoma Foundation to A. Dibas.

REFERENCES

1. Marmor MF. Mechanisms of fluid accumulation in retinal edema. Doc Ophthalmol 1999;97:239–249.
2. Finnemann SC, Bonilha VL, Marmorstein AD, Rodriguez-Boulan E. Phagocytosis of rod outer segments by retinal pigment epithelial cells requires alpha(v)beta5 integrin for binding but not for internalization. Proc Natl Acad Sci USA 1997;94:12932–12937.
3. Boulton M and Dayhaw-Baker P. The role of retinal pigment epithelium: topographical variation and ageing changes. Eye 2001;15:384–389.
4. Tanihara H, Inatani M, Honda Y. Growth factors and their receptors in the retina and pigment epithelium. Prog Retina Eye Res 1997;16:271–301.
5. Baehr W, Wu SM, Bird AC, Palczewski K. The retinoid cycle and retina disease. Vis Res 2003;43:2957–2958.
6. Pfeffer BA, Clark VM, Flannery JG, Bok D. Membrane receptors for retinal-binding protein in cultured human retinal pigment epithelium. Invest Ophthalmol Vis Sci 1986;27:1031–1040.
7. Bazan NG, Rodr Bazan NG, Bazan NG, Guez de Turco EB, Gordon WC. Pathways for the uptake and conservation of docosahexoenoic acid in photoreceptors and synapses: biochemical and autoradiographic studies. Can J Physiol Pharmacol 1993;71:690–698.
8. Miller SS and Steinberg RH. Active transport of ions across frog retinal pigment epithelium. Exp Eye Res 1977;25:235–248.
9. Gallemore RP, Hughes BA, Miller SS. Light-induced responses of the retinal pigment epithelium. In: Marmor MF and Wolfensberger TJ, eds, The Retinal Pigment Epithelium. New York: Oxford University Press; 1998. p.175–198.
10. Giebel SJ, Menicucci G, McGuire PG, Das A. Matrix metalloproteinases in early diabetic retinopathy and their role in alteration of the blood-retinal barrier. Lab Invest 2005;85:597–607.
11. Hageman GS and Mullins RF. Molecular composition of drusen as related to substructural phenotype. Mol Vis 1999;5:28.
12. Stalman P and Himpens B. Effect of increasing glucose concentrations and protein phosphorylation on intracellular communication in cultured rat retinal pigment epithelial cells. Invest Ophthalmol Vis Res 1997;38:1598–1609.
13. Rajasekaran SA, Hu J, Gopal J, Gallemore R, Ryazantsev S, Bok D, Rajasekaran AK. Na,K-ATPase inhibition alters tight junction structure and permeability in human retinal pigment epithelial cells. Am J Physiol Cell Physiol 2003;284:C1497–C1507.
14. Hu JG, Gallemore RP, Bok D, Lee AY, Frambach DA. Localization of Na/K ATPase on cultured human retinal pigment epithelium. Invest Ophthalmol Vis Sci 1994;35:3582–3588.
15. Sugasawa K, Deguchi J, Okami T, Yamamoto A, Omori K, Uyama M, Tashiro Y. Immunocytochemical analyses of distributions of Na, K-ATPase and GLUT1, insulin and transferrin receptors in the developing retinal pigment epithelial cells. Cell Struct Funct 1994;19:21–28.
16. Frambach DA, Valentine JL, Weiter JJ. Precocious retinal adhesion is affected by furosemide and ouabain. Curr Eye Res 1989;8:553–556.
17. Crider JY, Yorio T, Sharif NA, Griffin BW. The effects of elevated glucose on Na+/K(+)-ATPase of cultured bovine retinal pigment epithelial cells measured by a new nonradioactive rubidium uptake assay. J Ocul Pharmacol Ther 1997;13:337–352.
18. Chandy KG and Gutman GA. Nomenclature for mammalian potassium channel genes. Trends in Pharmacol Sci 1991;14:434–435.
19. Yuan LL and Chen X. Diversity of potassium channels in neuronal dendrites. Prog Neurobiol 2006;78:374–389.

20. Ghatta S, Nimmagadda D, Xu X, O'Rourke ST. Large-conductance, calcium-activated potassium channels: structural and functional implications. Pharmacol Ther 2006;110:103–116.
21. Doupnik CA, Davidson N, Lester HA. The inward rectifier potassium channel family. Curr Opin Neurobiol 1995;5:268–277.
22. Lu Z. Mechanism of rectification in inward-rectifier K+ channels. Annu Rev Physiol 2004;66:103–129.
23. Nichols CG. K_{ATP} channels as molecular sensors of cellular metabolism. Nature 2006;440:470–476.
24. Pinto LH and Klupmm DJ. Localization of potassium channels in the retina. Prog Ret Eye Res 1998;17:207–230.
25. Wollmann G, Lenzner S, Berger W, Rosenthal R, Karl MO, Strauss O. Voltage-dependent ion channels in the mouse RPE: comparison with Norrie disease mice. Vision Res 2006;46:688–698.
26. Takahira M and Hughes BA. Isolated bovine retinal pigment epithelial cells express delayed rectifier type and M-type K+ currents. Am J Physiol 1997;273:C790–C803.
27. Claydon TW, Boyett MR, Sivaprasadarao A, Ishii K, Owen JM, O'Beirne HA, Leach R, Komukai K, Orchard CH. J Physiol 2000;526:253–264.
28. Tao Q and Kelly ME. Calcium-activated potassium current in cultured rabbit retinal pigment epithelial cells. Curr Eye Res 1996;15:237–246.
29. Sheu SJ and Wu SN. Mechanism of inhibitory actions of oxidizing agents on calcium-activated potassium current in cultured pigment epithelial cells of the human retina. Invest Ophthalmol Vis Sci 2003;44:1237–1244.
30. Hughes BA and Takahira M. Inwardly rectifying K+ currents in isolated human retinal pigment epithelial cells. Invest Ophthalmol Vis Sci 1996;37:1125–1139.
31. Strauss O, Richard G, Wienrich M. Voltage-dependent potassium currents in cultured human retinal pigment epithelial cells. Biochem Biophys Res Commun 1993;191:775–781.
32. Tao Q, Rafuse PE, Kelly ME. Potassium currents in cultured rabbit retinal pigment epithelial cells. J Membr Biol 1994;141:123–138.
33. Yang D, Pan A, Swaminathan A, Kumar G, Hughes BA. Expression and localization of the inwardly rectifying potassium channel Kir7.1 in native bovine retinal pigment epithelium. Invest Ophthalmol Vis Sci 2003;44:3178–3185.
34. Kusaka S, Horio Y, Fujita A, Matsushita K, Inanobe A, Gotow T, Uchiyama Y, Tano Y, Kurachi Y. Expression and polarized distribution of an inwardly rectifying K+ channel, Kir4.1, in rat retinal pigment epithelium. J Physiol 1999;520:373–381.
35. Yuan Y, Shimura M, Hughes BA. Regulation of inwardly rectifying K+ channels in retinal pigment epithelial cells by intracellular pH. J Physiol 2003;549(Pt 2):429–438.
36. Hughes BA and Takahira M. ATP-dependent regulation of inwardly rectifying K+ current in bovine retinal pigment epithelial cells. Am J Physiol 1998;275:C1372–C1383.
37. Ettaiche M, Heurteaux C, Blondeau N, Borsotto M, Tinel N, Lazdunski M. ATP-sensitive potassium channels (K_{ATP}) in retina: a key role for delayed ischemic tolerance. Brain Res 2001;890:118–129.
38. Botchkin LM and Matthews G. Voltage-dependent sodium channels develop in rat retinal pigment epithelium cells in culture. Proc Natl Acad Sci USA 1994;91:4564–4568.
39. Eggermont J. Calcium-activated chloride channels. Unknown, unsolved. Proc Am Thorac Soc 2004;1:22–27.
40. Jentsch TJ, Valentin S, Weinreich F, Zdebik AA. Molecular structure and physiological function of chloride channels. Physiol Rev 2001;82:503–568.
41. Weng TX, Godley BF, Jin GF, Mangini NJ, Kennedy BG, Yu AS, Wills NK. Oxidant and antioxidant modulation of chloride channels expressed in human retinal pigment epithelium. Am J Physiol Cell Physiol 2002;283:C839–C849.

42. Wills NK, Weng T, Mo L, Hellmich HL, Yu A, Wang T, Buchheit S, Godley BF. Chloride channel expression in cultured human fetal RPE cells: response to oxidative stress. Invest Ophthalmol Vis Sci 2000;41:247–255.
43. Bosl MR, Stein V, Hubner C, Zdebik AA, Jordt SE, Mukhopadhyay AK, Davidoff MS, Holstein AF, Jentsch TJ. Embo J 2001;20:1289–1299.
44. Kurz LL, Klink H, Jacob I, Kuchenbecker M, Benz S, Lehman-Horn F, Rudel R. Identification of three cysteines as targets for the Zn^{2+} blockade of the human skeletal muscle chloride channel. J Biol Chem 1999;274:11786–11692.
45. Silverstone BZ. Effects of zinc and copper metabolism in highly myopic patient. Ciba Found Symp 1990;155:210–217.
46. Anderson MP, Berger HA, Rich DP, Gregory RJ, Smith AE, Welsh MJ. Nucleoside triphosphate are required to open the CFTR chloride channel. Cell 1991;67:775–784.
47. Seibert FS, Chang XB, Aleksandrov AA, Clarke DM, Hanrahan JW, Riordan JR. Influence of phopshorylation by protein kinase A on CFTR at the cell surface and endoplasmic reticulum. Biochim Biophys Acta 1999;1461:275–283.
48. Beitz E, Volkel H, Guo Y, Schultz JE. Adenylyl cyclase type 7 is the predominant isoform in the bovine retinal pigment epithelium. Acta Anat (Basel) 1998;162:157–162.
49. Blaug S, Quinn R, Quong J, Jalickee S, Miller SS. Retinal pigment epithelial function: a role for CFTR? Doc Ophthalmol 2003;106;4350.
50. Miller SS, Rabin J, Storng T, Iannuzzi M, Adam A, Collins F. Cystic fibrosis (CF) gene product is expressed in retinas and retinal pigment epithelium (abs). Invest Ophthalmol Vis Sci 1992;33(suppl).
51. Tsunenari T, Sun H, Williams J, Cahill H, Smallwood P, Yau KW, Nathans J. Structure-function analysis of the bestrophin family of anion channels. J Biol Chem 2003;278:41114–41125.
52. O'Gorman S, Flaherty WA, Fishman GA, Berson EL. Histopathologic findings in Best's vitelliform macular dystrophy. Arch Ophthalmol 1988;106:1261–1268.
53. Sun H, Tsunenari T, Yau KW, Nathans J. The vitelliform macular dystrophy protein defines a new family of chloride channels. Proc Natl Acad Sci U S A 2002;99:4008–4013.
54. Eksandh L, Bakall B, Bauer B, Wadelius C, Anderson S. Best's vitelliform macular dystrophy caused by a news mutation (Val89Ala) in the VDM2 gene. Ophthalmic Genet 2001;22:107–115.
55. Kramer F, White K, Pauleikhoff D, Gehrig A, Passmore L, Rivera A, Rudloph G, Kellner U, Andrassi M, Lorenz B, Rohr-Schneider K, Blankenagel A, Jurklies B, Schilling H, Schutt F, Holz FG, Weber BH. Mutations in the VMD2 gene are associated with juvenile-onset vitelliform macular dystrophy (Best disease) and adult vitelliform macular dystrophy but not age-related macular degeneration. Eur J Hum Genet 2000;8:286–292.
56. Seddon JM, Afshari MA, Sharma S, Berstein PS, Chong S, Hutchinson A, Petrukhin K, Allikmets R. Assessment of mutations in the Best macular dystrophy (VMD2) gene in patients with adult-onset foveomacular vitelliform dystrophy, age-related maculopathy, and bull's-eye maculopathy. Ophthalmol 2001;108:2060–2067.
57. Marmorstein AD, Stanton JB, Yocom J, Bakall B, Schiavone MT, Wadelius C, Marmorstein LY, Peachey NS. A model of Best vitelliform macular dystrophy in rats. Invest Ophthalmol Vis Sci 2004;45:3733–3739.
58. Loewen ME, Smith NK, Hamilton DL, Grahn BH, Forsyth GW. CLCA protein and chloride transport in canine retinal pigment epithelium. Am J Physiol Cell Physiol 2003;285: C1314–C1321.
59. Qu Z, Fischmeister R, Hartzell C. Mouse bestrophin-2 is a bona fide Cl(-) channel: identification of a residue important in anion binding and conduction. J Gen Physiol 2004;123:327–340.

60. Fischmeister R and Hartzell HC. Volume sensitivity of the bestrophin family of chloride channels. J Physiol 2005;562:477–491.
61. Botchkin LM and Matthews G. Chloride current activated by swelling in retinal pigment epithelium cells. Am J Physiol 1993;265:C1037–C1045.
62. Loewen ME and Forsyth GW. Structure and function of CLCA proteins. Physiol Rev 2005;85:1961–10092.
63. Gruber BR, Elble RC, Ji HL, Schreur KD, Fuller CM, Pauli BU. Genomic cloning, molecular characterization, and functional analysis of human CLCA1, the first human member of the family of Ca^{2+}-activated Cl^- channel proteins. Genomics 1998;54:200–214.
64. Evans SR, Thoreson WB, Beck CL. Molecular and Functional Analyses of Two New Calcium-activated Chloride Channel Family Members from Mouse Eye and Intestine. J Biol Chem 2004;279:41792–41800.
65. Annunziato L, Pignataro G, Di Renzo GF. Pharmacology of brain Na^+/Ca^{2+} exchanger: from molecular biology to therapeutic perspectives. Pharmacol Rev 2004;56:633–654.
66. Schnetkamp PPM. The SLC24 Na^+/Ca^{2+}-K^+ exchanger family: vision and beyond. Pflugers Arch 2004;447:683–688.
67. Loeffler KU and Mangini NJ. Immunohistochemical localization of Na^+/Ca^{2+} exchanger in human retina and retinal pigment epithelium. Graefes Arch Clin Exp Ophthalmol 1998;236:929–933.
68. Mangini NJ, Haugh-Scheidt L, Valle JE, Cragoe EJJr, Ripps H, Kennedy BG. Sodium-calcium exchanger in cultured human retinal pigment epithelium. Exp Eye Res 1997;65:821–834.
69. Gamba G. Molecular physiology and pathophysiology of electroneutral cation-chloride cotransporters. Physiol Rev 2005;85:423–493.
70. Lauf PK and Theg BE. A chloride dependent K^+ flux induced by N-ethylmaleimide in genetically low K^+ sheep and goat erythrocytes. Biochem Biophys Res Commun 1980;92:1422–1428.
71. Hu JG, Gallemore RP, Bok D, Frambach DA. Chloride transport in cultured fetal human retinal pigment epithelium. Exp Eye Res 1996;62:443–448.
72. Rymer J, Miller SS, Edelman JL. Epinephrine-induced increases in $[Ca^{2+}]$(in) and KCl-coupled fluid absorption in bovine RPE. Invest Ophthalmol Vis Sci 2001;42:1921–1929.
73. Kennedy BG. Volume regulation in cultured cells derived from human retinal pigment epithelium. Am J Physiol 1994;266:C676–C683.
74. Gross E and Kurtz I. Structural determinants and significance of regulation of electrogenic Na^+-HCO_3^- cotransporter stoichiometry. Am J Renal Physiol 2002;283:F876–F887.
75. Pushkin A and Kurtz I. SLC4 base (HCO_3^-, CO_3^{2-}) transporters: classification, function, structure, genetic disease, and knockout models. Am J Physiol 2006;290:F580–F599.
76. Mount DB and Romero MF. The SLC26 gene family of multifunctional anion exchangers. Pflugers Arch 2004;477:710–721.
77. Alper SL, Chernova MN, Stewart AK. Regulation of Na^+-independent Cl^-/HCO_3^- exchangers by pH. JOP 2001;2(4Suppl):171–175.
78. Romero MF. The electrogenic Na^+/HCO_3^- cotransporter, NBC. JOP 2001;2(4Suppl): 182–191.
79. Kurtz I, Petrasek D, Tatishchev S. Molecular mechanisms of electrogenic sodium bicarbonate cotransport: structural and equilibrium thermodynamic considerations. J Memb Biol 2004;197:77–90.
80. Gross E, Fedotoff O, Pushkin A, Abuladze N, Newman D, Kurtz I. Phosphorylation-induced modulation of pNBC1 function: distinct roles for the amino and carboxy-termini. J Physiol 2003;549:673–682.

81. Bok D, Schibler MJ, Pushkin A, Sassani P, Abuladze N, Naser Z, Kurtz I. Immunolocalization of electrogenic sodium-bicarbonate cotransporters pNBC1 and kNBC1 in the rat eye. 2001;281: F920–F935.
82. Bok D, Galbraith G, Lopez I, Woodruff M, Nusinowitz S, BeltrandelRio H, Huan W, Zhao S, Geske R, Montgomery C, Sligtenhorst I, Friddle C, Platt K, Sparks M, Pushkin A, Abuladze N, Ishiyama A, Dukkipati R, Liu W, Kurtz I. Blindness and auditory impairment caused by loss of the sodium bicarbonate cotransporter NBC3. Nat Genet 2003;34:313–319.
83. Edelman JL, Lin H, Miller SS. Potassium-induced chloride secretion across the frog retinal pigment epithelium. Am J Physiol 1997;266:C957–C966.
84. Edelman JL, Lin H, Miller SS. Acidification stimulates chloride and fluid absorption across frog retinal pigment epithelium. Am J Physiol 1994;266:C946–C956.
85. Segawa Y, Shirao Y, Kawasaki K. Retinal pigment epithelial origin of bicarbonate response. Jpn J Ophthalmol 1997;41:231–234.
86. Lin H and Miller SS. pHi regulation in frog retinal pigment epithelium: two apical membrane mechanisms. Am J Physiol 1991;261:C132–C142.
87. Hughes BA, Adorante JS, Miller SS, Lin H. Apical electrogenic NaHCO3 cotransport. A mechanism for HCO3 absorption across the retinal pigment epithelium. J Gen Physiol 1989;94:25–50.
88. Kenyon E, Maminishkis A, Joseph DP, Miller SS. Apical and basolateral membrane mechanisms that regulate pHi in bovine retinal pigment epithelium. Am J Physiol 1997;273: C456–C472.
89. Yang Z, Alvarez BV, Chakarova C, Jiang L, Karan G, Frederick JM, Zhao Y, Sauve Y, Li X, Zrenner E, Wissinger B, Hollander AI, Katz B, Baehr W, Cremers FP, Casey JR, Bhattacharya SS, Zhang K. Mutant carbonic anhydrase 4 impairs pH regulation and causes retinal photoreceptor degeneration. Hum Mol Genet 2005;14:255–265.
90. Vince JW and Reithmeier RAF. Carbonic anhydrase II binds to the carboxy-terminus of human band 3, the erythrocyte Cl^-/HCO_3^- exchanger. J Biol Chem 1998;273:28430–28437.
91. Vince JW and Reithmeier RAF. Identification of the carbonic anhydrase II binding site in the Cl^-/HCO_3^- anion exchanger AE1. Biochem 2000;39:5527–5533.
92. Vince JW, Carlsson U, Reithmeier RAF. Localization of the Cl^-/HCO_3^- anion exchanger binding site to the amino-terminal carbonic anhydrase. Biochem 2000;39:13344–13349.
93. Reithmeier RAF. A membrane metabolon linking carbonic anhydrase with chloride/bicarbonate anion exchangers. Blood Cell Mol Dis 2001;27:85–89.
94. Korte GE and Smith J. Carbonic anhydrase type II in regenerating retinal pigment epithelium. A histochemical study in the rabbit. Experientia 1993;49:789–791.
95. Alvarez BV, Loiselle FB, Supuran CT, Schwartz GJ, Casey JR. Direct extracellular interaction between carbonic anhydrase IV and the human NBC1 sodium/bicarbonate co-transporter. Biochem 2003;42:12321–12329.
96. Eichhorn M, Schreckenberger M, Tamm ER, Lutjen-Drecoll E. Carbonic anhydrase activity is increased in retinal pigmented epithelium and choriocapillaris of RCS rats. Graefes Arch Clin Exp Ophthalmol 1996;234:258–263.
97. Esbaugh AJ and Tufts BL. The structure and function of carbonic anhydrase isoenzymes in the respiratory system of vertebrates. Res Physiol Neurobiol; 2006;154:185–198.
98. Sun MK and Alkon DL. Carbonic anhydrase gating of attention: memory therapy and enhancement. Trends in Pharmacol Sci 2002;23:83–89.
99. Pastroekova S, Parkkila A, Pastorek J, Supuran CT. Carbonic anhydrases: current state of the art, therapeutic applications and future prospects. J Enz Inhib Med Chem 2004;3:199–229.

100. Biskobing DM. Dihydroxyvitamin D3 and phorbol ester acetate synergistically increase carbonic anhydrase II expression in a human myelomonocytic cell line. Endocrinol 1994;134:1493–1498.
101. Narumi S and Miyamoto E. Activation and phopshorylation of carbonic anhydrase by 3′,5′-monophosphate-dependent protein kinases. Biochim Biophys Acta 1974;350:215–224.
102. Wolfensberger TJ, Chiang RK, Takeuchi A, Marmor MF. Inhibition of membrane-bound carbonic anhydrase enhances subretinal fluid absorption and retinal adhesiveness. Graefes Arch Clin Exp Ophthalmol 2000;238:76–80.
103. Wolfensberger TJ, Mahieu I, Jarvis-Evans J, Boulton M, Carter ND, Nogradi A, Hollande E, Bird AC. Membrane-bound carbonic anhydrase in human retinal pigment epithelium. Invest Ophthalmol Vis Sci 1994;35:3401–3407.
104. Wolfensberger, TJ. The role of carbonic anhydrase inhibitors in the management of macular edema. Doc Ophthalmol 1999;97:387–397.
105. Nagelhus EA, Mathiisen TM, Bateman AC, Haug FM, Ottersen OP, Grubb JH, Waheed A, Sly WS. Carbonic anhydrase XIV is enriched in specific membrane domains of retinal pigment epithelium, Muller cells, and astrocytes. Proc Natl Acad Sci USA 2005;102:8030–8035.
106. Ochrietor JD, Clamp MF, Moroz TP, Grubb JH, Shah GN, Waheed A, Sly WS, Linser PJ. Carbonic anhydrase XIV identified as the membrane CA in mouse retina: strong expression in Muller cells and the RPE. Exp Eye Res 2005;81:492–500.
107. Wolfensberger TJ, Dmitriev AV, Govardovskii VI. Inhibition of membrane-bound carbonic anhydrase decreases subretinal pH and volume. Doc Ophthalmol 1999;97:261–271.
108. Hediger MA, Romero MF, Peng J, Rolfs A, Takanga H, Bruford EA. The ABCs of solute carriers: physiological, pathological and therapeutic implications of human membrane transport proteins. Pflugers Arch 2004;477:465–468.
109. Zadunaisky JA, Kinne-Saffran E, Kinne RA. Na/H exchange mechanism in apical membrane vesicles of the retinal pigment epithelium. Invest Ophthalmol Vis Sci 1989;30:2332–2340.
110. Adler AJ and Southwick RE. Distribution of glucose and lactate in the interphotoreceptor matrix. Ophthalmol Res 1992;24:243–252.
111. Philp NJ, Ochrieto JD, Rudoy C, Muramatsu T, Linser PJ. Loss of MCT1, MCT3, and MCT4 expression in the retinal pigment epithelium and neural retina of the 5A11/basigin-null mouse. Invest Ophthalmol Vis Sci 2003;44:1305–1311.
112. Ochrietor JD and Linser PJ. 5A11/Basigin gene products are necessary for proper maturation and function of the retina. Dev Neurosci 2004;26:380–387.
113. Clamp MF, Ochrietor JD, Moroz TP, Linser PJ. Developmental analyses of 5A11/Basigin, 5A11/Basigin-2 and their putative binding partner MCT1 in the mouse eye. Exp Eye Res 2004;78:777–789.
114. Deora AA, Philp N, Hu J, Bok D, Rodriguez-Boulan E. Mechanisms regulating tissue-specific polarity of monocarboxylate transporters and their chaperone CD147 in kidney and retinal epithelia. Proc Natl Acad Sci USA 2005;102:16245–16250.
115. Philp NJ, Yoon H, Lombardi L. Mouse MCT3 gene is expressed preferentially in retinal pigment and choroid plexus epithelia. Am J Physiol Cell Physiol 2001;280: C1319–C1326.
116. Kenyon E, Yu K, La Cour M, Miller SS. Lactate transport mechanisms at apical and basolateral membranes of bovine retinal pigment epithelium. Am J Physiol 1994;267: C1561–C1573.
117. Philp NJ, Wang D, Yoon H, Hjelmeland LM. Polarized expression of monocarboxylate transporters in human retinal pigment epithelium and ARPE-19 cells. Invest Ophthalmol Vis Sci 2003;44:1716–1721.

118. Ito A, Yamaguchi K, Onogawa T, Unno M, Suzuki T, Nishio T, Suzuki T, Sasano H, Abe T, Tamai M. Distribution of organic anion-transporting polypeptide 2 (oatp2) and oatp3 in the rat retina. Invest Ophthalmol Vis Sci 2002;43:858–863.
119. Gao B, Wenzel A, Grimm C, Vavricka SR, Benke D, Meier PJ, Reme CE. Localization of organic anion transport protein 2 in the apical region of rat retinal pigment epithelium. Invest Ophthalmol Vis Sci 2002;43:510–514.
120. Ito A, Yamaguchi K, Tomita H, Suzuki T, Onogawa T, Sato T, Mizutamari H, Mikkaichi T, Nishio T, Suzuki T, Unno M, Sasano H, Abe T, Tamai M. Distribution of rat organic anion transporting polypeptide-E (oatp-E) in the rat eye. Invest Ophthalmol Vis Sci 2003;44: 4877–4884.
121. Garcia DM and Burnside B. Suppression of cAMP-induced pigment granule aggregation in RPE by organic anion transport inhibitors. Invest Ophthalmol Vis Sci 1994;35: 178–188.
122. Stamer DW, Snyder RW, Smith BL, Agre P, Regan JW. Localization of aquaporin CHIP is the human eye: implications in the pathogenesis of glaucoma and other disorders of ocular fluid balance. Invest Ophthalmol Vis Sci 1994;35:3867–3872.
123. Nielsen S, Smith BL, Christensen EI, Agre P. Distribution of the aquaporin CHIP in secretory and resorptive epithelia and capillary endothelia. Proc Natl Acad Sci USA 1993;90:7275–7279.
124. Hamann S, Zeuthen T, La Cour M, Nagelhus EA, Ottersen OP, Agre P, Nielsen S. Am J Physiol 1998;274:C1332–C1345.
125. Stamer DW, Bok D, Hu J, Jaffe GJ, McKay BS. Aquaporin-1 channels in human retinal pigment epithelium: role in transepithelial water movement. Invest Ophthalmol Vis Sci 2003;44:2803–2808.
126. Dibas A, Prasanna G, Yorio T. Characterization of endothelin system in Bovine optic nerve and retina. J Ocul Pharmacol Ther 2005;21:288–297
127. Yamamoto N, Sobue K, Miyachi T, Inagaki M, Miura Y, Katsuya H, Asai K. Differential regulation of aquaporin expression in astrocytes by protein kinase C. Brain Res Mol Brain Res 2001;95:110–116.
128. Khatami M, Stramm LE, Rockey JH. Ascorbate transport in cultured cat retinal pigment epithelial cells. Exp Eye Res 1986;43:607–615.
129. Karihaloo A, Kato K, Greene DA, Thomas TP. Protein kinase and Ca^{2+} modulation of myo-inositol transport in cultured retinal pigment epithelial cells. Am J Physiol 1997;273: C671–C678.
130. Yefimova MG, Jeanny JC, Guillonneau X, Keller N, Nguyen-Legros J, Sergeant C, Guillou F, Courtois Y. Iron, ferritin, transferrin, and transferrin receptor in the adult rat retina. Invest Ophthalmol Vis Sci 2000;41:2343–2351.
131. Hahn P, Dentchev T, Qian Y, Rouault T, Harris ZL, Dunaief JL. Immunolocalization and regulation of iron handling proteins ferritin and ferroportin in the retina. Mol Vis 2004;10:598–607.
132. Said HM, Wang S, Ma TY. Mechanism of riboflavin uptake by cultured human retinal pigment epithelial ARPE-19 cells: possible regulation by an intracellular Ca2+-calmodulin-mediated pathway. J Physiol 2005;566:369–377.
133. Naggar H, Fei YJ, Ganapathy V, Smith SB. Regulation of reduced-folate transporter-1 (RFT-1) by homocysteine and identity of transport systems for homocysteine uptake in retinal pigment epithelial (RPE) cells. Exp Eye Res 2003;77:687–697.
134. Naggar H, Van Ells TK, Ganapathy V, Smith SB. Regulation of reduced-folate transporter-1 in retinal pigment epithelial cells by folate. Curr Eye Res 2005;1:35–44.

135. Chancy CD, Kekuda R, Huang W, Prasad PD, Kuhnel JM, Sirotnak FM, Roon P, Ganapathy V, Smith SB. Expression and differential polarization of the reduced-folate transporter-1 and the folate receptor alpha in mammalian retinal pigment epithelium. J Biol Chem 2000;275:20676–20684.
136. Bridges CC, El-Sherbeny A, Ola MS, Ganapathy V, Smith SB. Transcellular transfer of folate across the retinal pigment epithelium. Curr Eye Res 2002;24:129–138.
137. Naggar H, Ola MS, Moore P, Huang W, Bridges CC, Ganapathy V, Smith SB. Downregulation of reduced-folate transporter by glucose in cultured RPE cells and in RPE of diabetic mice. Invest Ophthalmol Vis Sci 2002;43:556–563.
138. Peterson WM, Meggyesy C, Yu K, Miller SS. Extracellular ATP activates calcium signaling, ion, and fluid transport in retinal pigment epithelium. J Neurosci 1997;17:2324–2337.
139. Takahashi J, Hikichi T, Mori F, Kawahara A, Yoshida A, Peterson WM. Effect of nucleotide P2Y2 receptor agonists on outward active transport of fluorescein across normal blood-retina barrier in rabbit. Exp Eye Res 2004;78:103–108.
140. Maminishkis A, Jalickee S, Blaug SA, Rymer J, Yerxa BR, Peterson WM, Miller SS. The P2Y(2) receptor agonist INS37217 stimulates RPE fluid transport in vitro and retinal reattachment in rat. Invest Ophthalmol Vis Sci 2002;43:3555–3566.
141. Reigada D, Lu W, Zhang X, Friedman C, Pendrak K, McGlinn A, Stone RA, Laties AM, Mitchell CH. Degradation of extracellular ATP by the retinal pigment epithelium. Am J Cell Physiol 2005;289:C617–C624.
142. Mitchell CH. Release of ATP by a human retinal pigment epithelial cell line: potential for autocrine stimulation through subretinal space. J Physiol 2001;534:193–202.
143. Ammar DA, Hughes BA, Thompson DA. Neuropeptide Y and the retinal pigment epithelium: receptor subtypes, signaling, and bioelectrical responses. Invest Ophthalmol Vis Sci 1998;39:1870–1878.
144. Bragadottir, R., Kato, M., and Jarkman, S. Serotonin elevates the c-wave of the electroretinogram of the rabbit eye by increasing the transepithelial potential. Vis Res 1997:37:2495–2503.
145. Kawahara A, Hikichi T, Kitaya N, Takahashi J, Mori F, Yoshida A. Adenosine agonist regulation of outward active transport of fluorescein across retinal pigment epithelium in rabbits. Exp Eye Res 2005;80:493–499.
146. Koh SM. VIP enhances the differentiation of retinal pigment epithelium in culture: from cAMP and pp60(c-src) to melanogenesis and development of fluid transport capacity. Prog Retin Eye Res 2000;19:669–688.
147. Fujiseki Y, Omori K, Omori K, Mikami Y, Suzukawa J, Okugawa G, Uyama M, Inagaki C. Natriuretic peptide receptors, NPR-A and NPR-B, in cultured rabbit retinal pigment epithelium cells. Jpn J Pharmacol 1999;79:359–368.
148. Mikami Y, Hara M, Yasukura T, Uyama M, Minato A, Inagaki C. Artial natriuretic peptide stimulates Cl- transport in retinal pigment epithelial cells. Curr Eye Res 1995;14:391–397.

10
Glucose Transporters in Retinal Pigment Epithelium Development

Lawrence J. Rizzolo

CONTENTS

SUMMARY
INTRODUCTION
PROPERTIES OF GLUCOSE TRANSPORTERS
STRUCTURE OF THE OUTER BLOOD–RETINAL BARRIER
COORDINATE DEVELOPMENT OF RETINAL PIGMENT EPITHELIUM
  TIGHT JUNCTIONS AND TRANSCELLULAR GLUCOSE TRANSPORT
CONCLUSIONS
ACKNOWLEDGEMENTS
REFERENCES

SUMMARY

The retina relies on glucose for metabolic energy, but unlike systemic endothelia, the endothelia and epithelia of the blood–brain barrier allow little glucose to diffuse across their paracellular spaces. To compensate, these cells express high levels of glucose transporters to facilitate transcellular transport. One region, the outer blood–retinal barrier, has an unusual structure. The choroidal capillaries are fenestrated, and the barrier is formed by the overlying retinal pigment epithelium (RPE). The development of RPE is closely coordinated with the development of the neural retina and the choriocapillaris. Early in development, the RPE expresses basal levels of various members of the GLUT family of glucose transporters. Then, tight junctions begin to form and start to restrict paracellular diffusion. Late in development, as tight junctions become tighter still, the expression of GLUT1, GLUT11, and SGLT1 increases. This suggests that, early in development, a collection of GLUT family members serves the needs of the RPE for metabolic energy, but certain members are better suited for the level of transepithelial transport required to serve the needs of the neural retina. This change in expression also correlates with increased infolding of the RPE basal plasma membrane and the formation of fenestrae in the choriocapillaris.

From: *Ophthalmology Research: Ocular Transporters in Ophthalmic Diseases and Drug Delivery*
Edited by: J. Tombran-Tink and C. J. Barnstable © Humana Press, Totowa, NJ

INTRODUCTION

The retina is among the most metabolically active of tissues and derives most of its energy from glycolysis *(1)*. This is especially true of the chicken, where the retina loses most of its mitochondria during development *(2)*. There are two sources for retinal glucose, the capillary bed of the inner blood–retinal barrier and the choriocapillaris of the outer blood–retinal barrier. The primary difference between these two sources is that the inner vascular bed of mammals is intimately associated with the inner layers of the retina, but the choriocapillaris is separated from the neural retina by the retinal pigment epithelium (RPE). Like most endothelia associated with the central nervous system (CNS), endothelia of the inner retinal capillary bed must perform all the functions of the blood-brain barrier. In contrast, the choriocapillaris is fenestrated, and therefore very leaky. These endothelia must work in collaboration with the RPE to form a blood-retinal barrier. Because glucose readily diffuses out of choroidal capillaries, this chapter will focus on the coordinated formation of two opposing mechanisms within the RPE monolayer: the formation of tight junctions that block the diffusion of glucose through the paracellular spaces and the compensatory expression of glucose transporters that enable the transcellular diffusion of glucose.

I will discuss the properties of glucose transporters, the structure of the outer blood-retinal barrier and how structure and function change during the normal embryonic development of the tissue.

PROPERTIES OF GLUCOSE TRANSPORTERS

A family of transporters facilitates diffusion of glucose across the plasma membrane. The SGLT subfamily of sodium coupled transporters does not appear to be as prominent in the CNS as the facilitated-diffusion GLUT subfamily. Fourteen members of the GLUT family have been described *(3–14)*. Family members differ in their kinetic properties and tissue distribution. For example, GLUT4 is prominent in muscle and adipose tissue and its distribution between the plasma membrane and intracellular pools is rapidly regulated by insulin *(15)*. GLUT8 is the only other glucose transporter that is regulated by insulin *(6)*. The principal transporters found in the CNS are GLUT1 and GLUT3. These family members are widely distributed, but in the CNS, GLUT1 is found primarily in endothelia, glia, the RPE, the epithelium of the choroid plexus and the ependyma. GLUT3 is found primarily in neurons. GLUT1 is expressed at high levels in the apical and basolateral membranes of retinal endothelia and the RPE, which would facilitate transport across the inner and outer blood–retinal barriers *(16–18)*.

Although GLUT1 appears to be the major transporter in RPE, there is data that suggests the presence of others. Busik et al. *(19)* challenged primary cultures of human RPE with a 30-minute exposure to elevated glucose. There was an approximately a twofold increase in the V_{max} for glucose uptake. There was no effect on the intracellular localization or the steady-state levels of GLUT1. The increase in V_{max} could be inhibited by the disruption of microtubules, which would disrupt intracellular protein traffic. Possibly, elevated glucose levels affect the subcellular localization of a regulator of GLUT1 or of another glucose transporter. Later, I will provide direct evidence of additional transporters in chick RPE.

GLUT1 and GLUT3 are expressed in many cell types. In rodents, GLUT3 provides a basal level of glucose transport, but GLUT1 is upregulated in response to mitogens or oncogenes. In the studies described below, it is important to note one of the differences in the expression of GLUT1 and GLUT3 in chickens, as opposed to mammals: chickens upregulate GLUT3 in response to the oncogene, v-src *(20,21)*. The authors of these studies suggest that during evolution, the GLUT3 promoter assumed functions in avian species that are performed by the GLUT1 promoter in mammals. This raises the question of whether all functions of the mammalian GLUT1 promoter are expressed by the chick GLUT3 promoter, or whether oncogenes regulate GLUT expression by a mechanism that differs from the mechanisms of embryonic development. I will return to this when discussing the development of RPE tight junctions and transcellular glucose transport.

STRUCTURE OF THE OUTER BLOOD–RETINAL BARRIER

In contrast to the bulk of the blood–brain barrier, the outer blood–retinal barrier is formed by a collaboration of two tissues: the choriocapillaris and the RPE. Because the two tissues form a functional unit, they should be considered together (Fig. 1). The choriocapillaris is fenestrated, which prevents these capillaries from forming a tissue barrier. The interaction of the choriocapillaris with the RPE is reflected by the RPE's ability to induce the formation of the fenestrae, and the location of the fenestra on the side of the capillaries that face the RPE *(22,23)*. Interposed between the RPE and the capillaries, lies a thick basement

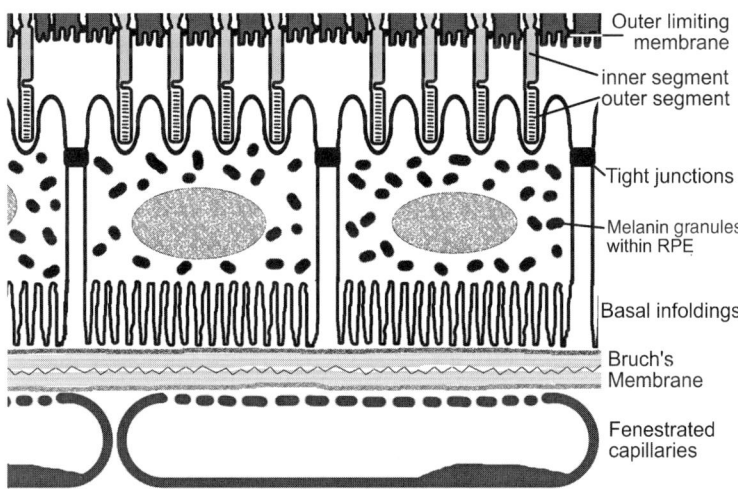

Fig. 1. Structure of the outer blood–retinal barrier. At the top, the inner and outer segments of photoreceptors are depicted interspersed with the apical pole of Muller cells. These cells are bound together by an adherens-containing apical junctional complex, the outer limiting membrane. The outer segments interdigitate with the microvilli of the retinal pigment epithelium (RPE), which includes tight junctions within its apical junctional complex. Pigment granules fill the RPE cytoplasm. The basal infoldings of the RPE sit on a pentilaminar, Bruch's membrane. Fenestrated capillaries underlie Bruch's membrane. The fenestrae are localized at the capillary face adjacent to Bruch's membrane.

membrane, called Bruch's membrane, which is produced by a collaboration of both tissues. The basal plasma membrane of RPE sits directly on Bruch's membrane. It is highly infolded to enhance the exchange of nutrients and wastes between the choriocapillaris and the retina. At the opposite, apical, pole of the RPE, long microvilli interdigitate with the outer segments of photoreceptors.

Typically, the apical membrane of an epithelium faces a lumen rather than a solid tissue. The unusual structure of the RPE is best understood by recalling its embryonic development. The RPE and neural retina form from a diverticulum of the neurotube called the optic vesicle. The neuroepithelium that lines the optic vesicle is a pseudostratified epithelium whose apical surface faces the lumen. To understand how the apical surface of the neural retina comes to appose the apical surface of the RPE, imagine a fist pushing into the optic vesicle, which pushes in until the surface covering the fist contacts the opposite surface. The result is a two-layered cup with the apical surfaces of each layer apposed to one another. The inner layer proliferates to form the multilayered neural retina, while the outer layer forms a simple monolayer, the RPE. This intimate association allows the neural retina to regulate the development and function of the RPE *(24,25)*.

Near the apical surface, each cell is bound to its neighbors by an apical junctional complex. The apical junctional complex encircles each cell much like the plastic rings that hold together a six-pack of canned beverage. In the neural retina, this apical junctional complex is the outer limiting membrane, which binds photoreceptors and Muller cells together (Fig. 1). The outer limiting membrane contains adherens junctions, but lacks tight junctions. Accordingly, it allows the diffusion of solutes smaller than serum albumin *(26)*.

The apical junctional complex of the RPE is actually a complex of three types of junctions (tight, adherens and gap) whose functions are intertwined. The adherens junctions bind neighboring cells together, as in the outer limiting membrane, while the tight junctions form a partially occluding seal that semiselectively retards diffusion through the paracellular spaces of the monolayer *(27–31)*. Both portions of the complex participate in signal transduction pathways that regulate cell size and proliferation and help polarize the distribution of plasma membrane proteins. To enable these diverse functions, the apical junctional complex is an extraordinarily complex assembly of proteins.

Tight junctions were thought to block diffusion through the paracellular spaces of neighboring epithelial or endothelial cells, which gave rise to the name zonula occludens or tight junction. Some epithelia, such as the urinary bladder, do form a nearly occluding seal, but most epithelia and endothelia require the junctions to be leaky to some degree in order to properly perform their function *(32,33)*. By retarding diffusion, tight junctions enable endothelia and simple, transporting epithelia to use active transport mechanisms to regulate transmonolayer transport and establish concentration gradients across the monolayer. Although a leaky tight junction would allow gradients to dissipate, some transport mechanisms rely on a semiselective 'leakiness'. Accordingly, the selectivity and the permeability of the tight junctions depend upon the physiologic role of the epithelium. For any given epithelium, it is controversial whether selectivity and permeability are regulated by normal physiologic changes within a tissue, although these properties can be altered pharmacologically *(25,34,35)*. Recent studies of kidney collecting tubules demonstrate that aldosterone regulates the phosphorylation and selectivity of tight junctions on a physiological time scale *(36)*. Because selectivity and permeability can be regulated independently, single measures of function alone, such as the transepithelial electrical resistance (TER) or permeability to a particular solute, e.g., glucose, fail to define this function of the tight junction fully *(37–41)*.

Unlike the capillaries of the blood–brain barrier, the RPE is moderately leaky. The TER depends upon the species and ranges from $138\,\Omega\text{-cm}^2$ in the chick to $426\,\Omega\text{-cm}^2$ in the frog *(42)*, as compared to the $1500–2000\,\Omega\text{-cm}^2$ observed in capillaries within the pia mater *(43)*. This difference reflects the different physiological roles played by the inner and outer blood–retinal barriers. The outer blood-retinal barrier is specialized to support the unique needs of the photoreceptors that abut its apical surface. From this perspective, it is not surprising that the tissue–tissue interactions that regulate the tight junctions of the RPE differ from those that regulate endothelial cells of the CNS *(35,44,45)*. The secretions of the neural retina that increase the tightness of the RPE barrier have different physical properties than the endothelial-active factors secreted by astrocytes, and they do not act via a cyclic adenosine monophosphate- (cAMP-) mediated pathway. Despite the differences in TER, the tight junctions of RPE and CNS endothelial cells strongly resist the diffusion of small organic tracers, including glucose, through the paracellular spaces of the epithelial or endothelial monolayer.

COORDINATE DEVELOPMENT OF RETINAL PIGMENT EPITHELIUM TIGHT JUNCTIONS AND TRANSCELLULAR GLUCOSE TRANSPORT

The development of RPE tight junctions in relation to the expression of GLUT transporters has been studied in chick. These studies were facilitated by the slow time course for the development of RPE tight junctions. In mammals and chick, regardless of the time of gestation, development of the retina and choroid can be tied to developmental milestones of the photoreceptors *(25,46)*. This observation suggests that basic interactions among these tissues are conserved among species. The early phase of development is defined as the period before the penetration of the outer limiting membrane by the inner segments of the photoreceptors. The late phase is defined as the period that follows the first appearance of the outer segments (Fig. 2). Because the development of the choroid and the neural retina are coordinated in chick and mammals, and because the RPE influences and is influenced by the development of its neighbors, it follows that the development of the RPE in chick should be a model for development in other species. In this section, I will discuss the data that support this hypothesis and relate it to the expression of glucose transporters.

Briefly, the expression of glucose transporters rises during the late phase of development and correlates with the formation of fenestra in the choriocapillaris. In the RPE, this increase correlates with the formation of basal membrane infoldings and maturation of tight junctions.

Development of Tight Junctions in Chick Retinal Pigment Epithelium

There is a substantial delay between assembly of the adherens junction and assembly of the tight junction. Although they will be remodeled, adherens junctions are already present in the neuroepithelium that formed the RPE on embryonic day 3 (E3), days before rudimentary tight junctions began to form on E7 *(47–51)*. In this early period,

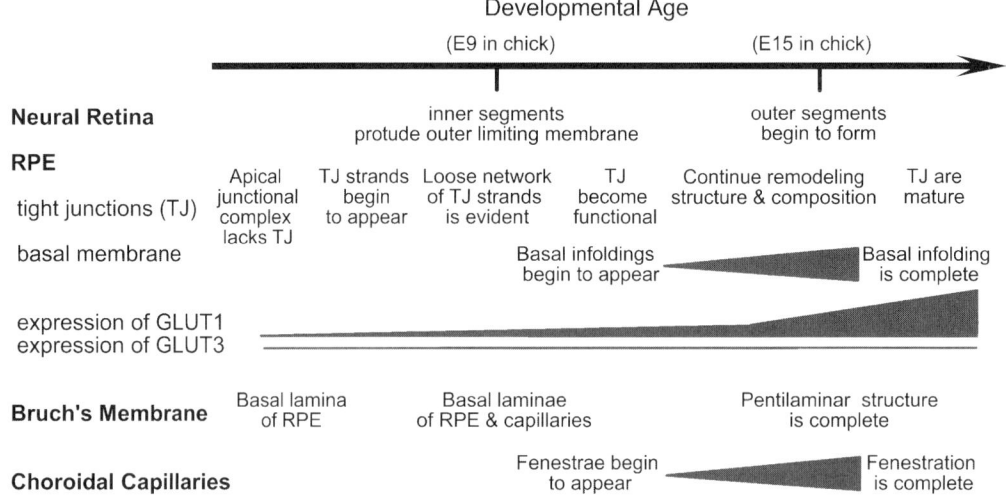

Fig. 2. Three stages in the development of the outer blood-retinal barrier. Two milestones of retinal development subdivide developmental events into early, intermediate and late stages. These temporal relationships among the neural retina, RPE and choroid also hold for mammalian and chick development. (For a review of this subject see *25,46*)

there is the presence of the tight junctional proteins, ZO-1 and occludin, and assembly proteins, such as AF-6, JAM-A, PAR3 or PAR6 *(52)*. Nonetheless, tight junctional strands were absent until claudins were expressed on E7 *(47,53)*. This is an important point, because the expression of ZO-1 and occludin is often taken as evidence of the presence of tight junctions *(28,30)*, and although their presence is necessary, it is insufficient. ZO-1 is found in all cells, where it participates in different types of cell junction *(48,54–57)*. Cogent examples are the apical junctional complex of the ependyma and outer limiting membrane, which are also derived from the neuroepithelium. Here, ZO-1 is found in an apical junctional complex that lacks tight junctions *(48,58,59)*. Similarly, occludin is expressed in some cells that lack tight junctions, such as primordial RPE cells and the precursors of astrocytes and neurons in culture *(47,48,60)*. In these circumstances, occludin might function as a regulator of the TGF-β receptor *(61)*. In cultures of chick RPE and some strains of the ARPE-19 cell line, zonular rings of ZO-1 and occludin are evident even when the tight junctions are discontinuous and therefore, nonfunctional *(47,62)*.

Tight junctions require the incorporation of claudins into the apical junctional complex. The 24 claudin family members form the strands of the tight junction and determine its selectivity and permeability *(31)*. Slowly, between E7 and E14, tight junctional strands grow in number and length to gradually coalesce into a complex, continuous network that encircle the cell *(47)*. The tight junction first becomes functional between E10 and E12 *(48)*. This event correlated with a switch in the expression of ZO-1 isoforms, an event that was also observed during tight junction formation in pre-implantation embryos *(63)*. After the active period of strand formation, structural modifications continue between

E14 and E18 *(64)*. As the fine structure of the tight junctions was remodeled during development, there was a gradual change in the relative expression of different claudin family members. This remodeling of composition would be expected to change selectivity and permeability of the junctions to different ionic solutes *(31)*. This slow process allows us to explore the assembly of the adherens junction and tight junction semi-independently.

A parallel development is observed in the pecten, a tuft of capillaries at the optic nerve head that serves as the avian inner blood–retinal barrier. In the pecten, junctional strands increase in number and complexity after E15, when the permeability of lanthanum nitrate decreases *(65)*. In both the pecten and the RPE, the greatest increase in GLUT1 correlate most closely with the tightening of the blood-retinal barrier that occurs in the late phase of development.

Expression of Glucose Transporters In Vivo

As the neural retina develops, its demand for glucose increases. However, this occurs when the outer blood–retinal barrier is maturing and restricting diffusion through the paracellular pathway. To compensate, the RPE expresses glucose transporters to facilitate transcellular diffusion. Previous studies established that mammalian RPE expresses GLUT1, but other isoforms were not investigated *(4)*. Using the reverse-transcriptase polymerase chain reaction (RT-PCR), we amplified the mRNAs for GLUT1, GLUT2 and GLUT3 *(66)*. As expected, GLUT2, an intestinal family member, was not detected in the RPE, but the more-ubiquitous GLUT 1 and GLUT 3 were evident. Northern blot analysis indicated that GLUT1 is the predominant isoform on E14. These findings allowed us to address two fundamental questions. Which isoform is regulated during development, and how is its expression coordinated with the increase in barrier function of the RPE?

The regulation of GLUT 1 and GLUT 3 is a little unusual in chick. Studies using oncogenes indicate that the signaling pathways that upregulate GLUT1 expression in rodents upregulate GLUT3 instead in chick *(20,21)*. This suggests that the chick uses GLUT1 to provide a basal level of transport and GLUT3 when transport needs to be upregulated, as in actively dividing cells. It further implies that the GLUT3 promoter of chick evolved to assume functions that the GLUT1 promoter performs in mammals.

RPE development provided a test of the hypothesis that GLUT 3 replaces GLUT 1 in chick *(66)*. Total RNA was isolated from the RPE of E5, E7, E14, and E18 chick embryos. GLUT1 mRNA rose during development relative to 18S rRNA, but GLUT3 showed little change. There was a sixfold increase in RT-PCR product for GLUT1, as determined by semi-quantitative RT-PCR. The steady-state level of GLUT1 protein also increased 8-fold, as determined by immunoblotting. Both experiments indicate that expression increased most during the late phase of tight junction development. The upregulation of GLUT1 was also observed in the inner blood-retinal barrier of chick, the pecten *(65)*. These data demonstrate that GLUT1 and GLUT3 can each be regulated under different circumstances. Although v-src upregulates both GLUT3 and cell proliferation, the glucose transporters of RPE must be upregulated in quiescent cells. Mitotic figures in the RPE monolayer are greatly diminished by E6 and nearly absent by E12 *(67)*. In the absence of pathology, RPE cells no longer divide during the life of the animal. Therefore, the regulatory targets of certain proto-oncogenes

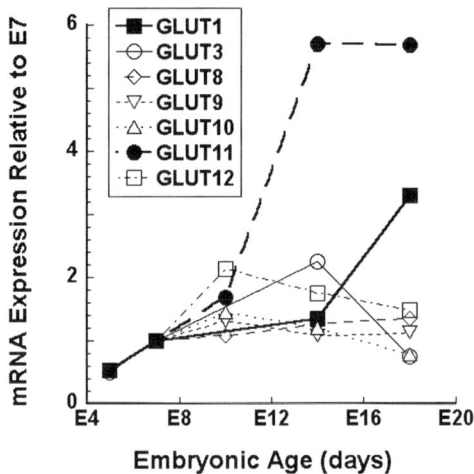

Fig. 3. Developmental regulation of the expression of GLUT family members. Total mRNA was isolated from the retinal pigment epithelium (RPE) of chicken embryos of the indicated age and used to probe Affymetrix microarrays of the chick genome. For comparison, semi-quantitative reverse transcriptase polymerase chain reaction (RT-PCR) data for GLUT1 and GLUT3 is included *(66)*.

may have diverged, but other regulatory features of the GLUT1 gene appear to be conserved between chick and rodents: these could include features of the 3′ and 5′ untranslated regions of the mRNA *(68,69)*.

Subsequent studies identified additional GLUT family members that were expressed in RPE. The entire chick genome was analyzed using Affymetrix gene microarrays. Changes in mRNA levels, relative to the expression on E7, are shown for each family member in Fig. 3. Like GLUT1, the expression of GLUT11 increased during development. The rise in GLUT11 preceded GLUT1, but followed the closure of discontinuities in the tight junctions (approximately E11, Fig.2). The other glucose transporters that were expressed in RPE either did not change during development or perhaps showed a slight, transient peak of expression. Besides GLUT3, these included GLUT8, 9, 10 and 12. Other transporters expressed in chick, but not detected in RPE included GLUT2, 4, 6, and 7.

One family member of the sodium-coupled glucose transporters, SGLT1, was identified in the chick genome. SGLT1 was not detected on the microarray until E10. Between E10 and E18 the expression of SGLT1 mRNA increased 2.5-fold. Although the microarray data needs to be verified by RT-PCR, it is notable that all the observed increases of glucose transporters occurred soon after functional tight junctions were formed. The presence of sodium-coupled glucose transporters has not been previously reported in RPE. The proposed localization of SGLT1 (Fig. 4) is based on an analogy with the distribution of the transporters in the intestine. In that case, the SGLT1 is in the apical membrane where glucose would be coported into the cell down a sodium gradient created by the Na,K-ATPase. The GLUT2 transporter facilitates the passive diffusion of

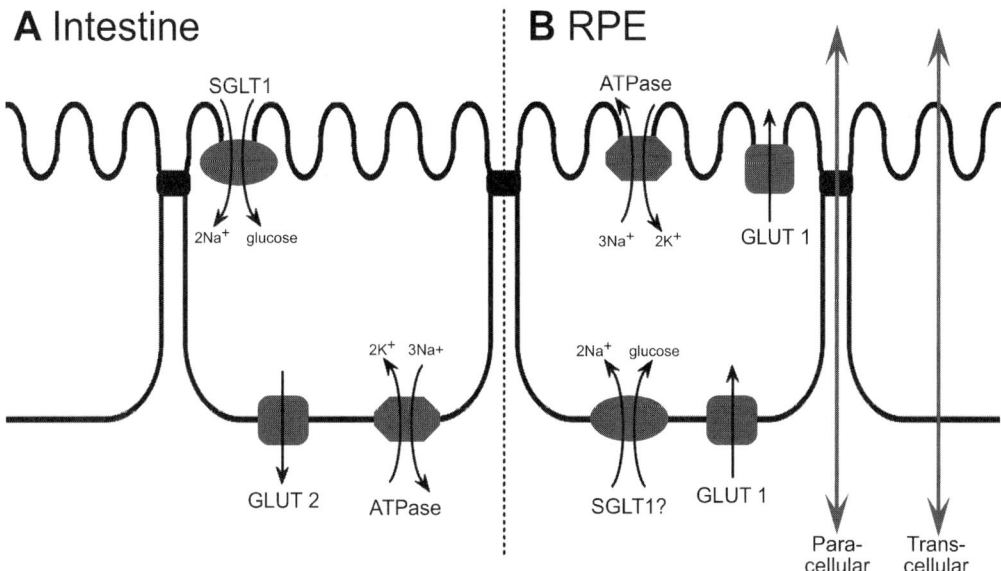

Fig. 4. Mediators of transepithelial glucose transport. The subcellular distribution of glucose transporters and the Na, K-ATPase are indicated for **(A)** intestine and **(B)** retinal pigment epitheilium (RPE). The localization of SGTL1 in the RPE is postulated, as described in the text. Double arrows indicate the paracellular and transcellular pathways.

glucose across the basolateral membrane and into the blood stream (4). In the RPE, the situation would need to be reversed with SGLT1 in the basolateral membrane. Although a number of sodium coporters and antiporters function in the apical membrane of RPE, the need for glucose derives from apical secretion, not apical absorption. The net secretion of sodium is in the apical direction, which requires a mechanism for sodium uptake across the basolateral membrane (42). This basal membrane flux of sodium could drive glucose through SGLT1. Unlike GLUT2 in the intestine, the distribution of GLUT1 has been identified in the apical and basolateral membranes of RPE (16,17). This suggests the main force driving a basal to apical flux of glucose would be the rapid metabolism of glucose by the retina. This suggests the role of SGLT1 would be to augment the facilitated diffusion that could occur by GLUT1 and GLUT11 alone.

Effect of the Neural Retina on the Expression of Glucose Transporters

To examine the hypothesis that GLUT1 (and by extension GLUT11 and SGLT1) is involved in transepithelial transport rather than RPE cell maintenance, the expression of GLUT1 and GLUT3 was examined in a primary cell-culture model of RPE development (38,47,70). In this model, the permeability of tight junctions decreases when RPE is cultured in the presence of retinal secretory products. The rationale was that if the developing neural retina satisfied its increased need for glucose by regulating GLUT expression, then retinal factors should regulate expression the GLUT that increased during development. As noted earlier, this would be GLUT1 instead of GLUT3.

When RPE was isolated from the late–early (E7) or late–intermediate (E14) stages and cultured on laminin-coated filters, the cells formed a monolayer with an apical junctional complex consisting of adherens junctions and a rudimentary tight junction. Tight junctional strands were present, but like the intermediate stage of tight junction assembly in vivo, the network of strands was discontinuous. Consequently, the junctions were leaky to tracers as large as horseradish peroxidase. These discontinuities were sealed when the cultures were incubated in a medium that was conditioned by organ cultures of E14 neural retinas *(47)*. Further, these secretions of the neural retina, presented to the apical side of the RPE monolayer, acted synergistically with various factors presented to the basal membrane, which supports the hypothesis that communication between the RPE and surrounding tissues regulates each other's development *(25,46,70)*.

With retinal-conditioned medium, the fine structure of the tight junctions closely resembled that of RPE in vivo, as viewed by freeze-fracture electron microscopy *(47)*. The remarkable finding was that despite the similar structure, the E14 junctions had a higher TER than the E7 cultures. This finding was explained by examination of the members of the claudin family that were expressed. The relative expression of the various claudins differed in each culture and the effect of retinal-conditioned medium on claudin expression also varied. For example, retinal-conditioned medium induced the expression of claudin 1 in both cultures, but induced the expression of claudin 2, and decreased the expression of claudin 4L2, only in E14 cultures. These different responses by the two cultures may reflect the expression of different signaling pathways. For example, the E7 cultures respond to a labile factor that has a mass of less than 10 kDa, while the E14 cultures respond to a protein of 49 kDa. Notably, these active factors have different properties from the astrocyte factors that induce the formation of tight junctions in endothelial cells *(35,45,71)*.

Despite these differences in the permeability of ions, the paracellular pathways of E7 and E14 RPE cultures are equally impermeable to small organic tracers such as mannitol. Their permeability to these solutes is comparable to that of endothelial cells in the presence of astrocytes-conditioned medium plus cAMP analogues, even though the TER of the endothelial cells is fivefold higher. Several laboratories have demonstrated independent regulation of the flux of ions and uncharged organic compounds with a hydrodynamic radius similar to glucose *(37–40)*. Several models have been proposed to explain the differential regulation of ions and organic tracers *(31,39,46)*.

To determine whether retinal factors affect transcellular glucose transport, the diffusion of 3-O-methylglucose and mannitol was measured. Both compounds have similar hydrodynamic radii, but mannitol is restricted to the paracellular pathway. 3-O-methylglucose uses paracellular and transcellular pathways *(72,73)*. In both the E7 and E14 cultures, retinal-conditioned medium failed to affect the transepithelial diffusion of 3-O-methylglucose despite a decrease in the diffusion of mannitol. To block transcellular transport, glucose transporters were inhibited by phloretin *(74)*, which has no effect on the paracellular pathway of rabbit gall bladder *(75)*. In this circumstance, retinal-conditioned medium reduced the permeation of 3-O-methylglucose to that of mannitol *(66)*. By subtracting this measure of paracellular transport from total transport, we observed that the level of transcellular glucose transport increased by 40% in the presence of retinal conditioned medium ($p < 0.05$). To determine whether the increase in activity results from increased mRNA or protein expression, immunoblotting and semiquantitative RT-PCR were performed. In three experiments, no significant increases were observed, nor was there an observable change in the intracellular distribution of GLUT1.

CONCLUSIONS

The outer blood–retinal barrier needs to tightly regulate transepithelial flux by severely restricting the movement of some solutes, and actively or passively facilitating the movement of others. The tight-junction portion of the barrier restricts the paracellular flux of organic compounds, but is leaky to certain ions. This means that glucose must traverse the transcellular portion of the barrier. The neural retina regulates both portions of the barrier by regulating the structure and function of tight junctions, and presumably by inducing the expression of transporters that enhance a basal-to-apical transcellular flux of glucose. Notably, the tight junctions do not fully mature until other mechanisms are in place: infoldings of the basal plasma membrane are elaborated to enhance basal uptake, choroidal capillaries are fenestrated to enhance egress of serum components and retina-induced glucose transporters are expressed.

The limitation of our experimental model is that the major increase in the expression of GLUT1 occurs on E18, which may explain why no discernable effect on gene expression was observed. The assays for transcellular transport are more sensitive than those for gene expression. There are several reasons why the increase in transport activity was greater than was accounted for by changes in GLUT1 expression. As noted above, Busik et al. (19) found that an increase in glucose concentration increased the V_{max} of transepithelial glucose transport without any discernable effect on the expression or distribution of GLUT1. In their experiments and ours, the kinetic properties of GLUT1 could have been modulated. However, genomic analysis indicates that a number of transporters are present that could have been affected. We now know that GLUT11 and SGLT1 also increase during development and are candidates for regulation by retinal-conditioned medium. It appears that RPE constitutively expresses a number of GLUT family members that mediate a basal level of glucose uptake to meet the RPE's specific needs. The formation of a functional outer blood–retinal barrier is accompanied by the expression of additional transporters, including an energy-driven transporter, to serve the needs of the outer retina.

ACKNOWLEDGEMENTS

I thank Ru Sun and Matthew Weitzman for critically reviewing the manuscript. Work in the author's laboratory was supported by National Institutes of Health grant EY08694 (LJR) and CORE grant EY00785 (Department of Ophthalmology and Visual Science, Yale University).

REFERENCES

1. Krebs HA. The Pasteur effect and the relations between respiration and fermentation. Essays in Biochemistry 1972;8:1–35.
2. Ruggiero FP and Sheffield JB. The use of avidin as a probe for the distribution of mitochondrial carboxylases in developing chick retina. J Histochem Cytochem 1998;46:177–183.

3. Mueckler M. Facilitative glucose transporters. Eur J Biochem 1994;219:713–725.
4. Takata K. Glucose transporters in the transepithelial transport of glucose. J Electron Microsc 1996;45:275–284.
5. Waddell ID, Zomerschoe AG, Voice MW, Burchell A. Cloning and expression of a hepatic microsomal glucose transport protein. Comparison with liver plasma-membrane glucose-transport protein GLUT 2. Biochem J 1992;286 (Pt 1):173–177.
6. Doege H, Schurmann A, Bahrenberg G, Brauers A, Joost HG. GLUT8, a novel member of the sugar transport facilitator family with glucose transport activity. J Biol Chem 2000;275:16275–16280.
7. Doege H, Bocianski A, Scheepers A, Axer H, Eckel J, Joost HG, Schurmann A. Characterization of human glucose transporter (GLUT) 11 (encoded by SLC2A11), a novel sugar-transport facilitator specifically expressed in heart and skeletal muscle. Biochem J 2001;359:443–449.
8. Joost HG, Bell GI, Best JD, Birnbaum MJ, Charron MJ, Chen YT, Doege H, James DE, Lodish HF, Moley KH, Moley JF, Mueckler M, Rogers S, Schurmann A, Seino S, Thorens B. Nomenclature of the GLUT/SLC2A family of sugar/polyol transport facilitators. Am J Physiol Endocrinol Metab 2002;282:E974–E976.
9. Scheepers A, Joost HG, Schurmann A. The glucose transporter families SGLT and GLUT: molecular basis of normal and aberrant function. JPEN J Parenter Enteral Nutr 2004;28:364–371.
10. Joost HG and Thorens B. The extended GLUT-family of sugar/polyol transport facilitators: nomenclature, sequence characteristics, and potential function of its novel members (review). Mol Membr Biol 2001;18:247–256.
11. Kayano T, Burant CF, Fukumoto H, Gould GW, Fan YS, Eddy RL, Byers MG, Shows TB, Seino S, Bell GI. Human facilitative glucose transporters. Isolation, functional characterization, and gene localization of cDNAs encoding an isoform (GLUT5) expressed in small intestine, kidney, muscle, and adipose tissue and an unusual glucose transporter pseudogene-like sequence (GLUT6). J Biol Chem 1990;265:13276–13282.
12. McVie-Wylie AJ, Lamson DR, Chen YT. Molecular cloning of a novel member of the GLUT family of transporters, SLC2a10 (GLUT10), localized on chromosome 20q13.1: a candidate gene for NIDDM susceptibility. Genomics 2001;72:113–117.
13. Phay JE, Hussain HB, Moley JF. Cloning and expression analysis of a novel member of the facilitative glucose transporter family, SLC2A9 (GLUT9). Genomics 2000;66:217–220.
14. Rogers S, Macheda ML, Docherty SE, Carty MD, Henderson MA, Soeller WC, Gibbs EM, James DE, Best JD. Identification of a novel glucose transporter-like protein-GLUT-12. Am J Physiol Endocrinol Metab 2002;282:E733–E738.
15. Ishiki M, Klip A. Minireview: recent developments in the regulation of glucose transporter-4 traffic: new signals, locations, and partners. Endocrinology 2005;146:5071–5078.
16. Takata K, Kasahara T, Kasahara M, Ezaki O, Hirano H. Ultracytochemical localization of the erythrocyte/HepG2-type glucose transporter (GLUT1) in cells of the blood-retinal barrier in the rat. Invest Ophthalmol Vis Sci 1992;33:377–383.
17. Sugasawa K, Deguchi J, Okami T, Yamamoto A, Omori K, Uyama M, Tashiro Y. Immunocytochemical analyses of distributions of Na, K-ATPase and GLUT1, insulin and transferrin receptors in the developing retinal pigment epithelial cells. Cell Struct Funct 1994;19:21–28.
18. Tserentsoodol N, Shin BC, Suzuki T, Takata K. Colocalization of tight junction proteins, occludin and ZO-1, and glucose transporter GLUT1 in cells of the blood-ocular barrier in the mouse eye. Histochem Cell Biol 1998;110:543–551.

19. Busik JV, Olson LK, Grant MB, Henry DN. Glucose-induced activation of glucose uptake in cells from the inner and outer blood-retinal barrier. Invest Ophthalmol Vis Sci 2002;43:2356–2363.
20. Wagstaff P, Kang HY, Mylott D, Robbins PJ, White MK. Characterization of the avian GLUT1 glucose transporter: differential regulation of GLUT1 and GLUT3 in chicken embryo fibroblasts. Mol Biol Cell 1995;6:1575–1589.
21. Steane SE, Mylot D, White MK. Regulation of a heterologous glucose transporter promoter in chicken embryo fibroblasts. Biochem Biophys Res Commun 1998;252:318–323.
22. Burns MS and Hartz MJ. The retinal pigment epithelium induces fenestration of endothelial cells in vivo. Curr Eye Res 1992;11:863–873.
23. Korte GE, Burns MS, Bellhorn RV. Epithelium-capillary interactions in the eye: The retinal pigment epithelium and the choriocapillaris. Int Rev Cytol 1989;114:221–248.
24. Rizzolo LJ. Polarity and the development of the outer blood-retinal barrier. Histol Histopathol 1997;12:1057–1067.
25. Wilt SD and Rizzolo LJ. Unique Aspects of the Blood-Brain Barrier. In: Anderson JM, Cereijido M, eds. Tight Junctions. 2nd ed. Boca Raton: CRC Press; 2001:415–443.
26. Bunt-Milam AH, Saari JC, Klock IB, Garwin GG. Zonulae adherentes pore size in the external limiting membrane of the rabbit retina. Invest Ophthalmol Vis Sci 1985;26:1377–1380.
27. Cereijido M and Anderson JM, eds. Tight Junctions. 2nd ed. Boca Raton: CRC Press; 2001.
28. Gonzalez-Mariscal L, Betanzos A, Nava P, Jaramillo BE. Tight junction proteins. Prog Biophys Mol Biol 2003;81:1–44.
29. Matter K, Aijaz S, Tsapara A, Balda MS. Mammalian tight junctions in the regulation of epithelial differentiation and proliferation. Curr Opin Cell Biol 2005;17:453–458.
30. Schneeberger EE and Lynch RD. The tight junction: a multifunctional complex. Am J Physiol Cell Physiol 2004;286:C1213–C1228.
31. Van Itallie CM and Anderson JM. Claudins and Epithelial Paracellular Transport Annu Rev Physiol 2006;68:403–429.
32. Frömter E and Diamond JM. Route of passive ion permeation in epithelia. Nature (New Biology) 1972;235:9–13.
33. Powell DW. Barrier function of epithelia. Am J Physiol 1981;241:G275–G288.
34. Bentzel CJ, Palant CE, Fromm M. Physiological and pathological factors affecting the tight junction. Boca Raton: CRC Press; 1992.
35. Rubin LL and Staddon JM. The cell biology of the blood-brain barrier. Annu Rev Neurosci 1999;22:11–28.
36. Le Moellic C, Boulkroun S, Gonzalez-Nunez D, Dublineau I, Cluzeaud F, Fay M, Blot-Chabaud M, Farman N. Aldosterone and tight junctions: modulation of claudin-4 phosphorylation in renal collecting duct cells. Am J Physiol Cell Physiol 2005;289:C1513–1521.
37. Ban Y and Rizzolo LJ. A culture model of development reveals multiple properties of RPE tight junctions. Mol Vis 1997;3:18. http://www.molvis.org/molvis/v3/ban.
38. Ban Y and Rizzolo LJ. Differential regulation of tight junction permeability during development of the retinal pigment epithelium. Am J Physiol 2000;279:C744–C750.
39. Balda MS, Whitney JA, Flores C, González S, Cereijido M, Matter K. Functional dissociation of paracellular permeability and transepithelial electrical resistance and disruption of the apical-basolateral intramembrane diffusion barrier by expression of a mutant tight junction membrane protein. J Cell Biol 1996;134:1031–1049.
40. McCarthy KM, Skare IB, Stankewich MC, Furuse M, Tsukita S, Rogers RA, Lynch RD, Schneeberger EE. Occludin is a functional component of the tight junction. J Cell Sci 1996;109:2287–2298.

41. Tang VW and Goodenough DA. Paracellular ion channel at the tight junction. Biophys J 2003;84:1660–1673.
42. Gallemore RP, Hughes BA, Miller SS. Retinal pigment epithelial transport mechanisms and their contributions to the electroretinogram. Prog Retin Eye Res 1997;16:509–566.
43. Butt AM, Jones HC, Abbott NJ. Electrical resistance across the blood-brain barrier in anaesthetized rats: a developmental study. J Physiol 1990;429:47–62.
44. Reinhardt CA and Gloor SM. Co-culture blood-brain barrier models and their use for pharmatoxicological screening. Toxicol in vitro 1997;11:513–518.
45. Ban Y, Wilt SD, Rizzolo LJ. Two secreted retinal factors regulate different stages of development of the outer blood-retinal barrier. Brain Res Dev Brain Res 2000;119:259–267.
46. Rizzolo LJ. Development and role of tight junctions in the retinal pigment epithelium. Int Rev Cytol 2006;258:195–234.
47. Rahner C, Fukuhara M, Peng S, Kojima S, Rizzolo LJ. The apical and basal environments of the retinal pigment epithelium regulate the maturation of tight junctions during development. J Cell Sci 2004;117:3307–3318.
48. Williams CD and Rizzolo LJ. Remodeling of junctional complexes during the development of the blood-retinal barrier. Anat Rec 1997;249:380–388.
49. Sandig M and Kalnins VI. Morphological changes in the zonula adhaerens during embryonic development of chick retinal pigment epithelial cells. Cell Tissue Res 1990;259:455–461.
50. Liu X, Mizoguchi A, Takeichi M, Honda Y, Ide C. Developmental changes in the subcellular localization of R-cadherin in chick retinal pigment epithelium. Histochem. Cell Biol 1997;108:35–43.
51. Grunwald GB. Cadherin cell adhesion molecules in retinal development and Pathology. Prog Retin Eye Res 1996;15:363–392.
52. Luo Y, Fukuhara M, Weitzman M, Rizzolo LJ. Expression of JAM-A, AF-6, PAR-3 and PAR-6 during the assembly and remodeling of RPE tight junctions. Brain Res 2006;1110:55–63.
53. Fujisawa H, Morioka H, Watanabe K, Nakamura H. A decay of gap junctions in association with cell differentiation of neural retina in chick embryonic development. J Cell Sci 1976;22:585–596.
54. Itoh M, Nagafuchi A, Yonemura S, Kitani-Yasuda T, Tsukita S, Tsukita S. The 220-kd protein colocalizing with cadherins in non-epithelial cells is identical to ZO-1,a tight junction-associated protein in epithelial cells: cDNA cloning and immunoelectron microscopy. J Cell Biol 1993;121:491–502.
55. Itoh M, Nagafuchi A, Moroi S, Tsukita S. Involvement of ZO-1 in cadherin-based cell adhesion through its direct binding to alpha catenin and actin filaments. J Cell Biol 1997;138:181–192.
56. Howarth AG, Hughes MR, Stevenson BR. Detection of the tight junction-associated protein ZO-1 in astrocytes and other nonepithelial cell types. Am J Physiol 1992;262:C461–469.
57. Yonemura S, Itoh M, Nagafuchi A, Tsukita S. Cell-to-cell adherens junction formation and actin filament organization: similarities and differences between non-polarized fibroblasts and polarized epithelial cells. J Cell Sci 1995;108 (Pt 1):127–142.
58. Dermietzel R and Krause D. Molecular anatomy of the blood-brain barrier as defined by immunocytochemistry. Int Rev Cytol 1991;127:57–109.
59. Saitou M, Ando-Akatsuka Y, Itoh M, Furuse M, Inazawa J, Fujimoto K, Tsukita S. Mammalian occludin in epithelial cells: its expression and subcellular distribution. Eur J Cell Biol 1997;73:222–231.
60. Bauer H, Stelzhammer W, Fuchs R, Weiger TM, Danninger C, Probst G, Krizbai IA. Astrocytes and neurons express the tight junction-specific protein occludin in vitro. Exp Cell Res 1999;250:434–438.

61. Barrios-Rodiles M, Brown KR, Ozdamar B, Bose R, Liu Z, Donovan RS, Shinjo F, Liu Y, Dembowy. J, Taylor IW, Luga V, Przulj N, Robinson M, Suzuki H, Hayashizaki Y, Jurisica I, J.L. W. High-throughput mapping of a dynamic signaling network in mammalian cells. Science 2005;307:1621–1625.
62. Luo Y, Zhuo Y, Fukuhara M, Rizzolo LJ. Effects of culture conditions on heterogeneity and the apical junctional complex of the ARPE-19 cell line. Invest Ophthalmol Vis Sci 2006;47:3644–3655.
63. Sheth B, Fontaine JJ, Ponza E, McCallum A, Page A, Citi S, Louvard D, Zahraoui A, Fleming TP. Differentiation of the epithelial apical junctional complex during mouse preimplantation development: a role for rab13 in the early maturation of the tight junction. Mech Dev 2000;97:93–104.
64. Kniesel U and Wolburg H. Tight junction complexity in the retinal pigment epithelium of the chicken during development. Neurosci Lett 1993;149:71–74.
65. Wolburg H, Liebner S, Reichenbach A, Gerhardt H. The pecten oculi of the chicken: a model system for vascular differentiation and barrier maturation. Int Rev Cytol 1999;187:111–159.
66. Ban Y and Rizzolo LJ. Regulation of glucose transporters during development of the retinal pigment epithelium. Brain Res Dev Brain Res 2000;121:89–95.
67. Stroeva OG and Mitashov VI. Retinal pigment epithelium: proliferation and differentiation during development and regeneration. Int Rev Cytol 1983;83:221–293.
68. Boado RJ and Pardridge WM. Amplification of gene expression using both 5'- and 3'-untranslated regions of GLUT1 glucose transporter mRNA. Brain Res Mol Brain Res 1999;63:371–374.
69. Qi C and Pekala PH. The Influence of mRNA Stability on Glucose Transporter (GLUT1) Gene Expression. Biochem Biophys Res Commun 1999;263:265–269.
70. Peng S, Rahner C, Rizzolo LJ. Apical and basal regulation of the permeability of the retinal pigment epithelium. Invest Ophthalmol Vis Sci 2003;44:808–817.
71. Rubin LL, Hall DE, Porter S, Barbu K, Cannon C, Horner HC, Janatpour M, Liaw CW, Manning K, Morales J, Tanner LI, Tomaselli KJ, Bard F. A cell culture model of the blood-brain barrier. J Cell Biol 1991;115:1725–1735.
72. Madara JL. Regulation of the movement of solutes across tight junctions. Annu Rev Physiol 1998;60:143–159.
73. Schultz SG and Solomon AK. Determination of the effective hydrodynamic radii of small molecules by viscometry. J Gen Physiol 1961;44:1189–1199.
74. Regina A, Roux F, Revest PA. Glucose transport in immortalized rat brain capillary endothelial cells in vitro: transport activity and GLUT1 expression. Biochim Biophys Acta 1997;1335:1351–43.
75. van Os C, de Jong MD, Slegers JF. Dimensions of polar pathways through rabbit gallbladder epithelium. The effect of phloretin on nonelectrolyte permeability. J Membr Biol 1974;15:363–382.

11
Ca²⁺ Channels in the Retinal Pigment Epithelium
Modulators of Retinal Pigment Epithelium Function and Communication with Neighboring Tissues

Olaf Strauss

CONTENTS

CALCIUM
Ca^{2+}-REGULATED FUNCTIONS OF THE RETINAL PIGMENT EPITHELIUM
Ca^{2+} CHANNELS IN GENERAL
Ca^{2+} CHANNELS IN THE RETINAL PIGMENT EPITHELIUM
SUMMARY
REFERENCES

CALCIUM

Ca^{2+} has a high affinity for proteins *(1–4)*. Ca^{2+} binding to proteins subsequently results in three-dimensional changes in the protein structure and function *(1–4)*. With respect to these properties, Ca^{2+} ions differ from many other ions. Thus, Ca^{2+} can regulate protein and cell function, and can serve as a second messenger in regulatory signalling pathways *(2,3,5–7)*. To fulfil this function cells keep a large Ca^{2+} concentration gradient between intracellular and extracellular space. Using active transporters and metabolic energy, cells maintain an intracellular Ca^{2+} concentration that is 10,000 times smaller than that of the extracellular space *(5,6)*. To achieve regulatory changes in cell function, even small increases in intracellular free Ca^{2+} are sufficient *(5,6)*. This occurs either by release of Ca^{2+} from cytosolic Ca^{2+} stores or by influx of Ca^{2+} into intracellular space. Ca^{2+} channels play a central role in the latter process.

Ca^{2+}-REGULATED FUNCTIONS OF THE RETINAL PIGMENT EPITHELIUM

Retinal pigment epithelium (RPE) cells contain a high concentration of Ca^{2+}, of which most is bound to the melanin and stored in melanosomes *(8–14)*. The retinal pigment epithelium closely interacts with photoreceptors and fulfils many functions that are essential for visual function *(15–17)*: these are the absorption of light, control of the

From: *Ophthalmology Research: Ocular Transporters in Ophthalmic Diseases and Drug Delivery*
Edited by: J. Tombran-Tink and C. J. Barnstable © Humana Press, Totowa, NJ

ion homeostasis in the subretinal space, regeneration of all-*trans* retinal, phagocytosis of shed, destroyed photoreceptor outer segments, the establishment of a barrier to ensure the immune privilege of eye is maintained, and the secretion of a variety of growth factors. Most of these functions have been found to be regulated by increases in intracellular free Ca^{2+} (Fig. 1A).

Light energy focused by the lens causes the destruction of proteins, lipoproteins, and lipids in photoreceptor outer segments *(16–18)*. This necessitates constant renewal of photoreceptor outer segments. New photoreceptor outer segments are built from the base while the destroyed tips are shed. The shed outer segments are phagocytosed by the RPE. Activation of the inositol triphosphate (InsP3)/Ca^{2+} system seems to play a role in the initiation of phagocytosis, and increases in intracellular free Ca^{2+} with subsequent activation of protein kinase C were found to represent the shut-off signal for phagocytosis. Thus, both Ca^{2+} signalling pathways regulate phagocytosis in concert.

A major mechanism maintaining the ion homeostasis in the subretinal space is the transepithelial transport of ions and water *(15)*. Increases in intracellular free Ca^{2+}, activated by a variety of different extracellular stimuli, were found to increase transepithelial transport of ions and water across the RPE *(19–29)*. Comparable effects were achieved by usage of Ca^{2+} ionophores to increase intracellular free Ca^{2+} *(20)*. The modulation of transepithelial ion transport results in the subsequent activation of Ca^{2+}-dependent Cl^- channels and Ca^{2+}-dependent K^+ channels *(20,25,27,30–38)*.

The secretion of growth factors by the RPE is essential for the maintenance of the structural integrity of photoreceptors and the endothelium of the choriocapillaris *(15,39,40)*. In addition, the secretion of immune suppressive factors by the RPE is a major mechanism by which the RPE participates in the establishment of the immune privilege of the eye *(41–47)*. Furthermore, secretion is a process that is central to the coordination of all other functions of the RPE. Phagocytosis, transepithelial transport and regeneration of all-trans retinal require a close communication between the RPE and the adjacent tissues *(15)*. The receptor-controlled secretion of a variety of factors represents an important part of this communication. In many different tissues transmitter release or exocytotic processes are triggered by increases in intracellular free Ca^{2+} *(7,48–51)*. Thus, it is likely that the same applies to the regulation of secretory activity in RPE cells. Although this is not extensively studied, many factors which are known to stimulate secretion by the RPE cells are also known to increase intracellular free Ca^{2+}. For example, basic fibroblast growth factor (bFGF) causes an increase in intracellular free Ca^{2+} and the secretion of vascular endothelial growth factor (VEGF) *(52)*. The same applies for insulin-like growth factor 1 (IGF-1) *(53)*.

Thus, increases in intracellular free Ca^{2+} are not only essential in the regulation of the different functions, but also contribute to the coordination of these functions between the RPE and the adjacent tissues. Ca^{2+} channels contribute to increases in intracellular free Ca^{2+} by providing a pathway for Ca^{2+} to enter the cell along the favorable electrochemical gradient. Even a small increase in the Ca^{2+} conductance caused by activation of Ca^{2+} channels, results in efficient changes in the intracellular Ca^{2+} concentration. Thus the study of the types of ion channels present in RPE cells, their characteristics and their regulation, which permits the participation in signalling pathways, are essential to understanding RPE function in health and disease.

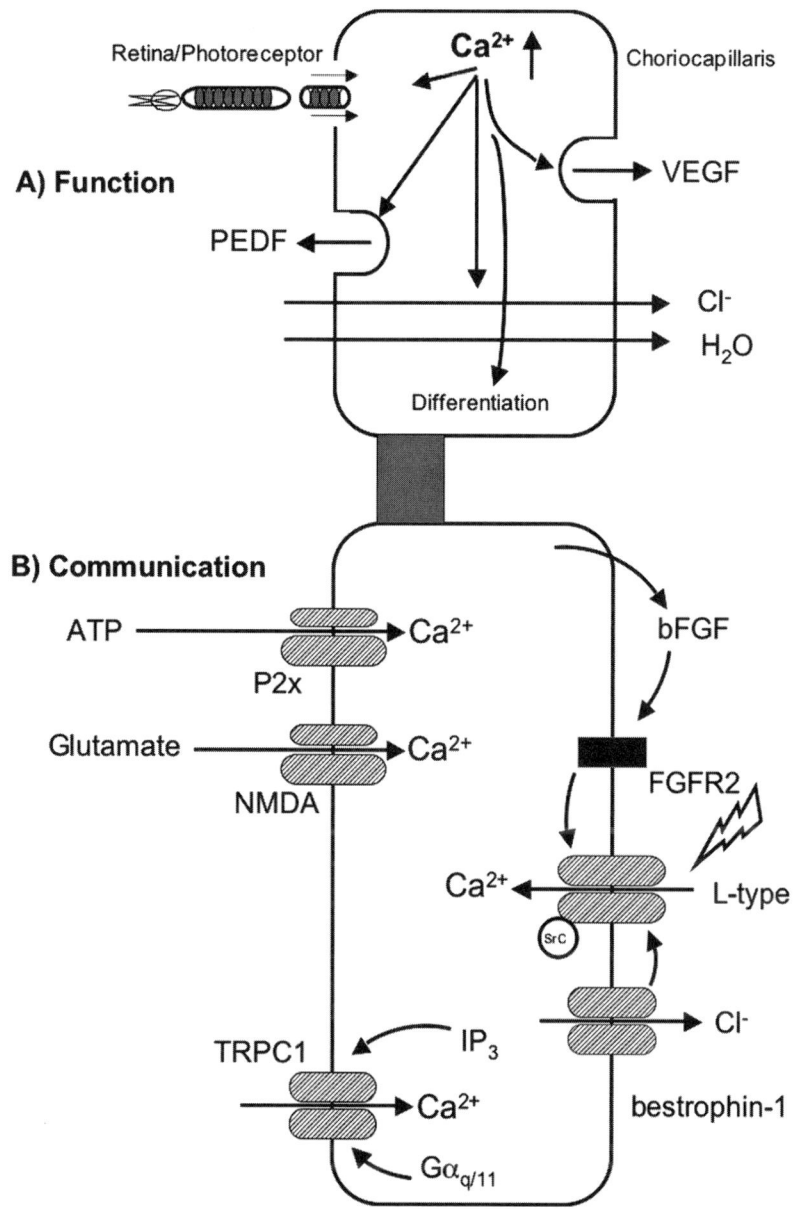

Fig. 1. Function of the retinal pigment epithelium (RPE) and Ca^{2+} channels. (**A**) Summary of different functions of the RPE that are known to be regulated by increases in intracellular free Ca^{2+}. (PEDF = pigment epithelium-derived factor; VEGF = vascular endothelial growth factor). (**B**) Ca^{2+} channels in the RPE and their regulation. Note that the localization within either the apical membrane or the basolateral membrane is not defined. ATP = adenosine triphosphate; bFGF = basic fibroblast growth factor; FGFR2 = fibroblast growth factor 2; $G\alpha_{q/11}$ = G protein α-subunit q/11; IP3 = inositol-1,4,5-trisphosphate; L-type = L-type voltage-dependent Ca^{2+} channel; NMDA = N-methyl-D-aspartate; P2X = purinergic receptor 2X; VEGF = vascular endothelial growth factor.

Ca^{2+} CHANNELS IN GENERAL

Ca^{2+} channels are membrane-spanning proteins that are highly specifically for the conduction of Ca^{2+} ions. They can be either activated by changes in the membrane potential, by intracellular or extracellular binding of specific ligands, or by a variety of different physical stimuli such as temperature.

The voltage-gated channels are members of a gene family of 10 homologous proteins *(50,54,55)*. All these channels are composed of four homologous domains, of which each consists of six membrane-spanning domains. The extracellular loop between the fifth and the sixth membrane-spanning domain dips into the membrane. All four homologous domains are arranged in such a way that the four loops which dip into the membrane form the membrane-spanning Ca^{2+} pore. The family of voltage-dependent Ca^{2+} channels contains three groups: L-type, P/N-type and T-type channels. The voltage-dependent Ca^{2+} channels differ in their voltage-dependence, time-dependent kinetics, and blocker sensitivity.

Another large group of Ca^{2+} channels are ligand-gated channels. This group is represented by the different ionotropic transmitter receptors. These include, for example, glutamate receptors *(56,57)* and adenosine triphosphate (ATP) receptors *(58,59)*. Extracellular binding of the specific ligand results in the activation of channel and an influx of Ca^{2+} into the cells. Other ligand-gated channels are activated by intracellular ligands. For example, cyclic guanidine monophosphate (cGMP)-gated channels are involved in the phototransduction cascade in the photoreceptor outer segments *(60)*. Some of the ligand-gated channels are not true Ca^{2+} channels, because they can also conduct other ions such as Na^+. Nevertheless, the Ca^{2+} that enters the cell serves to provide the major regulatory effects via the ligand-gated channel.

A third group of Ca^{2+}-conducting channels is the large family of TRP channels *(61–63)*. The gene was first identified in *Drosophila* where the mutant form leads to the transient receptor potential phenotype (trp). The family of the TRP channels in humans consists of six groups, whose members differ in the many ways through which they can be activated, e.g., via α-subunits of G-proteins, (cADP-ribox) cADP-ribose, intracellular Ca^{2+}, release of Ca^{2+} from cytosolic stores, acidic pH or changes in the temperature. With these many different activation mechanisms, TRP channels are involved in many important sensory and regulatory signalling pathways.

Ca^{2+} CHANNELS IN THE RETINAL PIGMENT EPITHELIUM

So far, voltage-dependent Ca^{2+} channels, a member of the TRP family of channels and ligand-gated Ca^{2+} channels have been identified in RPE cells (Fig. 1B). Of these channels, the voltage-dependent channels have been most extensively investigated.

Voltage-Dependent Ca^{2+} Channels

Early patch-clamp experiments with either freshly isolated or cultured RPE cells from rat and human indicated the functional presence of voltage-dependent Ca^{2+} channels in the RPE *(64–72)*. These were identified as Ba^{2+} currents in whole-cell patch-clamp experiments. The currents were activated at rather positive membrane potentials, showed in some cases slow inactivation kinetics and were influenced by dihydropyridine compounds such as nifedpine, a Ca^{2+} channel blocker, and BayK8644, a Ca^{2+} channel activator. Owing to these electrophysiological properties, these currents could be identified as currents through L-type Ca^{2+} channels *(73)*. Furthermore, systemic application of the dihydropyridine nimodipine, an L-type channel blocker, reduced the amplitude of the light-peak in the rat and mouse DC-electroretinogram (ERG) *(74)*. This indicates the functional presence of L-type channels in the intact eye. Further molecular biological analysis of the Ca^{2+} channel proteins and Ca^{2+} channel genes indicate that RPE cells primarily express a neuroendocrine subtype of L-type Ca^{2+} channels, which is represented by the presence of the α1D or $Ca_V 1.3$ subunit in RPE cells *(66–69)*. Reverse-transcriptase polymerase chain reaction (RT-PCR) analysis showed that RPE cells express, exclusively, a splicing variant of the neuroendocrine L-type channel that has not been reported in other studies of other tissues *(66)*.

The Functional Role of Neuroendocrine L-Type Ca^{2+} Channels in the Retinal Pigment Epithelium

To understand the functional role of these high-voltage-activated channels in retinal epithelial cells, which can be considered as nonexcitable cells, several questions have to be answered. The first is whether these Ca^{2+} channels can contribute to changes in intracellular free Ca^{2+}, because L-type channels are high-voltage-activated channels, which need rather positive membrane potentials to be activated, and RPE cells show a resting potential of between −40 and −50 mV *(25,33,66,70,72,75–80)*. An answer arises from heterologous expression studies in which currents through $Ca_V 1.3$ subunits were investigated. Here, these channels appeared to be subtype of L-type channels with the most negative activation thresholds among L-type channels, which can be as negative as −40 mV *(81–83)*. These characteristics found in heterologous expression systems were confirmed in studies with native cells which exclusively express this L-type channel subtype *(82)*. Furthermore, these channels can contribute to changes in intracellular free Ca^{2+} not only by altering the membrane potential, but also by shifts in their voltage-dependent activation towards more negative potentials by phosphorylation of the $Ca_V 1.3$ subunit *(66,67)*. In a combined study of patch-clamp measurements of the L-type channels currents together with changes in intracellular free Ca^{2+} in the same cell, it could be demonstrated that Ca^{2+} can enter the cell through L-type channels at a fixed membrane potential of −40 mV *(84)*. This is enabled by phosphorylation of the Ca^{2+} channel by tyrosine kinases. The cytosolic subtype of tyrosine kinase, $pp60^{c-src}$, appeared to be a housekeeping kinase which physically interacts with $Ca_V 1.3$ subunit and is responsible for the basic activity of the L-type channel in these cells *(68,69)*. However, this kinase also seems to integrate L-type channels into the InsP3/Ca^{2+} second messenger system. Intracellular application of InsP3 to cultured RPE cells resulted in an increase in cytosolic free Ca^{2+}, caused by an influx of Ca^{2+} through L-type channels activated by shifts of the voltage-dependent activation into a more negative voltage range *(84)*. This was dependent on activation of tyrosine kinase. Another tyrosine kinase which leads to activation of L-type channels through a shift in the voltage dependence is the bFGF receptor FGFR2 *(67)*. Application of bFGF leads to an increase in intracellular free Ca^{2+} by Ca^{2+} influx through L-type channels *(85)*. This is enabled by physical interaction of FGFR2 with the $Ca_V 1.3$ subunit, which also results in a shift in the voltage dependence towards more negative potentials *(67)*. The participation of $pp60^{c-src}$ could be excluded in this signalling pathway. Among serine/threonine kinases, protein kinase C appears to be an important regulator of L-type channels in the RPE. Protein kinase C is, on the one hand, an activator of L-type channels in the RPE *(69)*, however, activation of PKC seems to influence the tyrosine kinase-dependent activation of the L-type channels. In cells with stimulated PKC, pp60$^{c\text{-src}}$ then exerts an inhibitory effect on L-type channel activity. The nature of this process is not understood, but it is likely that different isoforms of PKC might contribute to these different effects. Finally, we could identify bestrophin-1 as a regulator of L-type channel activity, which also results in shifts in the voltage-dependence and in the kinetic behaviour of the L-type channels *(66)*. It is not clear whether this effect relies on a physical interaction or on an indirect stimulation based on phosphorylation-dependent processes.

In summary, L-type channels in RPE cells are capable of contributing to changes in intracellular free Ca^{2+}, not by changing the membrane potential, but by phosphorylation-dependent changes in the voltage-dependence of these channels. With these properties, L-type channels in the RPE are linked to many different signalling processes that can be initiated by tyrosine phosphorylation, for example by growth factors, or by stimulation of the InsP3/Ca^{2+} second-messenger system. Thus, the next question arises from the effects resulting from activation of L-type channels in RPE cells.

Intracellular application of InsP3 to rat RPE cells resulted in an increase in the membrane conductance for Cl^- *(36,86)*. This effect is based on the activation of Ca^{2+}-dependent Cl^- channels. The underlying increase in intracellular free Ca^{2+} partially results from activation of L-type channels *(36,86)*. A close regulation of Cl^- conductance and L-type channel activity is supported by a study in which the regulatory effect of bestrophin-1 on L-type channels was demonstrated *(66)*. Bestrophins were also identified as Cl^- channels, among which bestrophin-1 could represent a Ca^{2+}-dependent Cl^- channel in RPE cells *(37,87)*. Thus, the two ion channels are not only coupled by Ca^{2+}, but also by direct interactions that provide a close feedback loop in the control of transepithelial Cl^- transport. This hypothesis is supported by the observation that both L-type channels and bestrophin-1 are involved in a regulatory manner in the generation of the light-peak in the electro-oculogram *(74,88)*. This signal was shown to arise from activation of Cl^- channels in the basolateral membrane of the RPE in response to activation of a second-messenger cascade that is probably initiated by stimulation of purinergic receptors *(24,89,90)*. However, the exact contribution of the two proteins, L-type channels and bestrophin-1, to the light peak is not understood. As stated above, the L-type channels are regulated by protein tyrosine kinase *(68)*. This links these ion channels to growth-factor-dependent signalling. Even a growth factor receptor, the bFGF receptor FGFR2, is able to interact directly with the α1-subunit of the Ca^{2+} channel *(67)*. Since growth factors are able to change cell differentiation and behaviour by changing the gene expression profile, it might be possible that growth-factor-dependent activation of L-type channels can also contribute to changes in the gene expression of RPE cells. This idea is supported by observations in neurons, in which activation of L-type channels leads to increased expression of the immediate–early gene c-fos, whose product functions as a transcription factor *(91)*. Indeed, direct opening of L-type channels by the Ca^{2+}-channel opener dihydropyridine compound BayK8644 increased the expression rate of c-fos in RPE cells *(52)*. However, the bFGF-induced increase in the c-fos expression rate is independent of activation of FGFR2 and L-type channels *(52)*. Thus, only very distinct signalling pathways would be able to change c-fos expression via stimulation of L-type channels, while the L-type channel-dependent change in gene expression acts as a general mechanism.

The last effect L-type channels might have in RPE cells is implied by the subclass of L-type channels expressed in the RPE, the neuroendocrine channels. These L-type channels have been described in pancreatic beta islet cells *(55,83)* and were found to trigger the release of insulin by these cells *(48,92)*. Thus, L-type channels might have a comparable function in the RPE, where they may regulate secretory activity *(93)*. One important function of the RPE is the secretion of growth factors, but also of other substances, that are of importance to coordinate RPE function with that of adjacent tissues *(15,39)*. As stated above, bFGF can stimulate, via FGFR2, L-type channels in the RPE, resulting in a subsequent increase in intracellular free Ca^{2+} in these cells. In addition, bFGF is able to stimulate the secretion of VEGF. Indeed, the bFGF-stimulated increase in the VEGF secretion rate is dependent on the activation of L-type channels. It is very likely that L-type channels regulate the secretion of other important factors that represent the basis for the communication with adjacent tissues.

L-type Channels and Disease in the Retinal Pigment Epithelium

As regulators of secretion, which stands in the centre of the communication of the RPE with adjacent tissues, L-type channels are likely to play a role in pathophysiologic processes leading to loss of vision. One factor that is released under the control of L-type channels, and which might play a central role in pathophysiologic processes, is VEGF *(93)*. The effect of VEGF depends on its concentration *(94)*; at lower concentrations VEGF is important for maintenance of structural integrity of the endothelium of blood vessels, such as maintaining fenestrated structure, at higher concentrations VEGF initiates neovascularisation. In age-related macular degeneration, the major cause for loss of vision is choroidal neovascularisation *(95)*. It is generally accepted that this process is caused by increased levels of VEGF *(40,96)*. Anti-VEGF strategies have been recently established as a successful therapy to treat choroidal neovascularisation *(97)*. A considerable body of evidence demonstrates that the RPE is the major source of VEGF in this process *(94,96,98–101)*. Since L-type channels regulate the secretion of VEGF, they must play an important role in the initiation of choroidal neovascularisation. Indeed, in a study of freshly isolated RPE cells from surgically removed neovascular tissues we were able to demonstrate that these cells show functional presence of L-type channels *(93)*. Furthermore, these channels seem to have different tyrosine kinase-dependent regulatory properties, causing a higher VEGF secretion rate, even in cultured cells.

Another role of L-type channels in pathophysiological processes might be an involvement in a rare inherited form of macular degeneration, Best's vitelliform macular degeneration. As stated above, bestrophin-1, the product of the VMD2 gene, which causes Best's disease, can regulate the activity of L-type channels *(66)*. In an initial study that explored the function of bestrophin-1, it was shown that the mutant forms known to cause the disease display no Cl^- channel activity *(37)*. Thus, is it was concluded that the loss of Cl^- channel activity represents the major cause of the disease and explains the absence of the light-peak in the patients' electro-oculograms. However, several reports show that carriers of VMD2 mutations can have normal light-peaks, or decreases in the light-peak amplitude, that occur secondary to the onset of macular degeneration *(102–106)*. Furthermore, bestrophin-1 knockout mice show no decrease in the light peak amplitude and no retinal degeneration *(74)*. Thus, the loss of Cl⁻-channel activity in mutant bestrophin-1 is not necessarily the major cause of both macular degeneration and changes in the patients' electro-oculograms. Since the light-peak is also dependent on the activity of L-type channels, and since bestrophin-1 is able to change the activity of L-type channels, it could be that a mutation-dependent change in the L-type channel activity and a subsequent change in the Ca^{2+} homeostasis are involved in the chain of events leading to the disease *(66)*. This alternative hypothesis is supported by the observation that mutant bestrophin-1 displays different effects on the activity of L-type channels than the wild type *(66)*. These effects result in a prolonged activity of L-type channels or, by another mutant form, faster inactivation kinetics.

Finally, there are some indications that L-type channels possibly participate in the retinal degeneration of the Royal College of Surgeons (RCS) rat *(85)*. The retinal degeneration in this animal model for retinitis pigmentosa is caused by a mutation in a gene for a receptor tyrosine kinase, leading to a loss of the ability to phagocytose shed photoreceptor outer segments *(107–111)*. RPE cells of the RCS rat show increased activity of L-type channels due to altered regulation by tyrosine kinase, and subsequent faster activation of InsP3-dependent activation of Ca^{2+}-dependent Cl⁻ channels. It is not clear whether these changes are directly involved in the chain of events leading to the retinal degeneration or whether these are secondary effects.

TRP Channels

A recent study using ARPE-19 cells indicated the presence of the TRPC1 Ca^{2+} channel *(112)*. This is the first report of the expression of TRP channels in the RPE. The authors could demonstrate a physical interaction with proteins of the cytoskeleton. With such an interaction, the presence of TRP channels in cell membrane, and thus the Ca^{2+} permeability of the cell membrane, could be regulated. The TRPC1 channel is further known to be regulated by α-subunits of G proteins *(61–63,113)*. The presence of these channels could explain observations that the activation of a non-specific Ca^{2+}-conducting ion channel in RPE cells is initiated by purinergic stimulation *(27,114)*. The activation of the Ca^{2+} conductance was found to be G-protein dependent. Also an investigation of InsP3/Ca^{2+} second-messenger system in RPE cells indicated the participation of other Ca^{2+} channels than L-type channels *(84)*. Since TRP channels are known to be stimulated by metabolites of the InsP3 second-messenger system, it is likely that TRPC1 represents this additional Ca^{2+}-conducting channel *(61,63)*. TRP channel s are known to be activated by $G\alpha_{11/q}$ subunits. Thus it is likely that these channels contribute to changes in the RPE function, which result from stimulation of $G\alpha_{11/q}$ subunits coupled receptors. These changes in transepithelial ion transport are increased by stimulation of muscarinic P2Y ATP receptors, or adrenergic receptors. Others changes also influence phagocytic activity by P2Y or adrenergic receptor activity (113). Thus the TRPC1 channels in the RPE might represent another type Ca^{2+} channel of importance for the communication of the RPE with the adjacent tissues. However, the paper by Bollimuntha et al. *(112)* is a first initial report about TRP channels in the RPE and the definition of their functional role requires further experimental evidence.

Ionotropic Receptors

Purinergic Receptors

In many studies the importance of purinergic signalling in the RPE has been demonstrated *(22,24,25,27,29,115–117)*. Most of the studies describe the involvement of metabotropic ATP receptors. However, earlier studies indicated the presence of ionotropic ATP receptors, also, in the RPE *(27)*. Although not shown on a molecular basis, investigation of ATP-dependent Ca^{2+} signalling in the RPE, together with measurements of the membrane conductance, showed the ATP-dependent activation of a cation conductance in conjunction with rises in intracellular free Ca^{2+} as a second-messenger *(27)*. This cation conductance showed properties of ionotropic ATP receptors, the P2X receptors *(59)*.

The source of ATP for stimulation of purinergic signalling in RPE cells might be the retina or the RPE itself *(22,117)*. It is likely that stimulation of the retina by light results in the release of ATP by the retina, which diffuses to the RPE. Stimulation of RPE cells by bFGF or ATP leads to the release of ATP from RPE cells *(22,117)*. Stimulation of RPE cells by ATP results in a lot of different changes in RPE cell function, such as transepithelial transport of water and ions, or phagocytosis *(22)*. The presence of metabotropic receptors has been demonstrated on a functional and molecular basis. The additional presence of ionotropic receptors is likely. The presence of both types of ATP receptors might help us to understand how ATP can contribute to these different changes in RPE cells function.

Glutamate Receptors

Astonishingly, ionotropic glutamate receptors were described in RPE cells on both a functional and molecular basis *(118–120)*. These glutamate receptors were identified as N-methy-D-aspartate (NMDA) receptors and kainate receptors *(120)*, which are normally known to function as neurotransmitter receptors in the postsynaptic membrane of synaptic clefts. Activation of NMDA receptors results in a fast, inactivating depolarisation of the cell accompanied by fast, transient Ca^{2+} influx into the cell, which can be expected in RPE cells as well. However, it is unclear as to which changes in cell function glutamate-dependent stimulation of RPE cells results in. Only for kainate receptors, which cannot conduct Ca^{2+}, could it be shown that these can regulate phagocytosis by the RPE *(120)*.

SUMMARY

The RPE closely interacts with its adjacent tissues and helps to maintain their structural integrity. The interaction with photoreceptors is of importance for the maintenance of visual function. The coordinated interaction needed to fulfil these tasks requires a close communication between the RPE and the photoreceptors on one side, and the endothelium of the choriocapillaris on the other side.

REFERENCES

1. Carafoli E. The Symposia on Calcium Binding Proteins and Calcium Function in Health and Disease: an historical account, and an appraisal of their role in spreading the calcium message. Cell Calcium 2005;37:279–81.
2. Carafoli E. Calcium–a universal carrier of biological signals. Delivered on 3 July 2003 at the Special FEBS Meeting in Brussels. FEBS J 2005;272:1073–89.
3. Williams RJ. Calcium ions: their ligands and their functions. Biochem Soc Symp 1974; 133–8.
4. Williams RJ. Calcium-binding proteins in normal and transformed cells. Cell Calcium 1994;16:339–46.
5. Berridge MJ. Unlocking the secrets of cell signaling. Annu Rev Physiol 2005;67:1–21.
6. Berridge MJ, Lipp P, Bootman MD. The versatility and universality of calcium signalling. Nat Rev Mol Cell Biol 2000;1:11–21.
7. Shuttleworth TJ. Intracellular Ca2+ signalling in secretory cells. J Exp Biol 1997;200:303–14.
8. Bellhorn MB and Lewis RK. Localization of ions in retina by secondary ion mass spectrometry. Exp Eye Res 1976;22:505–18.
9. Boulton M. Ageing of the retinal pigment epithelium. In: Osborne NN and Chader GJ, eds. Progress in retinal research. New York: Pergamon, Oxford; 1991. p. 125–51.
10. Boulton M, Dayhaw-Barker P. The role of the retinal pigment epithelium: topographical variation and ageing changes. Eye 2001;15:384–389.
11. Drager UC. Calcium binding in pigmented and albino eyes. Proc Natl Acad Sci USA 1985;82:6716–20.
12. Fishman ML, Oberc MA, Hess HH, Engel WK. Ultrastructural demonstration of calcium in retina, retinal pigment epithelium and choroid. Exp Eye Res 1977;24:341–53.
13. Hess HH. The high calcium content of retinal pigmented epithelium. Exp Eye Res 1975;21:471–9.
14. Ulshafer RJ, Allen CB, Rubin ML. Distributions of elements in the human retinal pigment epithelium. Arch Ophthalmol 1990;108:113–7.
15. Strauss O. The Retinal Pigment Epithelium in Visual Function. Physiol Rev 2005; 85:845–851.
16. Bok D. The retinal pigment epithelium: a versatile partner in vision. J Cell Sci Suppl 1993;17:189–95.
17. Steinberg RH. Interactions between the retinal pigment epithelium and the neural retina. Doc Ophthalmol 1985;60:327–46.
18. Nguyen-Legros J and Hicks D. Renewal of photoreceptor outer segments and their phagocytosis by the retinal pigment epithelium. Int Rev Cytol 2000;196:245–313.
19. Edelman JL and Miller SS. Epinephrine stimulates fluid absorption across bovine retinal pigment epithelium. Invest Ophthalmol Vis Sci 1991;32:3033–40.
20. Joseph DP and Miller SS. Alpha-1-adrenergic modulation of K and Cl transport in bovine retinal pigment epithelium. J Gen Physiol 1992;99:263–90.
21. Maminishkis A, Jalickee S, Blaug SA, Rymer J, Yerxa BR, Peterson WM, Miller SS. The P2Y(2) receptor agonist INS37217 stimulates RPE fluid transport in vitro and retinal reattachment in rat. Invest Ophthalmol Vis Sci 2002;43:3555–66.
22. Mitchell CH. Release of ATP by a human retinal pigment epithelial cell line: potential for autocrine stimulation through subretinal space. J Physiol 2001;534:193–202.
23. Nash MS and Osborne NN. Agonist-induced effects on cyclic AMP metabolism are affected in pigment epithelial cells of the Royal College of Surgeons rat. Neurochem Int 1995;27: 253–62.

24. Peterson WM, Meggyesy C, Yu K, Miller SS. Extracellular ATP activates calcium signaling, ion, and fluid transport in retinal pigment epithelium. J Neurosci 1997;17:2324–37.
25. Quinn RH and Miller SS. Ion transport mechanisms in native human retinal pigment epithelium. Invest Ophthalmol Vis Sci 1992;33:3513–27.
26. Quinn RH, Quong JN, Miller SS. Adrenergic receptor activated ion transport in human fetal retinal pigment epithelium. Invest Ophthalmol Vis Sci 2001;42:255–64.
27. Ryan JS, Baldridge WH, Kelly ME. Purinergic regulation of cation conductances and intracellular Ca2+ in cultured rat retinal pigment epithelial cells. J Physiol 1999;520(3):745–59.
28. Rymer J, Miller SS, Edelman JL. Epinephrine-induced increases in [Ca2+](in) and KCl-coupled fluid absorption in bovine RPE. Invest Ophthalmol Vis Sci 2001;42:1921–9.
29. Sullivan DM, Erb L, Anglade E, Weisman GA, Turner JT, Csaky KG. Identification and characterization of P2Y2 nucleotide receptors in human retinal pigment epithelial cells. J Neurosci Res 1997;49:43–52.
30. Bialek S, Quong JN, Yu K, Miller SS. Nonsteroidal anti-inflammatory drugs alter chloride and fluid transport in bovine retinal pigment epithelium. Am J Physiol 1996;270: C1175–89.
31. Fischmeister R and Hartzell HC. Volume sensitivity of the bestrophin family of chloride channels. J Physiol 2005;562:477–91.
32. Hu JG, Gallemore RP, Bok D, Frambach DA. Chloride transport in cultured fetal human retinal pigment epithelium. Exp Eye Res 1996;62:443–8.
33. Miller SS and Edelman JL. Active ion transport pathways in the bovine retinal pigment epithelium. J Physiol 1990;424:283–300.
34. Ryan JS and Kelly ME. Activation of a nonspecific cation current in rat cultured retinal pigment epithelial cells: involvement of a G(alpha i) subunit protein and the mitogen-activated protein kinase signalling pathway. Br J Pharmacol 1998;124:1115–22.
35. Sheu SJ and Wu SN. Mechanism of inhibitory actions of oxidizing agents on calcium-activated potassium current in cultured pigment epithelial cells of the human retina. Invest Ophthalmol Vis Sci 2003;44:1237–44.
36. Strauss O, Wiederholt M, Wienrich M. Activation of Cl- currents in cultured rat retinal pigment epithelial cells by intracellular applications of inositol-1,4,5-triphosphate: differences between rats with retinal dystrophy (RCS) and normal rats. J Membr Biol 1996;151:189–200.
37. Sun H, Tsunenari T, Yau KW, Nathans J. The vitelliform macular dystrophy protein defines a new family of chloride channels. Proc Natl Acad Sci USA 2002;99:4008–13.
38. Tao Q and Kelly ME. Calcium-activated potassium current in cultured rabbit retinal pigment epithelial cells. Curr Eye Res 1996;15:237–46.
39. Tanihara H, Inatani M, Honda Y. Growth factors and their receptors in the retina and pigment epithelium. Prog Retin Eye Res 1997;16:271–301.
40. Campochiaro PA. Cytokine production by retinal pigmented epithelial cells. Int Rev Cytol 1993;146:75–82.
41. Crane IJ, Wallace CA, McKillop-Smith S, Forrester JV. CXCR4 receptor expression on human retinal pigment epithelial cells from the blood-retina barrier leads to chemokine secretion and migration in response to stromal cell-derived factor 1 alpha. J Immunol 2000;165:4372–8.
42. Enzmann V, Hollborn M, Kuhnhoff S, Wiedemann P, Kohen L. Influence of interleukin 10 and transforming growth factor-beta on T cell stimulation through allogeneic retinal pigment epithelium cells in vitro. Ophthalmic Res 2002;34:232–40.

43. Ishida K, Panjwani N, Cao Z, Streilein JW. Participation of pigment epithelium in ocular immune privilege. 3. Epithelia cultured from iris, ciliary body, and retina suppress T-cell activation by partially non-overlapping mechanisms. Ocul Immunol Inflamm 2003;11:91–105.
44. Matsumoto M, Yoshimura N, Honda Y. Increased production of transforming growth factor-beta 2 from cultured human retinal pigment epithelial cells by photocoagulation. Invest Ophthalmol Vis Sci 1994;35:4245–52.
45. Streilein JW. Ocular immune privilege: therapeutic opportunities from an experiment of nature. Nat Rev Immunol 2003;3:879–89.
46. Streilein JW, Ma N, Wenkel H, Ng TF, Zamiri P. Immunobiology and privilege of neuronal retina and pigment epithelium transplants. Vision Res 2002;42:487–95.
47. Wenkel H and Streilein JW. Evidence that retinal pigment epithelium functions as an immune-privileged tissue. Invest Ophthalmol Vis Sci 2000;41:3467–73.
48. Sher E, Giovannini F, Codignola A, Passafaro M, Giorgi-Rossi P, Volsen S, Craig P, Davalli A, Carrera P. Voltage-operated calcium channel heterogeneity in pancreatic beta cells: physiopathological implications. J Bioenerg Biomembr 2003;35:687–96.
49. Rossier MF, Burnay MM, Vallotton MB, Capponi AM. Distinct functions of T- and L-type calcium channels during activation of bovine adrenal glomerulosa cells. Endocrinology 1996;137:4817–26.
50. Catterall WA. Structure and regulation of voltage-gated Ca^{2+} channels. Annu Rev Cell Dev Biol 2000;16:521–55.
51. Catterall WA. Structure and function of neuronal Ca^{2+} channels and their role in neurotransmitter release. Cell Calcium 1998;24:307–23.
52. Rosenthal R, Malek G, Salomon N, Peill-Meininghaus M, Coeppicus L, Wohlleben H, Wimmers S, Bowes Rickman C, Strauss O. The fibroblast growth factor receptors, FGFR-1 and FGFR-2, mediate two independent signalling pathways in human retinal pigment epithelial cells. Biochem Biophys Res Commun 2005;337:241–7.
53. Rosenthal R, Wohlleben H, Malek G, Schlichting L, Thieme H, Bowes Rickman C, Strauss O. Insulin-like growth factor-1 contributes to neovascularization in age-related macular degeneration. Biochem Biophys Res Commun 2004;323:1203–8.
54. Catterall WA, Perez-Reyes E, Snutch TP, Striessnig J. International Union of Pharmacology. XLVIII. Nomenclature and structure-function relationships of voltage-gated calcium channels. Pharmacol Rev 2005;57:411–25.
55. Striessnig J, Hoda JC, Koschak A, Zaghetto F, Mullner C, Sinnegger-Brauns MJ, Wild C, Watschinger K, Trockenbacher A, Pelster G. L-type Ca^{2+} channels in Ca^{2+} channelopathies. Biochem Biophys Res Commun 2004;322:1341–6.
56. Kew JN and Kemp JA. Ionotropic and metabotropic glutamate receptor structure and pharmacology. Psychopharmacology (Berl) 2005;179:4–29.
57. Mayer ML. Glutamate receptor ion channels. Curr Opin Neurobiol 2005;15:282–8.
58. Burnstock G. Introduction: P2 receptors. Curr Top Med Chem 2004;4:793–803.
59. North RA. Molecular physiology of P2X receptors. Physiol Rev 2002;82:1013–67.
60. Kaupp UB and Seifert R. Cyclic nucleotide-gated ion channels. Physiol Rev 2002;82:769–824.
61. Inoue R. TRP channels as a newly emerging non-voltage-gated Ca^{2+} entry channel superfamily. Curr Pharm Des 2005;11:1899–914.
62. Clapham DE, Montell C, Schultz G, Julius D. International Union of Pharmacology. XLIII. Compendium of voltage-gated ion channels: transient receptor potential channels. Pharmacol Rev 2003;55:591–6.

63. Ramsey IS, Delling M, Clapham DE. An Introduction to TRP Channels. Annu Rev Physiol 2005.
64. Ueda Y and Steinberg RH. Voltage-operated calcium channels in fresh and cultured rat retinal pigment epithelial cells. Invest Ophthalmol Vis Sci 1993;34:3408–18.
65. Ueda Y and Steinberg RH. Dihydropyridine-sensitive calcium currents in freshly isolated human and monkey retinal pigment epithelial cells. Invest Ophthalmol Vis Sci 1995;36:373–80.
66. Rosenthal R, Bakall B, Kinnick T, Peachey N, Wimmers S, Wadelius C, Marmorstein A, Strauss O. Expression of bestrophin-1, the product of the VMD2 gene, modulates voltage-dependent Ca^{2+} channels in retinal pigment epithelial cells. FASEB J 2006;20:178–80.
67. Rosenthal R, Thieme H, Strauss O. Fibroblast growth factor receptor 2 (FGFR2) in brain neurons and retinal pigment epithelial cells act via stimulation of neuroendocrine L-type channels (Ca(v)1.3). FASEB J 2001;15:970–7.
68. Strauss O, Buss F, Rosenthal R, Fischer D, Mergler S, Stumpff F, Thieme H. Activation of neuroendocrine L-type channels (alpha1D subunits) in retinal pigment epithelial cells and brain neurons by pp60(c-src). Biochem Biophys Res Commun 2000;270:806–10.
69. Strauss O, Mergler S, Wiederholt M. Regulation of L-type calcium channels by protein tyrosine kinase and protein kinase C in cultured rat and human retinal pigment epithelial cells. FASEB J 1997;11:859–67.
70. Strauss O and Wienrich M. Cultured retinal pigment epithelial cells from RCS rats express an increased calcium conductance compared with cells from non-dystrophic rats. Pflugers Arch 1993;425:68–76.
71. Strauss O and Wienrich M. Ca(2+)-conductances in cultured rat retinal pigment epithelial cells. J Cell Physiol 1994;160:89–96.
72. Wollmann G, Lenzner S, Berger W, Rosenthal R, Karl MO, Strauss O. Voltage-dependent ion channels in the mouse RPE: Comparison with Norrie disease mice. Vision Res 2006;46:688–98.
73. McDonald TF, Pelzer S, Trautwein W, Pelzer DJ. Regulation and modulation of calcium channels in cardiac, skeletal, and smooth muscle cells. Physiol Rev 1994;74:365–507.
74. Marmorstein LY, Wu J, McLaughlin P, Yocom J, Karl MO, Neussert R, Wimmers S, Stanton JB, Gregg RG, Strauss O, Peachey NS, Marmorstein AD. The Light Peak of the Electroretinogram Is Dependent on Voltage-gated Calcium Channels and Antagonized by Bestrophin (Best-1). J Gen Physiol 2006;127:577–89.
75. Hughes BA, Takahira M, Segawa Y. An outwardly rectifying K^+ current active near resting potential in human retinal pigment epithelial cells. Am J Physiol 1995;269:C179–87.
76. Hughes BA and Steinberg RH. Voltage-dependent currents in isolated cells of the frog retinal pigment epithelium. J Physiol 1990;428:273–97.
77. la Cour M. The retinal pigment epithelium controls the potassium activity in the subretinal space. Acta Ophthalmol 1985;173(Suppl):9–10.
78. Miller SS and Steinberg RH. Active transport of ions across frog retinal pigment epithelium. Exp Eye Res 1977;25:235–48.
79. Miller SS and Steinberg RH. Passive ionic properties of frog retinal pigment epithelium. J Membr Biol 1977;36:337–72.
80. Oakley B 2nd, Steinberg RH, Miller SS, Nilsson SE. The in vitro frog pigment epithelial cell hyperpolarization in response to light. Invest Ophthalmol Vis Sci 1977;16:771–4.

81. Koschak A, Reimer D, Huber I, Grabner M, Glossmann H, Engel J, Striessnig J. alpha 1D (Cav1.3) subunits can form I-type Ca^{2+} channels activating at negative voltages. J Biol Chem 2001;276:22100–6.
82. Michna M, Knirsch M, Hoda JC, Muenkner S, Langer P, Platzer J, Striessnig J, Engel J. Cav1.3 (alpha1D) Ca^{2+} currents in neonatal outer hair cells of mice. J Physiol 2003; 553:747–58.
83. Scholze A, Plant TD, Dolphin AC, Nurnberg B. Functional expression and characterization of a voltage-gated CaV1.3 (alpha1D) calcium channel subunit from an insulin-secreting cell line. Mol Endocrinol 2001;15:1211–21.
84. Mergler S and Strauss O. Stimulation of L-type Ca(2+) channels by increase of intracellular InsP3 in rat retinal pigment epithelial cells. Exp Eye Res 2002;74:29–40.
85. Mergler S, Steinhausen K, Wiederholt M, Strauss O. Altered regulation of L-type channels by protein kinase C and protein tyrosine kinases as a pathophysiologic effect in retinal degeneration. FASEB J 1998;12:1125–34.
86. Strauss O, Steinhausen K, Mergler S, Stumpff F, Wiederholt M. Involvement of protein tyrosine kinase in the InsP3-induced activation of Ca^{2+}-dependent Cl- currents in cultured cells of the rat retinal pigment epithelium. J Membr Biol 1999;169:141–53.
87. Qu Z, Fischmeister R, Hartzell C. Mouse bestrophin-2 is a bona fide Cl(−) channel: identification of a residue important in anion binding and conduction. J Gen Physiol 2004;123:327–40.
88. Arden, G. B. & Constable, P. A. (2006). The electro-oculogram. Prog Retin Eye Res 25, 207–48.
89. Gallemore RP, Griff ER, Steinberg RH. Evidence in support of a photoreceptoral origin for the "light-peak substance". Invest Ophthalmol Vis Sci 1988;29:566–71.
90. Gallemore RP, Li JD, Govardovskii VI, Steinberg RH. Calcium gradients and light-evoked calcium changes outside rods in the intact cat retina. Vis Neurosci 1994;11:753–61.
91. Bito H, Deisseroth K, Tsien RW. CREB phosphorylation and dephosphorylation: a Ca^{2+}- and stimulus duration-dependent switch for hippocampal gene expression. Cell 1996;87: 1203–14.
92. Mears D. Regulation of insulin secretion in islets of Langerhans by Ca(2+)channels. J Membr Biol 2004;200:57–66.
93. Strauss O, Heimann H, Foerster MH, Agostini H, Hansen LL, Rosenthal R. Activation of L-type Ca^{2+} Channels is Necessary for Growth Factor-dependent Stimulation of VEGF Secretion by RPE Cells. Invest Ophthalmol Vis Sci 2003;44:(e-abstract)3926.
94. Witmer AN, Vrensen GF, Van Noorden CJ, Schlingemann RO. Vascular endothelial growth factors and angiogenesis in eye disease. Prog Retin Eye Res 2003;22:1–29.
95. Ambati J, Ambati BK, Yoo SH, Ianchulev S, Adamis AP. Age-related macular degeneration: etiology, pathogenesis, and therapeutic strategies. Surv Ophthalmol 2003;48:257–93.
96. Campochiaro PA. Retinal and choroidal neovascularization. J Cell Physiol 2000;184: 301–10.
97. Eyetech, Study & Group. Anti-vascular endothelial growth factor therapy for subfoveal choroidal neovascularization secondary to age-related macular degeneration: phase II study results. Ophthalmology 2003;110:979–86.
98. Frank RN. Growth factors in age-related macular degeneration: pathogenic and therapeutic implications. Ophthalmic Res 1997;29:341–53.
99. Kliffen M, Sharma HS, Mooy CM, Kerkvliet S, de Jong PT. Increased expression of angiogenic growth factors in age-related maculopathy. Brit J Ophthalmol 1997;81: 154–62.

100. Lopez PF, Sippy BD, Lambert HM, Thach AB, Hinton DR. Transdifferentiated retinal pigment epithelial cells are immunoreactive for vascular endothelial growth factor in surgically excised age-related macular degeneration-related choroidal neovascular membranes. Invest Ophthalmol Vis Sci 1996;37:855–68.
101. Slomiany, MG and Rosenzweig S A (2004). IGF-1-induced VEGF and IGFBP-3 secretion correlates with increased HIF-1 alpha expression and activity in retinal pigment epithelial cell line D407. Invest Ophthalmol Vis Sci 45, 2838–47.
102. Kramer F, White K, Pauleikhoff D, Gehrig A, Passmore L, Rivera A, Rudolph G, Kellner U, Andrassi M, Lorenz B, Rohrschneider K, Blankenagel A, Jurklies B, Schilling H, Schutt F, Holz FG, Weber BH. Mutations in the VMD2 gene are associated with juvenile-onset vitelliform macular dystrophy (Best disease) and adult vitelliform macular dystrophy but not age-related macular degeneration. Eur J Hum Genet 2000;8:286–92.
103. Marquardt A, Stohr H, Passmore LA, Kramer F, Rivera A, Weber BH. Mutations in a novel gene, VMD2, encoding a protein of unknown properties cause juvenile-onset vitelliform macular dystrophy (Best's disease). Hum Mol Genet 1998;7:1517–25.
104. Pollack K, Kreuz FR, Pillunat LE. Best's disease with normal EOG. Case report of familial macular dystrophy. Ophthalmologe 2005;102:891–4.
105. Renner AB, Tillack H, Kraus H, Kramer F, Mohr N, Weber BH, Foerster MH, Kellner U. Late onset is common in best macular dystrophy associated with VMD2 gene mutations. Ophthalmology 2005;112:586–92.
106. Wabbels BK, Demmler A, Preising M, Lorenz B. Fundus autofluorescence in patients with genetically determined Best vitelliform macular dystrophy: Evaluation of genotype-phenotype correlation and longitudinal course. Invest Ophthalmol Vis Sci 2004;45: (e-abstract)1762.
107. D'Cruz PM, Yasumura D, Weir J, Matthes MT, Abderrahim H, LaVail MM, Vollrath D. Mutation of the receptor tyrosine kinase gene Mertk in the retinal dystrophic RCS rat. Hum Mol Genet 2000;9:645–51.
108. Edwards RB and Szamier RB. Defective phagocytosis of isolated rod outer segments by RCS rat retinal pigment epithelium in culture. Science 1977;197:1001–3.
109. Gal A, Li Y, Thompson DA, Weir J, Orth U, Jacobson SG, Apfelstedt-Sylla E, Vollrath D. Mutations in MERTK, the human orthologue of the RCS rat retinal dystrophy gene, cause retinitis pigmentosa. Nat Genet 2000;26:270–1.
110. Goldman AI and O'Brien PJ. Phagocytosis in the retinal pigment epithelium of the RCS rat. Science 1978;201:1023–5.
111. Strauss O, Stumpff F, Mergler S, Wienrich M, Wiederholt M. The Royal College of Surgeons Rat: an animal model for inherited retinal degeneration with a still unknown genetic defect. Acta Anat (Basel) 1998;162:101–11.
112. Bollimuntha S, Cornatzer E, Singh BB. Plasma membrane localization and function of TRPC1 is dependent on its interaction with beta-tubulin in retinal epithelium cells. Vis Neurosci 2005;22:163–70.
113. Wettschureck N and Offermanns S. Mammalian G proteins and their cell type specific functions. Physiol Rev 2005;85:1159–204.
114. Poyer JF, Ryan JS, Kelly ME. G protein-mediated activation of a nonspecific cation current in cultured rat retinal pigment epithelial cells. J Membr Biol 1996;153:13–26.
115. Collison DJ, Tovell VE, Coombes LJ, Duncan G, Sanderson J. Potentiation of ATP-induced Ca^{2+} mobilisation in human retinal pigment epithelial cells. Exp Eye Res 2005;80:465–75.
116. Reigada D, Lu W, Zhang X, Friedman C, Pendrak K, McGlinn A, Stone RA, Laties AM, Mitchell CH. Degradation of extracellular ATP by the retinal pigment epithelium. Am J Physiol Cell Physiol 2005;289:C617–24.

117. Reigada D and Mitchell CH. Release of ATP from retinal pigment epithelial cells involves both CFTR and vesicular transport. Am J Physiol Cell Physiol 2005;288:C132–40.
118. Fragoso G and Lopez-Colome AM. Excitatory amino acid-induced inositol phosphate formation in cultured retinal pigment epithelium. Vis Neurosci 1999;16:263–9.
119. Lopez-Colome AM, Fragoso G, Wright CE, Sturman JA. Excitatory amino acid receptors in membranes from cultured human retinal pigment epithelium. Curr Eye Res 1994;13:553–60.
120. Besharse JC and Spratt G. Excitatory amino acids and rod photoreceptor disc shedding: analysis using specific agonists. Exp Eye Res 1988;47:609–20.

12
Taurine Transport Pathways in the Outer Retina in Relation to Aging and Disease

Ali A. Hussain and John Marshall

CONTENTS

INTRODUCTION
TAURINE: GENERALIZED DISTRIBUTION, INTRA-RETINAL LOCALIZATION
FUNCTIONAL IMPORTANCE OF TAURINE IN THE RETINA
TRANSPORT PATHWAYS FOR DELIVERY OF RETINAL TAURINE
AGEING OF THE TAURINE TRANSPORT SYSTEM: IMPLICATIONS FOR DISEASE
CONCLUSIONS
REFERENCES

INTRODUCTION

The crucial role played by taurine in maintaining the mammalian visual system has been highlighted by the observation that a nutritional deficiency in cats is associated with severe retinal degeneration. Other animal species were more resistant to dietary manipulation, but either drug-mediated inhibition of the taurine transporter or the construction of models with disrupted transporter genes resulted in severe retinal degeneration. In man also, taurine deficiency, as a consequence of pathological conditions, has been associated with abnormal ocular function. Abnormal taurine homeostasis has also been suggested in some forms of human retinitis pigmentosa, since the pattern of visual loss was very similar to that in the taurine-deficient cat. In the present analysis, the transport pathway for delivery of plasma taurine to photoreceptors of the retina has been examined and the effects of ageing of individual compartments on the integrity of the transport system assessed. Our results suggest that in diseases of aging such as macular degeneration, associated histopathological alterations of the transport pathway may precipitate a localized taurine deficiency that may act synergistically in the progression of the disease.

TAURINE: GENERALIZED DISTRIBUTION, INTRA-RETINAL LOCALIZATION

Taurine (2-aminoethane sulphonic acid) is widely distributed in mammalian species, and in the rat it has been estimated that taurine constitutes nearly 0.15% of its body weight *(1)*. This sulphonic amino acid is particularly enriched in excitable tissues, such as neuronal, contractile and secretory structures *(2)*.

The visual system maintains high concentrations of taurine, while the frontal and occipital lobes contain the highest levels in brain. In the retina, taurine accounts for nearly 40–50% of the total amino acid pool and retinal subfractionation techniques have localized most of this content to photoreceptor cells, the concentration in various species lying within the range 50 to 80 mM *(3,4)*. Since plasma levels of taurine are in the region of 40 to 60 µM, photoreceptor cells seem capable of maintaining a thousand-fold concentration gradient with respect to their extracellular environment *(5,6)*.

The major pathway for the biosynthesis of taurine in mammalian tissues, particularly liver and brain, is via cysteine sulphinic acid decarboxylase (CSD), a vitamin B6-dependent enzyme that catalyzes the conversion of cysteine sulphinic acid to hypotaurine *(7)*. Hypotaurine is then converted to taurine in a reaction catalyzed by hypotaurine oxidase. In the livers of cats, monkey and man, although the synthetic pathway is present, taurine biosynthesis is limited by low activity of CSD. In fact, human levels of CSD are particularly low, about a tenth of that in the cat *(1)*. The synthetic pathway has also been shown to be present in the retinas of various species *(8,9)*. Despite this presence, the activity of CSD is considerably low, requiring blood-borne import of the amino acid to maintain the high levels of retinal taurine.

Circulating taurine passes rapidly from the plasma to the retina. Autoradiographic studies have shown an initial rapid accumulation into the retinal pigment epithelium (RPE), followed by a slower exchange with retinal pools *(10,11)*. In rats, inhibition of the taurine transport mechanism by administration of the potent antagonist guanidoethane sulphonic acid (GES) decreased retinal taurine content by 75%, confirming that the transport of taurine across the blood-retinal barrier, rather than biosynthesis, is the major mechanism for maintaining taurine levels *(12,13)*.

Identifying the intra-retinal distribution of taurine uptake sites is one possibility in determining which of the retinal cell types relies on transport, rather than de novo synthesis, for maintaining its level of the amino acid. Generally, intravascular or intravitreal administration of ^{3}H-taurine, apart from labelling RPE, results in predominant accumulation by photoreceptor and Müller cells with diffuse labelling of inner retinal neurones *(10)*. Similarly, in many species, in vitro incubations show either a photoreceptor-Müller cell or photoreceptor-amacrine cell distribution, but in all these studies, ganglion cells were poorly labeled *(4,11,14)*.

FUNCTIONAL IMPORTANCE OF TAURINE IN THE RETINA

Given its universal distribution, it is not surprising that taurine has been associated with many functional attributes. Apart from the established role in bile acid conjugation, these include anti-arrhythmic effects on induced arrhythmias, modulation of sperm motility and anticonvulsant actions in experimentally induced epilepsy *(15–17)*. Many of the other protective effects against toxic insults have led to taurine being classified as a membrane stabilizer *(18,29)*. In the retina, taurine has depressant effects on the b-wave of the electroretinogram (ERG) *(19,20)*. However, uncertainties remain as to the specificity of this response, since similar inhibition can be obtained with glycine and gamma-amino butyric acid (GABA). Synaptosomal localization and release following depolarizing stimuli have demonstrated a role for taurine in neurotransmission in the inner retina *(21,22)*.

Roles attributed to taurine that may be very important for maintaining the visual system include membrane stabilization, anti-oxidant properties, and osmoregulation. Briefly, photoreceptors operate in an environment high in polyunsaturated fatty acids, oxygen tension and light, an explosive mixture for the generation of free radicals, leading to membrane damage *(23,24)*. In the long term, the resulting damage is dealt with by shedding of the outer segment tips and renewal at the base, with full renewal of the rod outer segment taking about 10 days, a process that is active for the entire lifetime of an individual *(25–27)*. However, taurine may protect outer segment membranes directly against radical-induced damage in the following sequence. Absorption of light results in release of taurine from outer segments such that the level in the interphotoreceptor matrix (IPM) surrounding outer segments is transiently elevated. Taurine as a taurine-zinc or taurine-calcium complex appears to protect against this type of oxidative damage by attaching to the membrane and quenching peroxidative products *(28,29)*.

The antioxidant properties of taurine have been documented in many tissues *(18)*. Toxic oxidants such as the superoxide radical are rendered slightly less reactive by superoxide dismutase, resulting in the production of hydrogen peroxide, which is then reduce to water by catalase. However, the myeloperoxidase of neutrophils and other tissue peroxidases catalyze the formation of hypochlorous acid from hydrogen peroxide and the chloride anion. Hypochlorous acid is a strong oxidizing agent that will oxidize carbohydrates, nucleic acids, peptide linkages, and amino acids. Hypochlorous acid will react primarily with α-amino acids to form mono- or dichloroamines. These are unstable and spontaneously deaminate, decarboxylate, and dechlorinate to form the respective aldehydes, which are highly toxic. Taurine can compete for chlorination by the myeloperoxidase/H_2O_2/Cl^- system to form stable monochlorotaurine *(30)*. Monochlorotaurine is a less-severe oxidant than hypochlorous acid, and will oxidize sulphahydryls such as glutathione to regenerate taurine. Taurine can also abstract the chlorine from other amino acid N-chloroamines, thereby preventing the formation of toxic aldehydes *(31)*. This is an important role for taurine in the highly oxidative environment of the retina since RPE cells are known to produce and release chloroamines.

Most cells maintain their volume by accumulating or releasing small organic osmolytes such as taurine and myo-inositol *(32,33)*. During the light response for example, due to ionic changes, receptor outer segments swell. The compensatory response is the release of taurine. Similarly, nursery cells such as RPE and Müller cells that need to regulate their extracellular microenvironments utilise taurine as an osmolyte *(34)*. Taurine, as an osmolyte, therefore plays an important role in the maintenance of the neural retina, and researchers into disease states indicative of abnormal osmotic regulation such as central serous retinopathy, some diabetic complications and macular degeneration have often queried the involvement of this amino acid in the pathophysiological process *(34–36)*.

Mammalian Taurine Deficiency

As outlined above, the retina is highly dependent on adequate provision of taurine for normal visual function and structural maintenance. Since endogenous synthesis plays a relatively small part, the retina is dependent on an adequate supply of this amino acid from the circulation.

Photoreceptor cells are particularly susceptible to taurine deficiency and in cats, a diet lacking this amino acid causes tissue depletion with progressive reduction in ERG amplitudes and severe photoreceptor degeneration *(37,38)*. Cone photoreceptors are the first to show signs of abnormality and are also less capable of recovery following return to a taurine-supplemented diet. Thus, in the early stages, when rod implicit times are normal, cone ERGs show both lowered amplitudes and delays in a- and b-wave implicit times. As the degeneration advances, the rod dark-adapted response-intensity profile becomes abnormal together with the delayed implicit times. During the early stages of the deficiency, when retinal taurine is reduced by 70–80%, retinal DNA, protein and funduscopic appearance remain normal. The early receptor potential (ERP) remains normal at a time when the ERG a- and b-waves are barely detectable, suggesting that photoreceptors, though still present (at least 90%), are functionally decoupled. Progression results in the shortening and disappearance of outer segments followed by loss of photoreceptor nuclei.

Most mammalian species, however, are resistant to dietary manipulation. In rats and mice, tissue taurine depletion has been achieved by use of the taurine transport inhibitor guanidoethane sulphonic acid (GES). Administration of GES caused a rapid depletion of retinal taurine resulting in decreased ERG amplitudes and photoreceptor degeneration *(12,13,39)*.

In human infants, pathologically induced reductions in plasma taurine were associated with abnormalities of the ERG *(40,41)*. In this study, eight patients with non-functioning bowels or extensive bowel resection were placed on home total parenteral nutrition (TPN) for an average of 2.3 years. Of the eight patients, six were children with an average age of two years. Five of the six had minimal food intake and their taurine levels decreased to about 17 µM, compared to control levels of 58 ± 19 µM. Although four showed diffuse mild granularity of the RPE and retina, all showed prolonged rod b-wave implicit times. The adult patients showed no ocular changes even though their plasma taurine levels were reduced. These results demonstrate the greater susceptibility of infants to taurine deficiency. Administration of 1.5–2.25 g/day taurine to three children returned their plasma taurine levels and ERG parameters to normal within five months.

The importance of the taurine transporter in visual function has been highlighted by a study that developed a mouse model with a disrupted gene coding for the transporter *(42)*. The model was characterized by severe retinal degeneration, suggesting that transport of taurine was important not only for visual function but also for retinal development.

TRANSPORT PATHWAYS FOR DELIVERY OF RETINAL TAURINE

The human retina is supplied by two vascular networks. Branches of the central retinal artery permeate the inner retina and choriocapillaris of the choroidal circulation supply the outer retina. From measurements of blood flow and oxygen extraction, it has been shown that in those regions of the fundus with a dual supply, the choroidal contribution accounts for over 90% of the metabolic requirements of photoreceptor cells *(43,44)*. In the foveolar region of the human fundus, the absence of inner retinal capillaries means that the choroidal circulation must provide metabolic support to the whole retina. The importance of the choroidal circulation in the macular area is emphasized by the nearly eightfold higher blood flow in this region compared with the peripheral fundus *(45)*.

The outer retinal transport system for the delivery of blood-borne taurine to photoreceptors is presented in schematic form in Fig. 1. Characteristics of individual steps within the pathway are discussed.

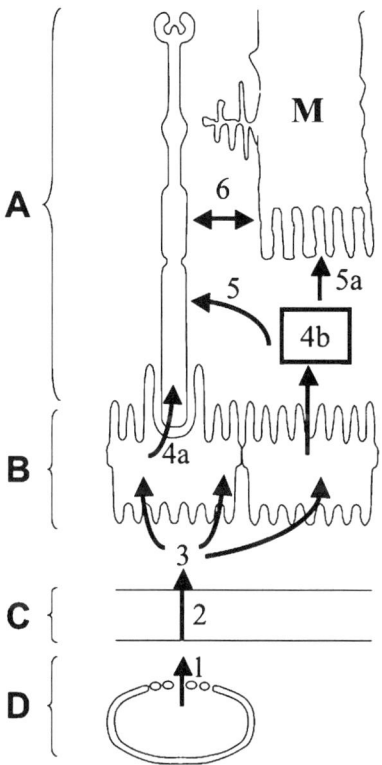

Fig. 1. Schematic representation of the transport pathway for the delivery of plasma taurine to photoreceptor cells of the retina. Following release from the fenestrated endothelium of the choriocapillaris (1), taurine diffuses passively across Bruch's membrane down its concentration gradient (2). It is then taken up by a Na^+- and Cl^--dependent high-affinity transporter on the basolateral surfaces of the retinal pigment epithelium (RPE) cell (3). Little is known about the release mechanism at the apical surface of the RPE, but taurine released in the region ensheathed by apical processes is taken up rapidly by high-affinity carriers located on the outer segment membranes (4a). Similarly, taurine released into the interphotoreceptor matrix (4b) is taken up by both photoreceptor and Müller (M) cells (5 and 5a). There is also the potential for intra-retinal distribution of the amino acid (6). A, photoreceptor layer; B, retinal pigment epithelium (RPE) layer; C, Bruch's membrane and D, the choriocapillaris.

The endothelium of the choriocapillary network (1) displays numerous fenestrations and therefore allows easy passage of taurine out for presentation to the outer aspects of Bruch's membrane *(46)*. Having exited the choriocapillary network, the first barrier encountered by the amino acid is Bruch's membrane (2). Bruch's is an acellular pentalaminated extracellular matrix (ECM) and can therefore only sustain passive transport down concentration gradients with rates being subject to Fick's laws of diffusion. The presence of glycosaminoglycans may also impart ion-selective properties to the membrane, but their effects on the movement of the zwitterionic taurine molecule have yet to be evaluated *(47–49)*.

Further inward movement of taurine is then dependent on the functional integrity of the RPE (3). Adjacent RPE cells are coupled by extensive junctional complexes called zonulae occludentes and these constitute the anatomical site of the outer blood-retinal barrier *(50,51)*. The basolateral aspect of the RPE cell rests in intimate contact with Bruch's membrane and thereby minimizes the diffusional distance, allowing rapid movement of released taurine. More importantly, convolutions of the RPE basal membrane greatly increase the surface area over which transport can occur. High affinity, (Na^+/Cl^-)-dependent taurine carriers have been identified in RPE cells to facilitate rapid uptake of the amino acid *(52–54)*. The human (Na^+/Cl^-)-dependent taurine transporter has been cloned and sequence homology has shown it to be a member of the (Na^+/Cl^-)-dependent neurotransmitter transporter family *(55,56)*. Its polypeptide chain comprises 619 amino acids and has 12 transmembrane domains.

Having entered the RPE cell, taurine may be sequestered by binding to melanin granules or may diffuse down its concentration gradient to reach the apical surface of the cell. Little is known of the release processes at the apical surface of the RPE cell, but the finding of a low-affinity sodium-independent carrier *(53)*, suggests that taurine may exit by passive carrier-mediated diffusion (4a,b). The intimate relationship between the RPE cytoplasmic extensions (microvilli) and the distal part of the outer segment may promote rapid uptake by minimizing the diffusional distance (4a). In addition, taurine may diffuse across the interphotoreceptor matrix for final uptake into photoreceptor and Müller cells (5, 5a). High-affinity uptake of taurine by these and RPE cells has been shown by autoradiographic studies outlined earlier and the presence of corresponding taurine transporters has also been demonstrated *(35,54,57–61)*. The figure also shows the considerable scope for intercompartmental redistribution of taurine within the retina. Thus taurine released by photoreceptor cells following light stimulation can be taken up by both the RPE and Müller (M) cells (6). More importantly, Müller and RPE cells, because of their ability to sequester and release taurine, can maintain a homeostatic environment for the neuronal elements of the retina.

Taurine Transport by Retinal Pigment Epithelium

Many anomalies exist regarding the transport of taurine across the RPE. Little is known of the release mechanism at the apical surface, and the possibility of bidirectional transport across the RPE monolayer has yet to be decided. We have previously elucidated the entire transportation pathway for translocation of the sulphur amino acid taurine across the bovine RPE cell layer and a schematic showing the mechanism of vectorial transport is given in Fig. 2 *(53)*.

Taurine Transport Pathways in the Outer Retina

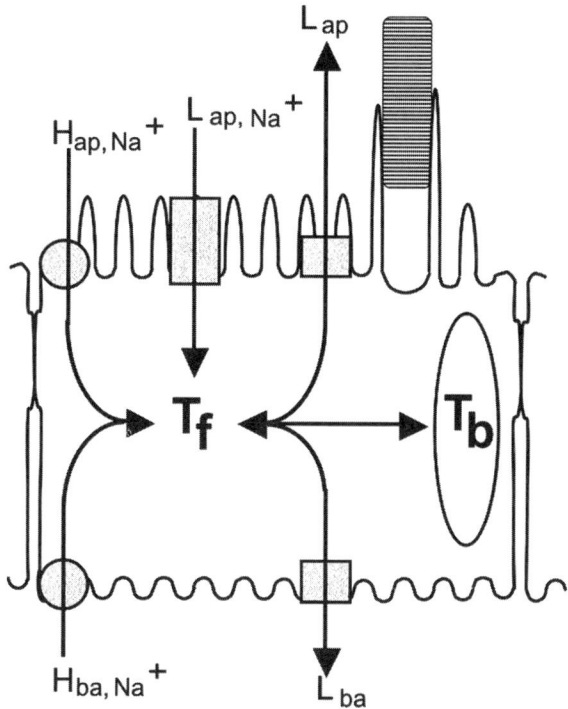

Fig. 2. The vectorial transport system for translocation of taurine across the bovine retinal pigment epithelium (RPE) (after Kundaiker et al. 1996 53). Values of V_{max} are given as pmol/5 min/4 mm disc of tissue. The basolateral membrane houses a Na$^+$-dependent high-affinity transporter (H_{ba}, Na$^+$) that delivers plasma taurine into the cell and is characterized by a K_m of 29 µM and V_{max} of 54.7. Also housed at this surface is a low-affinity Na$^+$-independent carrier with a K_m of 507 µM and V_{max} of 234 and its activity is governed by the taurine gradient at the basolateral surface, i.e., T_f-T_{plasma}. The apical surface contains one Na$^+$-dependent high-affinity carrier (K_m and V_{max} of 23.2 µM and 34, respectively) and one Na$^+$-dependent low-affinity carrier (K_m and V_{max} of 507 µM and 55, respectively). Both carriers can only transport taurine into the RPE cell. In addition, the apical surface possesses a Na$^+$-independent low-affinity carrier (Lap, with K_m of 507 µM, and V_{max} of 55) that can move taurine in response to the trans-apical gradient of taurine. Such a polarized distribution of carriers means that taurine can be transported in either direction depending on the concentration gradient across the RPE cell. Calculations have shown the presence of a large pool of bound taurine (T_b) that is in equilibrium with the free pool (T_f).

Using Ussing chambers to isolate apical and basal surfaces, and short radiochemical incubations to minimise loss of accumulated label, we were able to characterise the transporters on the two surfaces. The transport process was governed by the differential presence of high- and low-affinity, active and passive carriers on the polarized surfaces of the RPE cell. Since the sodium electrochemical gradient is always directed inwards, sodium-dependent carriers can only transport taurine into the RPE cell. Thus the high-affinity carrier on the basal surface transports plasma-derived taurine into the cell. Similarly, the high- and low-affinity sodium-dependent carriers on the apical surface transport taurine into the cell from the IPM.

The low-affinity sodium-independent carriers can only transport taurine down its concentration gradient. If the free intracellular concentration of taurine is higher than that in the extracellular space, then these sodium independent transporters can move taurine out of the RPE in either direction.

The kinetics of the transporters have been characterized (see Fig. 2), and by equating the input and output elements for given concentrations of extracellular taurine, it is possible to calculate vectorial fluxes and the intracellular concentration of free taurine. Thus, at physiological taurine concentrations of 62 µM for plasma (5), and 6 µM for IPM (62), the free concentration of intracellular taurine was determined to be 147 µM (about threefold higher than in plasma) with net transport in the basal-to-apical direction. High-performance liquid chromatography showed the level of taurine in the RPE to be 19.5 ± 3.6 mM, nearly 300 times higher than in plasma. However, the free concentration of 147 µM implies that most of the taurine in the RPE remains bound. Having the ability to sequester an osmolyte means that the cell can regulate its external environment.

AGEING OF THE TAURINE TRANSPORT SYSTEM: IMPLICATIONS FOR DISEASE

After release from the choriocapillaris, taurine must traverse Bruch's membrane by passive diffusional processes prior to uptake and delivery by active transporters in the RPE to primarily the photoreceptors of the retina. Ageing-related changes within these compartments are therefore likely to influence the trafficking of this amino acid, and the degree of modulation is expected to impact on both age-related diseases such as macular degeneration and other pathological alterations where the requirements for taurine are particularly exaggerated.

An analysis of the taurine transport system in the ageing human eye is complicated by uncontrollable post-mortem delays prior to analysis. For a transport system that relies on the energy status of the cell, such delays are expected to alter its functional characteristics. Since Bruch's membrane is an extracellular matrix, short post-mortem delays are expected to cause minimal damage and many studies have now confirmed the utility of such a preparation for transport studies (63–65). The stability of the high affinity taurine transport system in the retina/RPE of baboon and man has been investigated and results displayed in Fig. 3.

Baboon eyes were obtained fresh and stored on ice immediately after enucleation. Human donor eyes were enucleated at the earliest opportunity, but those within six hours were obtained from surgical enucleations for melanomas. Although some stability of transport was observed in baboon retinas, human retinal preparations showed a rapid fall in transport capability soon after death of donor. Kinetic analyses of human retinas showed a stable K_m for the high affinity carrier of 47 ± 5 µM but the V_{max} of the transport process diminished rapidly as a function of post-mortem time. As Fig. 3 demonstrates, the RPE was resistant to post-mortem delays and the tissue compartment was viable for the 48 hours of experimentation shown. Probing with a substrate concentration of 1 to 60 µM taurine,

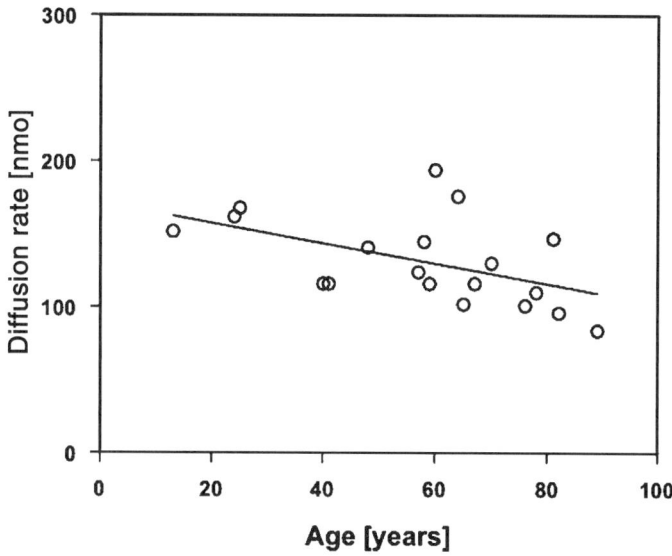

Fig. 3. The effect of age on the diffusion of taurine across the isolated human Bruch's-choroid complex. In these experiments, isolated samples of Bruch's-choroid were clamped in an Ussing chamber apparatus, and taurine, at a concentration of 10 mM, was added to one half-compartment. The opposing half-compartment was sampled at regular intervals and the amount of taurine traversing the preparation quantified by high-performance liquid chromatography. The respective diffusional fluxes were calculated and plotted as a function of age of donor. This showed a significant age-related decline in transport of taurine across Bruch's membrane ($p < 0.05$).

a high-affinity carrier was identified in human RPE cells with kinetic characteristics of $K_m = 50 \pm 10 \mu M$ and $V_{max} = 267 \pm 48$ pmol taurine/10 mins/5 mm disc of tissue. Thus, the status of taurine transport across Bruch's membrane and the RPE can be evaluated in relation to the ageing process.

Bruch's membrane undergoes age-related morphological and compositional alterations, with its thickness increasing two- to threefold over the life span of an individual *(66,67)*. The increase in the thickness of Bruch's membrane alone is expected to decrease the driving concentration gradients across the membrane for diffusional transport of taurine. Compositional alterations include the deposition of lipids *(68)*, proteins, and carbohydrate-associated moieties such as glycolipids, glycoproteins, and proteoglycans *(69)*. The insoluble collagen content of Bruch's membrane was shown to increase linearly with age *(70)* and over a 10-decade period accounted for nearly 50% of membrane collagen. The insolubility is indicative of denaturation and subsequent chemical modification including greater cross-linking and formation of advanced glycation endproducts (AGEs). All these age-related changes compromise membrane porosity and their effects on the diffusion of taurine are shown in Fig. 4. Ageing was associated with a significant decline in diffusional transport of taurine ($p < 0.05$). This has ramifications not only for normal ageing, but more so for advanced changes associated with macular degeneration or in Sorsby's fundus dystrophy where Bruch's membrane is considerably thickened *(71)*.

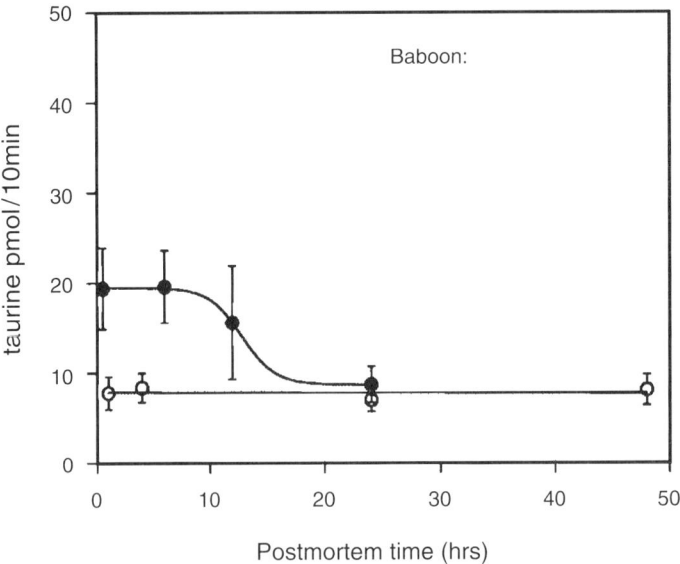

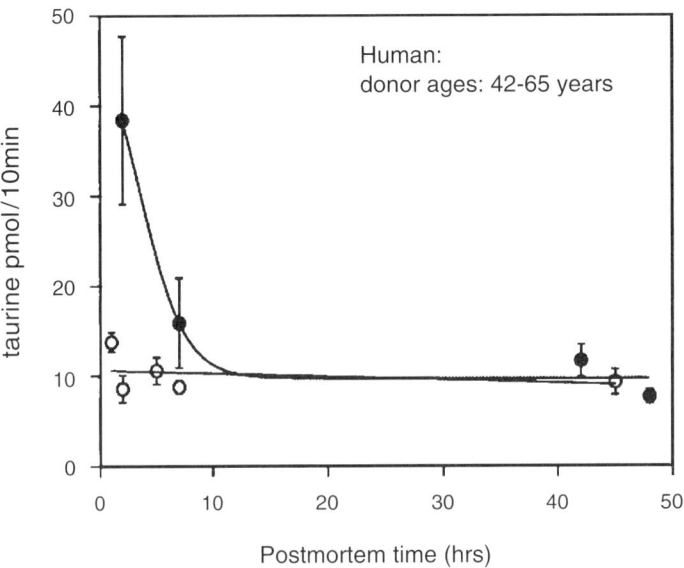

Fig. 4. Post-mortem survival of the Na⁺-dependent high-affinity taurine transport system in the retina and retinal pigment epithelium (RPE) of baboon and human donors. Donor samples of retina and RPE were obtained as free-floating 5 mm discs. Preliminary studies had established linearity of radiolabel uptake for at least 20 minutes of the incubation period at 1 µM taurine substrate concentration. Samples were then routinely incubated for 10 minutes followed by rapid quenching in ice-cold, taurine-free medium. They were then processed for scintillation counting to determine the amount of amino acid taken up. Filled circles, retina; open circles, RPE. Retinal tissue from baboons showed some stability with regards to high-affinity taurine transport and may have resulted from optimum storage conditions following enucleation. Human retinal samples degenerated rapidly highlighting the difficulties in obtaining reliable transport data. However, the taurine transport system in the RPE was very resilient to post-mortem deterioration and maintained stability for at least 72 hours (full data not shown).

The RPE also undergoes considerable age-related alterations in its structural organization and function. The most obvious morphological alteration capable of directly affecting transport processes is the age-dependent decrease in the number, size, and organization of membrane convolutions at both apical and basolateral surfaces of the RPE *(72)*. The decreased surface area for exchange will be associated with a corresponding reduction in the total number of membrane carriers, resulting in a lowered maximum capacity for transport. It is highly unlikely that the reduced area would maintain its original content of carriers since this would drastically reduce membrane fluidity, a parameter essential for normal operation of a host of carriers, receptors, and ionic channels. The decreased number of active transporters is perhaps reflected in the reduced concentration of mitochondria in the basal region of the RPE cell. Another important parameter is the accumulation of lipofuscin. Lipofuscin granules are discernable by the age of 10 years and occupy nearly 20% of cytoplasmic volume by the age of 80 years *(73)*. Decreased cytoplasmic volume may have ramifications for the storage potential for diffusible metabolites. Despite these gross morphological alterations, our analysis of a restricted ageing population (62 to 93 years) did not show any significant changes in RPE capacity for uptake of taurine *(54)*.

If we know the taurine diffusional characteristics of Bruch's membrane and the kinetic constants of the taurine transporters on the basal surface of the RPE cell, then plotting these transport processes as a function of substrate concentration (i.e., plasma taurine) can help to identify the rate-limiting compartments for delivery of plasma taurine to photoreceptor cells. In the case of the bovine eye, Bruch's membrane parameters were evaluated by Hillenkamp et al. *(74)* and the kinetic parameters of the sodium-dependent high-affinity taurine transporter by Kundaiker et al. *(53)*. The corresponding plot over plasma taurine concentrations of 0 to $-100\,\mu M$ is shown in Fig. 5.

When taurine is transported in the plasma-to-retina direction, the concentration at the basal surface of the RPE will always be lower than that in plasma because of the concentration gradient across Bruch's membrane and clearance of the amino acid by the RPE. Using the constants of the diffusion process through Bruch's and the kinetics of the transporter at the basal surface of the RPE, a sophisticated analysis could calculate the quiescent taurine concentration at the junction of Bruch's and RPE for a given concentration in the plasma. Since this concentration will be lower than in plasma, for a correct comparison, the taurine transport curve for the RPE in Fig. 5 needs to be translated to the right, i.e., the intersection of the two plots is shifted to the left. For the purposes of simplicity, and to explain the idea of a rate-limiting compartment, this correction has been ignored. Thus in Fig. 5, for taurine concentrations below the intersection of about $42\,\mu M$, transport through Bruch's would be the limiting step, whereas above this concentration, uptake by the RPE becomes the limiting step in the overall transport to the retina.

A similar analysis can be undertaken for the human Bruch's–RPE compartment, but the data is scanty. Age-related changes in the diffusion of taurine across human Bruch's-choroid preparations can be utilized to calculate diffusion rates at various concentrations of plasma taurine, and these have been plotted in Fig. 6 *(54,65)*. An analysis of the kinetics of the taurine transporter at the basal surface of the RPE has yet to be attempted. Many studies have previously characterized the kinetics of the sodium-dependent taurine transporter in human cell lines and have determined K_m values in the range 2.0–$8.9\,\mu M$

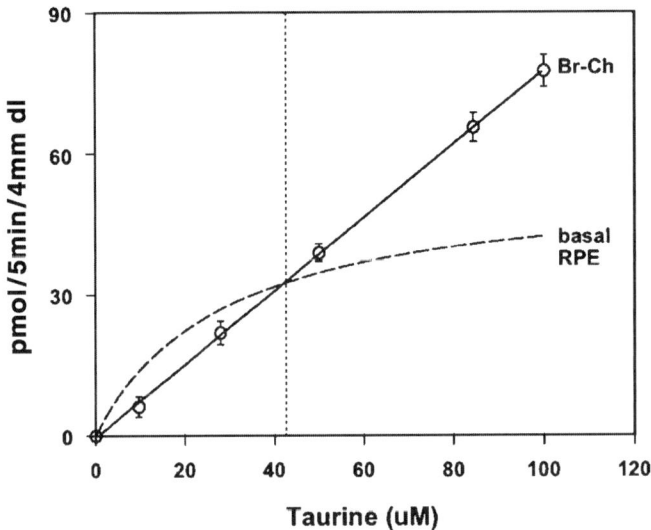

Fig. 5. Comparative analysis of transport rates for taurine across bovine Bruch's-choroid (Br-Ch) and the basal surface of the retinal pigment epithelium (basal RPE) as a function of plasma taurine concentration. Diffusional transport data for Bruch's-choroid was obtained from Hillenkamp et al. 2004 *(74)* and the Michaelis–Menten plot for the RPE was calculated from the kinetics of the high-affinity basal taurine transporter (K_m = 29 µM, V_{max} = 54.7 pmol/5 min/4 mm disc) characterized by Kundaiker et al. 1996 *(53)*. The data shows that above a plasma taurine concentration of about 42 µM (marked by the vertical dotted line), the basal membrane of the RPE cell is the rate limiting compartment for delivery of taurine to the retina.

(35,58,61,75). Our previous determination in intact donor tissue produced K_m values of 50 ± 10 µM and a V_{max} of 267 ± 48 pmol taurine/10 mins/5 mm disc, a value nearly fivefold higher than the results from the cell lines. This determination sampled both apical and basal surfaces of the preparation and therefore the composite V_{max} obtained would be much higher than that just for the basal surface. In the absence of more-specific data, we have plotted this higher transport function in Fig. 6 to compare with diffusion rates across ageing Bruch's membrane. Given the reservations that the plotted RPE rates are higher (apical plus basal) and the understanding that the RPE profile should be shifted to the right (analogous to the situation in the bovine eye explained earlier), it is obvious that, even with an age-related decline in transport across Bruch's membrane, the RPE remains the rate-limiting compartment for the delivery of plasma taurine to the photoreceptors of the retina (Fig. 6).

These results relate to the normal ageing human eye. Under pathophysiological conditions where there may be oxidative stress, the expression of the taurine transporter can be upregulated in RPE cells *(35,58,76)*. Upregulation is characterized by an increase in the number of transporters and hence in V_{max}, shifting the RPE curves upwards. Simultaneously, gross alterations in Bruch's membrane in patients with macular disease will severely compromise the diffusional transport of taurine across the Bruch's membrane compartment, shifting the diffusional profiles across Bruch's membrane downwards.

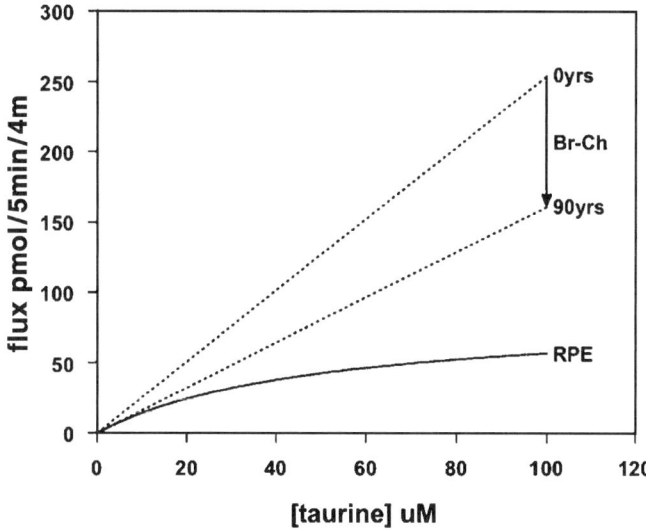

Fig. 6. Comparative analysis of transport rates for taurine across human Bruch's-choroid and the retinal pigment epithelium (RPE) as a function of plasma taurine concentration. The diffusional fluxes for Bruch's-choroid at various substrate concentrations together with the age-related changes were calculated from Hussain et al. 2002 *(65)* and are represented by the dotted lines. Transport data for taurine uptake was obtained from the kinetics of the Na^+-dependent high-affinity carrier given by Hillenkamp et al. 2004 *(54)*. Since the V_{max} represents both basal and apical contributions, the RPE plot (continuous line) is an overestimation of the transport across the basal surface. Despite this, the RPE is observed to be the rate-limiting step in the retinal delivery of taurine. As explained in the text, pathophysiological conditions can upregulate the RPE transport process and reduce transport across the Bruch's-choroid complex. Under these circumstances, Bruch's could dictate the rate at which taurine is transported from the plasma.

The possibility then exists for transport across Bruch's becoming the rate-limiting factor in the retinal delivery of taurine. Thus, despite normal plasma levels of the amino acid, the retina can undergo a nutritional deficiency, an insult that will compound the already degenerative status of photoreceptor cells.

CONCLUSIONS

Considerable work is required to clarify the existence of a vectorial transport system for taurine in human RPE. The presence of a (Na^+/Cl^-)-dependent taurine transporter of high affinity has been demonstrated, but the distribution between apical and basal surfaces of the RPE cell needs quantification. Little attention has been paid to the possibility of low-affinity carriers. All the high-affinity carriers characterized to date are sodium dependent and therefore can only transport taurine into the RPE cell. Since taurine is a charged molecule at physiological pH, it is highly unlikely to cross the lipid-rich plasma membrane of the RPE cell. It requires a membrane carrier for release by the RPE. The nature of this carrier needs urgent clarification. Finally, the RPE taurine carrier is amenable to up- or down-regulation, but the physiological signals that modulate this process are poorly understood. More importantly, the status of the taurine carrier in various disease states needs to be determined. The scanty data available at present suggests that Bruch's membrane in age-related macular degeneration may compromise the delivery of the much-required taurine, leading to a localized deficiency, augmenting the degenerative process in this disease.

REFERENCES

1. Jacobson JG and Smith LH. Biochemistry and physiology of taurine and taurine derivatives. Physiol Rev 1968;48:424–511.
2. Rubin RP, ed. Calcium and secretory processes. New York: Plenum; 1974. p. 1–124.
3. Orr HT, McIntyre N, Lowry OH. The distribution of taurine in the vertebrate retina. J Neurochem 1976;26:609–612.
4. Voaden MJ, Lake N, Marshall J, Morjaria B. Studies on the distribution of taurine and other neuroactive amino acids in the retina. Exp Eye Res 1977;25:249–257.
5. Hussain AA and Voaden MJ. Some observations on taurine homeostasis in patients with retinitis pigmentosa. In: Hollyfield JG, Anderson RE, LaVail MM, editors. Degenerative retinal disorders: Clinical and laboratory investigations. New York: Alan R. Liss; 1987. p.119–129.
6. Inoue H, Fukunaga K, Tsuruta Y. Determination of taurine in plasma by high-performance liquid chromatography using 4-(5,6-dimethoxy-2-phthalimidinyl)-2-methoxyphenylsulfonyl chloride as a fluorescent labelling agent. Anal Biochem 2003;319:138–142.
7. Sturman JA and Hayes KC. The biology of taurine in nutrition and development. Adv Nutr Res 1980;3:231–299.
8. Voaden MJ, Oraedu ACI, Marshall J, Lake N. Taurine in the retina. In: Schaffer SW, Baskin ST, Kocsis JJ, editors. The effects of taurine on excitable tissues. New York: Spectrum; 1981. p.145–160.
9. Lin C-T, Li H-Z, Wu J-Y. Immunocytochemical localisation of L-glutamate decarboxylase, GABA transaminase, cysteine sulfinic acid decarboxylase, aspartate aminotransferase and somatostatin in rat retina. Brain Res 1983;270:273–283.
10. Pourcho RG. Distribution of (^{35}S)-taurine in mouse retina after intravitreal and intramuscular injection. Exp Eye Res 1977;25:119–127.
11. Lake N, Marshall J, Voaden MJ. High affinity uptake sites for taurine in the retina. Exp Eye Res 1978;27:713–718.
12. Lake N. Is taurine an essential amino acid? Retina 1982;2:261–262.
13. Lake N. Depletion of taurine in the adult rat retina. Neurochem Res 1982;7:1385–1390.
14. Pasantes-Morales H, Bonaventure N, Wioland N, Mandel P. Effect of intravitreal injection of taurine and GABA on chick ERG. Int J Neurosci 1973;5:235–241.
15. Read WD and Welty JD. Effect of taurine on epinephrine and digoxin induced irregularities of dog heart. J Pharmacol Exp Ther 1963;139:283.
16. Mrsny RJ, Werman L, Meizel S. Taurine maintains and stimulates motility of hamster sperm during capacitation in vitro. J Exp Zool 1979;210:123–128.
17. Van Gelder N, Sherwin AL, Rasmussen T. Amino acid content of epileptogenic human brain: Focal versus surrounding regions. Brain Res 1972;40:385–393.
18. Wright CE, Tallan HH, Lin YY. Taurine: Biological update. Annu Rev Biochem 1986;55: 427–453.
19. Pasantes-Morales H, Kleith J, Urban PF, Mandel P. The physiological role of taurine in retina: uptake and effect on electroretinogram (ERG). Physiol Chem Physic 1972;4:339–348.

20. Urban F, Dreyfus H, Mandel P. Influence of various amino acids on the bioelectric response to light stimulation of a superfused frog retina. Life Sci 1976;18:473–480.
21. Cunningham R and Miller RE. Taurine: its selective action on neuronal pathways in rabbit retina. Brain Res 1976;117:341–345.
22. Pourcho RG. (^{3}H)-taurine accumulating neurones in the cat retina. Exp Eye Res 1981;32: 11–20.
23. Fridovich I. Oxygen is toxic. Bioscience 1977;27:462–466.
24. Fliesler DJ and Anderson RE. Chemistry and metabolism of lipids in the vertebrate retina. Prog Lipid Res 1983;22:79–131.
25. Feeny-Burns L, Berman ER, Rothman H. Lipofuscin of human RPE. Am J Ophthalmol 1980;90:783–791.
26. Young RW. Biological renewal, applications to the eye. Trans Ophthalmol Soc UK 1982;102:42–75.
27. Bok D. Retinal photoreceptor disc shedding and pigment epithelial phagocytosis. In: Ryan S, editor. Retina, 2nd edition. St Louis, USA: Mosby; 1994. p. 81–94.
28. Barbeau A and Donaldson J. Zinc, taurine, and epilepsy. Arch Neurol 1974;30:52–58.
29. Pasantes-Morales H and Cruz C. Taurine: A physiological stabiliser of photoreceptor membranes. Prog Clin Biol Res 1985;179:371–381.
30. Zgliczynski JM, Stelmaszynska T, Domanski J, Ostrowski W. Chloramines as intermediates of oxidation reaction of amino acids by myeloperoxidase. Biochem Biophys Acta 1971;235:419–424.
31. Wright CE, Lin TT, Lin YY, Sturman JA, Gaull GE. Taurine scavenges oxidised chlorine in biological systems. Prog Clin Biol Res 1985;179:137–147.
32. Pasantes-Morales H, Schousboe A. Volume regulation in astrocytes: a role for taurine as an osmoeffector. J Neurosci Res 1988;20:503–509.
33. Bitoun M and Tappaz M. Gene expression of taurine transporter and taurine biosynthetic enzymes in hyperosmotic states: a comparative study with the expression of the genes involved in the accumulation of other osmolytes. Adv Exp Med Biol 2000;483:239–248.
34. El-Sherbeny A, Naggar H, Miyauchi S, Ola MMS, Maddox DM, Martin PM, Ganapathy V, Smith SB. Osmoregulation of taurine transporter function and expression in retinal pigment epithelium, ganglion, and Muller cells. Invest Ophthalmol Vis Sci 2004;45:694–701.
35. Ganapathy V, Ramamoorthy JD, Del Monte MA, Leibac FH, Ramamoorthy S. Cyclic AMP-dependent up-regulation of the taurine transporter in a human retinal pigment epithelium cell line. Curr Eye Res 1995;14:843–850.
36. Yonemura D, Kawasaki K, Madachi-Yamamoto S. Hyperosmolarity response of ocular standing potential as a clinical test for retinal pigment epithelium activity: chorioretinal dystrophies. Doc Ophthalmol 1984;57:163–173.
37. Rabin AR, Hayes KC, Berson EL. Cone and rod responses in nutritionally induced retinal degeneration in the cat. Invest Ophthalmol 1973;12:694–704.
38. Hayes KC, Carey SY, Schmidt SY. Retinal degeneration associated with taurine deficiency in the cat. Science 1975;188:949–951.
39. Parmer R, Sheikh KH, Dawson WW, Toskes PP A parallel change of taurine and ERG in the developing rat retina. Comp Biochem Physiol 1982;72:109–111.
40. Geggel HS, Heckenlively JR, Martin DA, Ament ME, Kopple JD. Human retinal dysfunction and taurine deficiency. Doc Ophthalmol 1982;31:199–207.
41. Geggel HS, Ament HE, Heckenlively JR, Martin DA, Kopple JD. Nutritional requirement for taurine in patients receiving long-term parenteral nutrition. New Eng J Med 1985;312:142–146.

42. Heller-Stilb B, van Roeyen C, Rascher K, et al. Disruption of the taurine transporter gene (TauT) leads to retinal degeneration in mice. FASEB J. 2002;16:231–233.
43. Wilson TM, Strang R, Wallace J, Horton PW, Johnston NF. The measurement of the choroidal blood flow in the rabbit using Krypton 85. Exp Eye Res 1973;16:421–455.
44. Tornquist P and Alm A. Retinal and choroidal contribution to retinal metabolism in vivo: a study in pigs. Acta Physiol Scand 1979;106:351.
45. Alm A and Bill A. Ocular and optic nerve blood flow at normal and increased intraocular pressures in monkeys (*Macaca irus*): a study with radioactively labelled microspheres including flow determinations in brain and some other tissues. Exp Eye Res 1973;15:15–29.
46. Pino RM and Essner E. Permeability of rat choriocapillaris to hemeproteins: Restriction of tracers by a fenestrated endothelium. J Histochem Cytochem 1981;29:281–290.
47. Pino RM, Essner E, Pino LC. Localisation and chemical composition of anionic sites in Bruch's membrane of the rat. J Histochem Cytochem 1982;30:245–252.
48. Hewitt AT, Nakazawa K, Newsome DA. Analysis of newly synthesised Bruch's membrane proteoglycans. Invest Ophthalmol Vis Sci 1989;30:478–486.
49. Call T and Hollyfield J. Sulphated proteoglycans in Bruch's membrane of the human eye: localisation and characterisation using cupromeronic blue. Exp Eye Res 1990;51:451–462.
50. Shakib M, Rutkowski P, Wise GE. Fluorescein angiography and the retinal pigment epithelium. Am J Ophthalmol 1972;74:206–218.
51. Taniquchi Y. Ultrastructure of newly formed vessels in diabetic retinopathy. Jpn J Ophthalmol 1976;20:19–31.
52. Miyamoto Y and Del Monte MA. Na^+-dependent glutamate transporter in human retinal pigment epithelial cells. Invest Ophthalmol Vis Sci 1994;35:3589–3598.
53. Kundaiker S, Hussain AA, Marshall J. Component characteristics of the vectorial transport system for taurine in isolated bovine retinal pigment epithelium. J Physiol 1996;492(2):505–516.
54. Hillenkamp J, Hussain AA, Jackson TL, Cunningham JR, Marshall J. Taurine uptake by human retinal pigment epithelium: Implications for the transport of small solutes between the choroid and the outer retina. Invest Ophthalmol Vis Sci 2004;45:4529–4534.
55. Jhaing SM, Fithian L, Smanik P, McGill J, Tong Q, Mazzaferri EL. Cloning of the human taurine transporter and characterization of taurine uptake in thyroid cells. FEBS Lett 1993;318:139–144.
56. Ramamoorthy S, Leibach FH, Mahesh VB, et al. Functional characterisation and chromosomal localization of a cloned taurine transporter from human placenta. Biochem J 1994;300:893–900.
57. Vinnakota S, Qian X, Egal H, Sarthy H, Sarkar HK. Molecular characterization and in situ localization of a mouse retinal taurine transporter. J Neurochem 1997;69:2238–2250.
58. Bridges CC, Ola MS, Prasad PD, El-Sherbeny A, Ganapathy V, Smith SB. Regulation of taurine transporter gene expression by nitric oxide in cultured human retinal pigment epithelial cells. Am J Physiol 2001;281:C1825–C1836.
59. Miller SS and Steinberg RH. Potassium modulation of taurine transport across the frog retinal pigment epithelium. J Gen Physiol 1979;74:237–259.
60. Miyamoto Y, Kulanthaivel P, Leibach FH, Ganapathy V. Taurine uptake in apical membrane vesicles from the bovine retinal pigment epithelium. Invest Ophthalmol Vis Sci 1991;32:2542–2551.
61. Leibach JW, Cool DR, Del Monte DL, Ganapathy V, Leibach FH, Miyamoto Y. Properties of taurine transport in a human retinal pigment epithelium cell line. Curr Eye Res 1993;12:29–36.

62. Perry TL, Hansen S, Kennedy J. CSF amino acids and plasma-CSF amino acid ratios in adults. J Neurochem 1975;24:587–589.
63. Starita C, Hussain AA, Pagliarini S, Marshall J. Hydrodynamics of ageing Bruch's membrane: Implications for macular disease. Exp Eye Res 1996;62:565–572.
64. Moore DJ and Clover GM. The effect of age on the macromolecular permeability of human Bruch's membrane. Invest Ophthalmol Vis Sci 2001;42;2970–2975.
65. Hussain AA, Rowe L, Marshall J. Age related alterations in the diffusional transport of amino acids across the human Bruch's-choroid complex. J Opt Soc Am A 2002;19:166–172.
66. Sarks SH. Ageing and degeneration in the macular region: a clinico-pathological study. Br J Ophthalmol 1976;251:4062–4070.
67. Okuba A, Rosa RH, Bunce CV, Alexander RA, Fan JT, Bird AC, Luthert PJ. The relationships of age changes in retinal pigment epithelium and Bruch's membrane. Invest Ophthalmol Vis Sci 1999;40:443–449.
68. Holz FG, Sheraidah GS, Pauleikhoff D, Bird AC. Analysis of lipid deposits extracted from human macular and peripheral Bruch's membrane. Arch Ophthalmology 1994;112:402–406.
69. Mullins RF, Johnson LV, Anderson DH, Hageman GS. Characterization of drusen associated glycoconjugates. Ophthalmology 1997;104:288–294.
70. Karwatowski WSS, Jefferies TE, Duance VC, Albon J, Bailey AJ, Easty DL. Preparation of Bruch's membrane and analysis of the age related changes in the structural collagens. Br J Ophthalmol 1995;79:944–952.
71. Capon MR, Marshall J, Krafft JI, Alexander RA, Hiscott PS, Bird AC. Sorsby's fundus dystrophy. A light and microscopic study. Ophthalmology 1989;96(12):1769–77.
72. Garner A, Sarks S, Sarks JP. Degenerative and related disorders of the retina and choroid. In: Garner A and Klintworth GK, editors. Pathobiology of Ocular Disease: A Dynamic Approach, 2nd edition. New York: Marcel Dekker; 1994. p. 631–674.
73. Feeney-Burns L, Hilderbrand ES, Eldridge S. Ageing human RPE: morphometric analysis of macular, equatorial and peripheral cells. Invest Ophthalmol Vis Sci 1984;25:195–200.
74. Hillenkamp J, Hussain AA, Jackson TL, Constable PA, Cunningham JR, Marshall J. Compartmental analysis of taurine transport to the outer retina in the bovine eye. Invest Ophthalmol Vis Sci 2004;45:4099–4105.
75. Stevens MJ, Hosaka Y, Masterson JA, Jones SM, Thomas TP, Larkin DD. Downregulation of the human taurine transporter by glucose in cultured retinal pigment epithelium cells. Am J Physiol 1999;277:C760–C771.
76. Nakashima E, Pop-Busui R, Towns R, Thomas TP, Hosaka Y, Nakamura J, Greene DA, Killen PD, Schroader J, Larkin DD, Ho YL, Stevens MJ. Regulation of the human taurine transporter by oxidative stress in retinal pigment epithelial cells stably transformed to overexpress aldose reductase. Antioxid Redox Signal 2005;7:1530–1542.

13
P-Glycoprotein Expression and Function in the Retinal Pigment Epithelium

Paul A. Constable, John G. Lawrenson, and N. Joan Abbott

Contents

INTRODUCTION
THE RETINAL PIGMENT EPITHELIUM
P-GLYCOPROTEIN STRUCTURE
DRUG TRANSPORT
P-GLYCOPROTEIN SUBSTRATES
LOCALIZATION OF P-GLYCOPROTEIN IN THE RETINAL PIGMENT EPITHELIUM
OCULAR SIDE-EFFECTS OF DRUGS
DRUG DELIVERY
FUTURE DIRECTIONS
REFERENCES

INTRODUCTION

The retinal pigment epithelium (RPE) forms the major impediment to drug access to the outer retina. The basis of this outer blood–retinal barrier is the tight junctions between the pigment epithelial cells, in conjunction with an array of active efflux transporters that prevent the entry of xenobiotics into the sub-retinal space and the pigment epithelium's cytosol. However, these transporters also act as an impediment to successful delivery of potentially therapeutic drugs to the photoreceptors and RPE. One such transporter is P-glycoprotein, which actively removes a broad range of substrates from the plasma membrane and is the subject of this chapter.

THE RETINAL PIGMENT EPITHELIUM

The RPE forms a pigmented monolayer posterior to the photoreceptors. The RPE forms tight intercellular junctions that create a barrier between the outer retina and the underlying highly vascular and leaky choriocapillaris. The transepithelial resistance of the RPE is of the order of 200–500 $\Omega.cm^2$, depending upon the species. The outer retina depends upon the RPE for survival and the maintenance of visual function. The RPE participates in the transportation of essential amino acids and fatty acids, and is responsible for the isomerization of all-trans-retinal into 11-cis-retinal that is used by the photoreceptors in phototransduction. The RPE also contains, at the apical and basal membranes, numerous ion channels, pumps and co-transporters that regulate pH, fluid transportation and the ionic composition of the subretinal space. For a review of the RPE's physiological roles see Strauss (2005) *(1)*.

The flow of the choroidal circulation is high, owing to the large metabolic demand of the outer retina. Consequently, the basal membrane of the RPE is exposed to a large variety of endogenous and exogenous substances that must be either excluded from, or transported to, the outer retina. Transporters whose function is to transport taurine *(2)*, monocarbocylic acids *(3)*, glucose *(4)* and folic acid *(5)* have all been identified in the RPE of different species, as well as RPE cell lines. Associated with the RPE are several drug transporters whose substrates overlap, but serve to protect the RPE and the outer retina from assault from potentially harmful substances. Recent findings have demonstrated a novel organic cation transporter in the apical region of rat RPE cells *(6)*. Furthermore, an organic anion transporter has also been demonstrated to occur in human RPE subcultures *(7)*. The multidrug resistance-associated protein (MRP), which shares some substrates with P-glycoprotein (P-gp), has also been localized to the RPE *(8)*. For a review of ocular drug transporters see Mannermaa et al. (2006) *(9)*.

P-glycoprotein is an efflux pump that confers multidrug resistance (MDR) upon a tissue by actively removing a large range of compounds from the plasma membrane. However, the presence of P-gp can also prevent potentially therapeutic drugs from reaching the outer retina. In particular P-gp confers chemoresistance to tumors with the upregulation of MDR1 transcript *(10)*. In normal tissue, P-gp is expressed in the gut, testes, liver, kidney, adrenal gland, choroid plexus, and brain endothelium forming the blood-brain barrier, which is consistent with P-gp's role as an efflux pump designed to remove xenobiotics *(11–14)*.

P-GLYCOPROTEIN STRUCTURE

P-glycoprotein belongs to the adenosine triphosphate (ATP)-binding cassette (ABC) family of proteins *(15,16)* that utilize the hydrolysis of ATP to transport substances across the plasma membrane. The multidrug resistance gene (MDR1 or ABCB1) that encodes P-gp is located on the long arm of chromosome 7 *(17)*. P-glycoprotein consists of two repeated six-transmembrane domains (TMDs), with one nucleotide-binding domain (NBD) for ATP hydrolysis on each repeat *(18–20)*. These two repeated units of 610 amino-acids sections are joined by a 60 amino-acid linker region *(21)*. P-gp has multiple sites for threonine and serine phosphorylation by protein kinases *(22–24)*. The level of protein phosphorylation may alter the rate of hydrolysis of ATP at the NBDs of P-gp and affect drug transport. Therefore P-gp may be regulated by circulating levels of second messengers and G-protein-coupled receptors that modulate the level of protein phosphorylation. Drug transport is reduced in acidic or hypoxic environments, or when there is lowered intracellular protein kinase C activity *(25)*. Each of the two six-TMDs share 65% homology with a highly conserved approximately 25 amino-acid sequence present upstream of the Walker A motif that is critical for binding to the aromatic adenine ring of ATP *(26)*. P-gp has a molecular weight of 170–180 kDa. The N-terminal glycosylation product of P-gp (around 15 kDa) is required for functional P-gp activity, but is not involved directly with drug binding, and confers stability to the P-gp within the plasma membrane *(27,28)*.

Recent developments in imaging and computer modelling have built upon the earlier crystallography work that produced the first low-resolution images of P-gp *(29,30)*. These studies revealed an elliptical protein with a central pore whose conformation changed upon binding of ATP. In the absence of ATP binding, the two sets of transmembrane domains (1–6) and (7–12) form a single barrel 5 to 6 nm in diameter and about 5 nm deep. The central pore is open to the extracellular space and forms an inverted funnel. Upon binding of ATP, there is a conformational change in the transporter, which opens the central pore to enable access of substrates from the lipid bilayer into the pore, before being transported out into the extracellular space. See Fig. 1 for a schematic of the structure P-gp.

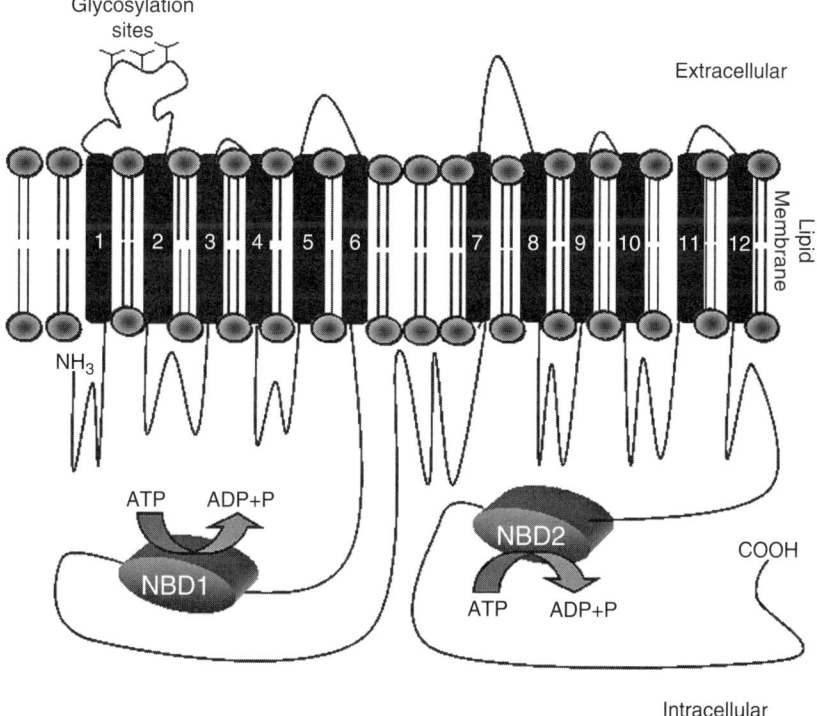

Fig. 1. Schematic representation of P-glycoprotein. P-gp consists of two sets of six transmembrane-domain (TMD) repeats each 610 amino acids long and linked by a 60 amino-acid sequence. There are three glycosylation sites on the extracellular face between the first and second trans-membrane domains. Two intracellular nucleotide-binding domains (NBD1 and NBD2) provide the sites for adenosine triphosphate (ATP) hydrolysis and the energy for extrusion of substrates. Figure adapted from Ramakrishnan (2003) *(31)*.

DRUG TRANSPORT

Substrates of P-gp are largely hydrophobic, which means that they either partition within the plasma membrane, or bind to proteins or chromosomes, and have a very low cytosolic concentration. P-gp is expressed in the plasma membrane and also along the nuclear envelope, in caveolae, cytoplasmic vesicles, the Golgi complex and rough endoplasmic reticulum *(32)*. P-gp is therefore an important transporter that removes xenobiotics from the site of protein synthesis, glycosylation, and membrane trafficking, as well as playing a role in the plasma membrane where it prevents entry of certain drugs into the cytosol.

P-gp has been proposed to resemble a hydrophobic vacuum cleaner *(33)* that removes substrates from the plasma membrane *(34)*. The broad range of substrates that P-gp is able to transport from the gut, liver and blood–brain barrier provides the body with an advantage in evolutionary terms, by protecting the body from harmful xenobiotics. It is not surprising that many of P-gp's substrates are floral derivatives, and therefore P-gp would have originally provided a selective advantage with respect to potential environmental toxins, such as digitalis and vinca alkaloids.

Mechanism of Drug Transport

The mechanism of drug capture and transport into the central pore is becoming clearer as the structure of P-gp emerges. Photoaffinity labeling of P-gp has identified the domains likely to be involved in the initial capture of substrates from the plasma membrane. The TMDs 3, 11 and 5, 8 provide the first point at which ligands are bound; they are then transported into the pore between the transmembrane interfaces, which may explain the large number of substrates that P-gp can transfer into the central pore through these interfaces *(35)*. This process requires ATP binding to the NBDs to open the cleft between the TMDs of the protein. The TMDs identified to date that are involved in drug binding within the pore are TMDs 1 and 7 *(36,37)*. Hydrolysis of ATP closes the plasma membrane clefts and the drug substrates are then removed from the pore to the extracellular space. Figure 2 shows the current model of P-gp, and how substrates enter the central pore through lateral channels between TMDs 3 and 11, and 5 and 8, upon ATP binding to the cytoplasmic NBDs.

P-GLYCOPROTEIN SUBSTRATES

P-glycoprotein was first observed in the Chinese hamster ovary (CHO) cell line, where a subset of cells showed a higher resistance to the antineoplastic drug, colchicine *(38,39)*. Similar resistance was found in CHO cells following increasing doses of daunorubicin and was associated with the upregulation of P-gp *(40)*.

Resistance to chemotherapeutic drugs has been observed in orbital tumors. P-glycoprotein has been found in retinoblastoma, and when coexpressed with multi drug-resistance associated protein-1 (MRP1) there is poor prognosis with chemotherapy *(41)*. P-glycoprotein has been identified in 12% of retinoblastomas following treatment by primary enucleation without any chemotherapy *(42)*. P-gp has also been detected in ocular melanoma (42%), and when present there is also a poor therapeutic outcome *(43)*.

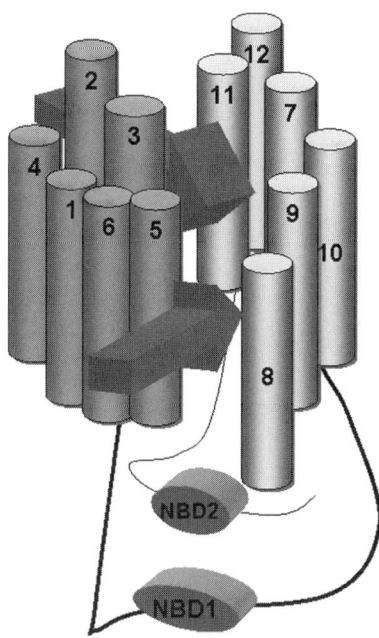

Fig. 2. The current model predicts that upon binding of adenosine triphosphate (ATP) to the nucleotide-binding domains (NBDs) located intracellularly, there is a conformational change in P-glycoprotein (P-gp) so that substrates enter the central pore in the spaces created between transmembrane domains (TMDs) 3 and 11, and between TMDs 5 and 8. The drug-binding pockets are located on TMDs 1 and 7 within the pore. Hydrolysis of ATP results, at NBD1 and NBD2, in the transport of the drugs out of the plasma membrane and back into the circulation.

The ability of P-gp to transport chemotherapeutic agents is a particular barrier to treating malignant tumors of the central nervous system. Colchicine, vinblastine, and doxorubicin were initially identified as substrates for P-gp in lung tissue from Chinese hamster *(44)*. Colchicine and vinblastine disrupt microtubule assemble by binding to β-tubulin, and therefore decrease cell motility and mitosis *(45)*. Doxorubicin acts to decrease blood flow to tumors *(46)*. Given that P-gp is able to remove these chemotherapeutic drugs, then the presence of P-gp often inhibits effective treatment of tumors *(47)*. P-gp can be upregulated with the antimitotic agent, daunomycin *(48)*.

P-gp actively extrudes a variety of compounds with mass in the range of 300 to 2000 Da that are typically hydrophobic, and are organic cationic compounds or organic bases. However, P-gp is also capable of transporting some organic anions, polypeptides, and uncharged compounds. P-gp is capable of transporting drugs of the alkaloid family. These substances derive from flora and comprise the vinca alkaloids, such as vinblastine whose binding site is distinct from that of the pyrimidine-based drugs *(49)* (including the cardiac glycoside, digoxin *(50)*).

P-gp is also able to transport a variety of other drugs that are used to combat a variety of diseases, such as HIV protease inhibitors (saquinavar and ritonavir) *(51)*, corticosteroids (dexamethasone, aldosterone and hydrocortisone) *(52,53)*, the immunosuppressant (cyclosporin A) *(54)*, the α_1-adrenergic antagonist prazosin *(55)*, and the antibiotics erythromycin, gramicidin D and valinomycin *(56–59)* and opioids (morphine-6-glucuronide) *(60)*. P-glycoprotein also transports some fluorescent indicators such as rhodamine 123 *(61)* and calcein, which enable fluorescence techniques to be used to determine the presence or absence of P-gp in vitro. Furthermore, P-gp also transports the hydrophobic acetoxymethyl ester of intracellular fluorescence indicators, such as fura- and fluo-compounds, that can impede the amount of the cytoplasmic fluorophore present *(62)*. Table 1 lists some of the main substrates of P-gp.

The presence of P-gp in the RPE may also have consequences for drug delivery to the outer retina. Novel treatments currently targeting neovascularization in the choroid require intravitreal injections *(77)*. This is in part due to the systemic side-effects of such drugs. Other routes, including oral, trans-scleral or sub-RPE injection still require

Table 1 List of substrates that are transported by P-glycoprotein

	Substrates transported by P-glycoprotein	
Ritonavir	HIV protease inhibitors	Lee et al. (1998) *(63)*
Saquinavir		
Indinavir		
Morphine-6-glucuronide	Opioids	Huwyler et al. (1996) *(60)*
Methadone		Callaghan and Riordan (1993) *(64)*
Fentanyl		Henthorn et al. (1999) *(65)*
Celiprolol	β-adrenergic agonist	Karlsson et al. (1993) *(66)*
Colchicine	Immunosuppressants	Ueda et al. (1987) *(67)*
Vinblastine		
Doxorubicin		
Taxotere		Janice et al. (1993) *(68)*
Daunorubicin		Kartner et al. (1983) *(40)*
Imatinib (STI571)		Mahon et al. (2000) *(69)*
Prazosin	α_1-agonist	Safa et al. (1990) *(55)*
Digoxin	Cardiac glycoside	Tanigawara et al. (1992) *(70)*
Midazolam	Benzodiazepine	Sanna et al. (2003) *(71)*
Ranitidine	Histamine (H2) receptor agonists	Collett et al. (1999) *(72)*
Cimetidine		
Methotrexate	Dihydrofolate reductase inhibitor	Rice et al. (1987) *(73)*
Ivermectin	Anti-parasitic	Schinkel et al. (1994) *(74)*
Octreotide	Somatostatin analogue	Gutmann et al. (2000) *(75)*
Rhodamine 123	Fluorescent indicators	Efferth et al. (1989) *(61)*
Calcein		Hollo et al. (1994) *(76)*
Acetoxymethyl ester (AM)		Homolya et al. (1993) *(62)*
Cortisol	Steroids	Ueda et al. (1992) *(52)*
Aldosterone		
Dexamethasone		

the drug to be transported across the RPE, and therefore knowledge of the barriers to drug delivery is important.

The question of what physiological role this transporter plays when not engaged in the transport of xenobiotics has led to some proposed further actions of P-gp that may or not be true of P-gp in the RPE. For reviews see Johnstone et al. (2000) *(78,79)*. These roles include the transport of phospholipids *(80,81)* and inhibiting apoptosis *(82,83)*. P-gp has also been associated with modulation of Cl⁻ channel activity and cell volume *(84,85)*. Furthermore, P-gp has been implicated in cholesterol esterification at the plasma membrane *(86)*. P-gp transcription is also up-regulated by cytokines in response to inflammation *(87)*. There is some evidence that statins may be substrates of P-gp or interfere with its activity *(88)*. In vitro studies suggest that simvastatin, atorvastin, and pravastatin inhibit P-gp activity, but pravastatin was ineffective *(89)*.

Inhibition of P-glycoprotein

One method for increasing the effectiveness of chemotherapeutic drugs is to co-administer drugs that inhibit the activity of P-gp. The calcium channel blockers, verapamil, and diltiazem inhibited the activity of P-gp in cultures of tumor cells *(90)*. Verapamil acts as a competitive inhibitor for the binding sites of adriamycin, vincristine, and colchicine *(90)*. The detection of P-gp in body tissues has exploited the specific transport of certain drugs using fluorescent or radioactively labeled compounds, in conjunction with a specific inhibitor of P-gp. Verapamil is used as a relatively specific inhibitor of P-gp activity and has been used as an effective inhibitor of P-gp in vivo *(91)* and in vitro in a number of systems *(8,92–94)*. Verapamil acts as a competitive inhibitor of P-gp and is used clinically to reverse the multidrug-resistance profile of P-gp by competing with chemotherapeutic drugs such as vinblastine *(90,95–97)*. Cyclosporine A may also be used to reduce drug resistance associated with P-gp in patients undergoing chemotherapy *(98)*.

The fluorescent probes, such as rhodamine-123 or calcein, are commonly used to determine functional P-gp activity. Rhodamine-123 enters cells passively, but can be extruded from the cytosol by P-gp. By adding a competitive inhibitor with rhodamine-123, it is possible to determine whether P-gp is functional or not. When functional P-gp is present, the cytosolic accumulation of rhodamine-123 is greater on addition of a P-gp inhibitor than in the absence of the inhibitor. Radiolabeled substrates of P-gp can also be used, such as colchicine.

Inhibitors of P-gp are generally classified as belonging to the first-, second-, third- or the current fourth-generation drugs. First-generation drugs, such as verapamil and cyclosporine A, and the second generation analogues of the first generation drugs, such as R-verapamil and cinchonine, have not been completely successful in reversing drug resistance. The first-generation drugs also had adverse side-effects and were used owing to their low affinity for P-gp transport. The second generation analogues, although designed to inhibit P-gp and reduce adverse side-effects, were not successful in clinical trials *(47,99,100)*. Third-generation drugs such as biricodar and R101933 that were designed specifically to inhibit P-gp are currently in clinical trials *(101,102)*.

The fourth-generation compounds, comprising monoclonal antibodies as well as high-affinity hydrophobic polypeptides (reversins), have also been used to demonstrate

Table 2 List of first- and second-generation inhibitors of P-glycoprotein (P-gp) activity

Cyclosporine A		Foxwell et al. (1989) *(110)*
FK-506 (fujimycin)	Immunosuppressants	Pourtier-Manzanedo et al. (1991) *(111)*
Sirolimus (rapamycin)		Arceci et al. (1992) *(112)*
Lonafarnib (SCH66336)		Wang et al. (2001) *(113)*
Verapamil	Calcium-channel inhibitors	Cornwell et al. (1987) *(90)*
Diltiazem Nifedipine		Safa et al. (1987) *(114)*
Carvedilol	β-adrenergic agonist	Jonsson et al. (1999) *(115)*
Erythromycin	Antibiotics	Hofsli and Nissen-Meyer (1989) *(116)*
Rifampicin		Fardel et al. (1995) *(117)*
Amitriptyline	Antidepressant	Varga et al. (1996) *(118)*

functional P-gp *(103,104)*. Further developments, following a greater understanding of the structure and function of P-gp, have led to the development of novel synthetic drugs targeted at inhibiting P-gp *(105)*. Phenoxazine derivatives show promise in inhibiting P-gp *(106)*, along with modified reversins, and are reported to show greater P-gp inhibition than cyclosporine A without side-effects *(107)*. Down-regulating MDR1 transcription by inhibiting messenger RNA (mRNA) is also promising, although to date only in vitro studies have been performed *(108,109)*. See Table 2 for a list of known inhibitors of P-gp. Strategies designed around inhibiting MDR1 transcription, or the delivery of drugs directly to tumor cells in liposomes, as well as the development of P-gp specific antibodies and antisense oligonucleotides, should provide better therapeutic outcomes in the future. For a review see Nobili et al. (2006) *(99)*.

LOCALIZATION OF P-GLYCOPROTEIN IN THE RETINAL PIGMENT EPITHELIUM

One study involved patients who had proliferative vitreoretinopathy, following either trauma or retinal detachment, which required vitrectomy for management. In such cases there is damage to the underlying RPE, and as such these cells can migrate, this contributes to the development of an epiretinal membrane. The expression of P-gp was examined in the surgically-removed epiretinal membranes in patients who had received intravitreal injections of daunomycin to reduce the risk of subsequent reproliferation. The authors found that P-gp was detected in all of the epiretinal membranes from the 10 patients who received injections, compared with just two out of 13 patients who had not received pre-treatment with intravitreal daunomycin. Colocalization with cytokeratin staining confirmed the epithelial origin of cells staining for P-gp. Furthermore, cultures of human RPE cells, in which MDR1 was detected, required the addition of a clinical dose (13 µM for 10 minutes) or subclinical (0.1–0.2 µM for two weeks) of daunomycin to translate the message into detectable levels of P-gp protein *(119)*.

Interestingly, P-gp has been detected at both the apical and basal membranes of native human RPE in frozen formaldehyde-fixed sections in the absence of any inducing agents

(103). These findings were confirmed by culturing human RPE cells and using two inhibitors of P-gp, a monoclonal antibody and reversin, to reveal an increased influx of rhodamine 123 in tissues treated with these inhibitors. The presence of P-gp at the apical membrane of the RPE may be related to a different function of P-gp at this membrane, given that any drugs transported would be returned to the subretinal space and photoreceptors, which would not be advantageous. However, the localization of P-gp at the basal membrane is in-keeping with its role as a drug transporter protecting the RPE and outer retina from circulating xenobiotics. Further work will be required to determine the significance of P-gp at the apical membrane of human RPE. In porcine RPE, P-gp is predominantly expressed at the basal membrane, based upon analysis of the relative flux of calcein from the apical-to-basal and basal-to-apical directions in the presence and absence of verapamil *(8)*.

There are several human RPE cell lines available to conduct in vivo experiments on drug transport across the RPE. However, cell lines have a varied phenotype depending upon substrate and culture conditions, and do not establish a tight epithelial barrier unless they are grown in carefully defined media *(120,121)*. Foetal RPE cells, whether from chick or human, can be manipulated to achieve a suitable RPE barrier for experimental work, if grown under carefully controlled conditions *(122,123)*. Of the human RPE cell lines commonly in use, only D407 expresses functional P-gp activity *(103,124)*.

P-gp has also been localized in cultured rat retinal vascular endothelium, where it would exclude harmful substances from the inner retina *(125)*. P-gp is also expressed in bovine ciliary epithelium where it inhibits Cl⁻ currents, where it may play a role in regulating cell volume *(126,127)*. It may be that P-gp located at the apical membrane of the RPE also transports Cl⁻ or is involved in the processing of cholesterol *(86)*, given that the RPE is involved in processing shed photoreceptor outer segments on a daily basis. Figure 3 shows a schematic of the RPE with the localization of P-gp and other drug transporters that have been identified in the RPE.

OCULAR SIDE-EFFECTS OF DRUGS

The presence of drug transporters in the RPE does not confer resistance to all drugs. Chloroquine retinopathy is well known. Recent clinical findings of constricted visual fields and reduced electro-oculograms associated with loss of vision in patients prescribed vigabatrin therapy for epilepsy, have been well documented *(128)*. However, there is little evidence for P-gp transporting anti-epileptic drugs (vigabatrin, gabapentin, phenobarbitone, lamotrigine, carbamazepine, phenytoin, and valproate) at physiologically relevant concentrations. The exception is acetazolamide, which appears to be a weak P-gp substrate *(129,130)*. The lack of P-gp substrate specificity for vigabatrin may account for its ability to enter the retina and cause visual loss.

Digoxin is a substrate of P-gp, however, there have been case reports of color vision loss and visual disturbances in patients receiving digoxin therapy *(131,132)*. In experiments using MDR1 knock-out mouse models, the uptake of radiolabeled digoxin in the brain was 27 times higher than in wild-type mice *(133)*. The presence of P-gp in both the RPE and inner retinal vascular endothelium would suggest that digoxin and other drugs that are retinotoxic are able to enter the retina by bypassing the inner and outer blood–retinal

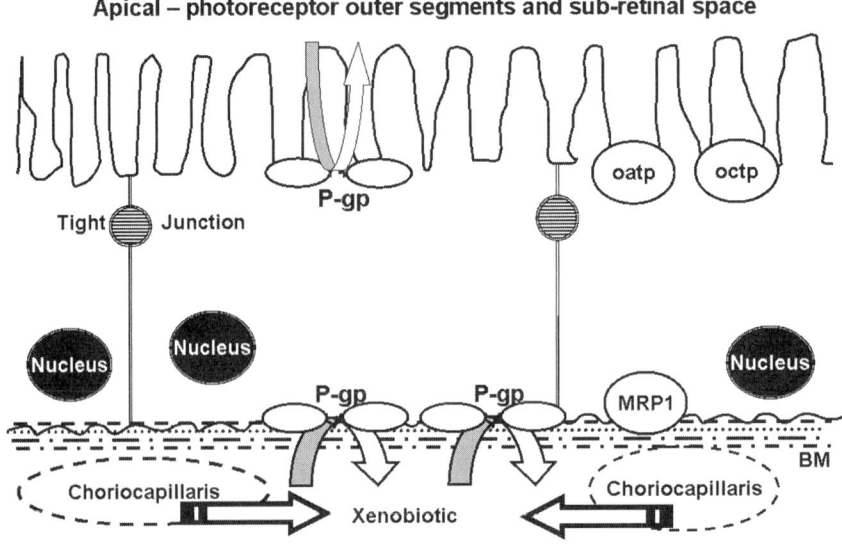

Fig. 3. The retinal pigment epithelium (RPE) forms the major barrier in the outer blood–retinal barrier. In this schematic the basal membrane of the RPE is continuous with Bruch's membrane (BM), beneath which the highly vascular and permeable choroid lies. The RPE acts as a barrier to xenobiotics by virtue of tight intercellular junctions that limit the diffusion of macromolecules and also an array of transporters that actively remove substrates from the plasma membrane. P-glycoprotein (P-gp) has been localized to the apical and basal membranes of the RPE. The function of P-gp at the apical membrane is uncertain, but may contribute to Cl⁻ transport and or transport of cholesterol. The localization of P-gp at the basal membrane is more in keeping with its role as a drug transporter, removing substrates that arrive through the fenestrations of the highly vascular choriocapillaris. Other drug transporters such as the organic anion (oatp) and organic cation transporters (octp) are also present along with multidrug resistance-associated protein 1 (MRP1) in the RPE and all serve to protect the outer retina and RPE from xenobiotics.

barriers. One potential region where drugs could obtain entry to the retina occurs at the junction between the optic nerve head and choroid. At this point there are no cellular impediments to drug diffusion from the choroid into the optic nerve head *(134,135)*.

DRUG DELIVERY

Recent success in delivering drugs to the RPE and outer retina for the management of neovascularization and inherited retinal degenerations requires intravitreal injections of specific oligonucleotides or antibodies. This is in part due to the need to apply the vector as near as possible to the cell of interest, and to avoid side-effects of the oligonucleotide. For review see Andrieu-Solar et al. (2006) *(136)*. It would be advantageous to be able to transfer drugs to the outer-retina/RPE by using an implant on the sclera to maintain long-term delivery of a drug, rather than repeated intravitreal injections.

To date, the successful delivery of betamethasone to the outer retina has been demonstrated in rabbit using an episcleral implant *(137)*. Pigment-epithelia-derived growth factor (PEDF) is also able to reach the RPE and retina following subconjunctival injection in rat, which raises the possibility of using this less-invasive route to deliver PEDF to the outer retina to reduce neovascularization *(138)*.

FUTURE DIRECTIONS

The expression of P-gp in the RPE is now established; however, several questions still remain as to how to reduce its activity with advances in drug therapies for diseases affecting the RPE and photoreceptors. Novel compounds are being developed that interfere with the transporter, and therefore in the future, the need for intravitreal injections will be reduced, as one of the RPE's defences is overcome. However, the role of P-gp at the apical membrane is still uncertain and so the full range of functions that P-gp performs are still to be fully understood with regard to lipid turnover and possible regulation of ionic channels and currents. The RPE remains the main impediment to drug delivery to the outer retina and the RPE itself, and yet we have still to fully characterize the drug transporters in this epithelium.

REFERENCES

1. Strauss O. The retinal pigment epithelium in visual function. Physiol Rev 2005;85:845–881.
2. Miyamoto Y, Kulanthaivel P, Leibach FH, Ganapathy V. Taurine uptake in apical membrane vesicles from the bovine retinal pigment epithelium. Invest Ophthalmol Vis Sci 1991;32:2542–2551.
3. Philp NJ, Wang D, Yoon H, Hjelmeland LM. Polarized expression of monocarboxylate transporters in human retinal pigment epithelium and ARPE-19 cells. Invest Ophthalmol Vis Sci 2003;44:1716–1721.
4. Kumagai AK, Glasgow BJ, Pardridge WM. GLUT1 glucose transporter expression in the diabetic and nondiabetic human eye. Invest Ophthalmol Vis Sci 1994;35:2887–2894.
5. Chancy CD, Kekuda R, Huang W, Prasad PD, Kuhnel J-M, Sirotnak FM, Roon P, Ganapathy V, Smith SB. Expression and differential polarization of the reduced-folate transporter-1 and the folate receptor alpha in mammalian retinal pigment epithelium. J Biol Chem 2000;275:20676–20684.
6. Han YH, Sweet DH, Hu DN, Pritchard JB. Characterization of a novel cationic drug transporter in human retinal pigment epithelial cells. J Pharmacol Exp Ther 2001;296:450–457.
7. Gao B, Wenzel A, Grimm C, Vavricka SR, Benke D, Meier PJ, Reme CE. Localization of organic anion transport protein 2 in the apical region of rat retinal pigment epithelium. Invest Ophthalmol Vis Sci 2002;43:510–514.
8. Steuer H, Jaworski A, Elger B, Kaussmann M, Keldenich J, Schneider H, Stoll D, Schlosshauer B. Functional characterization and comparison of the outer blood-retina barrier and the blood-brain barrier. Invest Ophthalmol Vis Sci 2005;46:1047–1053.
9. Mannermaa E, Vellonen KS, Urtti A. Drug transport in corneal epithelium and blood-retina barrier: Emerging role of transporters in ocular pharmacokinetics. Adv Drug Deliv Rev 2006;58:1136–1163.
10. Riordan JR, Deuchars K, Kartner N, Alon N, Trent J, Ling V. Amplification of P-glycoprotein genes in multidrug-resistant mammalian cell lines. Nature 1985;316:817–819.

11. Chin JE, Soffir R, Noonan KE, Choi K, Roninson IB. Structure and expression of the human MDR (P-glycoprotein) gene family. Mol Cell Biol 1989;9:3808–3820.
12. Fojo AT, Ueda K, Slamon DJ, Poplack DG, Gottesman MM, Pastan I. Expression of a multidrug-resistance gene in human tumors and tissues. Proced Natl Acad Sci USA 1987;84:265–269.
13. Thiebaut F, Tsuruo T, Hamada H, Gottesman MM, Pastan I, Willingham MC. Cellular localization of the multidrug-resistance gene product P-glycoprotein in normal human tissues. Proc Natl Acad Sci USA 1987;84:7735–7738.
14. Drion N, Risede P, Cholet N, Chanez C, Scherrmann JM. Role of P-170 glycoprotein in colchicine brain uptake. J Neurosci Res 1997;49:80–88.
15. Hyde SC, Emsley P, Hartshorn MJ, Mimmack MM, Gileadi U, Pearce SR, Gallagher MP, Gill DR, Hubbard RE, Higgins CF. Structural model of ATP-binding protein associated with cystic fibrosis, multidrug resistance and bacterial transport. Nature 1990;346:362–365.
16. Stefkova J, Poledne R, Hubacek JA. ATP-binding cassette (ABC) transporters in human metabolism and diseases. Physiol Rev 2004;53:235–243.
17. Callen DF, Baker E, Simmers RN, Seshadri R, Roninson IB. Localization of the human multiple drug resistance gene, MDR1, to 7q21.1. Hum Genet 1987;77:142–144.
18. Loo TW, Bartlett MC, Clarke DM. ATP hydrolysis promotes interactions between the extracellular ends of transmembrane segments 1 and 11 of human multidrug resistance P-glycoprotein. Biochemistry 2005;44:10250–10258.
19. Loo TW and Clarke DM. The human multidrug resistance P-glycoprotein is inactive when its maturation is inhibited: potential for a role in cancer chemotherapy. FASEB J 1999;13:1724–1732.
20. Azzaria M, Schurr E, Gros P. Discrete mutations introduced in the predicted nucleotide-binding sites of the mdr1 gene abolish its ability to confer multidrug resistance. Mol Cell Biol 1989;9:5289–5297.
21. Chen C-j, Chin JE, Ueda K, Clark DP, Pastan I, Gottesman MM, Roninson IB. Internal duplication and homology with bacterial transport proteins in the MDR1 (P-glycoprotein) gene from multidrug-resistant human cells. Cell 1986;47:381–389.
22. Chambers TC, McAvoy EM, Jacobs JW, Eilon G. Protein kinase C phosphorylates P-glycoprotein in multidrug resistant human KB carcinoma cells. J Biol Chem 1990;265: 7679–7686.
23. Hamada H, Hagiwara K, Nakajima T, Tsuruo T. Phosphorylation of the Mr 170,000 to 180,000 glycoprotein specific to multidrug-resistant tumor cells: effects of verapamil, trifluoperazine, and phorbol esters. Cancer Res 1987;47:2860–2865.
24. Chambers TC, Pohl J, Glass DB, Kuo JF. Phosphorylation by protein kinase C and cyclic AMP-dependent protein kinase of synthetic peptides derived from the linker region of human P-glycoprotein. Biochem J 1994;299:309–315.
25. Thews O, Gassner B, Kelleher DK, Schwerdt G, Gekle M. Impact of extracellular acidity on the activity of P-glycoprotein and the cytotoxicity of chemotherapeutic drugs. Neoplasia 2006;8:143–152.
26. Kim IW, Peng XH, Sauna ZE, FitzGerald PC, Xia D, Muller M, Nandigama K, Ambudkar SV. The conserved tyrosine residues 401 and 1044 in ATP sites of human P-glycoprotein are critical for ATP binding and hydrolysis: Evidence for a conserved subdomain, the A-loop in the ATP-Binding Cassette. Biochemistry 2006;45:7605–7616.
27. Loo TW and Clarke DM. The glycosylation and orientation in the membrane of the third cytoplasmic loop of human P-glycoprotein is affected by mutations and substrates. Biochemistry 1999;38:5124–5129.

28. Kramer R, Weber TK, Arceci R, Ramchurren N, Kastrinakis WV, Steele Jr G, Summerhayes IC. Inhibition of N-linked glycosylation of P-glycoprotein by tunicamycin results in a reduced multidrug resistance phenotype. Br J Cancer 1995;71:670–675.
29. Rosenberg MF, Velarde G, Ford RC, Martin C, Berridge G, Kerr ID, Callaghan R, Schmidlin A, Wooding C, Linton KJ, Higgins CF. Repacking of the transmembrane domains of P-glycoprotein during the transport ATPase cycle. EMBO J 2001;20:5615–5625.
30. Rosenberg MF, Kamis AB, Callaghan R, Higgins CF, Ford RC. Three-dimensional structures of the mammalian multidrug resistance p-glycoprotein demonstrate major conformational changes in the transmembrane domains upon nucleotide binding. J Biol Chem 2003;278:8294–8299.
31. Ramakrishnan P. The role of P-glycoprotein in the blood-brain barrier. Einstein Quart. J Biol Med 2003;19:160–165.
32. Bendayan R, Ronaldson PT, Gingras D, Bendayan M. In Situ localization of P-glycoprotein (ABCB1) in human and rat brain. J Histochem Cytochem 2006;54:1159–1167.
33. Del Moral RG, Olmo A, Aguilar M, O'Valle F. P glycoprotein: a new mechanism to control drug-induced nephrotoxicity. Exp Nephrol 1998;6:89–97.
34. Gottesman MM and Pastan I. Biochemistry of multidrug resistance mediated by the multidrug transporter. Annu Rev Biochem 1993;62:385–427.
35. Pleban K, Kopp S, Csaszar E, Peer M, Hrebicek T, Rizzi A, Ecker GF, Chiba P. P-Glycoprotein substrate binding domains are located at the transmembrane domain/transmembrane domain interfaces: a combined photoaffinity labeling-protein homology modeling approach. Mol Pharmacol 2005;67:365–374.
36. Loo TW, Bartlett MC, Clarke DM. Transmembrane segment 7 of human P-glycoprotein forms part of the drug-binding pocket. Biophys J 2006;399:351–359.
37. Loo TW, Bartlett MC, Clarke DM. Transmembrane segment 1 of human P-glycoprotein contributes to the drug-binding pocket. Biophys J 2006;396:537–545.
38. Juliano RL and Ling V. A surface glycoprotein modulating drug permeability in Chinese hamster ovary cell mutants. Biochimica Biophysica Acta 1976;455:152–162.
39. Riordan JR and Ling V. Purification of P-glycoprotein from plasma membrane vesicles of Chinese hamster ovary cell mutants with reduced colchicine permeability. J Biol Chem 1979;254:12701–12705.
40. Kartner N, Shales M, Riordan JR, Ling V. Daunorubicin-resistant Chinese hamster ovary cells expressing multidrug resistance and a cell-surface P-glycoprotein. Cancer Res 1983;43:4413–4419.
41. Chan HS, Lu Y, Grogan TM, Haddad G, Hipfner DR, Cole SP, Deeley RG, Ling V, Gallie BL. Multidrug resistance protein (MRP) expression in retinoblastoma correlates with the rare failure of chemotherapy despite cyclosporine for reversal of P-glycoprotein. Cancer Res 1997;57:2325–2330.
42. Wilson MW, Fraga CH, Fuller CE, Rodriguez-Galindo C, Mancini J, Hagedorn N, Leggas ML, Stewart CF. Immunohistochemical detection of multidrug-resistant protein expression in retinoblastoma treated by primary enucleation. Invest Ophthalmol Vis Sci 2006;47:1269–1273.
43. Mc Namara M, Clynes M, Dunne B, NicAmhlaoibh R, Lee WR, Barnes C, Kennedy SM. Multidrug resistance in ocular melanoma. Br J Ophthalmol 1996;80:1009–1112.
44. Safa AR, Mehta ND, Agresti M. Photoaffinity labeling of P-glycoprotein in multidrug resistant cells with photoactive analogs of colchicine. Biochem Biophys Res Commun 1989;162:1402–1408.
45. Roos E and Van de Pavert IV. Effect of tubulin-binding agents on the infiltration of tumour cells into primary hepatocyte cultures. J Cell Sci 1982;55:233–245.

46. Chaplin DJ and Horsman MR. Tumor blood flow changes induced by chemical modifiers of radiation response. Int J Radiat Oncol Biol Phys 1992;22:459–462.
47. Perez-Tomas R. Multidrug resistance: retrospect and prospects in anti-cancer drug treatment. Curr Med Chem 2006;13:1859–1876.
48. Fardel O, Lecureur V, Corlu A, Guillouzo A. P-glycoprotein induction in rat liver epithelial cells in response to acute 3-methylcholanthrene treatment. Biochem Pharmacol 1996;51:1427–1436.
49. Ferry DR, Russell MA, Cullen MH. P-glycoprotein possesses a 1,4-dihydropyridine-selective drug acceptor site which is alloserically coupled to a vinca-alkaloid-selective binding site. Biochem Biophys Res Commun 1992;188:440–445.
50. de Lannoy IAM and Silverman M. The MDR1 gene product, P-glycoprotein, mediates the transport of the cardiac glycoside, digoxin. Biochem Biophys Res Commun 1992;189:551–557.
51. Kim RB, Fromm MF, Wandel C, Leake B, Wood AJJ, Roden DM, Wilkinson GR. The drug transporter P-glycoprotein limits oral absorption and brain entry of HIV-1 protease inhibitors. J Clin Invest 1998;101:289–294.
52. Ueda K, Okamura N, Hirai M, Tanigawara Y, Saeki T, Kioka N, Komano T, Hori R. Human P-glycoprotein transports cortisol, aldosterone, and dexamethasone, but not progesterone. J Biol Chem 1992;267:24248–24252.
53. Van Kalken CK, Broxterman HJ, Pinedo HM, Feller N, Dekker H, Lankelma J, Giaccone G. Cortisol is transported by the multidrug resistance gene product P-glycoprotein. Br J Cancer 1993;67:284–289.
54. Tsuji A, Tamai I, Sakata A, Tenda Y and Terasaki T. Restricted transport of cyclosporin A across the blood-brain barrier by a multidrug transporter, P-glycoprotein. Biochem Pharmacol 1993;46:1096–1099.
55. Safa AR, Agresti M, Tamai I, Mehta ND, Vahabi S. The alpha 1-adrenergic photoaffinity probe [^{125}I]arylazidoprazosin binds to a specific peptide of P-glycoprotein in multidrug-resistant cells. Biochem Biophys Res Commun 1990;166:259–266.
56. Schuetz EG, Yasuda K, Arimori K, Schuetz JD. Human MDR and mouse mdrla P-glycoprotein alter the cellular retention and disposition of erythromycin, but not of retinoic acid or benzo(a)pyrene. Arch Biochem Biophys 1998;350:340–347.
57. Takano M, Hasegawa R, Fukuda T, Yumoto R, Nagai J, Murakami T. Interaction with P-glycoprotein and transport of erythromycin, midazolam and ketoconazole in Caco-2 cells. Eur J Pharmacol 1998;358:289–294.
58. Loe DW and Sharom FJ. Interaction of multidrug-resistant Chinese hamster ovary cells with the peptide ionophore gramicidin D. Biochim Biophys Acta 1994;1190:72–84.
59. Goda K, Krasznai Z, Gaspar R, Lankelma J, Westerhoff HV, Damjanovich S, Szabo J, Gabor. Reversal of multidrug resistance by valinomycin is overcome by CCCP. Biochem Biophys Res Commun 1996;219:306–310.
60. Huwyler J, Drewe J, Klusemann C; Fricker G. Evidence for P-glycoprotein-modulated penetration of morphine-6-glucuronide into brain capillary endothelium. Br J Pharmacol 1996;118:1879–1885.
61. Efferth T, Lohrke H, Volm M. Reciprocal correlation between expression of P-glycoprotein and accumulation of rhodamine 123 in human tumors. Anticancer Res 1989;9:1633–1637.
62. Homolya L, Hollo Z, Germann U, Pastan I, Gottesman M, Sarkadi B. Fluorescent cellular indicators are extruded by the multidrug resistance protein. J Biol Chem 1993;268:21493–21496.

63. Lee CGL, Gottesman MM, Cardarelli CO, Ramachandra M, Jeang KT, Ambudkar SV, Pastan I, Dey S. HIV-1 protcase inhibitors are substrates for the *MDR1* multidrug transporter. Biochemistry 1998;37:3594–3601.
64. Callaghan R and Riordan JR. Synthetic and natural opiates interact with P-glycoprotein in multidrug- resistant cells. J Biol Chem 1993;268:16059–16064.
65. Henthorn TK, Liu Y, Mahapatro M, Ng K-Y. Active transport of fentanyl by the blood-brain barrier. J Pharmacol Exp Ther 1999;289:1084–1089.
66. Karlsson J, Kuo SM, Ziemniak J, Artursson P. Transport of celiprolol across human intestinal epithelial (Caco-2) cells: mediation of secretion by multiple transporters including P-glycoprotein. Br J Pharmacol 1993;110:1009–1016.
67. Ueda K, Cardarelli C, Gottesman MM, Pastan I. Expression of a full-Length cDNA for the Human "MDR1" gene confers resistance to colchicine, doxorubicin, and vinblastine. Proc Natl Acad Sci U S A 1987;84:3004–3008.
68. Janice H, Barry HH, Nicholas LS. Drug absorption limited by P-glycoprotein-mediated secretory drug transport in human intestinal epithelial CACO-2 cell layers. Pharm Res 1993;10:743–749.
69. Mahon FX, Deininger MWN, Schultheis B, Chabrol J, Reiffers J, Goldman JM, Melo JV. Selection and characterization of BCR-ABL positive cell lines with differential sensitivity to the tyrosine kinase inhibitor STI571: diverse mechanisms of resistance. Blood 2000;96:1070–1079.
70. Tanigawara Y, Okamura N, Hirai M, Yasuhara M, Ueda K, Kioka N, Komano T, Hori R. Transport of digoxin by human P-glycoprotein expressed in a porcine kidney epithelial cell line (LLC-PK1). J Pharmacol Exp Ther 1992;263:840–845.
71. Sanna T-S, Jarkko R, Steve W, Joseph WP, James EP. Midazolam exhibits characteristics of a highly permeable p-glycoprotein substrate. Pharma Res 2003;20:757–764.
72. Collett A, Higgs NB, Sims E, Rowland M, Warhurst G. Modulation of the permeability of H2 receptor antagonists cimetidine and ranitidine by P-glycoprotein in rat intestine and the human colonic cell line CACO-2. J Pharmacol Exp Ther 1999;288:171–178.
73. Rice GC, Ling V, Schimke RT. Frequencies of independent and simultaneous selection of Chinese hamster cells for methotrexate and doxorubicin (adriamycin) resistance. Proc Natl Acad Sci USA 1987;84:9261–9264.
74. Schinkel AH, Smit JJ, van Tellingen O, Beijnen JH, Wagenaar E, van Deemter L, Mol CA, van der Valk MA, Robanus-Maandag EC, te Riele HP. Disruption of the mouse mdr1a P-glycoprotein gene leads to a deficiency in the blood-brain barrier and to increased sensitivity to drugs. Cell 1994;77:491–502.
75. Gutmann H, Miller DS, Droulle A, Drewe J, Fahr A, Fricker G. P-glycoprotein- and mrp2- mediated octreotide transport in renal proximal tubule. Br J Pharmacol 2000;129:251–256.
76. Hollo Z, Homolya L, Davis CW, Sarkadi B. Calcein accumulation as a fluorometric functional assay of the multidrug transporter. Biochim Biophys Acta 1994;1191:384–388.
77. Nowak JZ. Age-related macular degeneration (AMD): pathogenesis and therapy. Pharmacol Rep 2006;58:353–363.
78. Johnstone RW, Ruefli AA, Smyth MJ. Multiple physiological functions for multidrug transporter P-glycoprotein? Trends Biochem Sci 2000;25:1–6.
79. Johnstone RW, Ruefli AA, Tainton KM, Smyth MJ. A role for P-glycoprotein in regulating cell death. Leuk Lymphona 2000;38:1–11.
80. Romsicki Y and Sharom FJ. Phospholipid flippase activity of the reconstituted P-glycoprotein multidrug transporter. Biochemistry 2001;40:6937–6947.

81. Eckford PD and Sharom FJ. The reconstituted P-glycoprotein multidrug transporter is a flippase for glucosylceramide and other simple glycosphingolipids. Biochem J 2005;389:517–526.
82. Tainton KM, Smyth MJ, Jackson JT, Tanner JE, Cerruti L, Jane SM, Darcy PK, Johnstone RW. Mutational analysis of P-glycoprotein: suppression of caspase activation in the absence of ATP-dependent drug efflux. Cell Death Differ 2004;11:1028–1037.
83. Toshiyuki S, Tsutomu N, Midori H, Takashi K, Atsushi W, Tatsurou Y, Hironao K, Shunji N, Noboru O, Takayoshi Y, Toshiro S, Akinobu G, Masafumi M, Katsuhiko O. MDR1 up-regulated by apoptotic stimuli suppresses apoptotic signaling. Pharm Res 2002;19:1323–1329.
84. Valverde MA, Diaz M, Sepulveda FV, Gill DR, Hyde SC, Higgins CF. Volume-regulated chloride channels associated with the human multidrug-resistance P-glycoprotein. Nature 1992;355:830–833.
85. Gill DR, Hyde SC, Higgins CF, Valverde MA, Mintenig GM, Sepulveda FV. Separation of drug transport and chloride channel functions of the human multidrug resistance P-glycoprotein. Cell 1992;71:23–32.
86. Luker GD, Nilsson KR, Covey DF, Piwnica-Worms D. Multidrug resistance (MDR1) P-glycoprotein enhances esterification of plasma membrane cholesterol. J Biol Chem 1999;274:6979–6991.
87. Ho EA, Piquette-Miller M. Regulation of multidrug resistance by pro-inflammatory cytokines. Curr Cancer Drug Targets 2006;6:295–311.
88. Holtzman CW, Wiggins BS, Spinler SA. Role of P-glycoprotein in statin drug interactions. Pharmacotherapy 2006;26:1601–1607.
89. Er-Jia W, Christopher NC, Robert PC, William WJ. HMG-CoA reductase inhibitors (statins) characterized as direct inhibitors of P-glycoprotein. Pharmaceutical Res 2001;18:800–806.
90. Cornwell M, Pastan I, Gottesman M. Certain calcium channel blockers bind specifically to multidrug- resistant human KB carcinoma membrane vesicles and inhibit drug binding to P-glycoprotein. J Biol Chem 1987;262:2166–2170.
91. Pötschka H, Fedrowitz M, Löscher W. P-glycoprotein and multidrug resistance-associated protein are involved in the regulation of extracellular levels of the major antiepileptic drug carbemazepine in the brain. Neuroreport 2001;12:3557–3560.
92. Chishty M, Reichel A, Siva J, Abbott NJ, Begley DJ. Affinity for the P-glycoprotein efflux pump at the blood-brain barrier may explain the lack of side-effects of modern antihistamines. J Drug Target 2001;9:223–228.
93. Hämmerle SP, Rothen-Rutishauser B, Krämer SD, Günthert M, Wunderli-Allenspach H. P-glycoprotein in cell cultures: a combined approach to study expression, localisation, and functionality in the confocal microscope. Eur J Pharm Sci 2000;12:69–77.
94. Omidi Y, Campbell L, Barar J, Connell D, Akhtar S, Gumbleton M. Evaluation of the immortalised mouse brain capillary endothelial cell line, b.End3, as an in vitro blood-brain barrier model for drug uptake and transport studies. Brain Res 2003;990:95–112.
95. Beck WT, Cirtain MC, Glover CJ, Felsted RL, Safa AR. Effects of indole alkaloids on multidrug resistance and labeling of P-glycoprotein by a photoaffinity analog of vinblastine. Biochem Biophys Res Commun. 1988;153:959–966.
96. Garrigos M, Belehradek J, Mir LM, Orlowski S. Absence of cooperativity for MgATP and verapamil effects on the ATPase activity of P-glycoprotein containing membrane vesicles. Biochem Biophys Res Commun 1993;196:1034–1041.
97. Horio M, Gottesman MM, Pastan I. ATP-dependent transport of vinblastine in vesicles from human multidrug-resistant cells. Proc Natl Acad Sci US A 1988;85:3580–3585.

98. List AF, Spier C, Greer J, Wolff S, Hutter J, Dorr R, Salmon S, Futscher B, Baier M, Dalton W. Phase I/II trial of cyclosporine as a chemotherapy-resistance modifier in acute leukemia. J Clin Oncol 1993;11:1652–1660.
99. Nobili S, Landini I, Giglioni B, Mini E. Pharmacological strategies for overcoming multidrug resistance. Curr Drug Targets 2006;7:861–879.
100. Thomas H and Coley HM. Overcoming multidrug resistance in cancer: an update on the clinical strategy of inhibiting p-glycoprotein. Cancer Control 2003;10:159–165.
101. Randall PR, Albert E, Richard L, Tomasz MB, Yoo-Joung K, Henner WD, Glenn B, Elizabeth AM, Varun G, Ene E, Matthew WH, William SD. Safety and efficacy of the MDR inhibitor Incel (biricodar, VX-710) in combination with mitoxantrone and prednisone in hormone-refractory prostate cancer. Can Chemo Pharmacol 2003;51:297–305.
102. Van Zuylen L, Sparreboom A, van der Gaast A, Nooter K, Eskens FALM, Brouwer E, Bol CJ, de Vries R, Palmer PA, Verweij J. Disposition of docetaxel in the presence of P-glycoprotein inhibition by intravenous administration of R101933. Eur J Cancer 2002;38:1090–1099.
103. Kennedy BG and Mangini NJ. P-glycoprotein expression in human retinal pigment epithelium. Mol Vis 2002;8:422–430.
104. Sharom FJ, Yu X, Lu P, Liu R, Chu JWK, Szabo K, Muller M, Hose CD, Monks A, Varadi A. Interaction of the P-glycoprotein multidrug transporter (MDR1) with high affinity peptide chemosensitizers in isolated membranes, reconstituted systems, and intact cells. Biochemical Pharmacology 1999;58:571–586.
105. Jekerle V, Klinkhammer W, Scollard DA, Breitbach K, Reilly RM, Piquette-Miller M, Wiese M. In vitro and in vivo evaluation of WK-X-34, a novel inhibitor of P-glycoprotein and BCRP, using radio imaging techniques. Int J Cancer 2006;119:414–422.
106. Wesolowska O, Molnar J, Westman G, Samuelsson K, Kawase M, Ocsovszki I, Motohashi N, Michalak K. Benzo[a]phenoxazines: a new group of potent P-glycoprotein inhibitors. In Vivo 2006;20:109–113.
107. Koubeissi A, Raad I, Ettouati L, Guilet D, Dumontet C, Paris J. Inhibition of P-glycoprotein-mediated multidrug efflux by aminomethylene and ketomethylene analogs of reversins. Bioorganic Med Chem Letts 2006;16:5700–5703.
108. Xing H, Wang S, Weng D, Chen G, Yang X, Zhou J, Xu G, Lu Y, Ma D. Knock-down of P-glycoprotein reverses taxol resistance in ovarian cancer multicellular spheroids. Oncol Rep 2007;17:117–122.
109. Rittierodt M, Tschernig T, Harada K. Modulation of multidrug-resistance-associated P-glycoprotein in human U-87 MG and HUV-ECC cells with antisense oligodeoxynucleotides to MDR1 mRNA. Pathobiology 2004;71:123–128.
110. Foxwell BM, Mackie A, Ling V, Ryffel B. Identification of the multidrug resistance-related P-glycoprotein as a cyclosporine binding protein. Mol Pharmacol 1989;36:543–546.
111. Pourtier-Manzanedo A, Boesch D, Loor F. FK-506 (fujimycin) reverses the multidrug resistance of tumor cells in vitro. Anticancer Drugs 1991;2:279–283.
112. Arceci RJ, Stieglitz K, Bierer BE. Immunosuppressants FK506 and rapamycin function as reversal agents of the multidrug resistance phenotype. Blood 1992;80:1528–1536.
113. Wang E-J, Casciano CN, Clement RP, Johnson WW. The farnesyl protein transferase inhibitor SCH66336 is a potent inhibitor of MDR1 product P-glycoprotein. Cancer Res 2001;61:7525–7529.
114. Safa AR, Glover CJ, Sewell JL, Meyers MB, Biedler JL, Felsted RL. Identification of the multidrug resistance-related membrane glycoprotein as an acceptor for calcium channel blockers. J Biol Chem 1987;262:7884–7888.

115. Jonsson O, Behnam-Motlagh P, Persson M, Henriksson R, Grankvist K. Increase in doxorubicin cytotoxicity by carvedilol inhibition of P-glycoprotein activity. Biochem Pharmacol 1999;58:1801–1806.
116. Hofsli E and Nissen-Meyer J. Reversal of drug resistance by erythromycin: erythromycin increases the accumulation of actinomycin D and doxorubicin in multidrug-resistant cells. Int J Cancer 1989;44:149–154.
117. Fardel O, Lecureur V, Loyer P, Guillouzo A. Rifampicin enhances anti-cancer drug accumulation and activity in multidrug-resistant cells. Biochem Pharmacol 1995;49:1255–1260.
118. Varga A, Nugel H, Baehr R, Marx U, Hever A, Nacsa J, Ocsovszky I, Molnar J. Reversal of multidrug resistance by amitriptyline in vitro. Anticancer Res 1996;16:209–211.
119. Esser P, Tervooren D, Heimann K, Kociok N, Bartz-Schmidt KU, Walter P, Weller M. Intravitreal daunomycin induces multidrug resistance in proliferative vitreoretinopathy. Invest Ophthalmol Vis Sci 1998;39:164–170.
120. Luo Y, Zhuo Y, Fukuhara M, Rizzolo LJ. Effects of culture conditions on heterogeneity and the apical junctional complex of the ARPE-19 cell line. Invest Ophthalmol Vis Sci 2006;47:3644–3655.
121. Tian J, Ishibashi K, Handa JT. The expression of native and cultured RPE grown on different matrices. Physiol Genomics 2004;17:170–182.
122. Hu J and Bok D. A cell culture medium that supports the differentiation of human retinal pigment epithelium into functionally polarized monolayers. Mol Vis 2001;7:14–19.
123. Peng S, Rahner C, Rizzolo LJ. Apical and basal regulation of the permeability of the retinal pigment epithelium. Invest Ophthalmol Vis Sci 2003;44:808–817.
124. Constable PA, Lawrenson JG, Dolman DEM, Arden GB, Abbott NJ. P-Glycoprotein expression in retinal pigment epithelium cell lines. Exp Eye Res 2006;83:24–30.
125. Greenwood J. Characterization of a rat retinal endothelial cell culture and the expression of P-glycoprotein in brain and retinal endothelium in vitro. J Neuroimmunol 1992;39:123–132.
126. Wang L, Chen L, Walker V, Jacob TJC. Antisense to MDR1 mRNA reduces P-glycoprotein expression, swelling-activated Cl⁻ current and volume regulation in bovine ciliary epithelial cells. J Physiol 1998;511:33–44.
127. Chen LX, Wang LW, Jacob T. The role of MDR1 gene in volume-activated chloride currents in pigmented ciliary epithelial cells. Acta Physiol Sinca 2002;54:1–6.
128. Harding GF, Wild JM, Robertson KA, Lawden MC, Betts TA, Barber C, Barnes PM. Electro-oculography, electroretinography, visual evoked potentials, and multifocal electroretinography in patients with vigabatrin-attributed visual field constriction. Epilepsia 2000;41:1420–1431.
129. Weiss J, Kerpen CJ, Lindenmaier H, Dormann S-M G, Haefeli WE. Interaction of antiepileptic drugs with human P-glycoprotein in vitro. J Pharmacol Exp Ther 2003;307:262–267.
130. Crowe A and Teoh YK. Limited P-glycoprotein mediated efflux for anti-epileptic drugs. J Drug Target 2006;14:291–300.
131. Lawrenson JG, Kelly C, Lawrenson AL, Birch J. Acquired colour vision deficiency in patients receiving digoxin maintenance therapy. Br J Ophthalmol 2002;86:1259–1261.
132. Castells DD, Teitelbaum BA, Tresley DJ. Visual changes secondary to initiation of amiodarone: a case report and review involving ocular management in cardiac polypharmacy. Optometry 2002;73:113–121.
133. Schinkel AH, Mayer U, Wagenaar E, Mol CAAM, van Deemter L, Smit JJM, van der Valk MA, Voordouw AC, Spits H, van Tellingen O, Zijlmans JMJM, Fibbe WE, Borst P.

Normal viability and altered pharmacokinetics in mice lacking mdr1-type (drug-transporting) P-glycoproteins. Proceed Natl Acad Sci U S A 1997;94:4028–4033.
134. Flage T. A defect in the blood-retina barrier in the optic nerve head region in the rabbit and the monkey. Acta Ophthalmol (Copenh) 1980;58:645–651.
135. Cohen AI. Is there a potential defect in the blood-retinal barrier at the choroidal level of the optic nerve canal? Invest Ophthalmol Vis Sci 1973;12:513–519.
136. Andrieu-Soler C, Bejjani RA, de Bizemont T, Normand N, BenEzra D, Behar-Cohen F. Ocular gene therapy: a review of nonviral strategies. Mol Vis 2006;12:1334–1347.
137. Kato A, Kimura H, Okabe K, Okabe J, Kunou N, Ogura Y. Feasibility of drug delivery to the posterior pole of the rabbit eye with an episcleral implant. Invest Ophthalmol Vis Sci 2004;45:238–244.
138. Amaral J, Fariss RN, Campos MM, Robison Jr WG, Kim H, Lutz R, Becerra SP. Transscleral-RPE permeability of PEDF and ovalbumin proteins: implications for subconjunctival protein delivery. Invest Ophthalmol Vis Sci 2005;46:4383–4392.

VI
Transporters in the Retina

14
The Retinal Rod NCKX1 and Cone/Ganglion Cell NCKX2 Na^+/Ca^{2+}-K^+ Exchangers

Paul P. M. Schnetkamp, Yoskiyuki Shibukawa,
Haider F. Altimimi, Tashi G. Kinjo, Pratikhya Pratikhya,
Kyeong Jing Kang, and Robert T. Szerencsei

CONTENTS

INTRODUCTION
THE Na^+/Ca^{2+}-K^+ EXCHANGER IN RETINAL ROD PHOTORECEPTORS
MOLECULAR ANALYSIS OF NCKX GENE PRODUCTS
Na^+/Ca^{2+}-K^+ EXCHANGER STRUCTURE–FUNCTION RELATIONSHIPS
CONCLUSIONS
ACKNOWLEDGEMENTS
REFERENCES

INTRODUCTION

Regulation of cytosolic $[Ca^{2+}]$ is of critical importance to all cells in view of the ubiquitous role played by changes in cytosolic $[Ca^{2+}]$ in regulating cellular processes, and the role of sustained and increased cytosolic $[Ca^{2+}]$ in mediating cell death. It has been known since the 1970s that changes in extracellular $[Ca^{2+}]$ have dramatic effects on rod photoreceptor physiology *(1)* and that isolated rod outer segments (ROS) show very dynamic Ca^{2+} fluxes *(2,3)*. We now know that rapid changes in $[Ca^{2+}]$ are mediated by specific channels and transporters exclusively located in the ROS plasma membrane, and it has been suggested that changes in cytosolic $[Ca^{2+}]$ in ROS are effectively segregated from those in the rest of the rod photoreceptor *(4)*. The plasma membrane of both mammalian and amphibian ROS has been shown to contain only two ion transport proteins, the cyclic guanosine monophosphate (cGMP)-gated channels that carry the light-sensitive current and a high-powered Na^+/Ca^{2+} exchanger, however there are no currents *(5)* or fluxes *(6)* mediated by other channels or transporters. Due to this rather unique circumstance, isolated ROS have been the preparation of choice for the initial characterization and purification of both the rod cGMP-gated channel *(7)* and the Na^+/Ca^{2+} exchanger *(8,9)*. The characterization of the rod Na^+/Ca^{2+} exchanger revealed significant functional similarities with Na^+/Ca^{2+} exchangers found in other tissues, most notably the well-studied exchanger present in cardiac sarcolemmal membranes. However, quite distinct effects of K^+ on the rod Na$^+$/Ca^{2+} exchanger were noted early on *(3,8)*. It was shown subsequently that the rod exchanger not only uses the Na$^+$ gradient for Ca^{2+} extrusion (similar to the Na$^+$/Ca^{2+} exchanger in the heart), but in addition, and unlike the exchanger found in the heart, uses the transmembrane K$^+$ gradient as well, at a stoichiometry of four Na$^+$ to one Ca^{2+} and one K$^+$ *(10,11)*. Molecular cloning revealed that the rod Na$^+$/Ca^{2+}-K$^+$ exchanger, or NCKX, was the first member of the novel *SLC24* gene family, with remarkably little sequence similarity to members of the *SLC8* gene family of Na$^+$/Ca^{2+}exchangers or NCX *(12–14)*. Here, we will review our studies on the functional and molecular properties of the NCKX1 and NCKX2 Na$^+$/Ca^{2+}-K$^+$ exchangers found in the retina.

THE Na$^+$/Ca^{2+}-K$^+$ EXCHANGER IN RETINAL ROD PHOTORECEPTORS

The Na$^+$/Ca^{2+}-K$^+$ Exchanger in Rod Physiology

The light-sensitive inward current through the cGMP-gated channels in ROS is carried by both Na$^+$ and Ca^{2+} *(15,16)*. Thus, in darkness and under non-saturating illumination, cGMP-gated channels mediate a sustained Ca^{2+} influx (around 15% of the total inward current) into the ROS. The principal role of the Na$^+$/Ca^{2+}-K$^+$ exchanger is to extrude this Ca^{2+} and prevent the outer segments filling up with excess Ca^{2+}. Due to this dynamic equilibrium between Ca^{2+} entry via cGMP-gated channels and Ca^{2+} extrusion via Na$^+$/Ca^{2+}-K$^+$ exchangers, cytosolic calcium concentration ([Ca^{2+}]) in ROS in darkness is significantly elevated above normal resting [Ca^{2+}] in cells and values of approximately 500 nM and 250 nM have been reported for ROS of the intact photoreceptors of the salamander and mouse retina, respectively *(17,18)*. Illumination causes a graded closure of cGMP-gated channels and associated Ca^{2+} entry, which is then accompanied by a graded lowering of cytosolic [Ca^{2+}] mediated by the Na$^+$/Ca^{2+}-K$^+$ exchanger, as the exchanger is not directly regulated by light. The light-induced lowering of cytosolic [Ca^{2+}] in ROS is a major contributor to the process of light adaptation through its effect on guanylyl cyclase activity *(19–21)* and perhaps also by modulating the cGMP sensitivity of the channels *(22,23)*.

Functional and Molecular Properties of the Rod Outer Segment Na$^+$/Ca^{2+}-K$^+$ Exchanger

The *SLC24* gene family of Na$^+$/Ca^{2+}-K$^+$ exchangers comprises five or six members (depending on whether one considers NCKX6 to belong to a distinct gene family) with transcripts found in many tissues *(13,24)*. Remarkably, detailed studies on the in situ physiology of NCKX proteins are still limited to the NCKX1 isoform in retinal ROS, although NCKX has now been shown to contribute significantly to Ca^{2+} homeostasis in several neurons *(25–28)*. Perhaps the most striking feature of NCKX1-mediated Ca^{2+} fluxes in vertebrate ROS is their sheer capacity, which can change total ROS Ca^{2+} content by 0.05 and 0.5 mM/s in amphibian and mammalian ROS, respectively *(2,8,29)*. The bovine rod Na$^+$/Ca^{2+}-K$^+$ exchanger was purified and functionally reconstituted as an integral membrane protein with an apparent molecular weight (MW) of 230 kDa that is clearly visible on a Coomassie-stained sodium dodecyl sulfate polyacrylamide gel electrophoresis (SDS-PAGE) separation of total ROS proteins *(9)*. Bidirectional transport is the second key distinguishing feature of $Na^+/Ca^{2+}-K^+$ exchangers (shared with Na^+/Ca^{2+} exchangers) with important physiological consequences. The $Na^+/Ca^{2+}-K^+$ exchanger can carry out both efficient Ca^{2+} extrusion and Ca^{2+} import depending on the transmembrane Na^+ and K^+ gradients *(8,30)*. Thus, if the transmembrane Na^+ gradient is compromised under pathophysiological conditions or reversed by experimental manipulations, the $Na^+/Ca^{2+}-K^+$ exchanger will cause a very rapid, large and sustained rise in cytosolic $[Ca^{2+}]$. In the rod outer segment this sustained rise in cytosolic $[Ca^{2+}]$ leads to the rapid inhibition of guanylyl cyclase, resulting in a lowering of cGMP levels and closure of cGMP-gated channels, while in whole photoreceptors or other neurons this may lead to cell death caused by sustained Ca^{2+} overload.

The apparent dissociation constants (K_d) have been determined for both the intracellular and extracellular cation-binding sites for the bovine rod $Na^+/Ca^{2+}-K^+$ exchanger. Comparable K_d values were obtained for the inward- and outward-facing binding sites, respectively, for each of the transported cations: 25 to 40 mM for Na^+ (Hill coefficient of 2), 1 to 10 mM for K^+, and 1 to 3 µM for Ca^{2+} *(8,11,31,32)*. These values were obtained under conditions that minimized competitive interactions between cations at their respective binding sites. Multiple and complex interactions between physiologically occurring cations (Na^+, Ca^{2+}, K^+, Mg^{2+}, and protons) have been described and, as a result the apparent K_d values, may vary significantly depending on the exact ionic conditions used in the experiment *(33–36)*. The similarity in the apparent cation dissociation constants observed for the external and intracellular cation binding sites is consistent with the so-called alternating access or flip-flop model for cation transport, in which a single set of residues on the $Na^+/Ca^{2+}-K^+$ exchanger protein alternates access between the extracellular and intracellular space, respectively. Alternating access is the generally accepted model of solute/ion transport for most members of the *SLC* solute carrier gene families. Perhaps the most compelling evidence for an alternating-access model of transport mediated by the $Na^+/Ca^{2+}-K^+$ exchanger is the demonstration of self-exchange fluxes ($Ca^{2+} + K^+$:$Ca^{2+} + K^+$ exchange and Na^+:Na^+ exchange) in bovine ROS *(3,6,37)*. The properties of the in situ bovine NCKX1 are summarized in Table 1 and compared with properties observed for heterologously expressed NCKX1 and NCKX2 as discussed below.

Regulation of the Rod $Na^+/Ca^{2+}-K^+$ Exchanger

Most ion channels and transporters are regulated in complex ways by accessory binding proteins or by post-translational modifications, most notably phosphorylation. Very little is known about regulation of the various $Na^+/Ca^{2+}-K^+$ exchanger isoforms. There is no evidence yet that phosphorylation or other post-translational modifications have any effect on NCKX function. However, we did obtain strong evidence that the NCKX1 Ca^{2+} extrusion mode in ROS is subject to strong inactivation by a mechanism yet to be elucidated. Inactivation of the exchanger may prevent lowering of cytosolic $[Ca^{2+}]$ to the undesirably low values predicted by the NCKX stoichiometry, when Ca^{2+} influx via the cGMP-gated channels is interrupted for prolonged periods of time after saturation of rod photoreceptors during daylight illumination *(38,39)*. Consistent with the this inactivation observed in isolated ROS, light-induced and NCKX1-mediated lowering of cytosolic

Table 1 Functional properties of the Na+/Ca^{2+}-K+ exchangers NCKX1 and NCKX2

	Heterologously expressed NCKX1	Heterologously expressed NCKX2	In situ bovine NCKX1
Human gene name	SLC24a1	SLC24a2	-
Tissue distribution and cellular/subcellular expression	Rod photoreceptors, platelets	Retinal cone photoreceptors, retinal ganglion cells, brain	Rod outer segments
Gene locus	15q22	9p22	10
Number of residues	1098	661	1216
Stoichiometry	4Na$^+$:Ca^{2+}:K$^+$ (4:1:1)	4Na$^+$:Ca^{2+}:K$^+$ (4:1:1)	4Na$^+$:Ca^{2+}:K$^+$ (4:1:1)
Splice variants	Several	Two	Several likely
Mechanism	Alternating access	Alternating access	Alternating access
Transport modes	Forward (Ca^{2+} efflux)	Forward (Ca^{2+} efflux)	Forward (Ca^{2+} efflux)
	Reverse (Ca^{2+} influx)	Reverse (Ca^{2+} influx)	
	Self exchange	Self exchange	Reverse (Ca^{2+} influx)
			Self exchange
K_d for Ca^{2+} (external)	1–2 µM	2 µM	1 µM
K_d for K$^+$ (external)	2–10 mM	2–10 mM	1.5 mM
K_d for Na$^+$ (external)	35–45 mM	38 mM	35 mM
Protein–protein interactions	Form homodimers, binds other NCKX isoforms, binds CNGA1, CNGA3	Form homodimers, binds other NCKX isoforms, binds CNGA1, CNGA3	Form homodimers, binds to CNGA1

[Ca^{2+}] in intact photoreceptors using a saturating flash of light did not lower cytosolic [Ca^{2+}] to the extremely low values predicted by the NCKX stoichiometry (17,18).

Interaction of the Na$^+$/Ca^{2+}-K$^+$ Exchanger with other ROS Proteins

Proteins involved in cellular signaling are often organized in multi-protein complexes that ensure the coordinated action of these proteins in the appropriate domain of the cell. In addition, many ion channels and ion transporters operate as homo- or hetero-multimers. The bovine rod NCKX1 protein has been shown to form a dimer (40) and associate with the CNGA1 subunit of the cGMP-gated channel (41), and it has been suggested that NCKX1 dimerization results in inhibition of Ca^{2+} transport via the exchanger (42). The cGMP-gated channels not only associate with NCKX1 via the CNGA1 subunit, but the CNGB1 subunit binds directly to two small proteins, Rom-1 and Peripherin/rds, located in the rims of the internal disk membranes (43). Thus, the CNG-NCKX complex spans two membranes, but the physiological significance of this arrangement remains to be explored. To explore the specificity of the NCKX1-NCKX1 and NCKX1–CNGA1 interactions, we used both thiol-based crosslinkers and co-immunoprecipitation experiments to examine NCKX dimerization and CNGA–NCKX interactions when rod and cone NCKX and CNGA cDNAs were expressed in cell lines in various combinations.

When expressed in cell lines, both rod NCKX1 and cone NCKX2 were found to specifically associate with each other, most likely by forming dimers, as well as with the respective CNGA1 or CNGA3 subunits of the rod and cone cGMP-gated channels *(44)*. However, no specificity was observed among rod and cone isoforms, suggesting that both the NCKX–NCKX and NCKX–CNGA interactions were mediated by the two sets of transmembrane segments, as these represent the only domains conserved between the NCKX1 and NCKX2 isoforms (see below).

MOLECULAR ANALYSIS OF NCKX GENE PRODUCTS

Cloning and Analysis of Rod NCKX1 cDNAs

When the first NCKX1 cDNA was cloned from bovine retina, remarkably little sequence similarity to the previously cloned NCX1 Na^+/Ca^{2+} exchanger was observed, although hydropathy analysis suggested a very similar topology for the two proteins *(12)*. When comparing the NCKX1 sequence with sequences of other plasma membrane ion transporters, three rather uncommon features stood out:

1. Bovine NCKX1 contains a large hydrophilic loop of around 400 residues located close to the N-terminus and predicted to be in the extracellular space, a rather unusual feature for plasma membrane ion transport proteins.
2. The large extracellular loop at the N-terminus is preceded by a putative signal peptide, again this is rather unusual for plasma membrane proteins, although a signal peptide at the N-terminus is also found in NCX1.
3. The second large, hydrophilic loop separates two sets of predicted transmembrane segments and is thought to be located in the cytosol. About half of this loop consists of mostly acidic residues including multiple repeats of a 17-amino-acid motif, again a very unusual feature not seen in any other ion transport protein with the exception perhaps of the CNGB1 subunit of the rod cGMP-gated channel. Perhaps this acidic domain could contain regulatory Ca^{2+} binding sites that might control the inactivation process described earlier.

Unfortunately, the first NCKX1 cDNA cloned from cow did not yield functional expression in cell lines, which hampered experimental analysis of the mentioned unusual features of bovine NCKX1. The lack of functional expression of bovine NCKX1 spurred the cloning of NCKX1 cDNAs from other vertebrate species including human *(45)*, dolphin *(46)*, rat *(47)*, and chicken *(48)*. Analysis and comparison of these different NCKX1 sequences revealed the following:

1. All mammalian NCKX1 possess equally large hydrophilic loops close to the N-terminus, although sequence similarity is rather modest, with about 50% identity. Curiously, the chicken NCKX1 lacks the large hydrophilic loop close to the N-terminus and more closely resembles the other NCKX isoforms, which typically have N-terminal hydrophilic loops of 40 to 70 residues, not counting the signal peptide *(24,49)*.
2. All the NCKX1 sequences contained a putative signal peptide at the far N-terminus, and this is a feature shared with nearly all NCKX sequences in the data base with the notable exception of *Drosophila* NCKX-X *(49)*.

3. The acidic domain of the large cytosolic loop of bovine NCKX1 is found in other mammalian NCKX1 as well, although the specific sequence is not conserved and remarkably variable, e.g., rat NCKX1 has repeats of an acidic motif, but the sequence is different from that of bovine NCKX1, while the human NCKX1 has a long stretch of predominantly acidic residues at this position, but lacks clear repeats *(45)*. As observed with the large, extracellular hydrophilic loop, the acidic domain of the large cytosolic loop appears unique to mammalian NCKX as it is absent in chicken NCKX1.

As a result of the properties unique to mammalian NCKX1 genes, mammalian NCKX1 encode large proteins of 1100 to 1200 residues, whereas the chicken NCKX1 gene, as well as other NCKX genes, encode considerably smaller proteins of 600 to 700 residues. The significance of the divergent sequences found for rod NCKX1 proteins from different species is currently unclear. Analysis of functional correlates to the sequence variations observed is hampered the fact that, when transfected into cell lines, many of the NCKX1 cDNAs do not yield functional proteins that reach the plasma membrane.

Cloning and Analysis of a Second Retinal NCKX cDNA

In contrast to the extensive literature on the in situ retinal rod NCKX1 Na$^+$/Ca^{2+}-K$^+$ exchanger using both biochemical and electrophysiological methods (for reviews, see *37,50,51*), very little is known about Ca^{2+} fluxes in cone outer segments. Na$^+$-dependent Ca^{2+} release has been demonstrated in the cone outer segments of salamander *(52)* and zebra fish *(53)*, but it was not determined whether this reflected Na$^+$/Ca^{2+} or Na$^+$/Ca^{2+}-K$^+$ activity. We used homology cloning to identify a second NCKX isoform in the cone-rich chicken retina and localized these NCKX2 transcripts to the inner segments of the majority, if not all, cone photoreceptors, as well as most retinal ganglion cells *(48)*. Based on the chicken NCKX2 sequence, we used homology cloning to identify a second human NCKX from retinal cDNA and found that its transcripts were localized to cone photoreceptors and ganglion cells in the human retina *(54)*, similar to observations for chicken NCKX2 transcripts *(48)*. The NCKX2 transcripts are also abundantly present in various parts of the rat brain *(55)*. Curiously, no cone dysfunction was apparent from electroretinogram (ERG) recordings of NCKX2-knockout mice *(28)*. Clearly, more work needs to be carried out on characterization of Ca^{2+} fluxes and the role of NCKX2 in the outer segments of cone photoreceptors, and such studies are currently underway.

Chicken and human NCKX2 cDNAs encode proteins of about 660 residues, very similar in size to the chicken NCKX1 and also homologous to a rat NCKX cDNA cloned from brain *(56)*. The brain and retina NCKX2 proteins lack the variable domains found in mammalian NCKX1, and they are highly conserved over most of their sequence. The overall organization of NCKX2 proteins is identical to that of NCKX1 proteins as illustrated in Fig. 1. Two sets of putative TMS are separated by a large hydrophilic loop, thought to be located in the cytosol. The first set of TMS is preceded by an N-terminal domain with a putative signal peptide at the N-terminus. Very little sequence similarity is obvious between the N-terminal loops of NCKX1 and NCKX2 proteins, while only a small part of the large cytosolic loop has a domain conserved between NCKX1 and

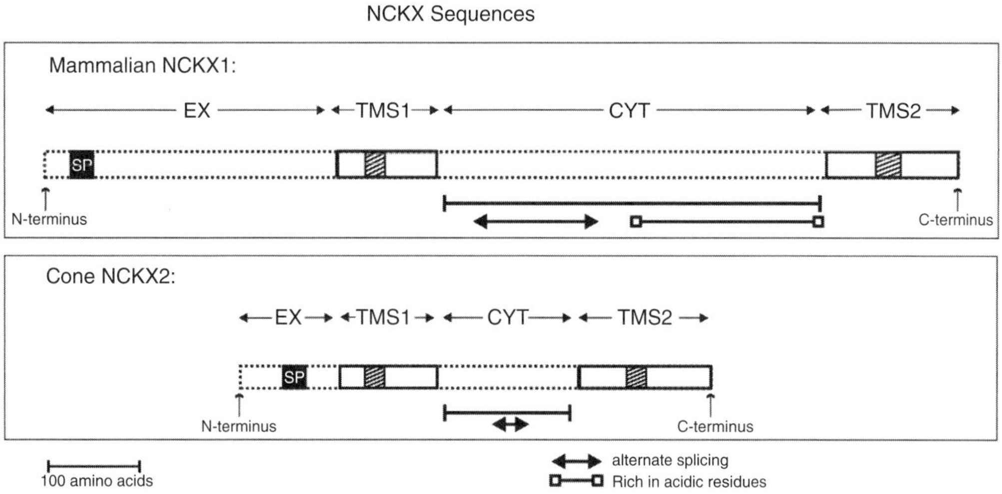

Fig. 1. Domain analysis of the mammalian Na$^+$/Ca^{2+}-K$^+$ exchanger NCKX1 and NCKX2 proteins. The shaded boxes close to the N-terminus indicate the hydrophobic segment that represents a putative and cleavable signal peptide. The relative position of the two sets of transmembrane segments (TMS) containing the α repeats (striped boxes) is shown, as are the location of alternate splicing and the domain rich in acidic residues (mammalian NCKX1 only).

NCKX2 *(48)*. In contrast, the 11 hydrophobic segments making up the two sets of TMS show a high degree of sequence identity between the various NCKX1 and NCKX2 proteins (Fig. 2). The central sections of the two TMS show internal sequence similarity suggested to have arisen from an ancient gene duplication event *(57)*. These two sections, the so-called α1 and α2 repeats, have subsequently been shown to contain many of the residues important for cation transport (see below).

Post-translational Modification: Alternative Splicing, N-terminal Signal-peptide Cleavage and Glycosylation of NCKX1 and NCKX2

Sequence comparison of the first three mammalian NCKX1 cDNAs cloned revealed gaps in the alignment of the otherwise-conserved N-terminal part of the large cytosolic loop: human NCKX1 lacked 18 residues found in bovine NCKX1, while dolphin NCKX1 lacked 114 residues found in the bovine sequence *(46)*. Analysis of the genomic organization of the human NCKX1 gene revealed that the different NCKX1 clones likely represent three different splice variants: bovine NCKX1 represents the full-length mammalian NCKX1, human NCKX1 lacks exon III, whereas our dolphin NCKX1 clone lacks exons III, IV, V, and VI *(58)*. Another study described the cloning of rat NCKX1 cDNA and as many as five different NCKX1 splice variants were found *(47)*. Curiously, only the full-length chicken NCKX1 transcript was found with no evidence for the presence of other splice variants, despite the fact that the amino-acid sequence encoded for by exon IV, the site of alternate splicing in mammalian NCKX1, is well conserved between mammals and chicken *(48)*.

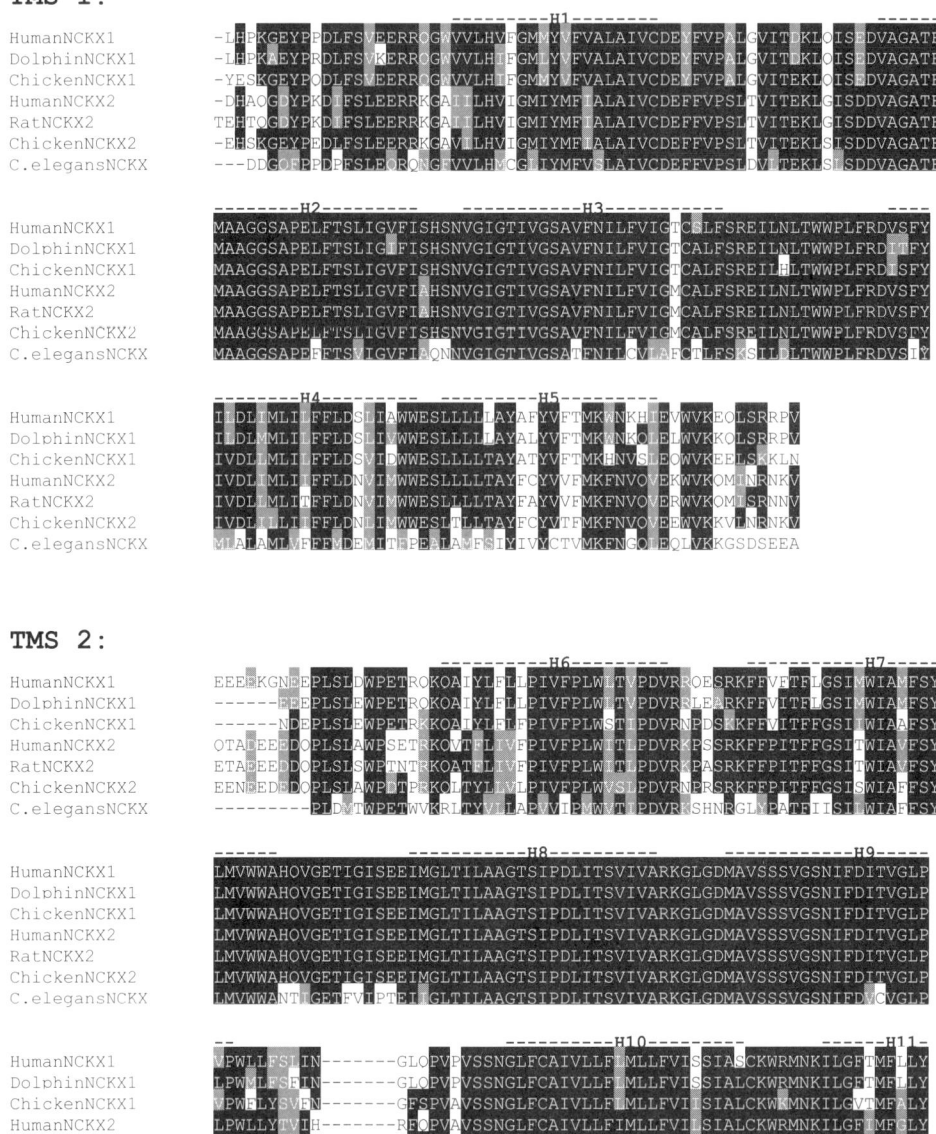

Fig. 2. Sequence alignment of the transmembrane segments (TMS) domains of the Na$^+$/Ca^{2+}-K$^+$ exchanger (NCKX) from different species. The sequences of NCKX1 and NCKX2 from various species were compared and aligned with an NCKX paralog cloned from *C. Elegans*. The alignment is limited to the two sets of hydrophobic segments containing a total of eleven putative transmembrane segments (labeled H1 through H11) that are conserved between all members of the NCKX gene family. H5 is separated from H6 by a large hydrophilic loop predicted to be located in the cytosol. H1 is preceded by an N-terminal loop that contains a putative cleavable signal peptide.

N-terminal sequencing of the NCKX1 protein purified from bovine retinal ROS showed that the N-terminal residue was Asp66, deduced from the full-length bovine NCKX1 cDNA *(12)*, suggesting that the putative signal peptide was cleaved off, with additional processing of the N-terminus as well. In contrast, analysis of the synthesis of the bovine NCKX1 protein in a cell-free system revealed no evidence of N-terminal processing and it was suggested that the hydrophobic domain at the N-terminus was essential for the NCKX1 protein to adopt the native topology; furthermore, an antibody against the N-terminal sequence suggested the presence of uncleaved NCKX1 in bovine retinal extracts, from which the authors conclude that even in the retina bovine NCKX1 is not cleaved *(59)*. In our laboratory, we have examined putative signal-peptide cleavage of dolphin NCKX1 by inserting the FLAG tag either before, or after, the putative cleavage site; we used dolphin NCKX1 since it is the only NCKX1 cDNA that results in strong and consistent functional expression in cell lines, and since the N-terminal sequence is well conserved with respect to that of bovine NCKX1. Our results revealed that the signal peptide of a fraction of the dolphin NCKX1 proteins was indeed cleaved at the predicted cleavage site after residue 38; moreover, signal peptide cleavage was critical for correct plasma membrane targeting *(60)*. Curiously, the signal peptide of the chicken NCKX1 was not cleaved in either HEK 293 cells or in insect High Five cells, despite the fact that algorithms (http://www.cbs.dtu.dk/services/SignalP/) predict that signal peptide cleavage is most likely in the case of chicken NCKX1, while the longer signal peptides predicted for mammalian NCKX1, and in particular NCKX2, suggest the presence of a uncleaved signal anchor *(60)*. The observation that the NCKX1 protein purified from bovine ROS begins at residue 66, almost thirty residues beyond the predicted cleavage point, suggests that additional processing (not seen in HEK293 cells) occurs in rod photoreceptors.

The bovine NCKX1 protein in retinal ROS is primarily glycosylated via O-linked sialo-carbohydrate chains, giving rise to an often-broad high-MW band at around 230 kDa on SDS-PAGE and there are often several bands at lower MW as well *(61)*. In contrast, the heterologously expressed dolphin and bovine NCKX1 proteins appear less glycosylated, as on SDS-PAGE sharp bands were observed at apparent MWs of 180 and 210 kDa, respectively *(46,60)*. This suggests that glycosylation of NCKX1 expressed in cell lines differs significantly from in situ glycosylation of NCKX1 in ROS.

In addition to full-length NCKX2, a single major splice variant has been found for human, rat and chicken NCKX2, in which 17 residues are missing from the middle of the large cytosolic loop *(48,55)*. Although functional expression levels were found to be lower for the full-length human and chicken NCKX2 when compared with the shorter splice variant, the significance of the occurrence of two NCKX2 splice variants is unclear. The NCKX2 contains a single and conserved N-glycosylation site in the extracellular loop, which is glycosylated in the heterologously expressed NCKX2 protein *(60,62)*. As observed for NCKX1, NCKX2 contains a putative signal peptide at the N-terminus, although in this case the predicted cleavage site would result in an unusually large signal peptide of 53 residues, more consistent with an uncleaved signal anchor (http://www.cbs.dtu.dk/services/SignalP/). Using the same strategy as described above for the NCKX1 protein, we determined that the signal peptide of part of the NCKX2 protein population is cleaved when NCKX2 is expressed in cell lines and, as observed for NCKX1, signal-peptide cleavage is critical for plasma membrane targeting *(60)*. To date, no studies on post-translational modification of the in situ NCKX2 protein have been published.

Mutated Alleles of the Rod NCKX1 and Cone NCKX2 Genes in Patients with Retinal Disease

The rod NCKX1 and cone NCKX2 genes are located on chromosomes 15q22 and 9p22, respectively, while the genomic organization of both genes has been shown to be very similar, consisting of 10 and 11 exons, respectively *(58,63)*. The large second exon of both genes codes for the N-terminus, the extracellular loop and the first set of transmembrane segments (TMS), while the NCKX2 gene contains an additional exon coding for 17 residues in the large cytosolic loop, and this exon is missing in the short splice variant that is the predominant form of NCKX2 found in the retina. In the only study to date to address mutated alleles of NCKX genes in human disease, 815 patients with a variety of retinal disorders and 166 patients with cone-rod degeneration and cone dysfunction syndromes were screened for sequence changes in the NCKX1 and NCKX2 genes, respectively. Relatively few sequence variants were observed for the cone NCKX2 gene; in contrast, many sequence variants were observed for the rod NCKX1 gene, some of which were considered to be pathogenic although this could not be proved *(63)*. Several of the sequence variants concerned mis-sense mutations in codons of residues that are conserved between NCKX1 and NCKX2, and in some cases even conserved among all five NCKX isoforms. Several of these sequence variants were introduced into our human NCKX2 cDNA and analyzed for changes in protein expression and transport function *(63)*. At least one of the rod NCKX1 mutations (Ile992Thr) resulted in a large (80%) decrease in NCKX function when this mutation was introduced into our human NCKX2 cDNA, although protein expression was not affected. Another mutation observed creates a putative novel splice site that may result in the deletion of the N-terminal 29 residues, including most of the signal peptide. As discussed above, this may prevent correct targeting to the plasma membrane. A reduction of NCKX1 function is likely to result in an increase in the dark cytosolic $[Ca^{2+}]$ in ROS, which could affect long-term survival of rods photoreceptors. No transgenic animals with altered NCKX1 proteins have been described yet, but an increase in dark $[Ca^{2+}]$ and photoreceptor degeneration was observed in a transgenic mouse model containing a mutation in the GCAP1 protein *(18)*. Therefore, it is likely that mutations causing greatly altered NCKX1 function will lead to retinal degeneration.

Na^+/Ca^{2+}-K^+ EXCHANGER STRUCTURE–FUNCTION RELATIONSHIPS

Residues Important for Cation Transport

Functional analysis of a series of NCKX1 and NCKX2 cDNAs from different species, as well as a NCKX paralog from *C. Elegans*, revealed that the apparent dissociation constants for each of the respective substrates (i.e. Na^+, Ca^{2+}, and K^+) were very similar across the species *(48,64,65)*. Moreover, these values were very similar to the apparent dissociation constants obtained with a bovine NCKX1-deletion mutant, from which the two large hydrophilic loops were deleted, suggesting that the residues required for cation transport were all contained in the two sets of TMS *(66)*. Maximal transport activity of the various NCKX proteins in heterologous systems was found to be more variable when compared with K_d values. The short splice variant of the human and chicken NCKX2 clones displayed the highest and most consistent activity of all NCKX cDNAs *(48,65)*, and therefore all the structural work subsequently carried out in our laboratory was based on the short splice variant of the human NCKX2. Comparing the various NCKX1 and NCKX2 sequences with that of *C. Elegans* NCKX revealed two large blocks of nearly completely-conserved sequence centered around the α1 and α2 repeats that form the core of the first and second sets of TMS, respectively (Fig. 2). These two blocks of conserved residues were selected for the first NCKX scanning mutagenesis study and a total of 96 residues were substituted, including every acidic residue found in the TMS. Twenty-five residues were identified for which substitution lowered the transport capacity of the mutant NCKX2 proteins to less than 20% of that found for wild-type NCKX2, while protein expression levels and plasma membrane targeting were not altered *(67)*. These residues included four acidic residues (Glu188, Asp258, Asp548, and Asp575) proposed to be important for Ca^{2+} binding, and several glycine residues that could provide for points of flexibility needed for the large conformational change implied in the alternating access model (these residues are highlighted in Fig. 3). Furthermore, several serine and threonine residues were found to be important for transport function, many of which did not even tolerate a highly conservative substitution (e.g., Ser to Thr) *(35)*. The sensitivity of NCKX2 function for such a significant number of residues is consistent with the notion that transport is associated with the large conformational change implicit in the alternating-access model of NCKX function.

These studies used an insect cell expression system that produced the most consistent functional expression for a large range of NCKX clones and mutants *(48,66,67)*. High-throughput functional analysis of sufficient resolution to permit detailed investigations of even low activity NCKX mutants is best carried out with fluorescent Ca^{2+}-indicating dyes, and the insect High Five cell system used above proved to be unsuitable for this purpose. A HEK293 cell expression system was used to obtain high-resolution recordings of the Ca^{2+} and K^+ dependencies of wild-type NCKX2 and large number of single-residue substitution NCKX2 mutants. Nine residues were identified in which substitutions caused a decrease in the apparent affinity for Ca^{2+}. This was surprisingly always accompanied by a decrease in apparent affinity for K^+ *(35)*. The position of some of these residues is illustrated in Fig. 3. Of the four critical acidic residues identified earlier *(67)*, the conservative Glu188Asp and Asp548Glu substitutions resulted in the largest shifts in Ca^{2+} and K^+ dependencies, whereas the charge-eliminating, but size-conservative, Glu188Gln and Asp548Asn substitutions yielded nonfunctional NCKX2 mutant proteins *(35)*. Glu188 and Asp548 are conserved in all NCKX1-5 and NCX1-3 isoforms, and these are the only two acidic residues in the NCKX2 protein that yield nonfunctional proteins upon charge-eliminating substitutions. Together, the above observations suggest that Glu188 and Asp548 of human NCKX2 and the corresponding residues in the other NCKX and NCX isoforms are the two most critical residues for Ca^{2+} binding *(35)*. No clear results were obtained with Asp258 substitutions, whereas the Asp575 residue was shown to be the key residue that imparts K^+ dependence on NCKX2, as the Asp575Cys and Asp575Asn substitutions yielded mutant NCKX2 proteins that had lost the unique K^+-dependence characteristic for NCXK proteins, whereas the more-conservative Asp575Glu substitution yielded a nonfunctional NCKX2 mutant protein *(68)*. Asp575 is conserved in all NCKX1-5 isoforms while NCX1-3 have an asparagine residue at the corresponding position.

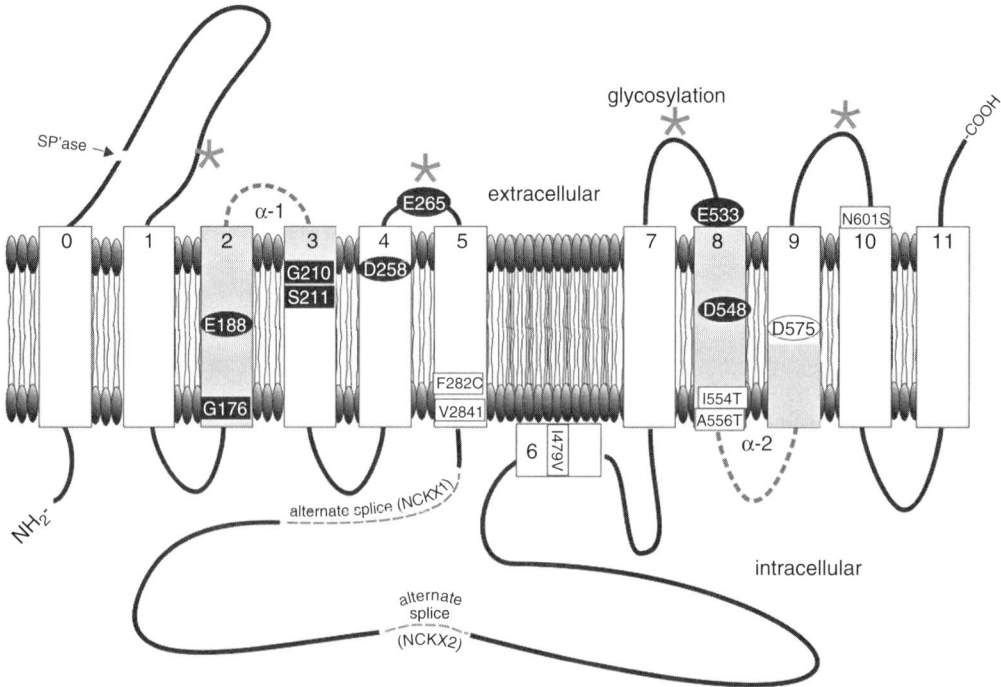

Fig. 3. Topological model of the Na$^+$/Ca^{2+}-K$^+$ exchanger 2 (NCKX2) protein. The current topological model of the NCKX2 protein is illustrated with the position and orientation of the twelve hydrophobic segments common to all NCKX isoforms. SP'ase indicates the signal peptidase shown to cleave heterologously expressed NCKX1 and NCKX2 proteins (60). The asterisks indicate glycosylation of (inserted) glycosylation sites, indicating exposure to the extracellular space (62). Substitution of residues indicated by black rectangles results in >90% loss of transport function (67). Substitutions of residues indicated by black ovals show large shifts in Ca^{2+} and K$^+$ dependencies (35). Substitution of D575 (clear oval) eliminates the K$^+$-dependence characteristic for all NCKX proteins (68). Sequence variants found in patients with retinal disease are shown as white rectangles (63).

Membrane Topology and Helix Packing of the NCKX2 Protein

Hydropathy analysis of the various NCKX isoforms results in a remarkable range of different topological models for different NCKX isoforms, or even the same NCKX isoform, when different algorithms are used. We used a combination of modifications of inserted cysteine residues and insertion of glycosylation sites to experimentally determine which of the hydrophilic loops that connect TMS are exposed to the extracellular space, yielding the complete topological model for the NCKX2 protein illustrated in Fig. 3 (62). Given the high degree of sequence similarity between the NCKX1 and NCKX2 isoforms, we believe the same model applies to NCXK1 and also very likely to the NCKX3-5 isoforms. In our topological model both the N-terminus (after signal-peptide cleavage) and the C-terminus are located in the extracellular space, and this model differs from the current topological model of the NCX1 Na$^+$/Ca^{2+} exchanger with respect to the three

C-terminal TMS *(69,70)*. However, in both topological models the two α repeats are in an inverted or antiparallel orientation, as is commonly observed for similar internal repeats in transporter proteins.

When residues important for cation transport are positioned within our NCKX topological model, it is clear that residues from both sets of TMS contribute to Ca^{2+} binding and that these residues are clustered in the two α repeats (Fig. 3). Presumably, within the three-dimensional structure of NCKX2, these TMS come together and form the cation-binding pocket. To address proximity of residues important for cation transport we used site-directed disulfide mapping in a functional and cysteine-free NCKX2 mutant. Although all NCKX1 and NCKX2 sequences in the database share five cysteine residues in the TMS, replacement with serine residues yielded a functional and cysteine-free NCKX2 mutant available for cysteine scanning mutagenesis and site-directed disulfide mapping *(71)*. Site-directed disulfide mapping involves insertion of pairs of cysteine residues followed by formation of a disulfide bond by copper-phenanthroline, which requires that the two cysteine residues making up the pair are within about 3 Å of each other. Site-directed disulfide mapping showed that the critical acidic residues Glu188, Asp548, and Asp575 are in close proximity, consistent with their critical role in the Ca^{2+} and K^+ binding pocket, and also confirmed the antiparallel orientation of the α1 and α2 repeats *(72)*. Moreover, these residues were shown to be in an area of considerably increased flexibility, perhaps indicative of the conformational change that results in alternating access of the cation binding site(s) to the intracellular and extracellular space, respectively.

CONCLUSIONS

Since the discovery in the late eighties of the first Na^+/Ca^{2+}-K^+ exchanger shown to be present in isolated retinal rod outer segments, much progress has been made in characterizing the in situ NCKX1 exchanger found in vertebrate rod photoreceptors, many NCKX1 and NCKX2 cDNAs have been cloned and the molecular analysis of the NCKX1 and NCKX2 gene products expressed in cell lines is proceeding. The mammalian rod NCKX1 protein contains two large hydrophilic domains, of remarkable sequence variability, in addition to two sets of highly conserved TMS. Many sequence variants were found in the human rod NCKX1 protein, some of which might be linked to retinal disease *(63)*. The NCKX1 occurs as a dimer in rod photoreceptors and is part of a multiprotein complex that includes the heteromultimeric cGMP-gated channels. It is now clear that the NCKX2-5 isoforms are found in many different tissues including retina, brain and skin, and in the latter case a single-nucleotide polymorphism (SNP) in the *SLC24A5* gene appears to be an important determinant of skin pigmentation in humans *(73)*. However, very little is yet known about the physiology of NCKX proteins in cells other than retinal rod photoreceptors, but the emerging importance of NCKX in tissues as diverse as hippocampal neurons and skin suggests that NCKX physiology should be an exciting field of study in the coming years. Structure–function analyses of the NCKX2 isoform expressed in cell lines have already identified a set of residues important for cation transport *(35,67,68)*, and resulted in a topological model of NCKX2 that shows an antiparallel orientation of the two α repeats *(62,72)*. Yet, many aspects of NCKX structure-function relationships remain to be elucidated and we lack fundamental insight in the workings of members of this gene family of intricate cation counter transporters.

ACKNOWLEDGEMENTS

This work was supported by an operating grant from the Canadian Institutes for Health Research to P.P.M.S., who is a scientist of the Alberta Heritage Foundation for Medical Research. T.G.K. is a recipient of a studentship from the Alberta Heritage Foundation for Medical Research. H.A. and K.J.K. are recipients of a studentship from the Foundation Fighting Blindness, Canada.

REFERENCES

1. Hagins WA and Yoshikami S. Ionic mechanisms in excitation of photoreceptors. Ann NY Acad Sci 1975;314–325.
2. Schnetkamp PPM. Calcium translocation and storage in isolated intact cattle rod outer segments in darkness. Biochim Biophys Acta 1979;554:441–459.
3. Schnetkamp PPM. Ion selectivity of the cation transport system of isolated cattle rod outer segments: evidence of a direct communication between the rod plasma membrane and the rod disk membranes. Biochim Biophys Acta 1980;598:66–90.
4. Krizaj D and Copenhagen DR. Compartmentalization of calcium extrusion mechanisms in the outer and inner segments of photoreceptors. Neuron 1998;21:249–256.
5. Lagnado L, Cervetto L, McNaughton PA. Ion transport by the Na:Ca exchange in isolated rod outer segments. Proc Natl Acad Sci U S A 1988;85:4548–4552.
6. Schnetkamp PPM, Szerencsei RT, Basu DK. Unidirectional Na^+, Ca^{2+} and K^+ fluxes through the bovine rod outer segment Na-Ca-K exchanger. J Biol Chem 1991;266:198–206.
7. Cook N J, Hanke W, Kaupp UB. Identification, purification, and functional reconstitution of the cyclic GMP-dependent channel from rod photoreceptors. Proc Natl Acad Sci USA 1987;84:585–589.
8. Schnetkamp PPM. Sodium-Calcium exchange in the outer segments of bovine rod photo receptors. J Physiol 1986;373:25–45.
9. Cook NJ and Kaupp UB. Solubilization, Purification, and Reconstitution of the Sodium-Calcium Exchanger from Bovine Retinal Rod Outer Segments. J Biol Chem 1988;263:11382–11388.
10. Schnetkamp PPM, Szerencsei RT, Basu DK. Na-Ca exchange in bovine rod outer segments requires potassium. Biophys J 1988;53:389a.
11. Schnetkamp PPM, Basu DK, Szerencsei RT. Na-Ca exchange in the outer segments of bovine rod photoreceptors requires and transports potassium. Am J Physiol (Cell Physiol) 1989;257:C153–C157.
12. Reilander H, Achilles A, Friedel U, Maul G, Lottspeich F, Cook NJ. Primary structure and functional expression of the Na/Ca,K-exchanger from bovine rod photoreceptors. EMBO J 1992;11:1689–1695.
13. Schnetkamp PPM. The SLC24 Na^+/Ca^{2+}-K^+ exchanger family: vision and beyond. Eur J Physiol 2004;447:683–688.
14. Quednau BD, Nicoll DA, Philipson KD. The sodium/calcium exchanger family-SLC8. Eur J Physiol 2004;447:543–548.
15. Picones A and Korenbrot JI. Permeability and interaction of Ca^{2+} with cGMP-gated ion channels differ in retinal rod and cone photoreceptors. Biophys J 1995;69:120–127.
16. Dzeja C, Hagen V, Kaupp UB, Frings S. Ca^{2+} permeation in cyclic nucleotide-gated channels. EMBO J 1999;18:131–144.
17. Sampath AP, Matthews HR, Cornwall MC, Fain GL. Bleached Pigment Produces a Maintained Decrease in Outer Segment Ca^{2+} in Salamander Rods. J Gen Physiol 1998;111:53–64.

18. Olshevskaya EV, Calvert PD, Woodruff ML, Peshenko IV, Savchenko AB, Makino CL, Ho YS, Fain GL, Dizhoor AM. The Y99C mutation in guanylyl cyclase-activating protein 1 increases intracellular Ca^{2+} and causes photoreceptor degeneration in transgenic mice. J Neurosci 2004;24(27):6078–6085.
19. Matthews HR, Murphy RLW, Fain GL, Lamb TD. Photoreceptor light adaptation is mediated by cytoplasmic calcium concentration. Nature 1988;334:67–69.
20. Nakatani K and Yau K-W. Calcium and light adaptation in retinal rods and cones. Nature 1988;334:69–71.
21. Koch K-W and Stryer L. Highly cooperative feedback control of retinal rod guanylate cyclase by calcium ions. Nature 1988;334:64–66.
22. Molday RS. Calmodulin regaulation of cyclic-nucleotude-gated channels. Curr Top Dev Biol 1996;6:445–452.
23. Rebrik TI and Korenbrot JI. In intact mammalian photoreceptors, Ca^{2+}-dependent modulation of cGMP-gated ion channels is detectable in cones but not in rods. J Gen Physiol 2004;123:63–75.
24. Cai X and Lytton J. The cation/Ca(2+) exchanger superfamily: phylogenetic analysis and structural implications. Mol Biol Evol 2004;21:1692–1703.
25. Kiedrowski L. High activity of K^+-dependent plasmalemmal Na^+/Ca^{2+} exchangers in hippocampal CA1 neurons. Neuroreport 2004;15:2113–2116.
26. Kim MH, Lee SH, Park KH, Ho WK, Lee SH. Distribution of K^+-dependent Na^+/Ca^{2+} exchangers in the rat supraoptic magnocellular neuron is polarized to axon terminals. J Neurosci 2003;23:11673–11680.
27. Kim MH, Korogod N, Schneggenburger R, Ho WK, Lee S-H. Interplay between Na^+/Ca^{2+} exchangers and mitochondria in Ca^{2+} clearance at the calyx of Held. J Neurosci 2005;25:6057–6065.
28. Li XF, Kiedrowski L, Tremblay F, Fernandez FR, Perizzolo M, Winkfein RJ, Turner RW, Bains JS, Rancourt DE, Lytton J. Importance of K^+-dependent Na^+/Ca^{2+}-exchanger 2, NCKX2, in motor learning and memory. J Biol Chem 2006.
29. Schnetkamp PPM and Bownds MD. Sodium and cGMP-induced Ca^{2+} fluxes in frog rod photoreceptors. J Gen Physiol 1987;89:481–500.
30. Schnetkamp PPM and Szerencsei RT. Intracellular Ca^{2+} sequestration and release in intact bovine retinal rod outer segments. J Biol Chem 1993;268:12449–12457.
31. Schnetkamp PPM. Optical measurements of Na-Ca-K exchange currents in intact outer segments isolated from bovine retinal rods. J Gen Physiol 1991;98:555–573.
32. Schnetkamp PPM, Tucker JE, Szerencsei RT. Ca^{2+} influx into bovine retinal rod outer segments mediated by Na-Ca+K exchange. Am J Physiol (Cell Physiol) 1995;269:c1153–c1159.
33. Cordes FS, Bright JN, Sansom MS. Proline-induced distortions of transmembrane helices. J Mol Biol 2002;323:951–960.
34. Tieleman DP, Sansom MS, Berendsen HJ. Alamethicin helices in a bilayer and in solution: molecular dynamics simulations. Biophys J 1999;76:40–49.
35. Kang K-J, Kinjo TG, Szerencsei RT, Schnetkamp PPM. Residues contributing to the Ca^{2+} and K^+ binding pocket of the NCKX2 Na^+/Ca^{2+}-K^+ exchanger. J Biol Chem 2005;280:6823–6833.
36. Kang TM and Hilgemann DW. Multiple transport modes of the cardiac Na^+/Ca^{2+} exchanger. Nature 2004;427:544–548.
37. Schnetkamp PPM. Na-Ca or Na-Ca-K exchange in the outer segments of vertebrate rod photoreceptors. Prog Biophys Mol Biol 1989;54:1–29.
38. Schnetkamp PPM, Basu DK, Li XB, Szerencsei RT. Regulation of intracellular free Ca^{2+} concentration in the outer segments of bovine retinal rods by Na-Ca-K exchange measured with Fluo-3. II. Thermodynamic competence of transmembrane Na^+ and K^+ gradients and inactivation of Na^+-dependent Ca^{2+} extrusion. J Biol Chem 1991;266:22983–22990.

39. Schnetkamp PPM. How does the retinal rod Na-Ca + K exchanger regulate free cytosolic Ca^{2+}? J Biol Chem 1995;270:13231–13239.
40. Schwarzer A, Kim TSY, Hagen V, Molday RS, Bauer PJ. The Na/Ca-K exchanger of rod photoreceptor exists as dimer in the plasma membrane. Biochemistry 1997;36:13667–13676.
41. Schwarzer A, Schauf H, Bauer PJ. Binding of the cGMP-gated channel to the Na/Ca-K exchanger in rod photoreceptors. J Biol Chem 2000;275:13448–13454.
42. Bauer PJ and Schauf H. Mutual inhibition of the dimerized Na/Ca-K exchanger in rod photoreceptors. Biochim Biophys Acta 2002;1559:121–134.
43. Poetsch A, Molday LL, Molday RS. The cGMP-gated channel and related glutamic acid-rich proteins interact with peripherin-2 at the rim region of rod photoreceptor disc membranes. J Biol Chem 2001;276:48009–48016.
44. Kang K-J, Bauer PJ, Kinjo TG, Szerencsei RT, Bonigk W, Winkfein RJ, Schnetkamp PPM. Assembly of retinal rod or cone Na^+/Ca^{2+}-K^+ exchangers oligomers with cGMP-gated channel subunits as probed with heterologously expressed cDNAs. Biochemistry 2003;42:4593–4600.
45. Tucker JE, Winkfein RJ, Cooper CB, Schnetkamp PPM. cDNA cloning of the human retinal rod Na/Ca+K exchanger: Comparison with a revised bovine sequence. Invest Ophthalmol Vis Sci 1998;39:435–440.
46. Cooper CB, Winkfein RJ, Szerencsei RT, Schnetkamp PPM. cDNA-cloning and functional expression of the dolphin retinal rod Na-Ca+K exchanger NCKX1: Comparison with the functionally silent bovine NCKX1. Biochemistry 1997;38:6276–6283.
47. Poon S, Leach S, Li XF, Tucker JE, Schnetkamp PP, Lytton J. Alternatively spliced isoforms of the rat eye sodium/calcium+potassium exchanger NCKX1. Am J Physiol Cell Physiol 2000;278:C651–C660.
48. Prinsen CFM, Szerencsei RT, Schnetkamp PPM. Molecular cloning and functional expression the potassium-dependent sodium-calcium exchanger from human and chicken retinal cone photoreceptors. J Neurosci 2000;20:1424–1434.
49. Winkfein RJ, Pearson B, Ward R, Szerencsei RT, Colley NJ, Schnetkamp PPM. Molecular characterization, functional expression and tissue distribution of a second NCKX Na^+/Ca^{2+}-K^+ exchanger from Drosophila. Cell Calcium 2004;36:147–155.
50. Lagnado L and McNaughton PA. Electrogenic Properties of the Na:Ca Exchange. J Membr Biol 1990;113:177–191.
51. Schnetkamp PPM. Calcium homeostasis in vertebrate retinal rod outer segments. Cell Calcium 1995;18:322–330.
52. Nakatani K and Yau K-W. Sodium-dependent calcium extrusion and sensitivity regulation in retinal cones of the salamander. J Physiol 1989;409:525–548.
53. Cilluffo MC, Matthews HR, Brockerhoff SE, Fain GL. Light-induced Ca^{2+} release in the visible cones of the zebrafish. Vis Neurosci 2004;21:599–609.
54. Prinsen CFM, Cooper CB, Szerencsei RT, Murthy SK, Demetrick DJ, Schnetkamp PPM. The retinal rod and cone Na^+/Ca^{2+}-K^+ exchangers. Adv Exp Med Biol 2002;514:237–251.
55. Tsoi M, Rhee K-H, Bungard D, Li XB, Lee S-L, Auer RN, Lytton J. Molecular cloning of a novel potassium-dependent sodium–calcium exchanger from rat brain. J Biol Chem 1998;273:4155–4162.
56. Holcman D and Korenbrot JI. Longitudinal diffusion in retinal rod and cone outer segment cytoplasm: the consequence of cell structure. Biophys J 2004;86:2566–2582.
57. Schwarz EM and Benzer S. Calx, a Na-Ca exchanger gene of Drosophila melanogaster. Proc Natl Acad Sci U S A 1997;94:10249–10254.
58. Tucker JE, Winkfein RJ, Murthy SK, Friedman JS, Walter MA, Demetrick DJ, Schnetkamp PPM. Chromosomal localization and genomic organization of the human retinal rod Na/Ca+K exchanger. Hum Genet 1998;103:411–414.

59. McNaughton PA and Friedlander M. The retinal rod Na/Ca,K exchanger contains a noncleaved signal sequence required fro translocation of the N terminus. J Biol Chem 1999;274:38177–38182.
60. Kang K-J and Schnetkamp PPM. Signal sequence cleavage and plasma membrane targeting of the rod NCKX1 and cone NCKX2 Na$^+$/Ca^{2+}-K$^+$ exchangers. Biochemistry 2003;42:9438–9445.
61. Kim TSY, Reid DM, Molday RS. Structure–function relationships and localization of the Na/Ca-K exchanger in rod photoreceptors. J Biol Chem 1998;273:16561–16567.
62. Kinjo TG, Szerencsei RT, Winkfein RJ, Kang K-J, Schnetkamp PPM. Topology of the retinal cone NCKX2 Na/Ca-K exchanger. Biochemistry 2003;42:2485–2491.
63. Sharon D, Yamamoto H, McGee TL, Rabe V, Szerencsei RT, Winkfein RJ, Prinsen CFM, Barnes CS, Andreasson S, Fishman GA, Schnetkamp PPM, Berson EL, Dryja TP. Mutated alleles of the rod and cone Na/Ca+K exchanger genes in patients with retinal diseases. Invest Ophthalmol Vis Sci 2002;43:1971–1979.
64. Sheng J-Z, Prinsen CFM, Clark RB, Giles WR, Schnetkamp PPM. Na$^+$-Ca^{2+}-K$^+$ Currents measured in insect cells transfected with the retinal cone or rod Na$^+$-Ca^{2+}-K$^+$ exchanger cDNA. Biophys J 2000;79:1945–1953.
65. Szerencsei RT, Prinsen CFM, Schnetkamp PPM. The stoichiometry of the retinal cone Na/Ca-K exchanger heterologously expressed in insect cells: Comparison with the bovine heart Na/Ca exchanger. Biochemistry 2001;40:6009–6015.
66. Szerencsei RT, Tucker JE, Cooper CB, Winkfein RJ, Farrell PJ, Iatrou K, Schnetkamp PPM. Minimal domain requirement for cation transport by the potassium-dependent Na/Ca-K exchanger: Comparison with an NCKX paralog from Caenorhabditis elegans. J Biol Chem 2000;275:669–676.
67. Winkfein RJ, Szerencsei RT, Kinjo TG, Kang K-J, Perizzolo M, Eisner L, Schnetkamp PPM. Scanning mutagenesis of the alpha repeats and of the transmembrane acidic residues of the human retinal cone Na/Ca-K exchanger. Biochemistry 2003;42:543–552.
68. Kang K-J, Shibukawa Y, Szerencsei RT, Schnetkamp PPM. Substitution of a single residue, Asp^{575}, renders the NCKX2 K$^+$-dependent Na$^+$/Ca^{2+} exchanger independent of K$^+$. J Biol Chem 2005;280:6834–6839.
69. Nicoll DA, Hryshko LV, Matsuoka S, Frank JS, Philipson KD. Mutation of amino acid residues in the putative transmembrane segments of the cardiac sarcolemmal Na$^+$-Ca^{2+} exchanger. J Biol Chem 1996;271:13385–13391.
70. Iwamoto T, Nakamura TY, Pan Y, Uehara A, Imanaga I, Shigekawa M. Unique topology of the internal repeats in the cardiac Na+/Ca2+ exchanger. FEBS Lett 1999;446:264–268.
71. Kinjo TG, Szerencsei RT, Winkfein RJ, Schnetkamp PPM. Role of cysteine residues in the NCKX2 Na$^+$/Ca^{2+}-K$^+$ exchanger: generation of a functional cysteine-free exchanger. Biochemistry 2004;43:7940–7947.
72. Kinjo TG, Kang K-J, Szerencsei RT, Winkfein RJ, Schnetkamp PPM. Site-directed disulfide mapping of residues contributing to the Ca^{2+} and K$^+$ binding pocket of the NCKX2 Na$^+$/Ca^{2+}-K$^+$ exchanger. Biochemistry 2995;44:7787–7795.
73. Lamason RL, Mohideen MA, Mest JR, Wong AC, Norton HL, Aros MC, Jurynec MJ, Mao X, Humphreville VR, Humbert JE, Sinha S, Moore JL, Jagadeeswaran P, Zhao W, Ning G, Makalowska I, McKeigue PM, O'Donnell D, Kittles R, Parra EJ, Mangini NJ, Grunwald DJ, Shriver MD, Canfield VA, Cheng KC. SLC24A5, a putative cation exchanger, affects pigmentation in zebrafish and humans. Science 2005;310:1782–1786.

15
Excitatory Amino Acid Transporters in the Retina

Vijay Sarthy and David Pow

CONTENTS

INTRODUCTION
EXCITATORY AMINO ACID TRANSPORTER (EAAT) LOCALIZATION
DEVELOPMENTAL EXPRESSION
GLUTAMATE TRANSPORTER FUNCTION
REGULATION OF EXCITATORY AMINO ACID TRANSPORTER FUNCTION
GLUTAMINE TRANSPORTERS
DISEASE INVOLVEMENT
CONCLUSIONS
ACKNOWLEDGEMENTS
REFERENCES

INTRODUCTION

Plasma membrane neurotransmitter transporters contribute to the clearance and recycling of neurotransmitters, and can have a profound impact on the extent of receptor activation during neuronal signaling. Studies of neurotransmitter transporters have been crucial in the development of treatments for major neuropsychiatric conditions. The dopamine, norepinephrine and serotonin transporters (DAT, NET and SERT, respectively) are well-established targets for addictive drugs including cocaine and amphetamines, and for therapeutic antidepressants *(1)*. Recently, glutamate transporters have been suggested as potential targets for stroke therapy *(2)*. Therefore, a sound knowledge of the roles of glutamate transporters in the retina is of great relevance in developing treatments for human ocular diseases such as diabetic retinopathy. An understanding of the mechanisms that regulate these transporters has the potential to impact on both the physiology and pathology of glutamate in the central nervous system (CNS).

Glutamate is the major excitatory neurotransmitter in the retina that mediates signal transmission at photoreceptor-bipolar cell and bipolar cell-amacrine cell synapses *(3)*. Glutamate, however, can turn into a neurotoxin if its extracellular level rises excessively in the retina, as happens during ischemia, resulting in excitotoxic damage, neuronal loss and finally, blindness *(4–6)*. Retinal extracellular glutamate must be kept at a low level to minimize glutamate receptor desensitization, as well as to avoid excitotoxicity. Plasma membrane glutamate transporters, expressed by retinal neurons and glial cells, perform

From: *Ophthalmology Research: Ocular Transporters in Ophthalmic Diseases and Drug Delivery*
Edited by: J. Tombran-Tink and C. J. Barnstable © Humana Press, Totowa, NJ

this function admirably *(7,8)*. In this chapter, we will review the cellular distribution, developmental expression, function and regulation of excitatory amino acid transporters (EAATs) in the retina, with the goal that this information will be useful in identifying their biological roles in the retina.

EXCITATORY AMINO ACID TRANSPORTER (EAAT) LOCALIZATION

Based on DNA sequence, pharmacology, and channel properties, at least five subtypes of sodium-dependent, excitatory amino acid transporters have been identified and characterized: GLAST (EAAT1), GLT-1 (EAAT2), EAAC1 (EAAT3), EAAT4, and EAAT5 *(7)*. EAAC1 is localized to neurons whereas GLT-1 and GLAST are the predominant glutamate transporters in CNS glia *(7)*. EAAT4 is found mostly in the cerebellum and retina, while EAAT5 appears to be restricted to the retina *(7)*.

Immunocytochemical studies suggest that all the EAAT subtypes are expressed in the retina (Fig. 1 Table 1). The cellular localization of the EAATs, however, appears to be different. EAAT1 (GLAST) is localized to Müller cells in all vertebrate retinas examined so far *(9–11)*. Ultrastructural studies further show that GLAST is present throughout the Müller cell including the fine processes that surround the rod terminals *(12)*. This pattern of GLAST localization suggests that the transporter might participate in glutamate clearance, both at synaptic regions and in other areas. Moreover, molecular genetic and electrophysiological studies suggest that GLAST is the primary regulator of extracellular glutamate level in the retina *(13,14)*. Although GLAST has not been found in retinal neurons, it has been reported in retinal astrocytes and the retinal pigment epithelium (RPE) *(12)*.

EAAT2 (GLT-1) is not a glial transporter in the mammalian retina, but is found localized to cones and bipolar cells *(9)*. GLT-1 has also been seen in some amacrine cells in the rat retina *(9,10)*. The gerbil retina appears to be an exception, because GLT-1 was detected in ganglion cells, but not in bipolar cells or amacrine cells in this retina *(15)*. An electron-microscopy immunocytochemical study found that GLT-1 is present in bipolar cells, as well as Müller cells, in the goldfish retina *(16)*. GLT-1 exists in at least three splice-variant isoforms, GLT-1a, b and c. GLT-1a and GLT-1b exist in all species examined, but show variations in terms of cellular distribution (Fig. 2). Also, each cell type appears to express only one splice variant *(17)*. GLT-1c is expressed by rod and cone terminals in human and rat retina *(18)*.

EAAT3 (EAAC1) is present in the bipolar, amacrine and ganglion cells in a large number of vertebrates *(19)*. In the cat and rat retinas, EAAC1 has also been reported in horizontal cells *(19,20)*. It is not found in photoreceptors, bipolar cells or Müller cells, although there is a single report that EAAC1 is present in Müller cells in the carp and bullfrog retinas.

EAAT4 was initially reported to be present only in the cerebellum. More recent studies, however, show that EAAT4 is also expressed in the retina. Its cellular distribution, however, remains controversial. Fyk-Kolodziej et al. *(20)*, found that EAAT4 was present in Müller cells and astrocytes in the cat retina, while Ward et al. *(21)* reported that EAAT4 was present in retinal astrocytes. In contrast, in the human retina, EAAT4 is localized to photoreceptor inner and outer segments, which are completely non-synaptic regions of the retina *(22)* (Fig. 3). This conclusion is based on the following evidence: reverse-transcriptase polymerase chain reaction (RT-PCR) data indicated that EAAT4

Excitatory Amino Acid Transporters in the Retina

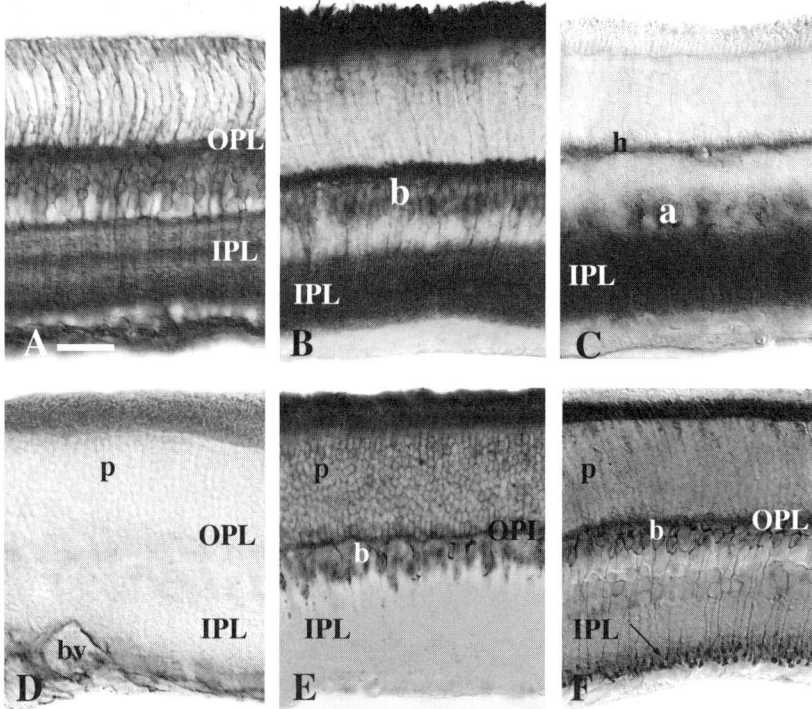

Fig. 1. Cellular distribution of excitatory amino acid transporters (EAATs) in the retina. (**A**) EAAT1 (GLAST) is abundant in Müller cells, which project radially through all retinal layers, and extend three layers of processes in the inner plexiform layer (IPL) and another layer in the outer plexiform layer (OPL). (**B**) EAAT2 (GLT-1) labeling using an antibody that detects all splice variants of GLT-1. This transporter is expressed by bipolar cells (b) as well as cone and rod photoreceptors, and also in some amacrine cell processes in the IPL. (**C**) EAAT3 (EAAC1) is expressed by a population of amacrine cells (a) and horizontal cells (h). (**D**) EAAT4 appears to be lightly expressed by photoreceptor inner segments and by astrocytes surrounding blood vessels (bv) on the vitreal surface. In rat (**E**) and mouse (**F**) retinas, EAAT5 is expressed by photoreceptors and by bipolar cells. The extent to which EAAT5 can be detected in bipolar cells varies between species, which probably reflects the presence of potentially cryptic epitopes. Scale bar = 25 µm.

Table 1 Exitatory amino acid transporter (EAAT) distribution in the mammalian retina

Transporter	Cell type
EAAT1 (GLAST)	Müller cell *(10,12,25)*
EAAT2 (GLT-1)	Rod terminals, cone terminals, bipolar cells *(9)*
EAAT3 (EAAC1)	Inner nuclear layer and ganglion cell layer neurons *(10,19)*; horizontal cells *(20)*
EAAT4	Rods and cones *(22)*, astrocytes *(20,21)*, Müller cells *(22)*
EAAT5	Rods *(23,24)*; Müller cells *(24)*

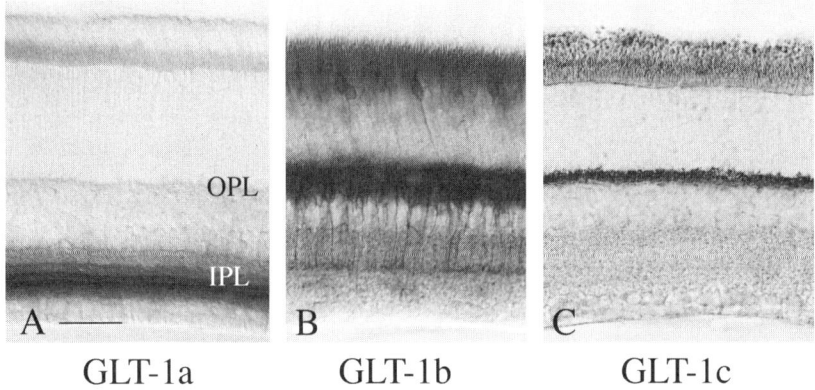

Scale bar 50 μm

Fig. 2. Differential distribution of GLT-1 splice variants in the rodent retina. **(A)** GLT-1a is present in amacrine cell processes in the inner plexiform layer (IPL), but is absent from the outer plexiform layer (OPL). **(B)** GLT-1b is expressed predominantly by cone photoreceptors and cone bipolar cells whereas **(C)** GLT-1c is expressed predominantly by rod photoreceptors. Scale bar = 50 μm.

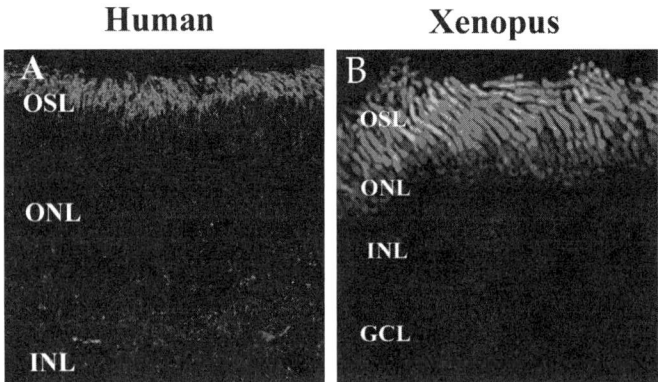

Fig. 3. Exciatory amino acid transporter 4 (EAAT4) localization in photoreceptors. The micrographs show EAAT4 localization in human and *Xenopus* retina, where it is found in the outer and the inner segment layers but not in other regions of the retina. OSL, outer segment layer; ONL, outer nuclear layer; INl, inner nuclear layer; GCL, ganglion cell layer.

transcripts are expressed at high levels in the human retina and in the cone-derived cell line, Y79. Immunoblotting showed the presence of EAAT4-immunoreactive bands in human and mouse retina homogenates, with molecular masses compatible with the transporter monomers and dimers. Immunohistochemistry revealed that EAAT4 was localized in the rod and cone outer segments in the human retina, and also in the inner segments in mouse and squirrel retinas. Immunostaining of isolated photoreceptor cultures was also consistent with EAAT4 expression in cone photoreceptors. Double-label immunohistochemistry with an opsin antibody supported the localization of EAAT4 to outer segments and this result was confirmed by immunoblotting of a rod outer segment preparation. Furthermore, the EAAT4-associated protein, GTRAP41, was found to colocalize with EAAT4 in the retina, and immunoprecipitation demonstrated the direct interaction of the two proteins in vivo. The reasons for this discrepancy in EAAT4 cellular localization among different species is not known, but it is quite clear that EAAT4 is present in the retina.

EAAT5 was originally isolated as a retina-specific glutamate transporter that was present in photoreceptors and bipolar cells (23). Subsequently, EAAT5 has been reported in amacrine cells, ganglion cells as well as in Müller cells (20,24). As in the case of EAAT4, the reasons behind the discrepancies in EAAT5 localization remain a mystery.

A fundamental question in retinal neurochemistry is why does the retina express a multitude of EAATs? Although we do not know the specific reason, several explanations have been advanced to explain the diversity: (i) distinct transporters may be expressed by different retinal cells, though this is definitely not the case because immuncytochemical localization studies show that many retinal cells express multiple transporter subtypes; (ii) the transporters may be localized to different subcellular compartments and might serve distinct functions. Although GLT-1 and EAAC-1 are present in synaptic regions, EAAT5 appears to be localized primarily to cell bodies where its function is not understood; (iii) there might be differences in developmental expression or gene regulation that correlate with function. This question has not been studied carefully; and finally, perhaps the most likely possibility is that (iv) because of differences in the kinetic, pharmacological and channel properties, the transporters may function optimally under different physiological conditions. In addition, the activity of EAAT subtypes might be controlled by distinct regulatory mechanisms.

DEVELOPMENTAL EXPRESSION

It is possible that different EAATs are expressed at different times during retinal development, thus providing rationale for their diversity. In the developing rat retina, GLAST is first seen in Müller cells at around postnatal day 1, but strong expression is not evident until the first postnatal week (25), when the outer plexiform layer begins to elaborate. GLT-1a appears late in postnatal development and is mainly seen in amacrine cells. GLT-1b, on the other hand, is seen in embryonic day 14 retinas and can be localized to cone photoreceptors by embryonic day 18. By first postnatal week, GLT-1b is seen in bipolar cell processes (17). Unlike these EAATs, EAAT5 is not expressed until the second postnatal week and is first seen in photoreceptors and later in bipolar cells only after the third postnatal week, when retinal histogenesis is complete. These observations show that different EAATs are expressed at different times in the developing retina, although there are overlaps in their expression patterns. A recent study in the human retina suggests that glutamine synthetase and GLAST are expressed first in the incipient fovea, and later in the peripheral retina at about 14 weeks gestation (26). It appears that during development, maximal GLAST expression coincides with synapse formation, suggesting a role for the transporter in synaptic transmission.

GLUTAMATE TRANSPORTER FUNCTION
Role in Glutamate Clearance

Glutamate transporters can concentrate glutamate more than 10^6-fold across cell membranes *(27)*. Because of this property, there is a general consensus that glutamate transporters play an important role in maintaining the extracellular glutamate concentration at low levels and in protecting neurons from the excitotoxic action of glutamate *(27,28–34)*. The involvement of GLAST in retinal glutamate homeostasis has been examined using antisense oligonucleotides, as well as GLAST knockout mice *(13,14,28,32,33)*. In rat eyes injected with antisense oligonucleotides to knock down GLAST, there was a reduction in D-aspartate uptake by Müller cells, as well as a decrease in the electroretinogram (ERG) b-wave. Surprisingly, there was no evidence of excitotoxic damage to retina and glutamate distribution was normal *(14)*. Studies with GLT-1 and GLAST knockout mice also show that loss of these transporters does not lead to excitotoxic damage to retinal neurons. GLT-1 knockout mice show normal ERGs and only mild damage following ischemia *(13)*. GLAST knockout mice, in contrast, show reduced b-wave amplitude and are more susceptible to ischemic damage compared to normal animals *(13)*. EAAC1 knockout mice show no abnormal retinal phenotype *(34)*. These data suggest that loss of any one EAAT does not affect the retinal function significantly, but alters retinal sensitivity to ischemic damage.

Although GLAST accounts for around 50% of glutamate uptake *(35)*, it is surprising that GLAST-KO mouse retinas look normal and show no signs of excitotoxic damage *(13,35)*. A priori, in the absence of GLAST, one would have expected some build up of extracellular glutamate leading to excitotoxic damage to retina. The benign retinal phenotype could be attributed to several factors. One possibility is that Müller cells might compensate for GLAST loss by expressing GLT-1 or another EAAT. Alternatively, glutamate clearance by EAAT5 (in Müller cells) might reduce glutamate toxicity *(20)*. Recent biochemical, immunocytochemical and electrophysiological studies, however, show that Müller cells from GLAST knockout mouse retinas do not express any electrogenic glutamate transporters *(35)*.

It is also possible that GLT-1 and/or EAAC1 can handle extracellular glutamate in the absence of GLAST. Because GLT1 is expressed by photoreceptors and bipolar cells, it could even be argued that the sites where the transporters are located (e.g., at photoreceptor synapses) may be more important than the aggregate quantity of transporter expressed in the retina as a whole. In support of this idea, a recent study found that glutamate is removed at rod synapses by presynaptic EAATs, rather than by postsynaptic or glial EAATs *(36)*. Future studies with double- and triple-knockout mice are necessary to determine the contributions of GLAST, GLT-1 and EAAC1 towards glutamate clearance.

Role in Synaptic Transmission

The biochemical properties and subcellular localization of glutamate transporters suggest that they are well positioned to regulate extracellular glutamate levels. However, there is controversy as to whether the transporters play an active role in determining the magnitude and time course of synaptic transmission at glutamatergic synapses *(37)*. Because kinetic studies show that the unitary transport rates are rather slow, e.g., 100–1000/sec *(38)*, it has been argued that diffusion itself is sufficient to account for glutamate clearance at the synaptic cleft. Moreover, such slow turnover rates mean that the transporters can affect signaling only at slow synapses, where signaling occurs on a time scale of hundreds of milliseconds and is mediated through G-coupled receptors. Transmitter uptake, however, may not be the only mechanism by which transporters function. There is evidence that transporters at 'fast' synapses may affect synaptic response by sequestering glutamate at the binding site on the transporter, serving as an effective diffusion sink *(38)*. Indeed, transporters appear to outnumber receptors at some glutamatergic synapses *(39)*. It is likely that presynaptic glutamate transporters such as GLT-1, control part of synaptic kinetics, whereas glial transporters serve to prevent extrasynaptic glutamate accumulation.

Experimental evidence suggests that L-glutamate released from rod terminals might provide a negative feedback signal to inhibit further L-Glu release *(40)*. However, the involvement of specific EAATs in regulating synaptic activity remains to be established. Because many EAATs are localized either in extrasynaptic regions or on glial processes separating synaptic terminals, it is likely that EAATs are involved in preventing lateral spread of L-glutamate to neighboring terminals.

It is worth pointing out that a significant question with regard to the presence of EAAT4 in the outer segments concerns its potential function *(22)*. One idea is that EAAT4 may have a scavenging function whereby it sucks up glutamate that escapes Müller cells, providing an additional mechanism to ensure that no glutamate is lost from the retina to the subretinal space or the RPE.

One of the intriguing features of EAATs, particularly with EAAT4 and EAAT5, is that these transporters have characteristics of both carriers and ligand-gated chloride channels *(7)*. The physiological function of the anion conductance is not understood, but it is possible that the anion conductance keeps glutamate transport more electroneutral *(41)*. Alternatively, the chloride permeability may influence the excitability of the membrane, acting as a feedback mechanism *(7)*.

Role in Gamma Aminobutyric Acid (GABA) Synthesis

Plasma membrane glutamate transporters maintain extracellular glutamate at a low level, thereby protecting neurons from excitotoxicity and also ensuring a high signal-to-noise ratio for glutamatergic neurotransmission *(1)*. Neuronal glutamate transporters have also been shown to play a role in GABA synthesis. For example, the glutamate transporter EAAC1 is found on the inhibitory terminals of GABAergic neurons where it might increase glutamate uptake and thus, elevate the glutamate pool available for GABA synthesis *(42–44)*. More-direct evidence for a role for glutamate transporters in GABA homeostasis is provided by antisense oligonucleotide-experiments in which suppression of EAAC1 expression in hippocampal slices was found to reduce new GABA synthesis, and decrease GABA levels by 50% *(45)*. Furthermore, it was recently reported that glutamate transporters regulate the vesicle pool of GABA in hippocampal CA1 interneurons, and that reduced glutamate uptake resulted

Table 2 Gamma aminobutyric acid (GABA) and glutamate levels in GLAST knockout mouse retina

	GLAST$^{-/-}$	Wild type
GABA	56.00 ± 4.44	26.90 ± 2.79
Glutamic acid	166.20 ± 9.89	76.40 ± 5.15
Aspartatic acid	27.40 ± 2.03	10.50 ± 1.07
Glutamine	62.60 ± 7.63	59.30 ± 5.61

in diminished evoked IPSCs and mIPSCs without affecting postsynaptic receptors *(46)*. These observations suggest that the neuronal glutamate transporter EAAC1 contributes to GABA formation.

A role for the glial glutamate transporter, GLAST, in GABA synthesis has come from biochemical studies in which glutamate and GABA levels were measured in the retinas of GLAST knockout mice *(47)*. It was reported that the GABA level was elevated twofold in the retinas of GLAST$^{-/-}$ mice (Table 2). Glutamate levels were also increased twofold, whereas the glutamine level was unaltered. The reason for an increase in GABA level is not understood, but is likely to be due to an increase in the availability of glutamate for GABA synthesis in the GLAST-knockout retina. Furthermore, because EAAC1 is localized to inner nuclear layer (INL) neurons and the inner plexiform layer (IPL), the role of EAAC1 in regulating retinal GABA levels could be examined using GLAST$^{-/-}$/EAAC1$^{-/-}$ double-knockout mice. Whether the GLT-1 has a similar role in GABA synthesis has not been studied.

REGULATION OF EXCITATORY AMINO ACID TRANSPORTER FUNCTION

Interacting Proteins

Glutamate uptake can be regulated by changes in transporter activity, and several mechanisms have been shown to kinetically regulate transporter activity. These include protein phosphorylation *(48)*, membrane trafficking *(49,50)*, interaction with zinc *(51)*, modulation by arachidonic acid *(52,53)*, multimerization *(54)* and a sulfhydryl-based redox mechanism *(55)*. Perhaps the most exciting development in the field is the recent identification of a group of proteins termed glutamate transporter-associated proteins (GTRAPs) that can bind to, and influence the activity of, glutamate transporters *(56–58)*. In two reports, Rothstein's group has described the identification of two proteins, GTRAP41 and GTRAP48, which specifically interact with the intracellular carboxyl-terminal domain of EAAT4 and modulate its glutamate transport activity *(56)*. Expression of GTRAP41 or 48 resulted in an increase in V_{max} without altering the K_m. Another protein, GTRAP3-18, binds to the intracellular carboxyl-terminal domain of EAAC1 and modulates EAAC1-mediated glutamate transport by decreasing the affinity of the transporter for glutamate *(57)*.

To examine the role of GLAST-interacting proteins, Marie and Atwell *(59)* perfused isolated retinal Müller cells with peptides corresponding to eight amino acids at the N- and C-terminals ends of GLAST, and determined their effect on transporter-mediated currents. The N-terminal peptide had no effect, whereas the C-terminal peptide increased the glutamate affinity for GLAST. These data suggest that interacting protein in Müller cells decreases the affinity of GLAST. The identity of the interacting protein remains unknown. Because the C-terminal amino acids have some similarity to the PDZ binding domain of ion channels, it is likely that GLAST might be anchored to the membrane through this domain.

The N-terminus of GLT-1 interacts with a cytoplasmic LIM protein, Ajuba, which is expressed in Bergmann glia and retinal neurons *(59)*. Co-expression of Ajuba and GLT-1 in COS cells did not affect the K_m or V_{max} for glutamate, which suggests that Ajuba might be involved in binding GLT-1 to the cytoskeleton or in intracellular signaling. Recently, the EAAT4-associated protein, GTRAP41, was found to colocalize with EAAT4 in the retina, and immunoprecipitation demonstrated the direct interaction of the two proteins in vivo *(22)*. From these studies it is clear that EAAT-interacting proteins are present in the retina, though their identity and role in regulating glutamate transport in retinal cells remains to be studied.

Excitatory Amino Acid Transporter Trafficking

The activity of many neurotransmitter transporters (dopamine, serotonin, norepinephrine, GABA and glutamate transporters) is regulated by protein kinase C (PKC) or mitogen-activated protein kinase (MAPK) through a mechanism involving the redistribution of transporter molecules between intracellular vesicles and the plasma membrane *(27,49,50,60)*. For example, PKC activation increases the activity and cell-surface expression of endogenous EAAC1 in C6 glioma cells *(49,50)*. In contrast, PKC activation in Müller cell cultures, following exposure to phorbol ester, causes a rapid decrease in transport activity and loss of GLAST protein *(61)*. In Y79 retinoblastoma cells, PKC activation also decreases GLT-1 activity by decreasing the K_m for glutamate five-fold, without changing V_{max} *(62)*. In rat retina, PKC inhibition using isoform-specific antagonists, suggest that blocking PKC delta in Müller cells leads to suppression of glutamate transport. Under these conditions, neuronal glutamate uptake was not affected *(63)*. The colocalization of EAAT3 with serum- and glucocorticoid-inducible kinase (SGK1) in retinal neurons, suggests that this kinase might modulate glutamate-induced current in retinal neurons *(64)*.

Another manner in which transporter action can be regulated is by selective targeting of transporter molecules to a subcellular compartment. In the case of EAAT3, there is evidence that a short peptide motif in the cytoplasmic C-terminal region is required for its targeting to dendrites in hippocampal neurons *(65)*. Recently, EAAT4 has been found in photoreceptor outer segments, but not in the synaptic regions *(39)*. How is EAAT4 targeted to the outer segments? In the case of rhodopsin and several transduction proteins, efficient outer segment localization depends on both membrane association through fatty acylation of conserved cysteines and a (V/I)XPX motif at the extreme C-terminus *(66–71)*. Furthermore, attaching the last eight amino acids of rhodopsin to green fluorescent protein (GFP) directs the tagged GFP to rod outer segment (ROS) in transgenic *Xenopus*. This motif is, however, absent in the C-terminus of human, rat or mouse EAAT4 (.......SRGRGGNESAM; NCBI/NIH protein database), which suggests that other sequences in EAAT4 are likely to be responsible for targeting the transporter to outer segments.

GLUTAMINE TRANSPORTERS

Glutamine is a key intermediary in the formation of glutamate by neurons, and is itself generated in glial cells from glutamate. The glutamate–glutamine cycle is a well-defined metabolic cycle that couples the metabolic activities of glial cells such as the Müller cells to the adjacent glutamatergic or GABAergic neurons *(72)*. Glutamine is synthesized in glial cells from glutamate by the glial enzyme, glutamine synthetase. The pools of glutamate in the glial cells may be generated by de novo synthesis, or by the glial accumulation of glutamate released from neurons, using the excitatory amino acid transporters. The release of glutamine from glial cells such as the Müller cells, occurs down a concentration gradient. Conversely the accumulation of glutamine into neurons normally occurs up a concentration gradient, indicating that transport needs to be driven by a different underpinning mechanism. Once accumulated in neurons, glutamine can be converted back into glutamate via the actions of phosphate-activated glutaminase.

Multiple families of transporters are involved in the translocation of glutamine *(73–75)*. The key candidate transport systems implicated in retinal and brain glutamine transport belong to what have been described as system A, system N, system L and system ASC. These historical system descriptions are based on their substrate preferences. System A transporters are characterized by their sensitivity to 2(methylamino)isobutyrate (MeAIB), system N transporters are characterized by their transport of asparagine (N), glutamine and in some cases histidine, system L transports branched-chain and aromatic amino acids in a sodium-independent manner, whilst system-ASC transporters are characterized by their preference for those amino acids with hydroxyl or sulfydryl-containing side chains. (Fig. 4).

System A Transporters

Solute carrier family 38, member 1 (normally referred to as SLC38A1) was the first member of the system A amino acid transporter subfamily to be cloned. Rat, human, and mouse sequences have been cloned and were formerly referred to as ATA1, SA2, SAT1, NAT2, or GlnT. The protein is now normally referred to as sodium-coupled neutral amino acid transporter 1 (SNAT 1), which is expressed abundantly by human cortical neurones and to a lesser extent by astrocytes *(76)*. SNAT another member of the SLC38 gene family, accounts for the majority of system A transport in many tissues.

SNAT 1 and SNAT 2 have been localized to neurons in the brain *(76,77)*. While this localization makes them attractive candidates as neuronal glutamine transporters involved in the glutamate-glutamine cycle for cycling synaptically released glutamate, their specific localization to dendrites and somata of neurons, rather than the synaptic terminals, suggests that this may not be their role. Instead they may well be involved in accumulating glutamine for other purposes such as protein synthesis.

System N Transporters

Three other SNATs (3–5) have been cloned. SNAT 4 does not transport glutamine. SNAT3 and SNAT 5 are system N transporters (also referred to as Na^+- and H^+-coupled amino-acid transport system N, or SN1 and SN2, respectively). SNAT 3 has been localized

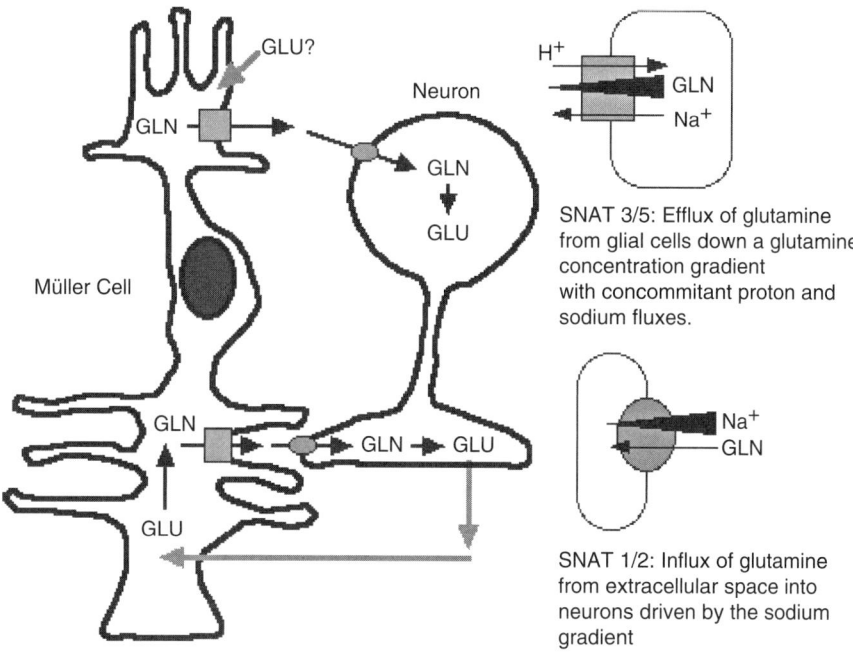

Fig. 4. Schematic diagram illustrating how sodium-coupled neutral amino acid transporters (SNATs) 1, 2, 3 and 5 may function to traffic glutamine out of glial cells down a concentration gradient and into neurons up a glutamine concentration gradient. GLN, glutamine; GLU, glutamate.

to glial cells (78) and appears to be the key mediator of glutamine efflux from such cells. Both SNAT 3 and SNAT 5 appear to be expressed by Müller cells (79). Conversely, other reports have suggested expression of system N transporters by neurons (80). Both SNAT 3 and SNAT 5 are characterized by the countertransport of protons and this has been proposed as part of a mechanism by which these two transporters might act in either direction, depending on the proton gradient (81,82).

Umapathy et al. (79) have analyzed the contributions of SNAT 3 and SNAT 5 to Müller cell glutamine uptake (rather than release), relative to contributions from system A (SNAT 1 and SNAT 2) and system L (LAT 1 and LAT 2). Whilst transcripts for all of these genes were expressed, they concluded that SNAT 3 and SNAT 5 probably accounted for the majority of glutamine uptake by Müller cells. Due to the reversibility of these transporters it is probable that these data indicate the predominance of SNAT 3 and SNAT 5 in mediating glutamine efflux from Müller cells under normal physiological conditions.

System L Transporters

Four proteins that induce system L (large, neutral amino-acid transport) activity have been identified: LAT 1, LAT 2, LAT 3, and LAT 4. The first two proteins belong to the solute carrier family 7 (SLC7), whereas the last two belong to SLC43 family. LAT 1 and LAT 2 exist as heterodimers, complexed with the 4F2hc/CD98 molecule, and can mediate the exchange of glutamine for cationic or neutral amino acids *(83,84)*. Whilst both LAT 1 and LAT 2 are expressed by cultured glia and neurons, their in vivo localization remains to be determined. It is plausible that in the eye, LAT 1 and LAT 2 may be associated predominantly with the blood-retinal barrier transport of amino acids, since both appear to be expressed by RPE cells *(85)*.

LAT 3 (previously known as POV1) mRNA is expressed mainly in pancreas and liver, but appears to be present at low levels in brain and has not been determined in retina. LAT 4 mRNA is detectable at low levels in brain, but again its expression has not been determined for retina *(86)*.

System ASC Transporters

ASCT1 and ASCT2 are transporters that exhibit ASC-like transport activities. ASCT2 is expressed by cultured astrocytes *(87)*, but the level of expression in brain is thought to be very low *(84,88)*. This transporter functions as an exchanger; the demonstrable efflux of glutamine from cultured astrocytes can be coupled to the influx of other neutral amino acids such as cysteine, which is probably required for astrocytes to synthesize glutathione. The presence of ASCT1 has been reported in Müller cells in the mouse retina *(35)*. The recent demonstration of D-serine fluxes in retina from Müller cells that are mediated by an ASCT-like transport system *(89)* also suggests that either ASCT1 or ASCT2 may be expressed by Müller cells, and thus may be able to participate to some extent in glutamine trafficking in the retina, though this remains to be determined.

DISEASE INVOLVEMENT

Retinal diseases such as diabetic retinopathy involve physiological conditions in which glutamate excitotoxicity is likely to be a serious problem. Because glutamate transporter dysfunction may be critical component in exacerbating glutamate toxicity and oxidative stress, several investigators have examined the role of glutamate transporters in the diabetic retina. In streptozotocin-treated rats, isolated Müller cells showed a decrease in glutamate transporter-induced currents four weeks after treatment; interestingly, transporter activity could be rapidly restored by treatment with disulfide-reducing agents, which suggests that the loss of GLAST activity is due to oxidation. In contrast, 3H-D-aspartate uptake experiments in the rat retina show that glutamate uptake is increased under diabetic conditions *(90)*.

In animal models of retinal ischemic disease, GLAST level was reported to be unaffected although there was a decrease in glutamate uptake *(91)*. In animals with optic nerve crush injuries, a decrease in GLT-1 mRNA levels has been reported *(92)*. GLAST has also been implicated in the GDNF-mediated photoreceptor rescue seen in rd1 mutant mice *(93)*. Perhaps the best evidence for a role for GLAST in retinal ischemic disease comes from experiments with GLAST knockout mice, which show that loss of GLAST makes the retina much more vulnerable to ischemic damage *(13)*. A recent study reported that GLT-1c, which is normally found only in rod photoreceptors, is additionally expressed by retinal ganglion cells in human glaucoma (Fig. 5). The authors suggested that the upregulation of GLT-1c might represent an attempt by

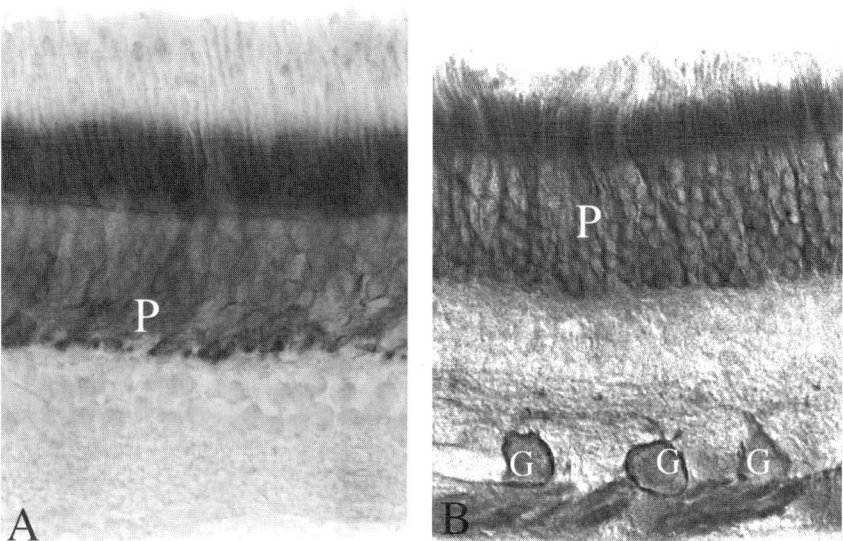

Fig. 5. GLT-1c localization in glaucoma. In the normal human retina, (**A**), GLT-1c is expressed by photoreceptors, particularly rods. In the glaucomatous retina, (**B**), GLT-1c is additionally expressed by retinal ganglion cells (G).

retinal ganglion cells to protect themselves from excitotoxic damage due to elevated glutamate levels *(94)*.

CONCLUSIONS

Although there is a tremendous amount of information on the structure, cellular distribution, transport properties and pharmacology of EAATs, far less is known about specific physiological roles of EAATs, or the mechanisms that regulate EAAT activity, in the vertebrate retina *(7,8)*. A major problem in understanding the physiological contributions of specific EAAT subtypes is the lack of selective inhibitors for different EAATs. Some clues to the physiological functions of EAATs, however, has come from studies with knockout mice *(13,32,34)*; but the existence of multiple glutamate transporters has complicated the interpretation of experimental findings. Therefore, careful studies with double- and triple-knockout mice are necessary to determine the contributions of GLAST, GLT-1, and EAAC1 in regulating excitotoxic damage to retinal neurons. It is expected that future studies in the field will utilize molecular genetic and cell biological techniques to elucidate the biochemical and cellular mechanisms that regulate EAAT function in the mammalian retina.

ACKNOWLEDGEMENTS

We wish to thank V. Joseph Dudley for help with manuscript preparation. We apologize to authors whose work could not be included because of limitations on the length of the article. Supported by NIH and NHMC (Australia).

REFERENCES

1. Amara SG and Kuhar MJ. Neurotransmitter transporters: recent progress. Ann Rev Neurosci 1993;16:73–93.
2. Lee J-M, Zipfel GJ, Choi DW. The changing landscape of ischaemic brain injury mechanisms. Nature 1999;399:A7–A14.
3. Massey SC. Cell types using glutamate as a neurotransmitter in the vertebrate retina. Prog Ret Res 1990;9:399–425.
4. Lucas DR and Newhouse JP. The toxic effect of sodium L-glutamate on the inner layers of the retina. Arch Ophthalmol 1957;58:193–201.
5. Romano C, Chen Q, Olney JW. The intact isolated (ex vivo) retina as a model system for the study of excitotoxicity. Prog Retin Eye Res 1998;17:465–483.
6. Ambati J, Chalam KV, Chawla DK, D'Angio CT, Guillet EG, Rose SJ, Vanderlinde RE, Ambati BK. Elevated gamma-aminobutyric acid, glutamate and vascular growth factor levels in the vitreous of patients with proliferative diabetic retinopathy. Arch Ophthalmol 1997;115:1161–1166.
7. Seal RP and Amara SG. Excitatory amino acid transporters: a family in flux. Ann Rev Pharmacol Toxicol 1999;39:431–456.
8. Pow DV. Amino acids and their transporters in the retina. Neurochem Int 2001;38:463–484.
9. Rauen T and Kanner BI. Localization of the glutamate transporter GLT-1 in rat and macaque monkey retinae. Neurosci Lett 1994;169:137–140.
10. Rauen T, Rothstein JD, Wassle H. Differential expression of three glutamate transporter subtypes in the rat retina. Cell Tissue Res 1996;286:325–336.
11. Lehre KP, Davanger S, Danbolt NC. Localization of the glutamate transporter protein GLAST in rat retina. Brain Res 1997;744:129–137.
12. Derouiche A and Rauen T. Coincidence of L-glutamate/aspartate transporter (GLAST) and glutamine synthetase (GS) immunoreactions in retinal glia: evidence for coupling of GLAST and GS in transmitter clearance. J Neurosci Res 1995;42:131–143.
13. Harada T, Harada C, Watanabe M, Inoue Y, Sakagawa T, Nakayama N, Sasaki S, Okuyama S, Watase K, Wada K, Tanaka K. Functions of two glutamate transporters GLAST and GLT-1 in the retina. Proc Natl Acad Sci USA 1998;95:4663–4666.
14. Barnett NL and Pow DV. Antisense knockdown of GLAST, a glial glutamate transporter, compromises retinal function. Invest Ophthalmol Vis Sci 2000;41:585–591.
15. Kang TC, Park SK, Jo SM, et al. Comparative studies on the distribution of glutamate transporters in the retinae of the Mongolian gerbil and the rat. Anat Histol Embryol 2000;29:381–383.
16. Vandenbranden CA, Yazulla S, Studholme KM, et al. Immunocytochemical localization of the glutamate transporter GLT-1 in goldfish (carassius auratus) retina. J Comp Neurol 2000;423:440–451.
17. Reye O, Sullivan R, Pow DV. Distribution of two splice variants of the glutamate transporter GLT-1 in the developing rat retina. J Comp Neurol 2002;447:323–330.
18. Rauen T, Wiessner M, Sullivan R, Lee A, Pow DV. A new GLT1 splice variant; Cloning and immunolocalization of GLT1c in the mammalian retina and brain. Neurochem Int 2004;45:1095–1106.
19. Schultz K and Stell WK. Immunocytochemical localization of the high affinity glutamate transporter, EAAC1, in the retina of representative vertebrate species. Neurosci Lett 1996;211:191–194.
20. Fyk-Kolodziej B, Qin P, Dzhagaryan A, Pourcho RG. Differential cellular and subcellular distribution of glutamate transporters in the cat retina. Vis Neurosci 2004;21:551–565.

21. Ward MM, Jobling AI, Puthuserry T, Foster LE, Fletcher EL. Localization and expression of the glutamate transporter, excitatory amino acid transporter 4, within astrocytes of the rat retina. Cell Tissue Res 2004;315:305–310.
22. Pignataro L, Sitaramayya A, Finnemann SC, Sarthy VP. Non-synaptic localization of EAAT4 in the outer segments of photoreceptors. MolCell Neurosci 2005;28:440–451.
23. Arriza JL, Eliasof S, Kavanaugh MP, Amara SG. Excitatory amino acid transporter 5, a retinal glutamate transporter coupled to a chloride conductance. Proc Natl Acad Sci USA 1997;94:4155–4160.
24. Eliasof S, Arriza JL, Leighton BH, Kavanaugh MP, Amara SG. Excitatory amino acid transporters of the salamander retina: identification, localization and function. J Neurosci 1998;18:698–712.
25. Pow DV and Barnett NL. Changing patterns of spatial buffering of glutamate in developing rat retinae are mediated by the Müller cell glutamate transporter GLAST. Cell Tissue Res 1999;297:57–66.
26. Georges P, Cornish EE, Provis JM, Madigan MC. Müller cell expression of glutamate cycle related proteins and anti-apoptotic proteins in early human retinal development. Br J Ophthalmol 2006;90;223–228.
27. Danbolt NC. Glutamate uptake. Prog Neurobiol 2001;65:1–105.
28. Rothstein JD, Dykes-Hoberg M, Pardo CA, Bristol LA, Jin L et al. Knockout of glutamate transporters reveals a major role for astroglial transport in excitotoxicity and clearance of glutamate. Neuron 1996;16:675–686.
29. Jabaudon D, Shimamoto K, Yasuda-Kamatani Y, Scanziani M, Gahwiler BH, Gerber U. Inhibition of uptake unmasks rapid extracellular turnover of glutamate of nonvesicular origin. Proc Natl Acad Sci USA 1999;96:8733–8738.
30. Szatkowski M and Attwell D. Triggering and execution of neuronal death in brain ischaemia: two phases of glutamate release by different mechanisms. Trends Neurosci 1994;17:359–365.
31. Li S, Mealing GAR, Morley P, Stys PK. Novel injury mechanism in anoxia and trauma of spinal cord white matter: Glutamate release via reverse Na^+-dependent glutamate transport. J Neurosci 1999;19(16):1–9.
32. Tanaka K, Watase K, Manabe T, Yamada K, Watanabe M, et al. Epilepsy and exacerbation of brain injury in mice lacking the glutamate transporter GLT-1. Science 1997;276: 1699–1702.
33. Watase K, Hashimoto K, Kano M, Yamada K, Watanabe M. Motor discordination and increased susceptibility to cerebellar injury in GLAST mutant mice. Eur J Neurosci 1998;10:976–988.
34. Peghini P, Janzen J, Stoffel W. Glutamate transporter EAAC1-deficient mice develop dicarboxylic aminoaciduria and behavioral abnormalities but no neurodegeneration. EMBO J 1997;16:3822–3832.
35. Sarthy VP, Pignataro L, Pannicke T, Weick M, Reichenbach A, Harada T, Tanaka K and Marc R. Glutamate transport by retinal Müller cells in glutamate/aspartate transporter knockout mice. Glia 2004;49:184–196.
36. Hasegawa J, Obara T, Tanaka K, Tachibana M. High-density presynaptic transporters are required for glutamate removal from the first visual synapse. Neuron 2006;50:63–74.
37. Bergles DE, Diamond JS, Jahr CE. Clearance of glutamate inside the synapse and beyond. Curr Opin Neurobiol 1999;9:293–298.
38. Diamond JS and Jahr CE. Transporters buffer synaptically released glutamate on a submillisecond time scale. J Neurosci 1997;17:4672–4687.

39. Lehre KP and Danblot NC. The number of glutamate transporter subtype molecules at glutamatergic synapses: chemical and stereological quantification in young and adult rat brain. J Neurosci 1998;18:8751–8757.
40. Rabl K, Bryson EJ, Thoreson WB. Activation of glutamate transporters in rods inhibits presynaptic calcium currents. Vis Neurosci 2003;20:557–566.
41. Grewer C and Rauen T. Electrogenic glutamate transporters in the CNS: Molecular mechanisms, pre-steady-state kinetics, and their impact on synaptic signaling. J Membrane Biol 2005;203:1–20.
42. Rothstein JD, Martin L, Levey AI, Dykes-Hoberg M et al. Localization of neuronal and glial glutamate transporters. Neuron 1994;13:713–725.
43. Conti F, DeBiasi S, Minelli A, Rothstein JD, Melone M. EAAC1, a high-affinity glutamate transporter, is localized to astrocytes and gabaergic neurons besides pyramidal cells in the rat cerebral cortex. Cereb Cortex 1998;8:108–116.
44. He Y, Janssen WG, Rothstein JD, Morrison JH. Differential synaptic localization of the glutamate transporter EAAC1 and glutamate receptor subunit GluR2 in the rat hippocampus. J Comp Neurol 2000;418:255–269.
45. Seputsky JP, Cohen AS, Eccles C, Rafiq A, et al. A neuronal glutamate transporter contributes to neurotransmittter GABA synthesis and epilepsy. J Neurosci 2002;22:6372–6379.
46. Matthews GC and Diamond JS. Neuronal glutamate uptake contributes to GABA synthesis and inhibitory synaptic strength. J Neurosci 2003;23:2040–2048.
47. Sarthy VP, Marc RE, Pignataro L, Tanaka K. Contribution of a glial glutamate transporter to GABA synthesis in the retina. Neuroreport 2004;15:1895–898.
48. Casado M, Bendahan A, Zafra F, Danbolt NC. Phosphorylation and modulation of brain glutamate transporter by protein kinase C J Biol Chem 1993;268:27313–27317.
49. Dowd LA and Robinson MB. Rapid stimulation of EAAC1 mediated Na$^+$-dependent L-glutamate transport activity in C6 glioma cells by phorbol ester. J Neurochem 1996;18:3603–3619.
50. Davis KE, Straff DJ, Weinstein EA, Bannerman PG, et al. Multiple signaling pathways regulate cell surface expression and activity of the excitatory amino acid carrier 1 subtype of glutamate transporter in C6 glioma. J Neurosci 1998;18:3603–3619.
51. Vandenberg RJ, Mitrovic AD, Johston GAR. Molecular basis for differential inhibition of glutamate transporter sub-types by zinc ions. Mol Pharmacol 1998;54:189–196.
52. Zerangue N, Arriza JL, Amara SG, Kavanaugh MP. Differential modulation of human transporter subtypes by arachidonic acid. J Biol Chem 1995;270:6433–6435.
53. Fairman WA, Sonders MS, Murdoch GH, Amara SG. Arachidonic acid elicits a substrate-gated proton current associated with the glutamate transpoorter EAAT4. Nature Neurosci 1998;1:105–113.
54. Eskandari S, Kreman M, Kavanaugh MP, Wright EM, Zamphigi GA. Pentameric assembly of a neuronal glutamate transporter. Proc Natl Acad Sci 2000;97:8641–8646.
55. Trotti D, Danbolt NC, Volterra A. Glutamate transporters are oxidant-vulnerable: a molecular link between oxidative and excitotoxic neurodegeneration? Trends Pharmacol Sci 1998;19:328–334.
56. Jackson M, Song W, Liu M, Jin L, Dyles-Hoberg M, et al. Modulation of the neuronal glutamate transporter EAAT4 by two interacting proteins. Nature 2001;410:89–93.
57. Lin CG, Orlov I, Ruggiero AM, Dykes-Hoberg M, et al. Modulation of the neuronal glutamate transporter EAAC1 by the interacting protein GTRAPS-18. Nature 2001;410:84–88.
58. Butchbach MER, Lai L, Lin C-I G. Molecular cloning, gene structure, expression profile and functional characterization of the mouse glutamate transporter (EAAT3) interacting protein GTRAP3-18. Gene 2002;292:81–90.

59. Marie H and Attwell D. C-terminal interactions modulate the affinity of GLAST glutamate transporters in salamander retinal glial cells. J Physiol 1999;520:393–397.
60. Moron JA, Zakharova I, Ferrer JV, Merrill GA, et al. Mitogen-activated protein kinase regulates dopamine transporter surface expression and dopamine transport capacity. J Neurosci 2003;23:8480–8488.
61. Gonzalez MI, Loez-Clome AM, Ortega A. Sodium-dependent glutamate transport in Müller cells: Regulation by phorbol esters. Brain Res 1999;831:140–145.
62. Ganel R and Crosson CE. Modulation of human glutamate transporter activity by phorbol ester. J Neurochem 1998;70:993–1000.
63. Bull ND and Barnett NL. Antagonists of protein kinase C inhibit rat retinal glutamate transporter activity in situ. J Neurochem 2002;81:472–480.
64. Schniepp R, Kohler K, Ladewig T, et al. Retinal localization and in vitro interaction of the glutamate transporter EAAT3 and the serum- and glucocorticoid-inducible kinase SGK1. Invest Ophthalmol Vis Sci 2004;45:1442–1449.
65. Cheng C, Grover G, Banker G, Amara SG. A novel sorting motif in the glutamate transporter excitatory amino acid transporter 3 directs its targeting in Madin–Darby canine kidney cells and hippocampal neurons. J Neurosci 2002;22:10643–10652.
66. Yokoyama S. Molecular evolution of vertebrate visual pigments. Prog Ret Eye Res 2000;19:385–419.
67. Moritz OL, Tam BM, Papermaster DS, Nakayama T. A functional rhodopsin-green fluorescent protein fusion protein localizes correctly in transgenic Xenopus laevis retinal rods and is expressed in a time-dependent manner. J Biol Chem 2001;276:28242–28251.
68. Tam BM, Moritz OL, Hurd LB, Papermaster DS. Identification of an outer segment targeting signal in the COOH terminus of rhodopsin using transgenic Xenopus laevis J Cell Biol 2000;151:1369–1380.
69. Luo W, Marsh-Armstrong N, Rattner A, Nathans J. An outer segment localization signal at the C terminus of the photoreceptor–specific retinol dehydrogenase. J Neurosci 2004;24:2623–2632.
70. Concepcion F, Mendez A, Chen J. The carboxy-terminal domain is essential for rhodopsin transport in rod photoreceptors. Vis Res 2002;42:417–426.
71. Sung CH, Makino C, Baylor D, Nathans J. A rhodopsin gene mutation responsible for autosomal dominant retinitis pigmentosa results in a protein that is defective in localization to the photoreceptor outer segment. J Neurosci 1994;14:5818–5833.
72. Hertz L and Zielke HR. Astrocytic control of glutamatergic activity: astrocytes as stars of the show. Trends Neurosci 2005;27:735–742.
73. Chaudhry FA, Reimer RJ, Edwards RH. The glutamine commute: take the N line and transfer to the A. J Cell Biol 2002;157:349–355.
74. Deitmer JW, Broer A, Broer S. Glutamine efflux from astrocytes is mediated by multiple pathways. J Neurochem 2003;87:127–35.
75. Broer S and Brookes N. Transfer of glutamine between astrocytes and neurons. J Neurochem 2001;77:705–719.
76. Melone M, Quagliano F, Barbaresi P, Varoqui H, et al. Localization of the glutamine transporter SNAT1 in rat cerebral cortex and neighboring structures, with a note on its localization in human cortex. Cereb Cortex 2004;14:562–574.
77. Melone M, Varoqui H, Erickson JD, Conti F. Localization of the Na(+)-coupled neutral amino acid transporter 2 in the cerebral cortex. Neuroscience 2006;140:281–292.
78. Boulland JL, Olsen KK, Levy LM, Danbolt NC, et al. Cell-specific expression of the glutamine transporter SN1 suggests differences in dependence on the glutamine cycle. Eur J Neurosci 2002;15:1615–1631.

79. Umapathy NS, Li W, Mysona BA, Smith SB, Ganapathy V. Expression and function of glutamine transporters SN1 (SNAT3) and SN2 (SNAT5) in retinal Müller cells. Invest Ophthalmol Vis Sci 2005;46:3980–3987.
80. Tamarappoo BK, Raizada MK, Kilberg MS. Identification of a system N-like Na(+)-dependent glutamine transport activity in rat brain neurons. J Neurochem 1997;68:954–960.
81. Boehmer C, Okur F, Setiawan I, Broer S, Lang F. Properties and regulation of glutamine transporter SN1 by protein kinases SGK and PKB. Biochem Biophys Res Commun 2003;306:156–62.
82. Mackenzie B and Erickson JD. Sodium-coupled neutral amino acid (System N/A) transporters of the SLC38 gene family. Pflugers Arch 2004;447:784–795.
83. Segawa H, Fukasawa Y, Miyamoto K, Takeda E, et al. Identification and functional characterization of a Na+-independent neutral amino acid transporter with broad substrate selectivity. J Biol Chem 1999;274:19745–19751.
84. Broer S and Wagner CA. Structure-function relationships of heterodimeric amino acid transporters. Cell Biochem Biophys 2002;36:155–168.
85. Nakauchi T, Ando A, Ueda-Yamada M, et al. Prevention of ornithine cytotoxicity by nonpolar side chain amino acids in retinal pigment epithelial cells. Invest Ophthalmol Vis Sci 2003;44:5023–28.
86. Bodoy S, Martin L, Zorzano A, Palacin M, et al. Identification of LAT4, a novel amino acid transporter with system L activity. J Biol Chem 2005;280:12002–12011.
87. Dolinska M, Zablocka B, Sonnewald U, Albrecht J. Glutamine uptake and expression of mRNA's of glutamine transporting proteins in mouse cerebellar and cerebral cortical astrocytes and neurons. Neurochem Int 2004;44:75–81.
88. Utsunomiya-Tate N, Endou H, Kanai Y. Cloning and functional characterization of a system ASC-like Na+-dependent neutral amino acid transporter. J Biol Chem 1996;271:14883–4890.
89. O'Brien KB, Miller RF, Bowser MT. D-Serine uptake by isolated retinas is consistent with ASCT-mediated transport. Neurosci Lett 2005;385:58–63.
90. Ward MM, Jobling AI, Kalloniatis M, Fletcher EL. Glutamate uptake in retinal glial cells during diabetes. Diabetologia 2005;48:351–60.
91. Barnett NL, Pow DV, Bull ND. Differential perturbation of neuronal and glial glutamate transport systems in retinal ischemia. Neurochem Int 2001;39:291–19.
92. Mawrin C, Pap T, Pallas M, et al. Changes of retinal glutamate transporter GLT-1 mRNA levels following optic nerve damage. Mol Vis 2003;13:10–13.
93. Delfyer MN, Simonutti M, Neveux, et al. Does GDNF exert its neuroprotective effects on photoreceptors in the rd1 retina through the glial glutamate GLAST? Mol Vis 2005;11:677–687.
94. Sullivan RKP, Woldemussie E, Macnab L, Ruiz G, Pow DV. Evoked expression of the glutamate transporter GLT-1c in retinal ganglion cells in human glaucoma and in a rat model. Invest Ophthalmol Vis Sci 2006;47:3853–3859.

16
Localization and Function of Gamma Aminobutyric Acid Transporter 1 in the Retina

Giovanni Casini

CONTENTS

SUMMARY
NEUROTRANSMITTER TRANSPORTERS
GABA TRANSPORTERS
THE GABAERGIC SYSTEM IN THE RETINA
GABA UPTAKE AND TRANSPORTERS IN THE RETINA
LOCALIZATION OF GAT-1 IN THE RETINA
FUNCTIONS OF GAT-1 IN THE RETINA
ANTI-EPILEPTIC DRUGS, RETINAL GABA, AND GAT-1
REFERENCES

SUMMARY

Plasma membrane transporters, located in the presynaptic terminal and/or surrounding glial cells, terminate synaptic transmission by facilitating rapid, high-affinity uptake of the neurotransmitter from the synaptic cleft. Pharmacological blockade of transporters increases extracellular neurotransmitter levels and prolongs transmitter exposure to the receptors. Gamma aminobutyric acid (GABA) transporters (GATs) belong to the Na^+- and Cl^--dependent transporter family. Four GATs have been isolated and cloned in mammals, of which GAT-1 and GAT-3 are expressed in the retina. The GAT-1 transporter has a widespread distribution to different retinal cell types, but it is prominently expressed in the amacrine cells of all vertebrate species studied to date. There are some species differences in the expression patterns of GAT-1 in the retina. It is expressed by horizontal cells in non-mammalian, but not in mammalian, retinas, and it is expressed in Müller glial cells of rats and guinea pigs, but not in those of rabbits and primates.

Functionally, GAT-1, together with GAT-3, regulates the extracellular GABA levels in the retina, thereby determining the level of inhibitory interactions and affecting visual processing in the retinal pathways. GAT-1 may interact with $GABA_C$ receptors on bipolar cell terminals and influence ganglion cell responses. It may also interact with $GABA_B$

receptors in the regulation of retinal waves of spontaneous activity, which are known to play critical roles during development of the visual system. Other important functional actions are exerted by GAT-1 through reversed GABA transport. These include GABA release by cholinergic/GABAergic starburst amacrine cells and GABA release during early retinal development.

NEUROTRANSMITTER TRANSPORTERS

Chemical neurotransmission is a highly complex process that involves the release of neurotransmitter from the presynaptic terminal, diffusion across the synaptic cleft and binding to receptors on the postsynaptic membrane, resulting in a response of the postsynaptic neuron. Overall, neurotransmitter transporters can be classified as intracellular vesicular transporters that are responsible for sequestering transmitters from the cytoplasm into synaptic vesicles, and plasma membrane transporters, located in the presynaptic terminal and/or surrounding glial cells, which terminate synaptic transmission by operating rapid, high-affinity uptake of the neurotransmitter from the synaptic cleft (1). The transmembrane transport of neurotransmitters is of fundamental importance for proper signaling between neurons, as it exerts a key role in controlling the neurotransmitter concentration in the synaptic cleft.

Plasma membrane transporters have long been recognized as important components of the machinery for neural signaling. Reuptake inhibitors increase the levels of neurotransmitter in the synapse, thus enhancing synaptic transmission, and provide important targets for therapeutic intervention. Indeed, the importance of neurotransmitter transporters is highlighted by the broad spectrum of drugs targeting these proteins, including those used to treat depression, anxiety, obesity and epilepsy, in addition to drugs of abuse, such as cocaine, amphetamine, and ecstasy (2). Furthermore, it is well established that neurotransmitter transporters have roles in several neurological and psychiatric diseases, including amyotrophic lateral sclerosis, severe orthostatic hypotension, obsessive–compulsive disorder, Asperger's syndrome, anorexia, and autism (3).

Cloning studies show that two distinct gene families encode neurotransmitter transporters. One family codes for the high-affinity glutamate transporters (SLC1 gene family), the other codes for the Na^+- and Cl^--coupled transporters (SLC6 gene family). The latter subclass is the largest and includes transporters of dopamine, serotonin, norepinephrine, glycine and gamma aminobutyric acid (GABA) (3). Additional proteins may associate with transporters in native systems. Indeed, transporters, in a similar manner to receptors and ion channels, are likely to be associated with other proteins that are necessary for their localization near the sites of transmitter release and/or for regulation of transporter function (4).

Important functions of plasma membrane transporters not only include the termination of the signaling process by neurotransmitter reuptake, but also replenishment of the neurotransmitter supply inside nerve terminals and, in the case of the glutamate transporters, keeping the extracellular concentration of glutamate below neurotoxic levels. In addition, following membrane depolarization, they can also work in a reverse mode and mediate nonvesicular, Ca^{2+}-independent transmitter release from presynaptic terminals *(5)*. Transmitter release by reverse transport can be modulated by other neurotransmitters: for instance, both GABA and glutamate may reciprocally influence their releases through plasma membrane transporters *(6)*.

GABA TRANSPORTERS

In a GABAergic synapse, about 80% of the released GABA is likely to be transported back into the GABAergic nerve ending, while the remaining 20% is taken up by the astrocytes surrounding the synapse. GABA transporters (GATs) belong to the Na^+- and Cl^--dependent transporter family, composed of several subfamilies including the choline, monoamine, taurine, glycine, and betaine amino-acid transporters. Most of these proteins are made up of approximately 600 amino acid residues, and display a molecular mass of around 80 kDa. Some of these transporter subfamilies show a common structural organization, characterized by 12 transmembrane segments, organized in dimers. A typical conserved feature is the wide extracellular loop localized between segments two and four. This structure acts as a site for glycosilation, which is important for the final stage of the insertion of transporters in the plasma membrane *(7)*. These proteins also contain three sites for phosphorylation by protein kinase C, and one for phosphorylation by protein kinase A, likely acting to influence the functional regulation of the transporter. Both the N- and C-terminal regions are located in the cytosol *(8)*.

The mechanism of the Na^+- and Cl^--dependent transport is complex, and it depends on ion cotransport directed by electrochemical gradients. GABA transport is electrogenic with a stoichiometry for $Na^+:Cl^-:GABA$ transport of 2:1:1 (all inwardly directed). In the presence of membrane depolarization and appropriate gradients, reversed transport of GABA may occur *(5,9)*. The GATs are influenced by a variety of physiological stimuli, including GABA *(10)*, brain-derived neurotrophic factor *(11)*, and hormones *(12)*, and they are regulated by multiple intracellular effectors including protein kinase C and syntaxin 1A *(4,10,11,13)*.

To date, three GATs and a betaine glycine transporter (BGT), which also transports GABA with high affinity, have been isolated and cloned in mammals *(14)*. Unfortunately, different nomenclatures have been used for the GAT subtypes in the mouse and in rat/human, and the resulting picture is rather misleading. The GAT1, GAT2, GAT3, and GAT4 transporters in the mouse nomenclature correspond to GAT-1, BGT-1, GAT-2, and GAT-3, respectively, in the rat and human nomenclatures *(15)*. The rat/human nomenclature will be used in this chapter.

GAT-1 is considered a neuronal transporter, while GAT-2 and GAT-3 are believed to be glial transporters. The role of BGT-1 in GABA uptake in the brain is not well understood *(16)*. GAT-1 is the most widely-expressed GAT in the central nervous system, mainly localized to presynaptic axon terminals forming symmetric synaptic contacts, and to a few astrocytic processes. GAT-2 is weakly expressed throughout the brain, however it is primarily present in the leptomeninges, and in ependymal and choroid plexus cells, and only to a minor extent in neuronal and non-neuronal elements. GAT-3 is almost exclusively localized to distal astrocytic processes *(14)*.

A number of inhibitors of GABA transport have been discovered *(17)*. At least two high-affinity systems were first identified, based on the effects of the GABA uptake

Table 1 Nomenclature and principal characteristics of GATs

Mouse nomenclature	Rat/human nomenclature	Characteristic inhibitors	Localization in the central nervous system
GAT1	GAT-1	ACHC, tiagabine	Axon terminals; a few astrocytic processes
GAT2	BGT-1	betaine	Astrocytes
GAT3	GAT-2	β-alanine	Neuronal and non-neuronal elements (low expression)
GAT4	GAT-3	β-alanine	Distal astrocytic processes

inhibitors cis-3-aminocyclohexanecarboxylic acid (ACHC) and β-alanine. In particular, ACHC, but not β-alanine, strongly inhibits GABA uptake by GAT-1. ACHC inhibition is considered a typical property of neuronal transporters. In contrast, GABA uptake by GAT-2 and GAT-3 is strongly inhibited by β-alanine, but not by ACHC, which is considered a property of glial transporters *(14)*. BGT-1 is competitively inhibited by betaine *(18)*. Nipecotic acid and guvacine, two additional inhibitors of GABA transport, have served as lead structures for the synthesis of a series of selective, high-affinity GAT-1 inhibitors, such as tiagabine, which penetrate the blood–brain barrier *(19)*. Some of the main characteristics of GATs are reported in Table 1.

THE GABAERGIC SYSTEM IN THE RETINA

GABA is the major inhibitory neurotransmitter in the retina. Inhibitory signaling pathways play a fundamental role in the shaping of visual information by modulating the flow of visual inputs from photoreceptors to bipolar cells and, subsequently, from bipolar cells to ganglion cells. The initial inhibitory interactions in the outer plexiform layer (OPL) are mediated by horizontal cells, while those in the inner plexiform layer (IPL) are mediated by amacrine cells (Fig. 1). Although the possibility of horizontal cell-cone photoreceptor interactions is plausible in mostly non-mammalian retinas (see *20* for review), generally GABA does not seem to contribute to lateral inhibitory interactions in the OPL *(21–23)*. It is well established, instead, that GABA mediates the lateral inhibitory phenomena in the IPL, which contribute to the surround of ganglion cell receptive fields *(24,25)*. The GABA-mediated inhibition in the IPL may also be involved in the temporal responses of ganglion cells *(26)* and plays a fundamental role in ganglion cell motion and direction sensitivity *(27)*. Recent papers have reviewed the localization patterns of GABA and its receptors as well as the physiology of the inhibitory GABAergic network in the vertebrate retina *(20,28–30)*.

GABA-containing retinal neurons are horizontal cells, amacrine cells, interplexiform cells (an amacrine cell variant), bipolar cells and ganglion cells. The issue of whether horizontal cells can be generally considered as GABAergic is destined to remain unresolved. Indeed, they appear to contain GABA or GAD in some vertebrate retinas but not in others, with high variability also depending on the developmental time and the retinal location (see *29* for review). Small subsets of bipolar cells may contain GABA in a number of vertebrate retinas. In primates, for instance, there are conflicting reports

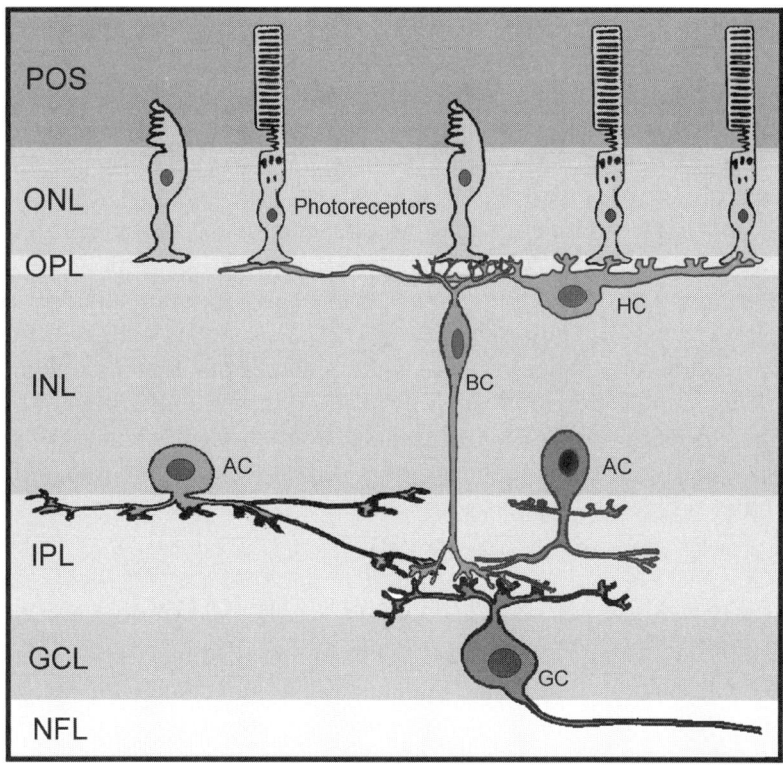

Fig. 1. Schematic representation of the vertebrate retina. Visual information passes from photoreceptors to bipolar cells (BC) through synapses in the outer plexiform layer (OPL). Horizontal cells provide inhibitory regulation of this flow, although there is uncertainty about the involvement of gamma aminobutyric acid (GABA). In the inner plexiform layer (IPL), bipolar cell axonal terminals contact ganglion cell (GC) dendrites. Here, prominent inhibitory actions provided by GABAergic amacrine cells (AC) importantly contribute to the physiological responses of ganglion cells, whose axons reach the retino-recipient nuclei in the brain. Other abbreviations: GCL, ganglion cell layer; INL, inner nuclear layer; NFL, nerve fiber layer; ONL, outer nuclear layer; POS, photoreceptor outer segments.

of the presence or the absence of GABA immunoreactivity in bipolar cells (see *31* for references). GABA-containing amacrine cells constitute one of the largest neuronal populations in the retina, and they account for 37–38% of the total amacrine cells in the rabbit retina (*32*). A number of subpopulations of GABA-containing amacrine and interplexiform cells can be identified on the basis of coexpressed transmitters/peptides (see *33* and *34* for references). In the ganglion cell layer (GCL), both displaced amacrine cells and ganglion cells containing GABA have been reported (*35*).

Three types of GABA receptors, with distinct structural and pharmacological characteristics, mediate inhibition in the retina: the ionotropic $GABA_A$ and $GABA_C$ receptors, and the metabotropic $GABA_B$ receptors (see *20* for a review). Activation of these receptors leads to neuron hyperpolarization, with subsequent reduction of transmitter release and/or action potential firing. $GABA_A$ receptors are pentameric and consist of α, β, γ, and δ subunits. The pharmacology of these receptors may vary depending on the assembled subunits. The most common combination is α1/β2/γ2, which is widely expressed in the central nervous system, including the retina. These receptors are directly coupled with Cl^- channels, can be specifically blocked by bicuculline and have modulatory binding sites for benzodiazepines, barbiturates, ethanol and neurosteroids. $GABA_B$ receptors are G-protein-coupled receptors, with second-messenger pathways leading to an increase in K^+ conductance, or a decrease in voltage-dependent Ca^{2+} currents. The $GABA_B$ receptor is activated by baclofen and it is insensitive to bicuculline. $GABA_C$ receptors are found predominantly in the retina. They are Cl^- pores, blocked by picrotoxin, but insensitive to bicuculline and baclofen. Structurally, $GABA_C$ receptors characteristically consist of ρ subunits, which form homoligomeric channels with $GABA_C$ receptor pharmacology. All of these receptors are widely expressed in vertebrate retinas *(20,28)*.

GABA UPTAKE AND TRANSPORTERS IN THE RETINA

High-affinity GABA uptake systems in the retina mediate GABA accumulation into neurons, Müller cells and retinal pigment epithelium cells, with complex kinetic and pharmacological properties *(36)*. In different vertebrate retinas, the uptake of GABA or GABA analogs has been extensively used in the past to map the retinal GABAergic system *(37–43)*. In rabbit retina, GABA, the GABA analog diaminobutyric acid, the GABA agonist isoguvacine and the GABA uptake blocker nipecotic acid, are taken up into neurons and Müller cells, while the GABA analogs γ-vinyl GABA and gabaculine are selectively accumulated, respectively, in amacrine cells and Müller cells *(44)*. These uptake patterns suggest that multiple GABA transporters operate in the retina. They do not seem to operate in horizontal cells of rabbit and rat retinas, as it has been demonstrated that these cells do not take up GABA or GABA analogs *(42,44–48)*.

A variety of studies have established that there is prominent expression of GATs in the retina. GAT-1, GAT-2 and GAT-3 mRNAs have been detected in rat retina using reverse transcriptase polymerase chain reaction (RT-PCR) and Northern blot analysis *(49,50)*. Furthermore, they have been reported in developing and adult rat optic nerve by Northern blotting *(51)*. GAT-1 is likely to be the principal GABA transporter in the retina (see below). GAT-2 does not seem to be expressed by cells of the neural retina. Indeed, GAT-2 immunoreactivity is predominantly distributed to the retinal pigment and ciliary epithelia *(52–54)*, congruent with findings that cultured bovine retinal pigment epithelia accumulate GABA *(55)*. GAT-3 mRNA has been detected mostly in the inner nuclear layer (INL) of rat retina by in situ hybridization *(56)*. Immunocytochemical analyses have shown that it is distributed throughout all retinal layers, with an expression pattern indicating that it is predominantly localized to Müller cells in rat, rabbit and guinea pig retinas *(52–54,57)*. In addition, GAT-3 is also expressed by a number of amacrine and displaced amacrine cells originating a dense plexus of GAT-3 immunoreactive fibers in the IPL *(53,54,57)*.

LOCALIZATION OF GAT-1 IN THE RETINA

Following the cloning of GAT-1 from the rat brain *(58)*, its expression in the rat retina, together with GAT-2 and GAT-3 expression, was first demonstrated with polymerase chain reaction analysis in 1992 *(49)*. Subsequently, the presence of GAT-1 in retinas of other vertebrates has been investigated. For instance, a skate GABA transporter displaying high homology with the mouse GAT-1 has been detected by Northern blot in the skate retina *(59)*, and GAT-1 immunoreactivity has been observed by immunoblot in goldfish retina (Klooster et al. 2004). In addition, studies of the uptake and release of GABA in the presence of the GAT-1 inhibitors nipecotic acid or NNC-711 provided evidence of the presence, in chicken dissociated retinal cells in culture, of a transporter with pharmacological properties similar, although not identical, to GAT-1 *(60)*. The expression patterns of GAT-1 in the retina have been investigated with a variety of techniques and the presence of this transporter in selected retinal cell types has been reported.

GAT-1 in Ganglion Cells and/or Displaced Amacrine Cells

The GCL contains both ganglion cells and displaced amacrine cells. The ganglion cells are the output neurons of the retina, and with their axons, contact the primary visual nuclei in the brain. In contrast, the displaced amacrine cells are similar to the normally-placed amacrine cells in the INL, their processes arborize in the IPL, and they do not exit the retina *(61)*. Using immunohistochemistry or in situ hybridization histochemistry, GAT-1 has been found to be expressed by cells in the GCL in tiger salamander *(62)*, salmon *(63)*, rat *(50,53,56,64)*, mouse *(65,66)*, rabbit *(67)*, bovine *(68)*, and primate *(31)* retinas. In most cases, these cells displaying GAT-1 immunoreactivity or GAT-1 mRNA have been interpreted as displaced amacrine cells mainly on the basis of the small soma size, suggesting these cells were unlikely to be ganglion cells. However, retrograde transport of fluorogold from the superior colliculus combined with situ hybridization histochemistry demonstrated that at least a few of the GAT-1 mRNA-expressing cells in the GCL of the rat retina are indeed ganglion cells *(50)*. In addition, the observation of some GAT-1 immunostained fibers in the nerve fiber layer (NFL, which contains the ganglion cell axons) of the rabbit retina could indicate the presence of some GAT-1-expressing ganglion cells in the rabbit retina as well *(67)*. This interpretation is not supported by observations in primate retinas, where GAT-1 immunopositive fibers have been observed in the NFL, but they did not reach the optic nerve, suggesting they are not ganglion cell axons *(31)*. Furthermore, in mouse retina, retrograde tracing from the superior colliculus fails to label neuropeptide Y- (NPY-) containing cells in the GCL, which also express GAT-1, indicating these cells are displaced amacrine cells *(66)*.

Taken together, these data are consistent with the presence of GAT-1 in displaced amacrine cells, although a small number of ganglion cells may also express this transporter. The possibility also exists that a higher number of ganglion cells possess GAT-1. For instance, in primates GABA-immunoreactive ganglion cells and ganglion cell axons have been identified *(69–72)*, but GAT-1-immunoreactive ganglion cells are unlikely *(31)*. These GABA-containing ganglion cells, if they express GAT-1, are therefore likely to express it at their pre-terminal axonal processes, and axonal terminals in retinal recipient nuclei, similar to the expression of GAT-1 on axonal terminals of hippocampal and cortical neurons *(73,74)*.

GAT-1 in Amacrine Cells

The majority of GAT-1-expressing cells in the retina are amacrine cells. The presence of GAT-1 in amacrine cells has been documented in most of the species studied to date *(31,50,52–54,56,57,62–68,75–77)*, while in skate and bullfrog retinas strong GAT-1 immunostaining is reported in the IPL *(78,79)*, which suggests the presence of GAT-1-expressing amacrine cells. With immunohistochemical labeling, the IPL appears heavily GAT-1 immunostained, displaying intensely stained puncta and varicose processes. In some instances, GAT-1-immunolabeled somata have been reported within the IPL (interstitial amacrine cells) *(31,50)*. The GAT-1-immunostained somata in the INL have often been observed to give rise to a process directed towards the OPL *(31,53,54,62,67)*, where a meshwork of GAT-1-immunoreactive fibers has been detected in the monkey *(31)* and in the rabbit retina (Casini G., Rickman D.W., Brecha N.C., unpublished data). The cells creating these processes are known as interplexiform cells, which can be considered an amacrine cell variant. The typical pattern of GAT-1 expression is represented in Fig. 2, which depicts an immunolabeling of the mouse retina (G. Casini, unpublished).

The predominant localization of GAT-1 to amacrine and interplexiform cells is generally congruent with the pattern of GABA- or GAD-containing cells and of GABA- or GABA analog-accumulating cells. Double-label immunohistochemical studies have addressed the presence of GABA in GAT-1-expressing amacrine cells (Fig. 3). Colocalization of the two markers has been observed in amacrine cells of the goldfish retina *(77)*, and quantitative data have been obtained in rabbit and in primate retinas. In one study of the rabbit retina, 99% of the GAT-1-immunoreactive amacrine cells were found to also contain GABA immunoreactivity, while 75% of the GABA-containing amacrine cells also expressed GAT-1 *(67)*. Our unpublished observations in rabbit retina give an estimate of 66% of GABA-immunostained amacrine cells also displaying GAT-1 immunoreactivity (Casini G., Rickman D.W., and Brecha N.C., unpublished data). Similar percentages

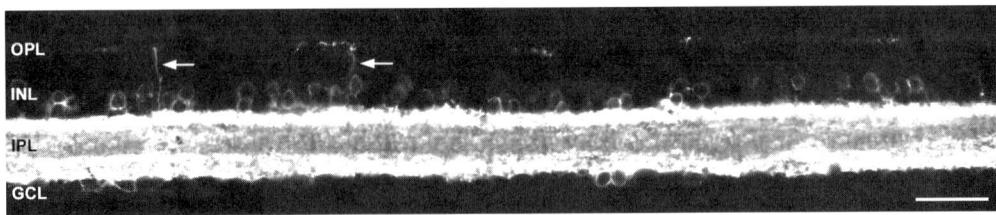

Fig. 2. Pattern of GAT-1 immunoreactivity in the mouse retina. Immunofluorescent somata are mainly amacrine cells located in the proximal inner nuclear layer (INL). Processes of GAT-1-immunolabeled interplexiform cells arise from the immunoreactive fiber plexus in the inner plexiform layer (IPL) or from immunostained cell bodies in the INL (arrows) and arborize in the outer plexiform layer (OPL). Displaced amacrine cells in the ganglion cell layer (GCL) can also be seen. GAT-1-immunoreactive processes are densely distributed in the IPL, with heavier labeling in the distal and in the proximal part of the layer. Paraformaldehyde-fixed cryostat section (10 µm thick). Immunofluorescence obtained with a rabbit primary antibody directed to GAT-1 (Chemicon) and a secondary goat anti-rabbit antibody conjugated with Alexa Fluor 488 (Molecular Probes). Calibration bar = 50 µm.

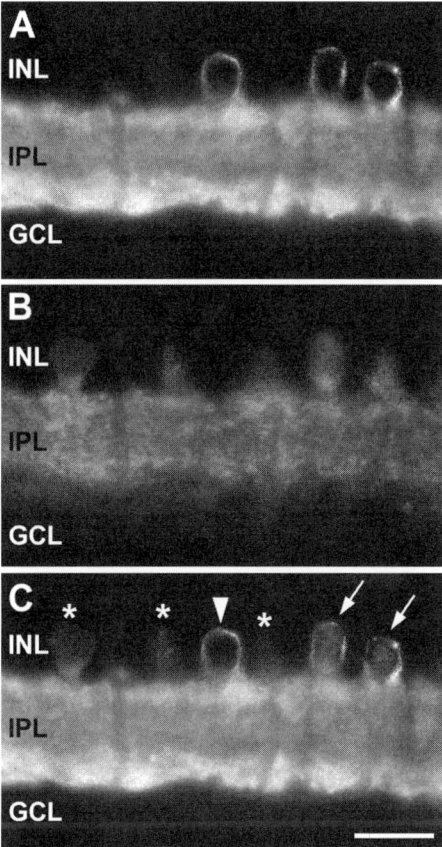

Fig. 3. Pattern of GAT-1 and GABA colocalization in the rabbit retina. (**A**) GAT-1 immunostaining; (**B**) GABA immunostaining; (**C**) overlay image of (A) and (B). Note that some GAT-1-expressing amacrine cells are also GABA-immunolabeled (arrows), while there are GABA-containing somata that do not express GAT-1 (asterisks). A GAT-1-expressing amacrine cell which does not contain GABA is also visible (arrowhead). GAT-1 immunostaining as in Fig. 2; the GABA antibody (Sigma) was made in mouse and visualized with goat anti-mouse secondary antibody conjugated with Alexa Fluor 546 (Molecular probes). Abbreviations as in Fig. 1. Calibration bar = 20 µm.

(99% of GAT-1 amacrine cells also containing GABA, and 66% of GABA-containing amacrine cells also displaying GAT-1) have been calculated in primate retinas *(31)*. Since about 38% of all amacrine cells in rabbit retinas contain GABA *(32)*, it follows that GAT-1 is expressed by 25–30% of all amacrine cells in these retinas. GABA-containing cells that do not express GAT-1 may take up GABA via another GAT, such as GAT-3, or they may lack a high-affinity GABA uptake system, similar to the Purkinje cells of the cerebellum, which do not express either GAT-1 or GAT-3 *(80)*.

Interestingly, some GAT-1-expressing amacrine cells that do not contain GABA immunoreactivity have been reported in the rabbit retina *(57)*. In particular, our quantitative studies of the rabbit retina (Casini G., Rickman D.W. and Brecha N.C., unpublished

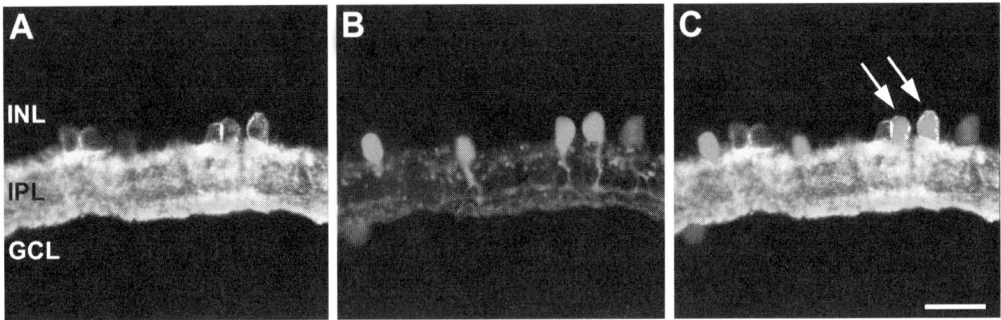

Fig. 4. GAT-1 and parvalbumin colocalization in the rabbit retina. (**A**) GAT-1 immunostaining; (**B**) parvalbumin immunostaining; (**C**) overlay image of (A) and (B). Parvalbumin antibodies label two different populations of amacrine cells in the rabbit retina, including a class of unidentified GABA-containing amacrine cells and the population of glycinergic AII amacrine cells. Here, two cells with the typical morphology of AII amacrine cells can be seen to display GAT-1 immunoreactivity at or near their plasma membrane (arrows). The parvalbumin antibody (Sigma) was made in mouse and visualized with goat anti-mouse secondary antibody conjugated with Alexa Fluor 546. Abbreviations as in Fig. 1. Calibration bar = 20 µm.

data) show that this subset of GAT-1/no-GABA cells amounts to 6.7% of all GAT-1-expressing amacrine cells, and a portion of them are likely to be represented by glycinergic AII amacrine cells (Fig. 4). The presence of GAT-1 immunoreactivity in cells that do not contain GABA is consistent with other reports that GAT-1 mRNA and immunoreactivity are also localized to non-GABA immunoreactive cells in other parts of the nervous system *(81)*. The expression of GAT-1 by a small percentage of AII amacrine cells may be the consequence of errors during retinal formation. Indeed, during the early part of postnatal retinal development, about 32% of the amacrine cells contain both GABA and glycine immunoreactivity *(82)*, and it is plausible that a small number of the cells destined to become glycinergic may maintain some GABAergic, nonfunctional characteristics.

Multiple subpopulations of GABA-containing amacrine cells have been detected in colocalization studies *(33,34)*, and the presence of GAT-1 in some of these subpopulations has also been tested. In the mouse retina, GAT-1 is expressed by the NPY-containing amacrine and displaced amacrine cells *(66)*, while in monkey retina it is in vasoactive intestinal peptide- (VIP-) containing amacrine cells *(31)*. In contrast, GAT-1 immunoreactivity has not been reported in rabbit cholinergic amacrine cells *(67)* and in monkey dopaminergic amacrine cells *(31)*. We have recently performed an extensive double-label quantitative immunohistochemical analysis of the composition of the population of GAT-1-expressing amacrine cells (Casini G., Rickman D.W., Brecha N.C., unpublished data). We found that the GAT-1-expressing amacrine cells include those containing VIP (Fig. 5), those accumulating indoleamines, and part of those containing substance P. In addition, about 20% of the GAT-1-immunoreactive amacrine cells also contain the calcium binding protein parvalbumin. GAT-1-immunostained amacrine cells containing the calcium binding protein calbindin (CaBP) were also observed (Fig. 6). Together, these identified subgroups of amacrine cells account for 57% of all GAT-1 immunoreactive

GAT-1 in the Retina

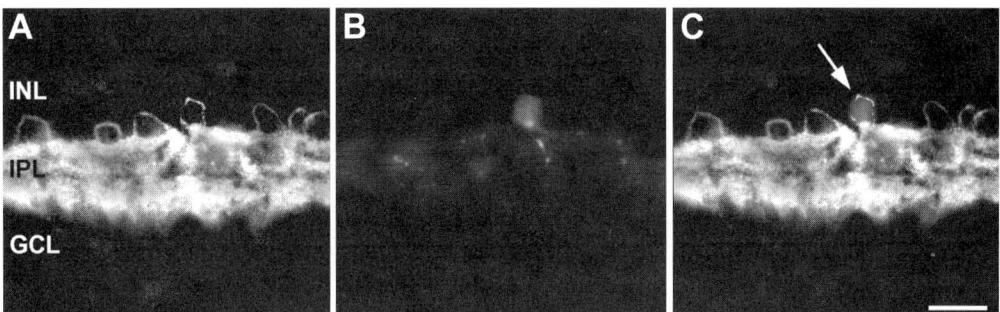

Fig. 5. GAT-1 and vasoactive intestinal peptide (VIP) colocalization in the rabbit retina. **(A)** GAT-1 immunostaining; **(B)** VIP immunostaining; **(C)** overlay image of (A) and (B). All the VIP-containing amacrine cells also express GAT-1 immunoreactivity (arrow). The VIP antibody (from H. Wong and Dr. J.H. Walsh, UCLA) was made in mouse and visualized with goat anti-mouse secondary antibody conjugated with Alexa Fluor 546. Abbreviations as in Fig. 1. Calibration bar = 20 µm.

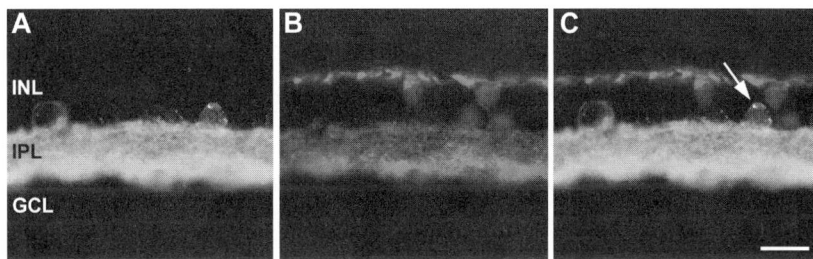

Fig. 6. GAT-1 and CaBP colocalization in the rabbit retina. **(A)** GAT-1 immunostaining; **(B)** CaBP immunostaining; **(C)** overlay image of **(A)** and **(B)**. None of the CaBP-immunostained horizontal cells, near the OPL, display GAT-1 immunoreactivity. In contrast, some of the CaBP-containing amacrine cells also express GAT-1 (arrow). The CaBP antibody (Sigma) was made in mouse and visualized with goat anti-mouse secondary antibody conjugated with Alexa Fluor 546. Abbreviations as in Fig. 1. Calibration bar = 20 µm.

amacrine cells. The remaining 43% are likely to be composed of other GABA-containing amacrine cell populations (Fig. 7). Consistent with previous observations in rabbit and in monkey retinas *(31,67)*, we did not observe GAT-1 expression in cholinergic or dopaminergic amacrine cells. The possibility exists that GAT-1 is expressed on the processes of these cells and not, or very rarely, on their cell bodies, as may be the case for some ganglion cells.

GAT-1 in Bipolar Cells

GAT-1-expressing bipolar cells have been demonstrated in tiger salamander retinas, where immunohistochemical studies have reported GAT-1 immunoreactivity in two different types of bipolar cells which also contained GABA *(62)*. In addition, whole-cell patch recordings

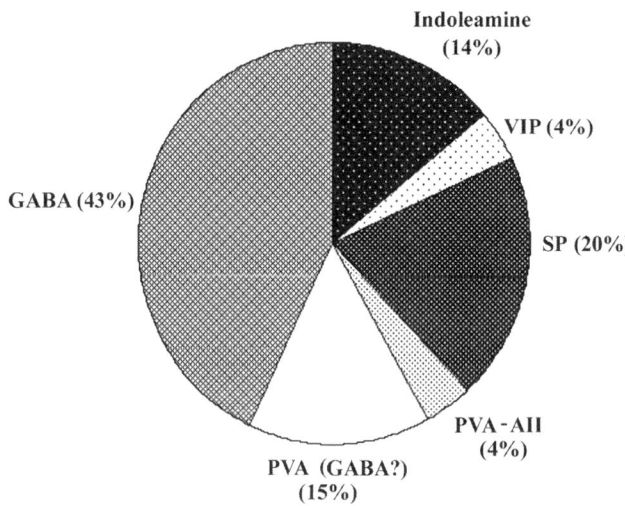

Fig. 7. Diagram summarizing the composition of the population of GAT-1 immunoreactive amacrine cells in the rabbit retina. Forty-three percent of these cells are unidentified gamma aminobutyric acid- (GABA-) containing amacrine cells, while 38% are identified subsets of GABA-containing amacrine cells, including all indoleamine-accumulating and vasoactive intestinal peptide- (VIP-) containing amacrine cells, and some of the substance P- (SP-) containing amacrine cells. In addition, 15% of the GAT-1-immunoreactive amacrine cells contain PVA immunoreactivity and are presumably GABAergic as well. Finally, 4% of the GAT-1 immunoreactive amacrine cells are likely to be AII amacrine cells.

from these cells revealed the presence of a GAT current (I_{GAT}) that was blocked by NO-711 *(83)*, a selective GAT-1 antagonist *(19)*. In contrast, bipolar cells expressing GAT-1 have not been reported in other vertebrate retinas. In particular, lack of colocalization of the immunoreactivities of GAT-1 and of a generic bipolar cell marker has been recently reported in the monkey retina *(31)*.

GAT-1 in Horizontal Cells

In the retina of several vertebrate species, GABA is used by horizontal cells as a neurotransmitter *(29)*. However, GAT-1 has not been localized to horizontal cells with in situ or immunohistochemical methods. For instance, horizontal cells identified by CaBP immunoreactivity in the rabbit retina are devoid of GAT-1 immunostaining (Casini G., Rickman D.W., Brecha N.C., unpublished data, Fig. 6). Only in non-mammalian retinas is there electrophysiological and pharmacological evidence of electrogenic GABA transport mechanisms consistent with the presence of GAT-1 in horizontal cells. Indeed, I_{GAT} values elicited by GABA uptake and/or release through a transporter have been detected in skate, catfish, goldfish, toad and tiger salamander horizontal cells, and these currents were sensitive to selective inhibitors of GABA transport through GAT-1, such as nipecotic acid, NO-711 or SKF89976A *(84–92)*.

The lack of GAT-1 expression in mammalian horizontal cells is congruent with studies showing that these cells do not accumulate exogenous GABA or GABA analogs

(42,44–48), and that they also lack GAT-1 or GAT-3 expression *(50,53,54,57)*. However, horizontal cells in several mammalian species express GABA and GAD *(93)*, suggesting that horizontal cells possess the machinery to synthesize and release GABA, but they do not have the cellular mechanisms to take up GABA from the extracellular space, similar to the Purkinje cell of the cerebellum *(80)*. We may hypothesize that in these cases GABA released in the OPL could be transported from the extracellular space by GAT-1 located on processes of interplexiform cells, or by GAT-3 expressed by Müller cells.

GAT-1 in Müller Cells

A highly efficient GABA uptake system by Müller cells has been described in the guinea pig retina, where both GAT-1 and GAT-3 are expressed by Müller cells *(54)*. In addition, the presence of GAT-1 in Müller cells of the rat retina has been demonstrated with in situ hybridization *(50)* and immunohistochemistry *(53)*. GAT-1, however, is only weakly expressed by these cells, which instead show a prominent GAT-3 expression *(50,53)*. In non-mammalian retinas, both GAT-1 and GAT-3 have been immunohistochemically detected in Müller cells of the skate retina *(78)*, while Müller cells of the bullfrog retina have been found to express GAT-1 and GAT-2 immunoreactivities, though they do not express GAT-3 *(79)*.

Summary of the Localization of GAT-1 in the Retina

Taken together, the data presented above show that GAT-1 is abundantly expressed in vertebrate retinas. Overall, the distribution of GAT-1 in the retina is similar to the distribution of GABA and GAD immunoreactivities, and of GABA or GABA analogue uptake. GAT-1-expressing cells are mainly amacrine cells, including displaced and interstitial amacrine cells, and interplexiform cells. A few ganglion cells may also express this transporter, while it is not expressed by photoreceptors.

Various subpopulations of GAT-1-immunoreactive amacrine cells have been identified, and many GABA-containing amacrine cells that do not express GAT-1 have also been detected. These cells could express other GATs, such as GAT-3, or they could express GAT-1 on their processes, but not on their cell bodies.

In mammalian retinas, GAT-1 is not expressed by horizontal or bipolar cells. The pattern of GAT-1 expression in mammalian retinas is consistent among mice, rats, guinea pigs, rabbits and primates, with the only exception that in rats and guinea pigs GAT-1 is weakly expressed in Müller cells, whereas it is not expressed in Müller cells of rabbits and primates.

The absence of GABA-uptake systems in horizontal cells of mammalian retinas raises the question of how GABA can be removed from the outer retina. The possibility exists that this function is operated by GAT-1 expressed on the distal processes of interplexiform cells and/or by GAT-1 and, mainly, by GAT-3 expressed by Müller cells.

In non-mammalian retinas, similar patterns of GAT-1 distribution have been reported, with the notable exceptions of the presence of GAT-1 in horizontal cells and, at least in the case of the tiger salamander retina, in bipolar cells.

FUNCTIONS OF GAT-1 IN THE RETINA

Both GAT-1 and GAT-3 Regulate GABA Levels in the Retina

The inhibition of GABA uptake mediated by GAT-1 in different regions of the central nervous system, including the retina, invariably results in increased tonic inhibition, suggesting that this transporter is a key player in the regulation of GABA neurotransmission. The widespread distribution of GAT-1 and GAT-3 in the retina indicates the presence of careful regulation of GABAergic signaling in the retina. For instance, GAT-1 and GAT-3 expressed by Müller cells would markedly influence GABA levels in all retinal regions, both by regulating GABA in the extracellular space and by limiting the spread of GABA from synapses *(94)*. Moreover, the GATs expressed by Müller cells may protect the synapses from inappropriate inhibition by excess GABA released elsewhere during physiological and/or pathological events.

The characteristic morphology of Müller cells, which span the entire neural retina, enables these cells to influence GABAergic transmission both in the outer and in the inner retina, by taking up GABA released by horizontal cells and by amacrine cells, respectively. In contrast, GATs expressed by neurons are likely to both remove GABA from the synaptic cleft following depolarization, and release GABA by reverse transport. Indeed, there is good evidence for carrier-mediated release of GABA by non-mammalian horizontal cells *(84,86,91)* and by the acetylcholine- and GABA-containing starburst amacrine cells of the rabbit retina, following depolarization *(95)*.

In summary, GAT-1, expressed by neurons, is likely to be involved in GABA removal from the synaptic cleft and extracellular space and perhaps in GABA release by a Ca^{2+}-independent mechanism, thus modulating GABA levels and GABA activity at GABA receptors. GAT-1 and GAT-3 expressed by Müller cells could also influence the levels of GABA, acting both in the inner and in the outer retina, and probably limit the spread of GABA within the retina. GAT-1 and GAT-3 have different functional and pharmacological properties, including different ionic dependences and inhibitor sensitivities, which influence GABA uptake. This pharmacological and functional heterogeneity is likely to provide considerable flexibility in the control of the extracellular levels of GABA in different physiological states.

Regulation of GABA Receptor Activation by GAT-1 at Retinal Synapses

The transmission of the visual signal from bipolar cells to ganglion cells is influenced by $GABA_C$ receptors located on the bipolar cell axon terminals and by $GABA_A$ receptors located both on bipolar cell terminals and, postsynaptic to the bipolar cell, on amacrine and ganglion cell dendrites *(28)*. The activation of $GABA_C$ receptors importantly contributes to the regulation of bipolar cell output, and dysfunction in $GABA_C$ receptor function results in altered ganglion cell responses (see 96 for references). The available evidence suggests that $GABA_C$ receptors may act to limit the extent of glutamate release from bipolar cell terminals. In the salamander retina, the selective GAT-1 blocker NO-711 determines an increase of the activation of $GABA_C$ receptors, with consequent reduction of the light-elicited signaling from bipolar cells to ganglion cells. This enhanced $GABA_C$-mediated response is likely to be due to activation of additional receptors by GABA spillover *(97)*. The suppression of bipolar cell-to-ganglion cell transmission, provoked by

GAT-1 blockade, affects the ganglion cell responses in a way similar to the reported effect of surround inhibition. These observations suggest that GAT-1 normally limits inhibitory signaling, acting particularly at $GABA_C$ receptors. If GAT-1 were nonfunctional, additional $GABA_C$ receptors would be activated by spillover transmission, with consequent impairment of both the spatial and temporal properties of ganglion cell responses.

Recently, recordings from isolated bipolar cell terminals in goldfish retinal slices have shown that a tonic current mediated by $GABA_C$ receptors is maintained by spontaneous GABA release and that the level of this current is tightly regulated by GAT-1. In addition, GAT-1 selectively limits the $GABA_C$ receptor-mediated reciprocal feedback between bipolar cell terminals and amacrine cells (96). These observations implicate GAT-1 as a major regulator of presynaptic excitability at the bipolar cell axonal terminal that acts by setting the concentration of extracellular GABA in the vicinity of bipolar cell $GABA_C$ receptors. Supporting this conclusion, GAT-1 inhibition significantly increases the amount of current necessary to depolarize the bipolar cell terminal to its threshold for generating action potentials (96).

Effects of GAT-1 Blockade on Electroretinographic Responses

In a recent study, the administration of the GAT-1 blocker NO-711 determined a pronounced increase of the b-wave of electroretinographic responses from isolated rabbit retinas (98). An explanation for this effect, proposed by the authors, is that as a consequence of the reduced GABA uptake, amacrine cells would become depleted of their GABA content, and their release of GABA upon light stimulation would be reduced. Therefore, the inhibition mediated by GABA receptors on the optic nerve-bipolar cell axonal terminals would also be reduced, and this would cause the increase of optic nerve-bipolar cell activity and the enlargement of the b-wave, which can increase to double its normal size.

GAT-1 and Retinal Development

In adult retinas, GABA is predominantly expressed by amacrine cells. During development, GABA is transiently expressed in additional cells, including ganglion and horizontal cells. GABA uptake and release mechanisms and GABA receptors are also expressed early in retinal development, well in advance of the onset of visual function. In addition, GABA transporter mRNAs have been detected in the developing rat optic nerve (51). These observations suggest that the GABAergic system may serve a developmental role, both in the establishment of retinal circuitries and in the organization of retinofugal projections (44,99). The GABA transporter is likely to play an extremely important role, because it seems to be responsible for GABA release, through reversed transport, early in development, prior to the establishment of vesicular synaptic transmission (99).

Spontaneous activity during development is essential for proper differentiation and refinement of the nervous system. In the retina such activity takes the form of rhythmic waves of intracellular $[Ca^{2+}]$ elevations (100). Recent observations have shown that endogenous GABA can modulate the frequency and the duration of such waves in the chick retina. This effect is mediated by $GABA_B$ receptors in a cAMP-independent manner, and this action is influenced by GAT-1. In particular, it has been observed that the GAT-1 inhibitor SKF89976A reduces the frequency of the transients presumably because it induces an increase in the extracellular concentration of GABA *(101)*.

Summary of the Functions of GAT-1 in the Retina

The experimental data reported in this section demonstrate that GABA transporters, and GAT-1 in particular, may affect multiple retinal functions including important aspects of retinal development, by regulating extracellular levels of GABA. By modulating the availability of GABA in the extracellular milieu and/or at specific synapses, GAT-1 and GAT-3 determine the overall level of inhibition in the retina or the amount of inhibitory signal at specific synapses. Mainly interacting with $GABA_C$ receptors, GAT-1 contributes to the shaping of ganglion cell responses, and by interacting with $GABA_B$ receptors, this transporter is likely to play important developmental functions. Other important functional actions are exerted through GABA release mediated by GAT-1 reversed transport. These include GABA release by cholinergic/GABAergic starburst amacrine cells and GABA release during early retinal development.

ANTI-EPILEPTIC DRUGS, RETINAL GABA, AND GAT-1

Anti-epileptic drugs have been developed with the aim of counteracting excess excitation by increasing GABAergic inhibition in the central nervous system. Two recently developed compounds, vigabatrin and tiagabine, increase GABA availability through two different mechanisms. Vigabatrin attenuates GABA metabolism by inhibiting the enzyme GABA-transaminase, whereas tiagabine blocks GABA uptake by GAT-1. As part of the central nervous system, the retina is also affected by treatment with these molecules. Vigabatrin accumulates in the retina and induces a greater increase in GABA concentration in the retina than in the brain. These abnormally high GABA levels in the retina often result in significant visual field constriction. In contrast, tiagabine does not accumulate in the retina, it does not overinhibit GAT-1 and does not cause visual field disturbances *(102)*. These observations indicate that interventions acting on the GABAergic system with anti-epileptic purposes should privilege the compounds, like tiagabine, directed at inhibiting GABA uptake, since these compounds do not seem to significantly impair visual function.

REFERENCES

1. Kanner BI, Schuldiner S. Mechanism of transport and storage of neurotransmitters. CRC Crit Rev Biochem 1987;22:1–38.
2. Chen NH, Reith ME, Quick MW. Synaptic uptake and beyond: the sodium- and chloride-dependent neurotransmitter transporter family SLC6. Pflugers Arch 2004;447:519–31.
3. Gether U, Andersen PH, Larsson OM, Schousboe A. Neurotransmitter transporters: molecular function of important drug targets. Trends Pharmacol Sci 2006;27:375–83.
4. Deken SL, Beckman ML, Boos L, Quick MW. Transport rates of GABA transporters: regulation by the N-terminal domain and syntaxin 1A. Nat Neurosci 2000;3:998–1003.
5. Attwell D, Barbour B, Szatkowski M. Nonvesicular release of neurotransmitter. Neuron 1993;11:401–7.

6. Bonanno G, Raiteri L, Paluzzi S, Zappettini S, Usai C, Raiteri M. Co-existence of GABA and Glu transporters in the central nervous system. Curr Top Med Chem 2006;6:979–88.
7. Jursky F, Tamura S, Tamura A, Mandiyan S, Nelson H, Nelson N. Structure, function and brain localization of neurotransmitter transporters. J Exp Biol 1994;196:283–95.
8. Worrall DM, Williams DC. Sodium ion-dependent transporters for neurotransmitters: a review of recent developments. Biochem J 1994;297(3):425–36.
9. Cammack JN, Rakhilin SV, Schwartz EA. A GABA transporter operates asymmetrically and with variable stoichiometry. Neuron 1994;13:949–60.
10. Quick MW. Substrates regulate gamma-aminobutyric acid transporters in a syntaxin 1A-dependent manner. Proc Natl Acad Sci USA 2002;99:5686–91.
11. Law RM, Stafford A, Quick MW. Functional regulation of gamma-aminobutyric acid transporters by direct tyrosine phosphorylation. J Biol Chem 2000;275:23986–91.
12. Herbison AE, Augood SJ, Simonian SX, Chapman C. Regulation of GABA transporter activity and mRNA expression by estrogen in rat preoptic area. J Neurosci 1995;15:8302–9.
13. Beckman ML, Bernstein EM, Quick MW. Multiple G protein-coupled receptors initiate protein kinase C redistribution of GABA transporters in hippocampal neurons. J Neurosci 1999;19:RC9.
14. Borden LA. GABA transporter heterogeneity: pharmacology and cellular localization. Neurochem Int 1996;29:335–56.
15. Sarup A, Larsson OM, Schousboe A. GABA transporters and GABA-transaminase as drug targets. Curr Drug Targets CNS Neurol Disord 2003;2:269–77.
16. Cherubini E, Conti F. Generating diversity at GABAergic synapses. Trends Neurosci 2001;24:155–62.
17. Dalby NO. Inhibition of gamma-aminobutyric acid uptake: anatomy, physiology and effects against epileptic seizures. Eur J Pharmacol 2003;479:127–37.
18. Liu QR, Lopez-Corcuera B, Mandiyan S, Nelson H, Nelson N. Molecular characterization of four pharmacologically distinct gamma-aminobutyric acid transporters in mouse brain [corrected]. J Biol Chem 1993;268:2106–12.
19. Borden LA, Murali Dhar TG, Smith KE, Weinshank RL, Branchek TA, Gluchowski C. Tiagabine, SK&F 89976-A, CI-966, and NNC-711 are selective for the cloned GABA transporter GAT-1. Eur J Pharmacol 1994;269:219–24.
20. Yang XL. Characterization of receptors for glutamate and GABA in retinal neurons. Prog Neurobiol 2004;73:127–50.
21. McMahon MJ, Packer OS, Dacey DM. The classical receptive field surround of primate parasol ganglion cells is mediated primarily by a non-GABAergic pathway. J Neurosci 2004;24:3736–45.
22. Kamermans M, Fahrenfort I, Schultz K, Janssen-Bienhold U, Sjoerdsma T, Weiler R. Hemichannel-mediated inhibition in the outer retina. Science 2001;292:1178–80.
23. Verweij J, Hornstein EP, Schnapf JL. Surround antagonism in macaque cone photoreceptors. J Neurosci 2003;23:10249–57.
24. Cook PB, McReynolds JS. Lateral inhibition in the inner retina is important for spatial tuning of ganglion cells. Nat Neurosci 1998;1:714–9.
25. Flores-Herr N, Protti DA, Wassle H. Synaptic currents generating the inhibitory surround of ganglion cells in the mammalian retina. J Neurosci 2001;21:4852–63.
26. Dong CJ, Werblin FS. Temporal contrast enhancement via GABAC feedback at bipolar terminals in the tiger salamander retina. J Neurophysiol 1998;79:2171–80.
27. Caldwell JH, Daw NW, Wyatt HJ. Effects of picrotoxin and strychnine on rabbit retinal ganglion cells: lateral interactions for cells with more complex receptive fields. J Physiol 1978;276:277–98.

28. Lukasiewicz PD, Eggers ED, Sagdullaev BT, McCall MA. GABAC receptor-mediated inhibition in the retina. Vision Res 2004;44:3289–96.
29. Marc RE. Retinal neurotransmitters. In: Chalupa LM, Werner JS, eds. The visual neurosciences. Cambridge (MA): MIT Press; 2003. p. 304–19.
30. Slaughter MM. Inhibition in the retina. In: Chalupa LM, Werner JS, eds. The visual neurosciences. Cambridge (MA): MIT Press; 2003. p. 355–68.
31. Casini G, Rickman DW, Brecha NC. Expression of the gamma-aminobutyric acid (GABA) plasma membrane transporter-1 in monkey and human retina. Invest Ophthalmol Vis Sci 2006;47:1682–90.
32. Strettoi E, Masland RH. The number of unidentified amacrine cells in the mammalian retina. Proc Natl Acad Sci USA 1996;93:14906–11.
33. Casini G, Brecha NC. Colocalization of vasoactive intestinal polypeptide and GABA immunoreactivities in a population of wide-field amacrine cells in the rabbit retina. Vis Neurosci 1992;8:373–8.
34. Marc RE. Structural organization of GABAergic circuitry in ectotherm retinas. Prog Brain Res 1992;90:61–92.
35. Marc RE, Jones BW. Molecular phenotyping of retinal ganglion cells. J Neurosci 2002;22:413–27.
36. Yazulla S. GABAergic mechanisms in the retina. In: Osborne NN, Chader GJ, eds. Progress in Retinal Research. Oxford: Pergamon; 1986:1–52.
37. Biedermann B, Eberhardt W, Reichelt W. GABA uptake into isolated retinal Muller glial cells of the guinea-pig detected electrophysiologically. Neuroreport 1994;5(4):438–40.
38. Yu BC, Watt CB, Lam DM, Fry KR. GABAergic ganglion cells in the rabbit retina. Brain Res 1988;439:376–82.
39. Yazulla S, Brecha N. Binding and uptake of the GABA analogue, 3H-muscimol, in the retinas of goldfish and chicken. Invest Ophthalmol Vis Sci 1980;19:1415–26.
40. Pourcho RG, Goebel DJ. Neuronal subpopulations in cat retina which accumulate the GABA agonist, (3H)muscimol: a combined Golgi and autoradiographic study. J Comp Neurol 1983;219:25–35.
41. Massey SC, Blankenship K, Mills SL. Cholinergic amacrine cells in the rabbit retina accumulate muscimol. Vis Neurosci 1991;6:113–7.
42. Hendrickson A, Ryan M, Noble B, Wu JY. Colocalization of [3H]muscimol and antisera to GABA and glutamic acid decarboxylase within the same neurons in monkey retina. Brain Res 1985;348:391–6.
43. Chun MH, Wassle H, Brecha N. Colocalization of [3H]muscimol uptake and choline acetyltransferase immunoreactivity in amacrine cells of the cat retina. Neurosci Lett 1988;94:259–63.
44. Pow DV, Baldridge W, Crook DK. Activity-dependent transport of GABA analogues into specific cell types demonstrated at high resolution using a novel immunocytochemical strategy. Neuroscience 1996;73:1129–43.
45. Brandon C, Lam DM, Wu JY. The gamma-aminobutyric acid system in rabbit retina: localization by immunocytochemistry and autoradiography. Proc Natl Acad Sci USA 1979;76:3557–61.
46. Agardh E, Ehinger B. Retinal GABA neuron labelling with [3H]isoguvacine in different species. Exp Eye Res 1983;36:215–29.
47. Pourcho RG. Uptake of [3H]glycine and [3H]GABA by amacrine cells in the cat retina. Brain Res 1980;198:33–46.

48. Freed MA, Nakamura Y, Sterling P. Four types of amacrine in the cat retina that accumulate GABA. J Comp Neurol 1983;219:295–304.
49. Borden LA, Smith KE, Hartig PR, Branchek TA, Weinshank RL. Molecular heterogeneity of the gamma-aminobutyric acid (GABA) transport system. Cloning of two novel high affinity GABA transporters from rat brain. J Biol Chem 1992;267:21098–104.
50. Brecha NC, Weigmann C. Expression of GAT-1, a high-affinity gamma-aminobutyric acid plasma membrane transporter in the rat retina. J Comp Neurol 1994;345:602–11.
51. Howd AG, Rattray M, Butt AM. Expression of GABA transporter mRNAs in the developing and adult rat optic nerve. Neurosci Lett 1997;235:98–100.
52. Honda S, Yamamoto M, Saito N. Immunocytochemical localization of three subtypes of GABA transporter in rat retina. Brain Res Mol Brain Res 1995;33:319–25.
53. Johnson J, Chen TK, Rickman DW, Evans C, Brecha NC. Multiple gamma-Aminobutyric acid plasma membrane transporters (GAT-1, GAT-2, GAT-3) in the rat retina. J Comp Neurol 1996;375:212–24.
54. Biedermann B, Bringmann A, Reichenbach A. High-affinity GABA uptake in retinal glial (Muller) cells of the guinea pig: electrophysiological characterization, immunohistochemical localization, and modeling of efficiency. Glia 2002;39:217–28.
55. Sivakami S, Ganapathy V, Leibach FH, Miyamoto Y. The gamma-aminobutyric acid transporter and its interaction with taurine in the apical membrane of the bovine retinal pigment epithelium. Biochem J 1992;283(2):391–7.
56. Durkin MM, Smith KE, Borden LA, Weinshank RL, Branchek TA, Gustafson EL. Localization of messenger RNAs encoding three GABA transporters in rat brain: an in situ hybridization study. Brain Res Mol Brain Res 1995;33:7–21.
57. Hu M, Bruun A, Ehinger B. Expression of GABA transporter subtypes (GAT1, GAT3) in the adult rabbit retina. Acta Ophthalmol Scand 1999;77:255–60.
58. Guastella J, Nelson N, Nelson H, et al. Cloning and expression of a rat brain GABA transporter. Science 1990;249:1303–6.
59. Qian X, Malchow RP, O'Brien J, al-Ubaidi MR. Isolation and characterization of a skate retinal GABA transporter cDNA. Mol Vis 1998;4:6.
60. Do Nascimento JL, Ventura AL, Paes de Carvalho R. Veratridine- and glutamate-induced release of [3H]-GABA from cultured chick retina cells: possible involvement of a GAT-1-like subtype of GABA transporter. Brain Res 1998;798:217–22.
61. Wassle H, Boycott BB. Functional architecture of the mammalian retina. Physiol Rev 1991;71(2):447–80.
62. Yang CY, Brecha NC, Tsao E. Immunocytochemical localization of gamma-aminobutyric acid plasma membrane transporters in the tiger salamander retina. J Comp Neurol 1997;389:117–26.
63. Ekstrom P, Anzelius M. GABA and GABA-transporter (GAT-1) immunoreactivities in the retina of the salmon (Salmo salar L.). Brain Res 1998;812:179–85.
64. Fletcher EL, Clark MJ, Furness JB. Neuronal and glial localization of GABA transporter immunoreactivity in the myenteric plexus. Cell Tissue Res 2002;308:339–46.
65. Ruiz M, Egal H, Sarthy V, Qian X, Sarkar HK. Cloning, expression, and localization of a mouse retinal gamma-aminobutyric acid transporter. Invest Ophthalmol Vis Sci 1994;35:4039–48.
66. Sinclair JR, Nirenberg S. Characterization of neuropeptide Y-expressing cells in the mouse retina using immunohistochemical and transgenic techniques. J Comp Neurol 2001;432:296–306.
67. Dmitrieva NA, Lindstrom JM, Keyser KT. The relationship between GABA-containing cells and the cholinergic circuitry in the rabbit retina. Vis Neurosci 2001;18:93–100.

68. Jones EM. Na(+)- and Cl(−)-dependent neurotransmitter transporters in bovine retina: identification and localization by in situ hybridization histochemistry. Vis Neurosci 1995;12(6):1135–42.
69. Wilson JR, Cowey A, Somogy P. GABA immunopositive axons in the optic nerve and optic tract of macaque monkeys. Vision Res 1996;36(10):1357–63.
70. Koontz MA, Hendrickson LE, Brace ST, Hendrickson AE. Immunocytochemical localization of GABA and glycine in amacrine and displaced amacrine cells of macaque monkey retina. Vision Res 1993;33(18):2617–28.
71. Da Costa BL, Hokoc JN, Pinaud RR, Gattass R. GABAergic retinocollicular projection in the New World monkey *Cebus apella*. Neuroreport 1997;8(8):1797–802.
72. Crooks J, Kolb H. Localization of GABA, glycine, glutamate and tyrosine hydroxylase in the human retina. J Comp Neurol 1992;315(3):287–302.
73. Minelli A, Brecha NC, Karschin C, DeBiasi S, Conti F. GAT-1, a high-affinity GABA plasma membrane transporter, is localized to neurons and astroglia in the cerebral cortex. J Neurosci 1995;15:7734–46.
74. Ribak CE, Tong WM, Brecha NC. GABA plasma membrane transporters, GAT-1 and GAT-3, display different distributions in the rat hippocampus. J Comp Neurol 1996;367:595–606.
75. Hu M, Bruun A, Ehinger B. Expression of GABA transporter subtypes (GAT1, GAT3) in the developing rabbit retina. Acta Ophthalmol Scand 1999;77:261–5.
76. Calaza Kda C, de Mello MC, de Mello FG, Gardino PF. Local differences in GABA release induced by excitatory amino acids during retina development: selective activation of NMDA receptors by aspartate in the inner retina. Neurochem Res 2003;28:1475–85.
77. Klooster J, Nunes Cardozo B, Yazulla S, Kamermans M. Postsynaptic localization of gamma-aminobutyric acid transporters and receptors in the outer plexiform layer of the goldfish retina: An ultrastructural study. J Comp Neurol 2004;474:58–74.
78. Birnbaum AD, Rohde SK, Qian H, Al-Ubaidi MR, Caldwell JH, Malchow RP. Cloning, immunolocalization, and functional expression of a GABA transporter from the retina of the skate. Vis Neurosci 2005;22:211–23.
79. Zhao JW, Du JL, Li JS, Yang XL. Expression of GABA transporters on bullfrog retinal Muller cells. Glia 2000;31:104–17.
80. Itouji A, Sakai N, Tanaka C, Saito N. Neuronal and glial localization of two GABA transporters (GAT1 and GAT3) in the rat cerebellum. Brain Res Mol Brain Res 1996;37(1–2):309–16.
81. Palacin M, Estevez R, Bertran J, Zorzano A. Molecular biology of mammalian plasma membrane amino acid transporters. Physiol Rev 1998;78:969–1054.
82. Crook DK, Pow DV. Analysis of the distribution of glycine and GABA in amacrine cells of the developing rabbit retina: a comparison with the ontogeny of a functional GABA transport system in retinal neurons. Vis Neurosci 1997;14:751–63.
83. Yang CY. gamma-aminobutyric acid transporter-mediated current from bipolar cells in tiger salamander retinal slices. Vision Res 1998;38:2521–6.
84. Kamermans M, Werblin F. GABA-mediated positive autofeedback loop controls horizontal cell kinetics in tiger salamander retina. J Neurosci 1992;12:2451–63.
85. Malchow RP, Andersen KA. GABA transporter function in the horizontal cells of the skate. Prog Brain Res 2001;131:267–75.
86. Schwartz EA. Depolarization without calcium can release gamma-aminobutyric acid from a retinal neuron. Science 1987;238:350–55.
87. Takahashi K, Miyoshi S, Kaneko A, Copenhagen DR. Actions of nipecotic acid and SKF89976A on GABA transporter in cone-driven horizontal cells dissociated from the catfish retina. Jpn J Physiol 1995;45:457–73.

88. Yang XL, Gao F, Wu SM. Modulation of horizontal cell function by GABA(A) and GABA(C) receptors in dark- and light-adapted tiger salamander retina. Vis Neurosci 1999;16:967–79.
89. Verweij J, Kamermans M, Negishi K, Spekreijse H. GABA sensitivity of spectrally classified horizontal cells in goldfish retina. Vis Neurosci 1998;15:77–86.
90. Haugh-Scheidt L, Malchow RP, Ripps H. GABA transport and calcium dynamics in horizontal cells from the skate retina. J Physiol 1995;488(3):565–76.
91. Cammack JN, Schwartz EA. Ions required for the electrogenic transport of GABA by horizontal cells of the catfish retina. J Physiol 1993;472:81–102.
92. Dong CJ, Picaud SA, Werblin FS. GABA transporters and GABAC-like receptors on catfish cone- but not rod-driven horizontal cells. J Neurosci 1994;14:2648–58.
93. Nguyen-Legros J, Versaux-Botteri C, Savy C. Dopaminergic and GABAergic retinal cell populations in mammals. Microsc Res Tech 1997;36(1):26–42.
94. Isaacson JS, Solis JM, Nicoll RA. Local and diffuse synaptic actions of GABA in the hippocampus. Neuron 1993;10:165–75.
95. O'Malley DM, Sandell JH, Masland RH. Co-release of acetylcholine and GABA by the starburst amacrine cells. J Neurosci 1992;12:1394–408.
96. Hull C, Li GL, von Gersdorff H. GABA transporters regulate a standing GABAC receptor-mediated current at a retinal presynaptic terminal. J Neurosci 2006;26:6979–84.
97. Ichinose T, Lukasiewicz PD. GABA transporters regulate inhibition in the retina by limiting GABA(C) receptor activation. J Neurosci 2002;22:3285–92.
98. Hanitzsch R, Kuppers L, Flade A. The effect of GABA and the GABA-uptake-blocker NO-711 on the b-wave of the ERG and the responses of horizontal cells to light. Graefes Arch Clin Exp Ophthalmol 2004;242:784–91.
99. Sandell JH. GABA as a developmental signal in the inner retina and optic nerve. Perspect Dev Neurobiol 1998;5:269–78.
100. Firth SI, Wang CT, Feller MB. Retinal waves: mechanisms and function in visual system development. Cell Calcium 2005;37(5):425–32.
101. Catsicas M, Mobbs P. GABAb receptors regulate chick retinal calcium waves. J Neurosci 2001;21:897–910.
102. Sills GJ. Pre-clinical studies with the GABAergic compounds vigabatrin and tiagabine. Epileptic Disord 2003;5:51–6.

VII
Genetic Variants of Ocular Transporters
Implication in Drug Metabolism and Eye Diseases

17
Biochemical Defects Associated with Genetic Mutations in the Retina-Specific ABC Transporter, ABCR, and Macular Degenerative Diseases

Esther E. Biswas-Fiss

Contents

INTRODUCTION
MACULAR DEGENERATIONS ASSOCIATED WITH THE ABCA4 GENE
ABCR IS A MEMBER OF THE ATP BINDING CASSETTE (ABC)
  TRANSPORTER FAMILY
CELLULAR TRANSPORT FUNCTION IN ABCR
MOLECULAR GENETICS OF HUMAN ABCA4
PROTEIN DOMAINS OF ABCR AND THEIR FUNCTIONS
THE NUCLEOTIDE BINDING DOMAINS OF ABCR
ABCR-RELATED DEGENERATIVE MACULOPATHIES REPRESENT COMPLEX
  ETIOLOGIES: POTENTIAL DIRECTIONS FOR FUTURE RESEARCH
ACKNOWLEDGMENTS
REFERENCES

INTRODUCTION

The retina-specific ABC transporter, ABCR, has been implicated in a wide spectrum of inherited macular degenerations including Stargardt disease (STGD), fundus flavimaculatus (FFM) and autosomal recessive cone-rod dystrophy, as well as increased susceptibility to age-related macular degeneration (AMD). Mutations in the ABCA4 gene appear to lead to defects in the energy-dependent transport of all-trans retinal by ABCR, thereby leading to the accumulation of cytotoxic lipofuscin fluorophores in the retinal pigment epithelium, characteristic of these diseases. Initial studies of the nucleotide binding domains (NBDs) of ABCR have provided insight towards understanding the coupling of ATP binding and hydrolysis to transport. These studies have also raised important questions surrounding the role of each NBD, their composition, and the mode of NBD dimerization during the transport reaction cycle. Recent studies have begun to address these concerns. This review summarizes the biochemical and structural characterizations of vertebrate ABCR proteins, with an emphasis on their nucleotide binding domains, and offers current perspectives on the functional mechanism of the ABCA subgroup of transporters.

From: *Ophthalmology Research: Ocular Transporters in Ophthalmic Diseases and Drug Delivery*
Edited by: J. Tombran-Tink and C. J. Barnstable © Humana Press, Totowa, NJ

MACULAR DEGENERATIONS ASSOCIATED WITH THE ABCA4 GENE

Advances in molecular genetics have led to the identification of genes and genetic mutations that are unequivocally linked to various visual diseases *(1–11)* and visual physiology *(12–16)*. Human genetic studies have correlated mutated forms of the ABCA4 gene with several inherited visual diseases, including STGD *(17–21)*, FFM *(21–23)*, AMD *(17,24–26)*, autosomal recessive retinitis pigmentosa (arRP) *(20,27–29)* and cone-rod dystrophy (arCRD) *(30–33)*. Stargardt disease and its adult onset variant, FFM, are autosomal recessive disorders that affect approximately 1 in 10,000 persons, while AMD is the leading cause of blindness in persons over the age of 75. These diseases are characterized by a progressive loss of central vision and atrophy of the retinal pigment epithelium (RPE). Histopathological studies in patients with STGD, FFM, and arCRD show an accumulation of lipofuscin fluorophores within the RPE cells *(34–36)*. Evidence suggests this accumulation of lipofuscin is cytotoxic, leading to loss of the RPE and photoreceptors, which ends in visual impairment *(36–40)*. Clinically and histopathologically, these symptoms are similar to those observed with AMD *(34,41)*.

ABCR IS A MEMBER OF THE ATP BINDING CASSETTE (ABC) TRANSPORTER FAMILY

The ABCA4 gene encodes the retina-specific ABC transporter commonly referred to as ABCR protein *(42–44)*. The ABC transporters have been shown to be involved in the active transport of a wide variety of hydrophobic substances across extra- and intracellular membranes, including drugs *(44,45)*, lipids *(46,47)*, metabolites *(48)*, peptides *(48,49)* and steroids *(44)*. The transmembrane domains (TMDs) contain 6 to 12 membrane-spanning α-helices and are important in substrate specificity, with each ABC transporter having its own unique substrate. Typically, eukaryotic ABC proteins are comprised of two tandem sets of six transmembrane helices followed by a Walker type A and a type B nucleotide-binding motif (Fig. 1) *(50)*. The two repeated NBD domains are joined by a linker region *(44,45)*. In addition, they posses a signature C' sequence located upstream of the Walker B motif. These functional domains are typically found within a single polypeptide, however, examples of bipartite transporters are present, but these must dimerize to form a functional transporter *(44,45)*. Sequence homology and the availability of DNA sequence databases for a variety of organisms, including humans, have aided in the identification of many other members of the ABC superfamily. To date, 48 members of the human ABC family have been identified, which, based on sequence homology, have been divided into seven subfamilies, ABC A to G *(43,51)*. Sequence alignment with other ABC transporters places ABCR in the ABCA gene family. The protein is homologous to the bovine and *Xenopus* Rim proteins previously identified in the rims of the rod outer segment discs *(52,53)*.

Human genetic studies have demonstrated that the different classes (A to G) of ABC transporter genes correlate with a variety of inherited diseases, all of which are commonly characterized by defects in the transport of specific substances. Cases in point, in addition to ABCA4, include the ABCG gene associated with accumulation of dietary cholesterol in sitosterolemia *(54–63)*, the cystic fibrosis transmembrane conductance regulator gene (CFTR) *(64–71)* and the P-glycoprotein multidrug resistance gene, which

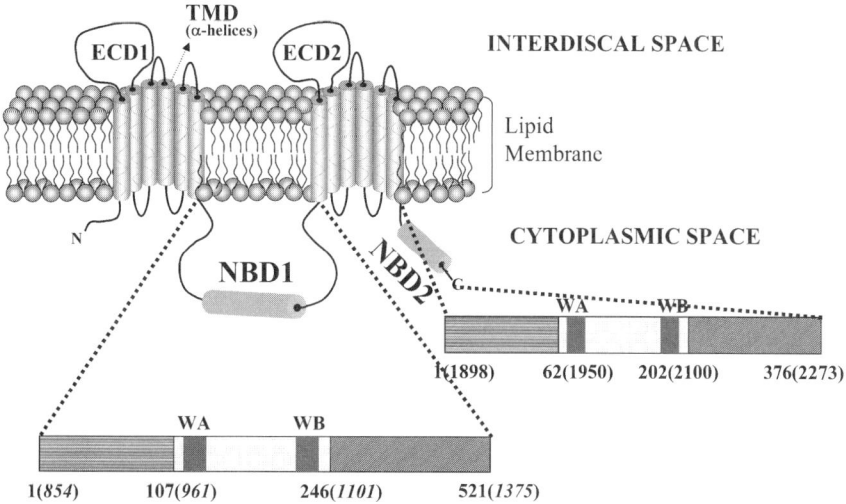

Fig. 1. Schematic representation of the ABCR protein showing important functional domains and regions of conservation. Schematic representation of the domains of ABCR depicting the transmembrane (TMD), extracellular (ECD) and nucleotide binding (NBD) domains. Also shown is a linear representation of the NBDs, where NBD1 corresponds to aa residues 854—1375 while NBD2 corresponds to residues 1898–2273 of the full-length protein. WA and WB refer to the Walker A and Walker B nucleotide binding motifs, respectively *(50)*. The numbers as they correspond to the individual, isolated NBD polypeptides are given in normal script, while those of ABCR are italicized.

can confer resistance to several chemotherapeutic agents upon cancer cells. Therefore, an understanding of the structure and function of these proteins has significant with ramifications in a number of physiological systems.

CELLULAR TRANSPORT FUNCTION IN ABCR

In vitro reconstitution studies carried out using purified bovine and recombinant human ABCR suggested that retinoids, most probably in the form of retinylidene-phosphatidylethanolamine, are the substrates of ABCR *(72)*. These findings were extended to ocular characterization of ABCR knockout mice that displayed delayed dark adaptation and increased levels of all-trans retinal and phosphatidylethanolamine in the outer segments following light exposure *(73–75)*. This has led to the hypothesis that ABCR acts as an outwardly directed flippase of the protonated Shiff's base complex of all-trans retinal and phosphatidylethanolamine (N-RPE) *(75)*. Accordingly, ABCR plays a pivotal role in the recycling of the 11-cis-retinal chromophore utilized by rhodopsin and cone pigments in the normal course of the visual transduction cycle (Fig. 2) *(14,15)*. A review of the visual cycle puts the pivotal role of ABCR in clear perspective (Fig. 2). Normally, following light capture by rhodopsin, 11-cis-retinal is converted to all-trans retinal, which leaves the disc membrane and enters the disc space, where it most likely combines with the phospholipid, phosphatidylethanolamine, forming the Shiff's base N-RPE. It is believed that ABCR transports N-RPE out of the disc space into the cytoplasm, where the Shiff's base is cleaved, allowing for further recycling of all-trans

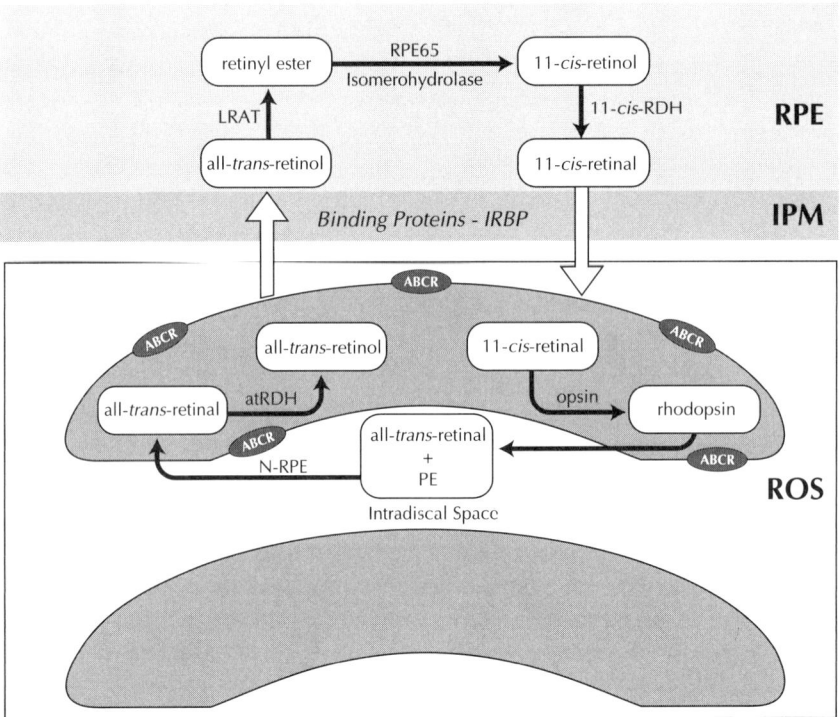

Fig. 2. Schematic diagram of the retinoid cycle. Shown are the various conversions of all-trans-retinol (Vitamin A) during the visual cycle. ABCR plays an essential role in this sequence by removal of all-trans retinal, most likely in the form of the complex lipid N-retinylidene- phosphatidylethanolamine (N-RPE) from the rod outer segment disks.

retinal in the visual cycle. In the absence of normal transport N-RPE is predicted to accumulate in the rod and cone disc spaces, where it can combine with another molecule of all-trans retinal to form N-retinylidene-N-retinyl-ethanolamine (A2E), which is a major component of lipofusin *(76,77)*. As a result, current pharmacological therapeutic strategies have been directed towards the development of drugs that impede the regeneration of 11 cis retinal by slowing the course of the visual cycle, such as Isotretinoin (Accutane®), which inhibits 11-cis-retinol dehydrogenase, and more recently small-molecule inhibitors of the RPE65 protein *(78,79)*.

Recent laboratory studies utilizing immobilized ABCR point towards N-retinylidene-phosphatidylethanolamine as the prefered substrate of this transporter *(80)*. Beyond this, the transport aspects of ABCR remain relatively uncharacterized. Currently, no in vitro or in vivo transport assay exists, nor have the specific domains of the ABCR protein that interact with the substrate been identified. From a biochemical perspective ABCR function has several components in addition to ATP binding and hydrolysis, which involve protein–protein and protein–ligand interactions, as well as the conformational changes which drive the retinal transport cycle. It is highly likely that the severity of ABCR dysfunction underlies the clinical presentation. Thus, at the present time it is difficult to assess those disease-associated mutations that interfere with substrate recognition and binding.

MOLECULAR GENETICS OF HUMAN ABCA4

The ABCA4 gene was identified in humans and localized to chromosomal position 1p22.1-p21 by fluorescence in situ hybridization *(19)*. The ABCA4 gene contains 50 exons and is estimated to span 150 kb. The exon sizes range from 33 to 266 base pairs. Initial studies have localized ABCR to the disc membrane of rod outer segments, but it may also be expressed in cones as well *(52,53,81,82)*.

Intensive studies in many labs and the establishment of the ABCR screening consortiums have aided in the identification of over 400 disease-associated mutations. Recently, genotyping microarray chips that can robustly identify more than 98% of the existing mutations have been developed for the ABCA4 gene *(83,84)*. Such technology provides an opportunity to generate comprehensive genotype-clinical phenotype profiles. Disease-associated ABCR alleles have shown extraordinary heterogeneity, and have been identified throughout the entire open reading frame. They include deletions and missense, as well as splice-site, mutations. Some have been shown to result in cellular mislocalization of the ABCR protein *(85)*. Many, but by no means all, of missense mutations are localized to the NBDs, suggesting that the underlying defect has a basis in nucleotide hydrolysis and/or aspects of energy transduction related to transport. What makes ABCR a more difficult diagnostic target is that the most frequent disease-associated ABCR alleles (G1961E, G863A, and A1038V) have been described in only around 10% of STGD patients in a distinct population, in contrast to the CFTR where the deletion F508 accounts for nearly 70% of all cases *(86)*. In order to fully understand the pathology of ABCR-related retinopathies, we need to understand the effect of a given mutation on the protein's structure and how this relates to the change in function.

PROTEIN DOMAINS OF ABCR AND THEIR FUNCTIONS

Membrane topology varies widely between different subclasses of ABC transporters *(87–92)*. In the ABCA family, each half transporter contains a TMD made up of six membrane-spanning units, which is followed by a cytoplasmic domain (Fig. 1). The greatest degree of commonality appears to be within the cytoplasmic regions that harbor the NBDs (in ABCR NBD1 corresponds to aa 854–1375 while NBD2 corresponds to aa 1898–2273). The NBDs of ABCR are described in detail in the following section. It is in the extracellular loops that each transporter seems to be distinctive, and based on the available literature the least amount of information is known about these. The long extracellular (EC) loops, or domains, are a unique feature of the topology of the ABCA family of transporters *(93)*. In the case of ABCR, several models of membrane topology have been proposed based on hydropathy profiles, and experimental data supports the notion that two large EC domains (ECDs) are present *(88)*. The EC loops project from TMD1 for ECD1 and from TMD7 for ECD2, and they represent significantly large polypeptide domains, with 603 residues for ECD1 (aa 43–646) and 285 residues for ECD2 (aa 1395–1680). The functions of these domains remain to be determined. ECD1 displays a unique and highly stable secondary structure, with a high beta sheet (β-barrel) content as determined by circular dichroism studies and supported by protein secondary structure prediction using sequence analysis (Biswas-Fiss, unpublished results). The high degree of sequence conservation observed in the ECDs of vertebrate ABCA4 proteins suggests an important physiological significance

(94). Two distinct possibilities are that these domains are (i) potential sites of protein–protein interaction, or (ii) that they may interact with the substrates, guiding them towards the outer leaflet of the outer segment membrane.

THE NUCLEOTIDE BINDING DOMAINS OF ABCR.

Clearly, ATP binding and hydrolysis play an important role in the function of ABCR, as well as other members of the ABC superfamily. To date, the most well-characterized biochemical aspect of ABCR remains its ATP hydrolysis. The ATPase activity of purified bovine or recombinant ABCR has been shown to be stimulated three- to fourfold by its putative retinoid substrates, a characteristic of many ABC transporters *(72,95)*. Several of the mutations associated with human retinal disease, are localized to the ABC (or nucleotide binding) domains *(20,30,32,33,96–107)*, which suggests that these mutations result in defective ABCR-mediated transport. In vitro analysis of the ATPase activity of disease-harboring mutants has demonstrated that several mutations can alter the enzymatic activity *(98,101,102,104,108–111)*. These alterations include increases, as well as decreases, in the rate of ATP hydrolysis, while others have no discernable effect *(104,108,112,113)*. Qualitative azido-ATP crosslinking and quantitative fluorescence analyses indicate that defects in ATP hydrolysis by mutant ABCR are in some cases due to defects in ATP binding *(98,104,108)*. Perhaps one of the most intriguing aspects of mutations in the ABCR gene is the wide range of clinical phenotypes. For example, the mutations P940R and R2038W are each localized to NBD1 and NBD2, respectively. The mutation P940R is associated with AMD, a late-onset, slowly progressing disorder. On the other hand, R2038W has been observed in individuals afflicted with STGD, an aggressive form of degeneration, striking individuals in early adulthood and rapidly leading to blindness. Yet, in each case, these mutations result in an around 50% diminution of ATPase activity *(104,108)*. Clearly, it is difficult to assess the full significance of ABCR ATPase activity data in the absence of a transport assay. To correlate disease-associated ABCR mutations with their role in macular degeneration, further studies on the structure, function, and mechanism of action of this protein need to be carried out.

The functional roles of the two NBDs of ABCR remain a subject of investigation. Using purified, recombinant NBDs, produced as individual functional domains, it has been demonstrated that the two nucleotide-binding domains of ABCR have distinct nucleotide specificity and activity *(112,113)*; NBD1 is a general ribonucleotidase, while NBD2 is specific for ATP hydrolysis. Quantitative analysis of the nucleotide binding and hydrolysis properties of these two domains has demonstrated that NBD1 binds nucleotides with higher affinity than NBD2, while NBD2 displays a greater rate of ATP hydrolysis. Functional asymmetry of the two NBDs has been described for other members of the ABC protein family *(114)*. It has been proposed that the two nucleotide binding domains of ABCR contribute differentially in the overall function of the protein *(102,115)*. Two conflicting models have been proposed concerning the roles of the NBDs of ABCR, as well as the coupling of ATP hydrolysis to transport. Based on studies involving coexpressed, and affinity-purified N- and C-terminal halves of ABCR, Ahn et al. have proposed that NBD2 is the principal site of ATP binding and hydrolysis, in the presence and absence of retinal substrate, and that NBD1 plays a noncatalytic role *(115)*. On the other hand, Sun et al. have proposed, on the basis of analysis of site-specific mutants, that NBD1 is responsible for basal-level ATPase, whereas NBD1 and NBD2 contribute to retinal-stimulated ATPase activity *(102)*.

Crystallographic studies of isolated NBDs or subunits show that nearly all NBDs share a conserved fold, underpinning the importance of protein confirmation in ATP hydrolysis and transport *(93,116)*. Spectroscopic analyses have shown that the NBDs of ABCR undergo conformational changes in response to nucleotide binding and subsequent hydrolysis *(108)*. In the absence of nucleotide binding, wild-type NBD2 protein maintains a taut conformation; upon ATP binding it adopts a relaxed conformation; while ADP binding leads to a taut conformation similar to that observed with the free protein *(108)*. Thus, hydrolysis of ATP to adenosine diphosphate (ADP) appears to lead to a significant conformational cycle of change of the protein. The observed conformational changes likely favor the ATP hydrolysis cycle, as the ATP-bound form is of lower free energy than the ADP-bound form. Proteins harboring disease-associated mutations, display a strikingly different cycle of conformational changes that correlates well with the observed changes in nucleotide hydrolysis, as well as disease severity associated with these mutations *(108)*.

An important aspect of ABC transporter studies has been aimed at understanding interdomain interaction as it relates to vectorial transport of the substrate. For members of the ABC family this could entail: (i) TMD–TMD interaction, affecting substrate binding and release, (ii) NBD–TMD interaction, coupling energy transduction to transport, and (iii) NBD–NBD interaction, allowing for communication between the two halves of the transporter. In the case of the ABCA protein family, the range of interactions could be extended to the ECDs as well. Studies carried out utilizing co-expressed half molecules of ABCR support a functional interaction between the two NBDs *(115)*. Using fluorescence anisotropy, a direct interaction of the two NBDs of ABCR has been demonstrated *(117)*. Quantitative analysis of the in vitro interaction as a function of nucleotide bound state, demonstrates that the interaction takes place in the absence of nucleotide as well as in the presence of ATP, and is attenuated in the ADP-bound state *(117)*. Analysis of the ATPase activities of the NBDs in the free and complex states indicated that the NBD1-NBD2 interaction significantly influences the ATPase activity. Further investigation, using site-specific mutants, showed that mutations in NBD2, but not NBD1 lead to alteration of the ATPase activity of the NBD1:NBD2 complex. These studies suggest that in ABCR, as in other ABC transporters, changes in the oligomeric state of the NBDs might represent a possible signal for the MSDs of ABCR to export the bound substrate. In addition, the data support a mechanistic model in which upon binding of NBD2, NBD1 binds nucleotide, but does not hydrolyze it or does so with significantly reduced rate (Fig. 3).

ABCR-RELATED DEGENERATIVE MACULOPATHIES REPRESENT COMPLEX ETIOLOGIES: POTENTIAL DIRECTIONS FOR FUTURE RESEARCH

Identification of over 400 disease-associated mutations in ABCR provides a unique road map to carry out detailed structure–function studies. Thus, in ABCR, there is no need to speculate which amino acids are critical to enzyme function, as they are already

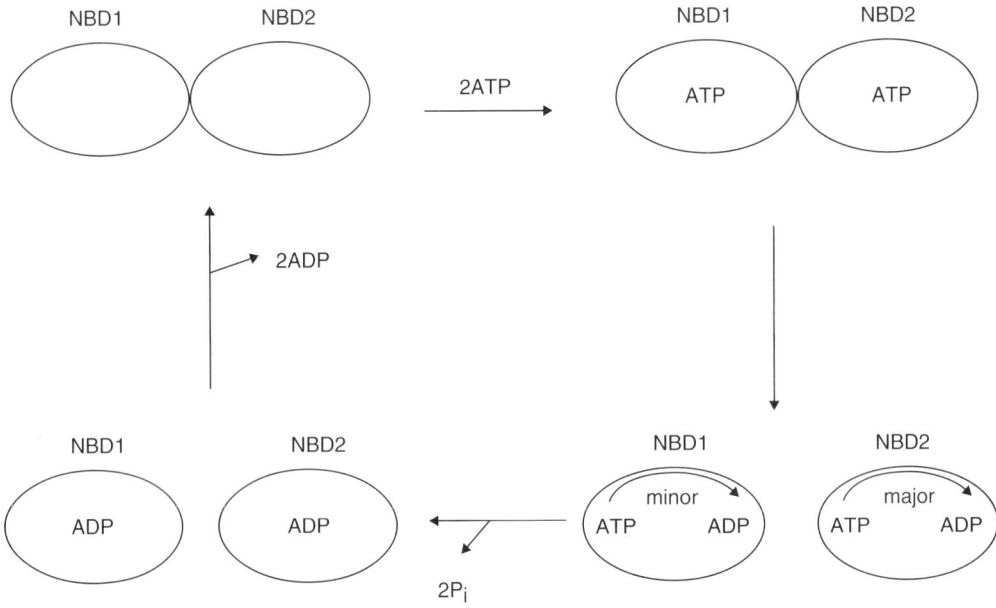

Fig. 3. Schematic model summarizing the relationship between nucleotide bound states and oligomerization of the nucleotide binding domains (NBDs) of ABCR. Interaction of the NBDs takes place in the absence of nucleotide, as well as in the presence of adenosine triphosphate (ATP), and only attenuates in the adenosine diphosphate (ADP)-bound state. Coupling between oligomeric state of the NBDs and ATP hydrolysis, support a mechanistic model in which upon binding of NBD2, NBD1 binds ATP, but does not hydrolyze it or does so with significantly reduced rate *(117)* P_i represents inorganic phosphate.

defined by the genetic mutations. Yet, these mutations represent a large spectrum of clinical phenotypes. The data clearly suggest that: (i) the function of the ABCR protein is critical in the pathogenesis of these diseases, and (ii) several of the observed mutations lead (or are likely to lead) to defects in the protein's mechanism of action. However, the mechanistic basis for alterations in function occurring as a result of these mutations remains poorly understood. Part of the challenge today is to understand the function of the ABCA4 gene product, including its network of functional interactions and how they are altered during the development of retinal degenerative disases such as STGD, FFM, arCRD, and AMD. Major issues remaining in unraveling the structure and function of ABCR include: (i) determing the function of the highly conserved ECD domains; (ii) identifying the nature of ABCR interactions with other proteins involved in the visual cycle, and perhaps most importantly (iii) the development of a transport assay. Progress in these areas will enable comprehensive biochemical characterization of the wide spectrum of ABCR4 mutations. From a clincal perspective, this in turn will enable the development of more-precise, perhaps ABCR-directed, threrapies and better-defined prognoses for patients harboring ABCA4 mutations. From a basic science perspective, such advancements are required to understand the nature of the transport process, including the mechanism of coupling of ATP hydrolysis to the transduction process, substrate binding, and conformational changes that take place during the transport cycle. These questions are significant not only as they relate to ABCR maculopathies, but to other members the ABCA protein family whose functions play a significant role in human health and disease.

ACKNOWLEDGMENTS

The author would like to thank Leelabati Biswas for her help with graphic design and S.B. Biswas for helpful discussions. This work was supported by a grant from the National Institutes of Health; National Eye Institute; EY 013113 to E. E. B.-F.

REFERENCES

1. Hinton DR, Spee C, He S, Weitz S, Usinger W, LaBree L, Oliver N, Lim JI. Accumulation of NH2-terminal fragment of connective tissue growth factor in the vitreous of patients with proliferative diabetic retinopathy. Diabetes Care 2004;27:758–764.
2. Katzmann DJ, Epping EA, Moye-Rowley WS. Mutational disruption of plasma membrane trafficking of Saccharomyces cerevisiae Yor1p, a homologue of mammalian multidrug resistance protein. Mol Cell Biol 1999;19:2998–3009.
3. Tucker CL, Pina AL, Loyer M, Dharmaraj S, Li Y, Maumenee IH, Hurley JB, Koenekoop RK. Functional analyses of mutant recessive GUCY2D alleles identified in Leber congenital amaurosis patients: protein domain comparisons and dominant negative effects. Mol Vis 2004;10:297–303.
4. Liu Q ZJ and EA Pierce. The retinitis pigmentosa 1 protein is a photoreceptor MAP. Neuroscience 2004;24:6427–6432.
5. Boissy RE, Krakowsky JM, Lamoreux ML, Lingrel JB, Nordlund JJ. Ocular pathology in mice with a transgenic insertion at the microphthalmia locus. J Submicrosc Cytol Pathol 1993;25:319–332.
6. Spooner PJ, Goodall SC, Bovee-Geurts PH, Verhoeven MA, Lugtenburg J, Pistorius AM, Degrip WJ, Watts A. The ring of the rhodopsin chromophore in a hydrophobic activation switch within the binding pocket. J Mol Biol 2004;345:719–730.
7. Rosenbaum EE, Hardie RC, Colley NJ. Calnexin is essential for rhodopsin maturation, Ca2+ regulation, and photoreceptor cell survival. Neuron 2006;49:229–241.
8. LaLonde MM, Janssens H, Rosenbaum E, Choi SY, Gergen JP, Colley NJ, Stark WS, Frohman MA. Regulation of phototransduction responsiveness and retinal degeneration by a phospholipase D-generated signaling lipid. J Cell Biol 2005;169:471–479.
9. Butler JM, Guthrie SM, Koc M, Afzal A, Caballero S, Brooks HL, Mames RN, Segal MS, Grant MB, Scott EW. SDF-1 is both necessary and sufficient to promote proliferative retinopathy. J Clin Invest 2005;115:86–93.
10. Stricker HM, Ding XQ, Quiambao A, Fliesler SJ, Naash MI. The Cys214-Ser mutation in peripherin/rds causes a loss-of-function phenotype in transgenic mice. Biochem J 2005;388:605–613.
11. Naash MI, Wu TH, Chakraborty D, Fliesler SJ, Ding XQ, Nour M, Peachey NS, Lem J, Qtaishat N, Al-Ubaidi MR et al. Retinal abnormalities associated with the G90D mutation in opsin. J Comp Neurol 2004;478:149–163.
12. Paskowitz DM, Lavail MM, Duncan JL. Light and inherited retinal degeneration. Br J Ophthalmol 2006;90:1060–1066.

13. El-Sherbeny A, Naggar H, Miyauchi S, Ola MS, Maddox DM, Martin PM, Ganapathy V, Smith SB. Osmoregulation of taurine transporter function and expression in retinal pigment epithelial, ganglion, and muller cells. Invest Ophthalmol Vis Sci 2004;45:694–701.
14. Maeda A, Maeda T, Imanishi Y, Kuksa V, Alekseev A, Bronson JD, Zhang H, Zhu L, Sun W, Saperstein DA et al. Role of photoreceptor-specific retinol dehydrogenase in the retinoid cycle in vivo. J Biol Chem 2005;280:18822–18832.
15. Baehr W, Wu SM, Bird AC, Palczewski K. The retinoid cycle and retina disease. Vision Res 2003;43:2957–2958.
16. Pappu KS, Chen R, Middlebrooks BW, Woo C, Heberlein U, Mardon G. Mechanism of hedgehog signaling during Drosophila eye development. Development 2003;130:3053–3062.
17. Allikmets R, Shroyer NF, Singh N, Seddon JM, Lewis RA, Bernstein PS, Peiffer A, Zabriskie NA, Li Y, Hutchinson A et al. Mutation of the Stargardt disease gene (ABCR) in age-related macular degeneration. Science 1997;277:1805–1807.
18. Kaplan J, Gerber S, Larget-Piet D, Rozet J-M, Dollfus HH, Dufier JL, Odent S, Postel-Vinary A, Janin N, Briard ML, Frezal J, Munnich A. A gene for Stargardt's disease (fundus flavimaculatus) maps to the short arm of chromosome 1. Nature Gen 1993;5:308–311.
19. Nasonkin I, Illing M, Koehler MR, Schmid M, Molday RS, Weber BH. Mapping of the rod photoreceptor ABC transporter (ABCR) to 1p21-p22.1 and identification of novel mutations in Stargardt's disease. Hum Genet 1998;102:21–26.
20. Rozet JM, Gerber S, Ghazi I, Perrault I, Ducroq D, Souied E, Cabot A, Dufier JL, Munnich A, Kaplan J. Mutations of the retinal specific ATP binding transporter gene (ABCR) in a single family segregating both autosomal recessive retinitis pigmentosa RP19 and Stargardt disease: evidence of clinical heterogeneity at this locus. J Med Genet 1999;36:447–451.
21. Lewis RA, Shroyer NF, Singh N, Allikmets R, Hutchinson A, Li Y, Lupski JR, Leppert M, Dean M. Genotype/Phenotype analysis of a photoreceptor-specific ATP-binding cassette transporter gene, ABCR, in Stargardt disease. Am J Hum Genet 1999;64:422–434.
22. Stone EM, Webster AR, Vandenburgh K, Streb LM, Hockey RR, Lotery AJ, Sheffield VC. Allelic variation in ABCR associated with Stargardt disease but not age-related macular degeneration. Nat Genet 1998;20:328–329.
23. Rozet JM, Gerber S, Souied E, Perrault I, Chatelin S, Ghazi I, Leowski C, Dufier JL, Munnich A, Kaplan J. Spectrum of ABCR gene mutations in autosomal recessive macular dystrophies. Eur J Hum Genet 1998;6:291–295.
24. Allikmets R. Further evidence for an association of ABCR alleles with age-related macular degeneration. The International ABCR Screening Consortium. Am J Hum Genet 2000;67:487–491.
25. Kuroiwa S, Kojima H, Kikuchi T, Yoshimura N. ATP binding cassette transporter retina genotypes and age related macular degeneration: an analysis on exudative non-familial Japanese patients. Br J Ophthalmol 1999;83:613–615.
26. Yates JR and Moore AT. Genetic susceptibility to age related macular degeneration. J Med Genet 2000;37:83–87.
27. Shroyer NF, Lewis RA, Yatsenko AN, Lupski JR. Null missense ABCR (ABCA4) mutations in a family with stargardt disease and retinitis pigmentosa. Invest Ophthalmol Vis Sci 2001;42:2757–2761.
28. Cremers FP, van de Pol DJ, van Driel M, den Hollander AI, van Haren FJ, Knoers NV, Tijmes N, Bergen AA, Rohrschneider K, Blankenagel A. Autosomal recessive retinitis pig-

mentosa and cone-rod dystrophy caused by splice site mutations in the Stargardt's disease gene ABCR. Hum Mol Genet 1998;7:355–362.
29. Martinez-Mir A, Paloma E, Allikmets R, Ayuso C, del Rio T, Dean M, Vilageliu L, Gonzalez-Duarte R, Balcells S. Retinitis pigmentosa caused by a homozygous mutation in the Stargardt disease gene ABCR. Nat Genet 1998;18:11–12.
30. Klevering BJ, Blankenagel A, Maugeri A, Cremers FP, Hoyng CB, Rohrschneider K. Phenotypic spectrum of autosomal recessive cone-rod dystrophies caused by mutations in the ABCA4 (ABCR) gene. Invest Ophthalmol Vis Sci 2002;43:1980–1985.
31. Birch DG, Peters AY, Locke KL, Spencer R, Megarity CF, Travis GH. Visual function in patients with cone-rod dystrophy (CRD) associated with mutations in the ABCA4(ABCR) gene. Exp Eye Res 2001;73:877–886.
32. Briggs CE, Rucinski D, Rosenfeld PJ, Hirose T, Berson EL, Dryja TP. Mutations in ABCR (ABCA4) in patients with Stargardt macular degeneration or cone-rod degeneration. Invest Ophthalmol Vis Sci 2001;42:2229–2236.
33. Maugeri A, Klevering BJ, Rohrschneider K, Blankenagel A, Brunner HG, Deutman AF, Hoyng CB, Cremers FP. Mutations in the ABCA4 (ABCR) gene are the major cause of autosomal recessive cone-rod dystrophy. Am J Hum Genet 2000;67:960–966.
34. Delori FC, Fleckner MR, Goger DG, Weiter JJ, Dorey CK. Autofluorescence distribution associated with drusen in age-related macular degeneration. Invest Ophthalmol Vis Sci 2000;41:496–504.
35. Delori FC, Staurenghi G, Arend O, Dorey CK, Goger DG, Weiter JJ. In vivo measurement of lipofuscin in Stargardt's disease–Fundus flavimaculatus. Invest Ophthalmol Vis Sci 1995;36:2327–2331.
36. Dorey CK, Wu G, Ebenstein D, Garsd A, Weiter JJ. Cell loss in the aging retina. Relationship to lipofuscin accumulation and macular degeneration. Invest Ophthalmol Vis Sci 1989;30:1691–1699.
37. Sparrow JR, Fishkin N, Zhou J, Cai B, Jang YP, Krane S, Itagaki Y, Nakanishi K. A2E, a byproduct of the visual cycle. Vision Res 2003;43:2983–2990.
38. Wolf G. Lipofuscin and macular degeneration. Nutr Rev 2003;61:342–346.
39. Fishkin NE, Sparrow JR, Allikmets R, Nakanishi K. Isolation and characterization of a retinal pigment epithelial cell fluorophore: an all-trans-retinal dimer conjugate. Proc Natl Acad Sci USA 2005;102:7091–7096.
40. Karan G, Lillo C, Yang Z, Cameron DJ, Locke KG, Zhao Y, Thirumalaichary S, Li C, Birch DG, Vollmer-Snarr HR et al. Lipofuscin accumulation, abnormal electrophysiology, and photoreceptor degeneration in mutant ELOVL4 transgenic mice: a model for macular degeneration. Proc Natl Acad Sci USA 2005;102:4164–4169.
41. Katz ML. Potential role of retinal pigment epithelial lipofuscin accumulation in age-related macular degeneration. Arch Gerontol Geriatr 2002;34:359–370.
42. Dean M, Rzhetsky A, Allikmets R. The human ATP-binding cassette (ABC) transporter superfamily. Genome Res 2001;11:1156–1166.
43. Dean M, Hamon Y, Chimini G. The human ATP-binding cassette (ABC) transporter superfamily. J Lipid Res 2001;42:1007–1017.
44. Higgins CF. ABC transporters: physiology, structure and mechanism–an overview. Res Microbiol 2001;152:205–210.
45. Higgins CF. ABC transporters: from microorganisms to man. Annu Rev Cell Biol 1992;8:67–113.
46. Hettema EH, van Roermund CW, Distel B, van den Berg M, Vilela C, Rodrigues-Pousada C, Wanders RJ, Tabak HF. The ABC transporter proteins Pat1 and Pat2 are required for

import of long-chain fatty acids into peroxisomes of Saccharomyces cerevisiae. Embo J 1996;15:3813–3822.

47. Ewart GD, Cannell D, Cox GB, Howells AJ. Mutational analysis of the traffic ATPase (ABC) transporters involved in uptake of eye pigment precursors in Drosophila melanogaster. Implications for structure-function relationships. J Biol Chem 1994;269:10370–10377.

48. Berkower C and Michaelis S. Mutational analysis of the yeast a-factor transporter STE6, a member of the ATP binding cassette (ABC) protein superfamily. Embo J 1991;10:3777–3785.

49. Ellis EA, Guberski DL, Hutson B, Grant MB. Time course of NADH oxidase, inducible nitric oxide synthase and peroxynitrite in diabetic retinopathy in the BBZ/WOR rat. Nitric Oxide 2002;6:295–304.

50. Walker JE, Saraste M, Runswick MJ, Gray N. Distantly related sequences in the alpha and beta subunits of ATP synthase, myosin, kinases, and other ATP requiring enzymes and a common nucleotide bindng fold. EMBO J 1982;1:945–951.

51. Higgins CF and Linton KJ. Structural biology. The xyz of ABC transporters. Science 2001;293:1782–1784.

52. Illing M, Molday LL, Molday RS. The 220-kDa rim protein of retinal rod outer segments is a member of the ABC transporter superfamily. J Biol Chem 1997;272:10303–10310.

53. Papermaster DS, Schneider BG, Zorn MA, Kraehenbuhl JP. Immunocytochemical localization of a large intrinsic membrane protein to the incisures and margins of frog rod outer segment disks. J Cell Biol 1978;78:415–425.

54. Albrecht C, Elliott JI, Sardini A, Litman T, Stieger B, Meier PJ, Higgins CF. Functional analysis of candidate ABC transporter proteins for sitosterol transport. Biochim Biophys Acta 2002;1567:133–142.

55. Berge KE, Tian H, Graf GA, Yu L, Grishin NV, Schultz J, Kwiterovich P, Shan B, Barnes R, Hobbs HH. Accumulation of dietary cholesterol in sitosterolemia caused by mutations in adjacent ABC transporters. Science 2000;290:1771–1775.

56. Annilo T, Tammur J, Hutchinson A, Rzhetsky A, Dean M, Allikmets R. Human and mouse orthologs of a new ATP-binding cassette gene, ABCG4. Cytogenet Cell Genet 2001;94:196–201.

57. Bodzioch M, Orso E, Klucken J, Langmann T, Bottcher A, Diederich W, Drobnik W, Barlage S, Buchler C, Porsch-Ozcurumez M et al. The gene encoding ATP-binding cassette transporter 1 is mutated in Tangier disease. Nat Genet 1999;22:347–351.

58. Berge KE, von Bergmann K, Lutjohann D, Guerra R, Grundy SM, Hobbs HH, Cohen JC. Heritability of plasma noncholesterol sterols and relationship to DNA sequence polymorphism in ABCG5 and ABCG8. J Lipid Res 2002;43:486–494.

59. Christiansen-Weber TA, Voland JR, Wu Y, Ngo K, Roland BL, Nguyen S, Peterson PA, Fung-Leung WP. Functional loss of ABCA1 in mice causes severe placental malformation, aberrant lipid distribution, and kidney glomerulonephritis as well as high-density lipoprotein cholesterol deficiency. Am J Pathol 2000;157:1017–1029.

60. Hong SH, Rhyne J, Zeller K, Miller M. ABCA1(Alabama): a novel variant associated with HDL deficiency and premature coronary artery disease. Atherosclerosis 2002;164:245–250.

61. Langmann T, Klucken J, Reil M, Liebisch G, Luciani MF, Chimini G, Kaminski WE, Schmitz G. Molecular cloning of the human ATP-binding cassette transporter 1 (hABC1): evidence for sterol-dependent regulation in macrophages. Biochem Biophys Res Commun 1999;257:29–33.

62. Oram JF and Lawn RM. ABCA1. The gatekeeper for eliminating excess tissue cholesterol. J Lipid Res 2001;42:1173–1179.
63. Rust S, Rosier M, Funke H, Real J, Amoura Z, Piette JC, Deleuze JF, Brewer HB, Duverger N, Denefle P et al. Tangier disease is caused by mutations in the gene encoding ATP-binding cassette transporter 1. Nat Genet 1999;22:352–355.
64. Cheng SH, Gregory RJ, Marshall J, Paul S, Suza DW, White GA, Riordan CR, Smith AE. Defective intracellular transport and processing of CFTR is the molecular basis of most cystic fibrosis. Cell 1990;63:827–834.
65. Massiah MA, Ko Y-H, Pederson PL, Mildvan AS. Cystic fibrosis transmembrane conductance regulator: solution structures of peptides based on the Phe508 region, the most common site of disease-causing deltaF508 mutation. 1999.
66. Riordan JR. Chloride impermeability in cystic fibrosis. Ann Rev Physiol 1993;55:609–630.
67. Wioland MA, Fleury-Feith J, Corlieu P, Commo F, Monceaux G, Lacau-St-Guily J, Bernaudin JF. CFTR, MDR1, and MRP1 immunolocalization in normal human nasal respiratory mucosa. J Histochem Cytochem 2000;48:1215–1222.
68. Sheppard DN and Welsh MJ. Structure and function of the CFTR chloride channel. Physiol Rev 1999;79:S23–45.
69. Schwiebert EM, Benos DJ, Egan ME, Stutts MJ, Guggino WB. CFTR is a conductance regulator as well as a chloride channel. Physiol Rev 1999;79:S145–166.
70. Ko YH and Pedersen PL. Cystic fibrosis: a brief look at some highlights of a decade of research focused on elucidating and correcting the molecular basis of the disease. J Bioenerg Biomembr 2001;33:513–521.
71. Klein I, Sarkadi B, Varadi A. An inventory of the human ABC proteins. Biochim Biophys Acta 1999;1461:237–262.
72. Sun H, Molday RS, Nathans J. Retinal stimulates ATP hydrolysis by purified and reconstituted ABCR, the photoreceptor-specific ATP-binding cassette transporter responsible for Stargardt disease. J Biol Chem 1999;274:8269–8281.
73. Flannery JG. Transgenic animal models for the study of inherited retinal dystrophies. Ilar J 1999;40:51–58.
74. Mata NL, Tzekov RT, Liu X, Weng J, Birch DG, Travis GH. Delayed dark-adaptation and lipofuscin accumulation in abcr+/− mice: implications for involvement of ABCR in age-related macular degeneration. Invest Ophthalmol Vis Sci 2001;42:1685–1690.
75. Weng J, Mata NL, Azarian SM, Tzekov RT, Birch DG, Travis GH. Insights into the function of Rim protein in photoreceptors and etiology of Stargardt's disease from the phenotype in abcr knockout mice. Cell 1999;98:13–23.
76. Eldred GE and Lasky MR. Retinal age pigments generated by self-assembling lysosomotropic detergents. Nature 1993;361:724–726.
77. Jang YP, Matsuda H, Itagaki Y, Nakanishi K, Sparrow JR. Characterization of peroxy-A2E and furan-A2E photooxidation products and detection in human and mouse retinal pigment epithelial cell lipofuscin. J Biol Chem 2005;280:39732–39739.
78. Maiti P, Kong J, Kim SR, Sparrow JR, Allikmets R, Rando RR. Small molecule RPE65 antagonists limit the visual cycle and prevent lipofuscin formation. Biochemistry 2006;45: 852–860.
79. Radu RA, Mata NL, Nusinowitz S, Liu X, Travis GH. Isotretinoin treatment inhibits lipofuscin accumulation in a mouse model of recessive Stargardt's macular degeneration. Novartis Found Symp 2004;255:51–63. Discussion 63–57,177–178.
80. Beharry S, Zhong M, Molday RS. N-retinylidene-phosphatidylethanolamine is the preferred retinoid substrate for the photoreceptor-specific ABC transporter ABCA4 (ABCR). J Biol Chem 2004;279:53972–53979.

81. Molday LL, Rabin AR, Molday RS. ABCR expression in foveal cone photoreceptors and its role in Stargardt macular dystrophy. Am J Ophthalmol 2000;130:689.
82. Sun H and Nathans J. Stargardt's ABCR is localized to the disc membrane of retinal rod outer segments. Nat Genet 1997;17:15–16.
83. Klevering BJ, Yzer S, Rohrschneider K, Zonneveld M, Allikmets R, van den Born LI, Maugeri A, Hoyng CB, Cremers FP. Microarray-based mutation analysis of the ABCA4 (ABCR) gene in autosomal recessive cone-rod dystrophy and retinitis pigmentosa. Eur J Hum Genet 2004;12:1024–1032.
84. Jaakson K, Zernant J, Kulm M, Hutchinson A, Tonisson N, Glavac D, Ravnik-Glavac M, Hawlina M, Meltzer MR, Caruso RC et al. Genotyping microarray (gene chip) for the ABCR (ABCA4) gene. Hum Mutat 2003;22:395–403.
85. Wiszniewski W, Zaremba CM, Yatsenko AN, Jamrich M, Wensel TG, Lewis RA, Lupski JR. ABCA4 mutations causing mislocalization are found frequently in patients with severe retinal dystrophies. Hum Mol Genet 2005;14:2769–2778.
86. Zielenski J and Tsui LC. Cystic fibrosis: genotypic and phenotypic variations. Annu Rev Genet 1995;29:777–807.
87. Bibi E and Beja O. Membrane topology of multidrug resistance protein expressed in Escherichia coli. N-terminal domain. J Biol Chem 1994;269:19910–19915.
88. Bungert S, Molday LL, Molday RS. Membrane topology of the ATP binding cassette transporter ABCR and its relationship to ABC1 and related ABCA transporters: identification of N-linked glycosylation sites. J Biol Chem 2001;276:23539–23546.
89. Pigeon RP and Silver RP. Topological and mutational analysis of KpsM, the hydrophobic component of the ABC-transporter involved in the export of polysialic acid in *Escherichia coli* K1. Mol Microbiol 1994;14:871–881.
90. Wyborn NR, Alderson J, Andrews SC, Kelly DJ. Topological analysis of DctQ, the small integral membrane protein of the C4-dicarboxylate TRAP transporter of Rhodobacter capsulatus. FEMS Microbiol Lett 2001;194:13–17.
91. Jones PM and George AM. Symmetry and structure in P-glycoprotein and ABC transporters what goes around comes around. Eur J Biochem 2000;267:5298–5305.
92. Chen M and Zhang JT. Membrane insertion, processing, and topology of cystic fibrosis transmembrane conductance regulator (CFTR) in microsomal membranes. Mol Membr Biol 1996;13:33–40.
93. Peelman F, Labeur C, Vanloo B, Roosbeek S, Devaud C, Duverger N, Denefle P, Rosier M, Vandekerckhove J, Rosseneu M. Characterization of the ABCA transporter subfamily: identification of prokaryotic and eukaryotic members, phylogeny and topology. J Mol Biol 2003;325:259–274.
94. Yatsenko AN, Wiszniewski W, Zaremba CM, Jamrich M, Lupski JR. Evolution of ABCA4 proteins in vertebrates. J Mol Evol 2005;60:72–80.
95. Ahn J, Wong JT, Molday RS. The effect of lipid environment and retinoids on the ATPase activity of ABCR, the photoreceptor ABC transporter responsible for Stargardt macular dystrophy. J Biol Chem 2000;275:20399–20405.
96. Cremers FP, Maugeri A, Klevering BJ, Hoefsloot LH, Hoyng CB. [From gene to disease: from the ABCA4 gene to Stargardt disease, cone-rod dystrophy and retinitis pigmentosa]. Ned Tijdschr Geneeskd 2002;146:1581–1584.
97. Bernstein PS, Leppert M, Singh N, Dean M, Lewis RA, Lupski JR, Allikmets R, Seddon JM. Genotype-phenotype analysis of ABCR variants in macular degeneration probands and siblings. Invest Ophthalmol Vis Sci 2002;43:466–473.

98. Sun H and Nathans J. Mechanistic studies of ABCR, the ABC transporter in photoreceptor outer segments responsible for autosomal recessive Stargardt disease. J Bioenerg Biomembr 2001;33:523–530.
99. Musarella MA. Molecular genetics of macular degeneration. Doc Ophthalmol 2001;102: 165–177.
100. Paloma E, Martinez-Mir A, Vilageliu L, Gonzalez-Duarte R, Balcells S. Spectrum of ABCA4 (ABCR) gene mutations in Spanish patients with autosomal recessive macular dystrophies. Hum Mutat 2001;17:504–510.
101. Sun H and Nathans J. ABCR, the ATP-binding cassette transporter responsible for Stargardt macular dystrophy, is an efficient target of all-trans-retinal-mediated photooxidative damage in vitro. Implications for retinal disease. J Biol Chem 2001;276:11766–11774.
102. Sun H, Smallwood PM, Nathans J. Biochemical defects in ABCR protein variants associated with human retinopathies. Nat Genet 2000;26:242–246.
103. Allikmets R. Simple and complex ABCR: genetic predisposition to retinal disease. Am J Hum Genet 2000;67:793–799.
104. Suarez T, Biswas SB, Biswas EE. Biochemical defects in retina-specific human ATP binding cassette transporter nucleotide binding domain 1 mutants associated with macular degeneration. J Biol Chem 2002;277:21759–21767.
105. Aldred MA. The ABCR of visual impairment. Mol Med Today 2000;6:417.
106. Simonelli F, Testa F, de Crecchio G, Rinaldi E, Hutchinson A, Atkinson A, Dean M, D'Urso M, Allikmets R. New ABCR mutations and clinical phenotype in Italian patients with Stargardt disease. Invest Ophthalmol Vis Sci 2000;41:892–897.
107. Souied EH, Ducroq D, Rozet JM, Gerber S, Perrault I, Munnich A, Coscas G, Soubrane G, Kaplan J. ABCR gene analysis in familial exudative age-related macular degeneration. Invest Ophthalmol Vis Sci 2000;41:244–247.
108. Biswas-Fiss EE. Functional analysis of genetic mutations in nucleotide binding domain 2 of the human retina specific ABC transporter. Biochemistry 2003;42:10683–10696.
109. Shroyer NF, Lewis RA, Yatsenko AN, Wensel TG, Lupski JR. Cosegregation and functional analysis of mutant ABCR (ABCA4) alleles in families that manifest both Stargardt disease and age-related macular degeneration. Hum Mol Genet 2001;10:2671–2678.
110. Zhang K, Garibaldi DC, Kniazeva M, Albini T, Chiang MF, Kerrigan M, Sunness JS, Han M, Allikmets R. A novel mutation in the ABCR gene in four patients with autosomal recessive Stargardt disease. Am J Ophthalmol 1999;128:720–724.
111. Allikmets R, Singh N, Sun H, Shroyer NF, Hutchinson A, Chidambaram A, Gerrard B, Baird L, Stauffer D, Peiffer A et al. A photoreceptor cell-specific ATP-binding transporter gene (ABCR) is mutated in recessive Stargardt macular dystrophy. Nat Genet 1997;15:236–246.
112. Biswas EE and Biswas SB. The C-terminal nucleotide binding domain of the human retinal ABCR protein is an adenosine triphosphatase. Biochemistry 2000;39:15879–15886.
113. Biswas EE. Nucleotide binding domain 1 of the human retinal ABC transporter functions as a general ribonucleotidase. Biochemistry 2001;40:8181–8187.
114. Proff C and Kolling R. Functional asymmetry of the two nucleotide binding domains in the ABC transporter Ste6. Mol Gen Genet 2001;264:883–893.
115. Ahn J, Beharry S, Molday LL, Molday RS. Functional interaction between the two halves of the photoreceptor-specific ATP binding assette protein ABCR (ABCA4). J Biol Chem 2003;278:39600–39608.

116. Lu G, Westbrooks JM, Davidson AL, Chen J. ATP hydrolysis is required to reset the ATP-binding cassette dimer into the resting-state conformation. Proc Natl Acad Sci USA 2005;102:17969–17974.
117. Biswas-Fiss EE. Interaction of the nucleotide binding domains and regulation of the ATPase activity of the human retina specific ABC transporter, ABCR. Biochemistry 2006;45:3813–3823.

18
Glutamate Transporters and Retinal Disease and Regulation

Nigel L. Barnett and Natalie D. Bull

CONTENTS

GLUTAMATE TRANSPORTERS
GLUTAMATE TRANSPORTERS AND RETINAL DISEASE
REGULATION OF GLUTAMATE TRANSPORT
ACKNOWLEDGMENTS
REFERENCES

GLUTAMATE TRANSPORTERS

Within the retina, extracellular glutamate concentrations must be strictly regulated to facilitate neurotransmission *(1)*. The outer retinal neurons respond to light stimulation with graded potentials, rather than action potentials; thus the concentration of glutamate in the synaptic cleft encodes the light signal. Furthermore, regulation of extracellular glutamate concentrations within the central nervous system is extremely important as over-stimulation of glutamate receptors causes neurotoxicity *(2)*. High-affinity glutamate transporters (excitatory amino acid transporters, EAATs) located on neuronal and glial cell membranes are responsible for cellular glutamate sequestration and the balancing of extracellular concentration between the level required for physiological neurotransmission and the level that would lead to pathological excitotoxicity *(3)*. The glial and neuronal high-affinity glutamate uptake systems allow the rapid removal of glutamate from the extracellular space, thereby terminating neurotransmission and reducing the possibility of excitotoxic neuronal damage. In addition to the membrane-bound EAATs, intracellular vesicular glutamate transporters (VGLUTs) facilitate the loading of glutamate into synaptic vesicles. VGLUTs will not be discussed in this chapter.

Five distinct human EAATs (EAAT1–5) have been cloned *(4–6)*. Rodent homologues of EAAT1–5 have also been cloned, these being named GLAST, GLT-1, EAAC1, EAAT4 and EAAT5, respectively. However, for the purpose of this review, the human transporter nomenclature of EAAT1 to EAAT5 will be used throughout. Sequence and hydrophobicity analysis has predicted that the EAATs consist of six α-helical transmembrane domains that form a pore through which glutamate, and various ions, pass *(7)*. The EAAT utilizes the energy generated by transmembrane electrochemical gradients to transport glutamate into the cell against a concentration gradient. Given that the cellular maintenance of the electrochemical gradient consumes adenosine triphosphate (ATP), this dictates that glutamate transporter activity is also energy dependent *(8)*. The stoichiometry for uptake of a single glutamate molecule is cotransport of three sodium ions and one proton into the cell (or possibly a hydroxyl ion countertransport), plus countertransport of a single potassium ion out of the cell. When open, the EAATs also facilitate varying degrees of chloride ion flux in either direction across the cell membrane, which may resist the consequent cellular depolarization generated by the transport of glutamate into the cell *(3,9)*. Alternatively, the glutamate transporter could act as a ligand-activated chloride channel (with glutamate and Na^+ acting as ligands); for further details and discussion of this role for high-affinity glutamate transporters within the central nervous system see the in depth reviews by Danbolt *(3)* and Vandenberg *(7)*.

All five subtypes of EAATs have been identified by immunohistochemistry in the rodent retina *(10–14)*. In addition, the existence of further, unidentified transporters has recently been suggested *(15)*. The function of each subtype appears to correlate with both its distribution within a particular cell type (e.g., neuronal versus glial) and its proximity to transmitter release. Retinal glutamate transporter expression differs from that of the brain in two major respects. Firstly, it appears that EAAT5 is unique to the retina. Secondly, within the retina, EAAT1 has been identified as the key high-affinity transporter responsible for maintaining glutamate homeostasis in order to facilitate neurotransmission and prevent glutamate excitotoxicity *(16)*. In contrast, this task is performed by EAAT2 in the brain. Following is an overview of the five known EAATs.

EAAT1

EAAT1 (GLAST) is expressed by both neurons and glia within the brain, but is localized exclusively to glia (Müller cells) within the retina *(13,17)*. Autoradiographic and immunohistochemical studies indicate that Müller cells dominate total retinal glutamate transport *(16,18–20)*: exogenous glutamate (or a transportable analogue) is accumulated predominantly by these cells. This dominance implies that EAAT1 has a role in the modulation of excitatory transmission. It has been suggested that EAAT1 operates as an early uptake mechanism, immediately reducing the level of glutamate as it floods out of synaptic vesicles into the synaptic cleft *(21)*. This removal of glutamate, combined with the secretion of neurotrophins, works to prevent excitotoxic damage to cells within the central nervous system (CNS), including retinal ganglion cells *(22,23)*. Furthermore, EAAT1 is critically involved in glial glutamate metabolism and plays a fundamental role in the glutamate-glutamine cycle *(24,25)*. Following EAAT1-mediated glutamate uptake from the synaptic cleft into Muller cells, the glial-specific enzyme glutamine synthetase (GS) amidates glutamate, thereby converting it into non-neuroactive glutamine. Glutamine is conveyed back into neurons, whereupon it is reconverted to glutamate. Inhibition of GS depletes retinal neurons of glutamate while EAAT1 inhibition suppresses retinal function *(18,24)*. Moreover, the activities of EAAT1 and glutamine synthetase appear to be coupled *(26,27)*. The importance of EAAT1 in the maintenance of retinal glutamate homeostasis is highlighted by evidence that reduction or loss of Müller-cell EAAT1 activity impairs second-order neurotransmission and retinal function *(17,18)*.

EAAT2

Protein for EAAT2 (GLT-1) has been localized to astrocyte processes in the brain, where it is the dominant glutamate transporter. Because the transporter is localized at sites away from the axon terminals, it has been suggested that this transporter acts to regulate the diffusion of glutamate that has escaped from the synaptic cleft *(28)*. A number of splice variants of EAAT2 have been described *(29)* and the current nomenclature for these variants can be found at www.eaats.org. Unlike those of the brain, immunohistochemical studies have determined that EAAT2 splice variants are predominantly localized to neurons in the retina *(13,30,31)*. GLT-1a is associated with neuronal processes of the inner plexiform layer *(31)* and has been described in cone, bipolar, amacrine and ganglion cells of the cat retina *(10)*. GLT-1v is associated with cone photoreceptors and a population of bipolar cells. GLT1c is expressed in the synaptic terminals of both rod and cone photoreceptors *(32)* and bipolar cells *(10)*. It is suggested that alternate splicing of EAAT2 is significant because it encodes the capacity for differential targeting of the transporter *(31)*.

EAAT3

EAAT3 (EAAC-1) is contained within the cytosol and trafficked to the plasma membrane *(33)*. EAAT3 has been identified in horizontal, amacrine and ganglion cells in the retina, at both synaptic and extrasynaptic sites *(10,34)*. The transporter has also been found in retinal pigment epithelium (RPE) cells in pig and human cell lines *(35)*. As EAAT3 is predominantly localized to non-glutamatergic neurons, roles for EAAT3 other than glutamate clearance have been proposed. While it appears that EAAT3 does not play a significant role in the GABA synthetic pathway *(34)*, it may have a role in the synthesis of glutathione, by mediating the uptake of cysteine into retinal neurons *(10,36)*.

EAAT4

EAAT4 is expressed by both neuronal and glial cells in the CNS. In the brain it has been identified on the membranes of cell bodies and dendrites *(37)*, and is the principal glutamate transporter in Purkinje cells. In the retina, EAAT4 has been localized to astrocytes *(14)*, Müller cells *(10)* and rod and cone photoreceptor outer segments *(38)*. It has been suggested that this restriction to the outer retina may provide a feedback mechanism for sensing the extracellular glutamate concentration surrounding photoreceptors, or act as a barrier to prevent glutamate escape from the retina *(38)*. Extensive colocalization of EAAT4 with EAAT1 has also been demonstrated *(14)*. This supports the proposition that EAAT4 acts as a sponge, mopping up glutamate spillage from other transporters and preventing cross-talk between synapses *(21)*. The transporter has also been found in human RPE and retinoblastoma cell lines, suggesting that EAAT4 may play a role in cell proliferation and migration *(35)*. EAAT4 appears to be a high-affinity/low-capacity uptake carrier that is more useful for high-affinity binding than rapid turnover *(39)*.

EAAT5

The existence of an electrogenic glutamate transporter linked to a chloride conductance in the outer retina has been demonstrated *(40)*, this being subsequently identified as EAAT5 *(41,42)*. EAAT5 appears to be exclusively expressed within the retina. It has been identified in photoreceptor terminals of the rat, cat, primate, and salamander

(10,12,42) where it provides for the rapid re-uptake of photoreceptor-released glutamate. In the tiger salamander retina, stimulation of EAAT5 activity inhibits calcium currents and limits the continued release of glutamate *(43)*. EAAT5 has also been found on bipolar cells, amacrine cells and ganglion cells *(10,12)*, but its distribution within the inner retina is species dependent.

Other Glutamate Transporters

Recent evidence also suggests the existence of another, as yet unidentified, sodium-dependent, high-affinity glutamate transporter in the retina. Müller cells of an EAAT1$^{-/-}$ knockout mouse have been shown to accumulate glutamate readily, by a mechanism independent of the currently known glutamate transporters *(15)*.

GLUTAMATE TRANSPORTERS AND RETINAL DISEASE

Overactivation of glutamate receptors, or glutamate excitotoxicity, has been implicated in many retinal diseases, including ischemic damage resulting from stroke, diabetic retinopathy, retinitis pigmentosa, glaucoma and retinal damage as a side effect of cancer treatment. This section will discuss recent investigations into the involvement of glutamate transporters in the pathogenesis of these diseases.

Ischemia

Interruption of adequate blood supply to the retina results in ischemic tissue damage. This can occur in central retinal artery or vein occlusion, carotid artery disease, stroke, heart disease, and possibly glaucoma (see below). Much of the retinal neuronal damage caused by ischemia is thought to be due to glutamate-mediated excitotoxicity *(44)*. Retinal neurons possess metabotropic and ionotropic glutamate receptors, and excessive activation of ionotropic glutamate receptors [primarily N-methyl D-aspartate (NMDA) receptors] leads to the entry of calcium ions into the cell and calcium overload *(45)*. This causes mitochondrial failure, excessive formation of free oxidative radicals and activation of complex cascades of caspases, proteases, nucleases, and lipases, which may trigger apoptotic neuronal cell death *(45)*. Endogenous glutamate has been shown to accumulate in the extracellular compartment of the retina under ischemic conditions *(46)*, but the sources of the subsequent rise in extracellular glutamate remain unclear.

It is accepted that neuronal vesicular release contributes greatly to the rapid rise in extracellular glutamate. Neuronal exocytosis of glutamate, under ischemic conditions, is promoted by the loss of membrane polarization. However, the effect of transport-mediated glutamate release is less apparent. Many studies have reported that glutamate transporter reversal actively contributes to the ischemia-induced elevation of extracellular glutamate *(47–49)*, whereas others have shown continued glutamate uptake under ischemic conditions *(50–52)*. It is clear, however, that retinal glutamate clearance is compromised under ischemic conditions *(50,52,53)*. The activity of the transporters is affected during ischemia by a combination of factors, including the disruption of the necessary sodium gradient, oxidative damage, acidosis, phosphorylation state changes and cleavage by caspase *(53–56)*, rather than the rapid down-regulation of transporter expression *(50,57,58)*.

Uptake studies in the retina have shown that EAAT1 dominates glutamate uptake. However, the activity of this transporter appears to be particularly susceptible to an acute ischemic insult, resulting in the failure of Müller cells to accumulate glutamate *(50)*, as illustrated in Fig. 1. When retinal glutamate transport is impaired, sub-millimolar

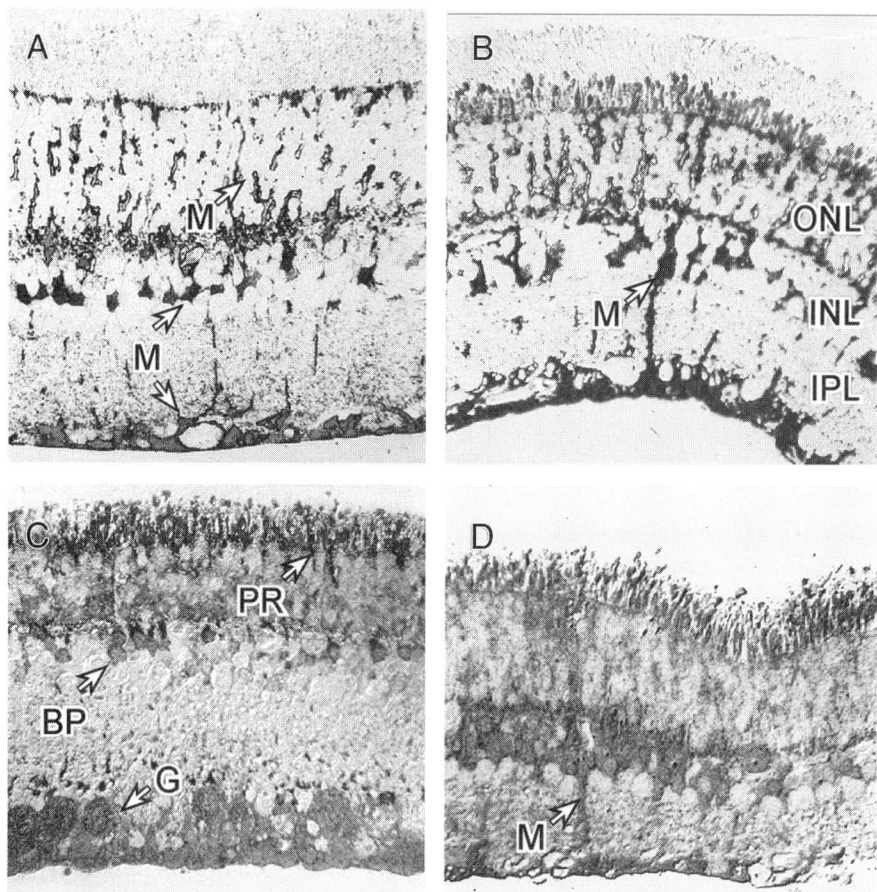

Fig. 1. Retinal glutamate transport is perturbed by an ischemic insult. Glutamate transporter activity in the retina reflected by the *in vivo* accumulation of the non-metabolizable glutamate analogue D-aspartate. D-Aspartate was injected intravitreally and revealed by immunocytochemistry. **(A)** D-Aspartate is taken up exclusively by Müller cells (M) in non-ischemic retinas. **(B)** Incomplete ischemia induced by carotid artery occlusion does not affect the glial cell uptake. **(C)** During an acute total ischemic insult caused by central retinal artery occlusion, D-aspartate is accumulated by photoreceptors (PR), bipolar cells (BP) and ganglion cells (G). **(D)** After five days of reperfusion following acute total ischemia, the Müller cells partially regain the ability to accumulate D-aspartate.

Modified from Barnett, N. L., Pow, D. V. and Bull, N. D. Differential perturbation of neuronal and glial glutamate transport systems in retinal ischaemia. Neurochem Int 2001;39:291–9. (Elsevier Science Ltd.).

concentrations of glutamate are neurotoxic *(59)*. Under normal, non-ischemic, experimental conditions, retinal neurons do not accumulate exogenous glutamate analogues *in vivo*, because glutamate is rapidly cleared by Müller cells *(16,18–20)*. However, under ischemic conditions, this situation is reversed such that exogenous glutamate analogues are accumulated by photoreceptors, bipolar, and ganglion cells *(50)*. This neuronal uptake, mediated by EAAT5, EAAT2 and EAAT3, respectively, is not sufficient to compensate for the failure of EAAT1 during ischemia. Although EAAT1 function recovers rapidly upon reperfusion, this does not prevent subsequent excitotoxic neurodegeneration *(57)*. The role of EAAT1 in the prevention of tissue damage during an ischemic attack is vital, as EAAT1-deficient mice are even more sensitive to ischemia than control animals, displaying exacerbated retinal damage following an ischemic insult *(17)*.

Interestingly, short periods of hypoxia, rather than complete ischemia, stimulate glutamate uptake in isolated rat retinal cells, possibly by the mobilization of transporters to the cell membrane *(60)*. This rapid upregulation of glutamate transport may be an adaptive neuroprotective process.

Glaucoma

Glaucoma is a progressive optic neuropathy with associated optic nerve damage and visual field defects. Elevated intraocular pressure is considered a primary risk factor for the initiation and progression of glaucomatous neuropathy.

The mechanisms underlying retinal degeneration in glaucoma are complex *(61)*. A number of biochemical processes have been implicated in the neuropathology of glaucoma, including glutamate excitotoxicity. The involvement of glutamate excitotoxicity in glaucoma is, however, controversial. Although it was initially claimed that glutamate levels in the vitreous are elevated in glaucomatous eyes *(62)*, this has subsequently been challenged *(63)*, and has not been confirmed in either human or animal models of glaucoma *(64)*. However, a large body of evidence suggests that glutamatergic mechanisms within the retina itself are involved in neuronal cell death. This evidence includes: (i) NMDA glutamate receptor antagonists offer protection to retinal ganglion cells in experimental glaucoma models *(65–67)* and clinical patients *(68)*, (ii) the expression of the endogenous glutamate receptor antagonist, kynurenic acid, is suppressed in an animal glaucoma model *(69)*, (iii) elevated intraocular pressure differentially regulates metabotropic glutamate receptors in the mouse retina *(70)*, (iv) pharmacological or antisense depression of retinal glutamate transporters (EAAT1 and EAAT2) can induce ganglion cell death *(71)*, and (v) glaucoma affects glutamate transporter expression, but experimental data are not consistent. For example, EAAT1 immunoreactivity is increased in monkey and rat models of glaucoma *(72,73)*, whereas decreased EAAT1 and EAAT2 expression has been reported in a rat model *(74)*, together with reduced EAAT3 immunoreactivity in human glaucomatous eyes *(75)*. Transient changes in EAAT2 expression following experimental optic nerve crush have also been reported *(76)*. Clearly, much evidence exists to suggest that glutamatergic mechanisms could be involved in glaucomatous neuropathy. However, whether glutamate is involved in the initial pathology or in secondary events subsequent to initial pressure-induced cell death (and hence glutamate release from dying neurons) is not known.

Diabetic Retinopathy

One of the complications of diabetes, both type I and type II, is the gradual development of retinal degeneration and loss of sight. The classic clinical hallmarks of diabetic retinopathy involve neovascularization, macular edema, microaneurysms and hemorrhages. Historically these changes were assumed responsible for the neurodegeneration and visual changes suffered, such as a depression of the pattern electroretinogram (ERG) *(77)*, a reduction in amplitude and delay of the ERG b-wave oscillatory potentials *(78)*, and a loss of contrast sensitivity *(79)*. In addition, diabetes triggers early apoptosis of inner retinal neurons *(80)*. However, it is now clear that glutamate excitotoxicity contributes to the inner retinal neuronal loss in this disease, and that this may begin well before the onset of clinical retinopathy. Within months of diabetes onset, glutamate levels rise above normal physiological levels in the retina of rats *(81)* and in the vitreous of patients *(82)*, possibly as a consequence of the breakdown of the blood-retinal barrier, which allows glutamate leakage from the blood. Furthermore, high glucose levels enhance the evoked release of glutamate analogues from the retina *(83)* and glutamate metabolism is altered in the diabetic retina. Glutamine synthetase activity and glutamate degradation via the citric-acid cycle are suppressed in diabetic rats, which could also contribute to raised extracellular glutamate concentrations *(81,84)*.

The involvement of glutamate transporters in diabetic retinopathy is less clear. Using a freshly isolated rat Müller cell preparation, it has been reported that the chemical induction of diabetes significantly decreases EAAT1-mediated glutamate uptake as a result of transporter oxidation, thereby contributing to the elevation of retinal glutamate levels and neuronal excitotoxicity *(85)*. Conversely, it has also been reported that diabetes stimulates glutamate transporter activity in rat Müller cells *in vivo*, where the expression of EAAT1 and EAAT4 is significantly elevated and the accumulation of a glutamate analogue by Müller cells is significantly enhanced *(86)*. These results suggest that Müller cells maintain their capacity to transport glutamate and their ability to regulate the extracellular concentrations of glutamate in the diabetic retina, implying that changes in glutamate transport during diabetes do not contribute to neuronal excitotoxicity and death.

Retinitis Pigmentosa

Glutamate excitotoxicity may also play a role in the death of photoreceptors in the heterogenous group of retinal degenerative diseases known collectively as retinitis pigmentosa. However, the involvement of glutamate transporters in the etiology of these conditions is not conclusive. In the Royal College of Surgeons' (RCS) rat model of retinitis pigmentosa, autoradiographical studies have revealed that Müller-cell glutamate uptake is maintained even after the onset of photoreceptor degeneration. Moreover, this uptake appears to be greater in adult RCS retinas than in control retinas *(87)*. However, glutamate degradation by Müller cells is slowed and these cells display an accumulation of glutamine *(88)*. This demonstrates that EAAT1 activity is maintained in the RCS retina, but implies a defect in Müller-cell-mediated recycling of neuronal glutamate via the intermediate glutamine. Dysfunctional Müller-cell metabolism in the RCS retina may be a causative factor in photoreceptor degradation.

Photoreceptor loss is directly linked to elevated extracellular glutamate levels and glutamate excitotoxicity in the *rd1* mouse model of retinal degeneration *(89)*. The degeneration of the rods is partially ameliorated by the application of an amino-3-hydroxy-5-methyl-4-isoxazolepropionic acid (AMPA)/kainate-type glutamate receptor antagonist. However, the onset of photoreceptor cell death is associated with the upregulation of both EAAT1 and glutamine synthetase expression, suggesting that Müller cell dysfunction (and hence EAAT1 dysfunction) is not responsible for the rise in extracellular glutamate. A possible cause is that the accumulation of cyclic guanosine monophosphate (cGMP), resulting from the genetic mutation in this model, induces prolonged depolarization of rod photoreceptors and enhanced glutamate transmitter release.

Leber Hereditary Optic Neuropathy

Leber hereditary optic neuropathy is an inherited condition resulting in ganglion cell degeneration. It is caused by mutations leading to mitochondrial dysfunction. Experimental evidence suggests that excessive production of reactive oxygen species by the faulty mitochondria compromises EAAT1 function, thus contributing to ganglion cell damage and optic neuropathy *(22)*.

REGULATION OF GLUTAMATE TRANSPORT

The search for specific pharmacological modulators of EAAT activity has only just begun, with most pharmacological research to date directed towards developing drugs capable of blocking glutamate receptor activation and, thus, excitotoxic neurodegeneration. However, given the central role of glutamate transporters in the (patho)physiological regulation of extracellular glutamate concentrations, investigations have begun into the manipulation of EAAT activity for the treatment (or prevention) of such maladies. This has led to the study of glutamate transporter regulation in the CNS, that is, which intracellular signaling pathways fine-tune glutamate transporter activity to maintain an appropriate extracellular milieu? Also, what is the respective physiological function of each transporter subtype? To date, most studies have focused on glutamate transporter activity in the brain, rather than the activity of EAATs in the retina. These studies will be outlined below. However, the findings have direct implications for the retina, which forms part of the CNS. *In vivo*, glutamate transporters are broadly regulated in three ways: (i) by the redistribution of protein to or from the cell surface, (ii) by modulating the rate of gene transcription, and (iii) by altering the catalytic efficiency of the protein at the cell membrane (for example, by altering phosphorylation state or by changing interactions with intracellular proteins).

Protein Kinases and Phosphatases

Protein kinases are potent endogenous regulators of glutamate transporter activity, with protein kinase C (PKC), protein kinase A (PKA), and phosphatidylinositol 3-kinase (PI3K) all implicated in the intracellular control of EAAT activity, expression and cell-surface trafficking.

PKC (of which there are a number of isoforms) is a diacylglycerol-activated, phospholipid-dependent enzyme, belonging to the protein kinase superfamily, which plays a central role in intracellular second-messenger signaling, thereby modulating a plethora of cellular events. The PKC signaling cascade is initiated by lipid second messengers, including diacylglycerol (DAG), fatty acids and phosphatidylinositol 3,4,5-triphosphate (PIP_3). In addition, some isoforms require calcium binding for activation. Isozyme localization and function can also be modulated via protein-protein interactions that refine the overlapping activity of PKC isoforms by restricting individual isoform control to specific cellular responses *(90,91)*. Pharmacologically, pan-PKC activity can be enhanced by a class of drugs known as phorbol esters *(92)*, while other compounds selectively activate isoforms of PKC, for example bistratene A, which stimulates PKCδ *(93,94)*. PKC activity can also be suppressed pharmacologically, for example, non-specific PKC activity is reduced with chelerythrine *(95)*, or in an isoform-dependent manner. The action of PKC on the activity of EAATs has been extensively studied (Fig. 2). However, conflicting results suggest that the effects are both PKC-isoform and EAAT-subtype dependent.

Modulation of EAAT1, which is expressed by retinal Müller cells, is of particular interest as this transporter is essential for maintaining glutamate homeostasis in the retina *(16,18,24)*. Three studies have reported that PKC activation decreases EAAT1 activity in a number of different expression systems, including transfected *Xenopus* oocytes and human embryonic kidney (HEK293) cells *(56)*, as well as cultured chick Bergmann glial cells and chick Müller cells *(96,97)*. This PKC-mediated suppression of EAAT1 glutamate uptake is due to transcriptional down-regulation *(98)*. Interestingly, PKC inhibition stimulates EAAT1-mediated glutamate uptake in cultured chick Bergmann glial cells *(97)*. However, in mammalian retina *in situ*, PKC inhibition suppresses EAAT1-mediated uptake *(55)*.

Reported effects of PKC modulation of EAAT2 activity are also conflicting. EAAT2 activity is either stimulated by PKC activation *(99)* or inhibited through internalization of the transporter protein by PKC activation *(100,101)*. In contradiction, neither the activation nor the inhibition of PKC alters EAAT2 activity in mammalian cell lines transfected to express moderate EAAT2 activity *(102)*. These conflicting effects therefore appear to be both tissue and species dependent.

PKC activation increases intracellular trafficking of EAAT3 to the cell membrane and enhances glutamate transport in C6 glioma cells *(33,103)*. This effect may be due to the formation of PKCα-EAAT3 complexes that aid movement to the cell membrane *(104)*. However, PKCα activation also inhibits transport by EAAT3 in other cell culture systems *(105)*. Interestingly, PKC activation inhibits EAAT3 activity in *Xenopus* oocytes and HEK293 cells, but stimulates activity in C6 glioma cells *(106)*. These results suggest that the modulatory effect of PKC on EAAT3 activity is cell-type specific; however, as none of these cell types endogenously express EAAT3, there is, as yet, no definitive answer.

PKC-mediated regulation of EAAT4 and EAAT5 has not been studied in detail. A single study found that PKC activation stimulated a chloride current via EAAT4 transfected into *Xenopus* oocytes, but failed to stimulate a concurrent rise in glutamate uptake by the transporter *(107)*.

Protein kinase A (PKA) plays a role in the regulation of glutamate transporter activity. PKA activation by dibutyryl cyclic adenosine monophosphate (dbcAMP) stimulates the relocation of EAAT2 to the cell surface in primary cortical astrocyte cultures (108). This trafficking of EAAT2 is dependent on the expression of the cytoskeletal protein, glial fibrillary acidic protein (GFAP) (109). Activation of PKA by cAMP also upregulates the expression of EAAT1 and EAAT2, and elicits a rise in transporter protein, which occurs concomitant with stimulation of glutamate transporter activity in cultured astrocytes (108). Conversely, inhibition of PKA enhances the EAAT1-mediated uptake of glutamate into Müller cells in the ischemic rat retina (110).

PKA activation also mediates the effects of growth factors on glutamate uptake. A number of growth factors appear to modulate EAAT-mediated glutamate uptake via the activity of PI3K. Inhibition of PI3K suppresses glutamate uptake by reducing cell surface expression of EAAT3 (33). Carbamazepine (an antiepileptic agent) stimulates EAAT3 activity in a PI3K-dependent manner in both transfected *Xenopus* oocytes and endogenously expressing C6 glioma cells (111). Furthermore, in cortical astrocyte cultures EAAT2, but not EAAT1, activity appears to be regulated by a signaling pathway involving Akt/protein kinase B, which can function downstream of PI3K (112).

It appears that the various kinases integrate signals to fine-tune glutamate transporter activity and that the different kinase-mediated signaling pathways converge to regulate the activity of each transporter differentially (113). In neuron-enriched cultures endogenously expressing EAAT1, EAAT2 and EAAT3, total glutamate uptake is suppressed by the inhibition of PKA and PI3K but is unaffected by PKC modulation. Specifically, the presence of EAAT1 on the cell membrane is reduced by PKA inhibition and PKC activation, but increased by the inhibition of PI3K; EAAT2 surface expression is stimulated by PKA inhibition, but diminished by both PKC activation and PI3K inhibition, while EAAT3 membrane expression is decreased by the inactivation of both PKA and PI3K but augmented by PKC activation.

GTRAPs

Recently, it has become apparent that glutamate transporter activity is sensitively regulated by intracellular protein-protein interactions. Glutamate transporter-associated proteins (GTRAPs) bind to EAATs and modify their activity. GTRAP3-18 binds to EAAT3, but not to EAAT1, EAAT2 or EAAT4, and decrease EAAT3 activity by reducing the binding efficiency of the transporter (114). It does not affect either the transcription or cellular distribution of the transporter. Antisense suppression of GTRAP3-18 expression results in an increase in EAAT3 activity, as GTRAP3-18 protein levels fall. Conversely, morphine can upregulate the expression of GTRAP3-18 and suppress EAAT3-mediated transport through a PKA-dependent pathway (115,116). GTRAP41 and GTRAP48 interact with the intracellular region of EAAT4, thereby modulating its activity (117). Both GTRAP41 and GTRAP48 stimulate glutamate uptake by stabilizing EAAT4 at the cell membrane and thereby increasing the active transporter pool. In the retina, GTRAP41 colocalizes and interacts with EAAT4 in photoreceptors (38).

Soluble factors

A large number of soluble factors that can influence glutamate transporter activity have been identified, some of which will be discussed here. The search for such factors

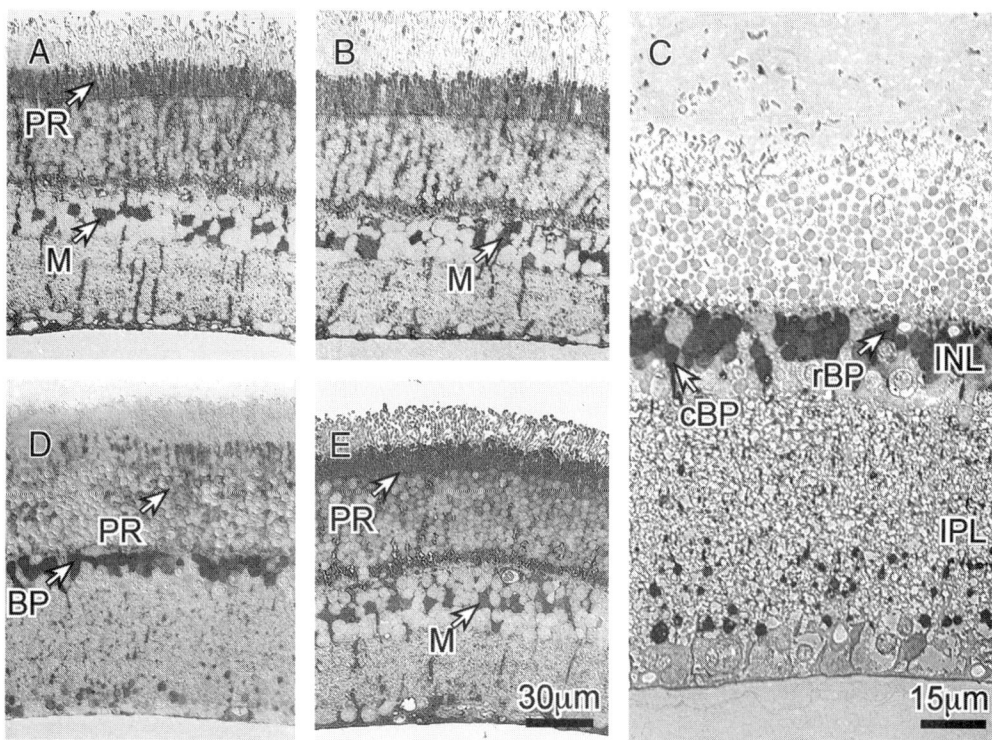

Fig. 2. Protein kinase C (PKC) modulates the activity of retinal glutamate transporters. Glutamate transporter activity in the retina reflected by the *in vitro* accumulation of the non-metabolizable glutamate analogue, D-aspartate. (**A**) D-Aspartate is predominantly accumulated by the radial glial Müller cells (M). The inner segments of the photoreceptors (PR) are also immunopositive. (**B**) Activation of retinal PKC with phorbol 12-myristate 13-acetate (PMA) does not change the pattern of D-Aspartate uptake within the retina. (**C**) The pan-isoform-specific inhibitor of PKC, chelerythrine, blocks D-aspartate uptake by Müller cells. D-Aspartate is accumulated by a population of rod (rBP) and cone (cBP) bipolar cell perikarya within the inner nuclear layer (INL), and their terminals within the inner plexiform layer (IPL). (**D**) Specific inhibition of PKCδ with rottlerin also prevents D-aspartate uptake by Müller cells. Positive D-aspartate immunoreactivity is observed in bipolar cells, as well as weak labeling of photoreceptor cell bodies. (**E**) Inhibition of PKC isoforms α, β and γ by Gö6976 does not alter retinal D-aspartate uptake.

Taken from Bull, N. D. and Barnett, N. L. Antagonists of protein kinase C inhibit rat retinal glutamate transport activity in situ. J Neurochem 2002;81:472–80. (International Society for Neurochemistry/Blackwell).

was stimulated by the observation that neurons could induce the expression of EAAT2 in cultured astrocytes independent of physical contact *(118)*. These soluble factors appear to regulate glutamate transporter activity through the aforementioned protein kinase signaling pathways. Epidermal growth factor (EGF) stimulates glutamate uptake via EAAT1 and EAAT2 *(119,120,121)*. The increase in glutamate uptake is mediated by an upregulation of gene expression, and is dependent on PI3K signaling and the transcription factor

NF-kappaB. These studies also identified transforming growth factor-alpha (TGFα) as an inducer of EAAT1 and EAAT2 expression and activity in cultured astrocytes, via the PI3K pathway. Fibroblast growth factor also upregulates EAAT1 and EAAT2 expression *(119)*. Interestingly, EGF and TGFα upregulate the expression of EAAT1 and EAAT2 in striatal astrocytic cultures, but not in astrocytic cultures established from the cerebellum, mesencephalon or spinal cord, hinting that glial glutamate transporter expression may be differentially regulated across CNS regions *(120)*. Platelet-derived growth factor (PDGF) stimulates EAAT3 relocation from intracellular stores to the cell surface *(122)* and promotes the upregulation of EAAT2, but not EAAT1, expression in cortical astroglial cultures *(119)*. Brain-derived neurotropic factor (BDNF) upregulates astrocytic transcription of EAAT2, via NF-kappaB *(123)*, while glial-derived neurotropic factor (GDNF) increases the production of both EAAT1 and EAAT2 in organotypic hippocampal slice cultures *(124)*. However, the effects of these neurotropic factors are far from clear, as in cortical astrocytic cultures both BDNF and GDNF fail to affect the expression of either EAAT1 or EAAT2 *(119)*. In addition, NF-kappaB signaling, triggered by exposure of brain slice cultures to tumor necrosis factor-alpha, has also been found to inhibit glutamate uptake *(125)*.

The hormone estrogen, acting upon nuclear estrogen receptors, also stimulates EAAT1 and EAAT2 transcription and translation, resulting in a rise in cellular glutamate uptake *(126)*. This mechanism may underlie the deleterious effects of the anti-estrogenic chemotherapeutic compounds tamoxifen and toremifene upon the retina, which may be mediated by reduced glutamate uptake *(127)*.

Transporter Substrates

Glutamate itself, as well as other substrates of the glutamate transporters, can also regulate EAAT activity. It has been demonstrated that long-term exposure of astrocytic cultures to glutamate can boost transporter activity by upregulating EAAT1, possibly through the action of glutamate on kainate receptors *(128)*. Glutamate release by neurons in mixed cultures has also been shown to stimulate astrocytic EAAT1 and EAAT2 activity by triggering redistribution of transporter protein to the cell surface; however, this is not associated with a concomitant increase in transporter synthesis and is not reliant upon glutamate receptor activation *(129)*. In contrast, glutamate receptor activation can suppress EAAT1 transcription, via PKC-mediated signaling, in cultured chick cerebellar Bergmann glia *(130)*. However, it is possible that this effect is mediated by metabotropic glutamate receptors (mGluRs), which activate intracellular signaling cascades, rather than by the previously mentioned stimulation of activity through kainate receptor activity *(128)*. Indeed, mGluR5a activation stimulates EAAT2 activity in cultured reactive astrocytes via intracellular signaling pathways involving phospholipase C and PKC *(131)*. Furthermore, long-term stimulation of group I mGluRs can suppress EAAT1 activity, while group II mGluR agonists can upregulate the expression of this transporter in cultured astrocytes *(118)*. Glutamate has also been shown to prompt the cell-surface expression of EAAT4 in a glioma cell line, independent of glutamate receptor activation *(118)*.

Transporter Blockers

A number of competitive and non-transportable blockers of EAATs have been developed as tools for elucidating the functions of these transporters *(132)*. The most potent of these to date are DL-threo-beta-benzyloxyaspartate (DL-TBOA) and its derivatives, including L-threo-beta-methoxyaspartate (L-TMOA). While DL-TBOA is a non-transportable blocker of all EAAT subtypes (1 to 5), L-TMOA exerts a differential effect upon each transporter subtype *(133)*.

Antibiotics

It has recently been discovered that β-lactam antibiotics are neuroprotective in CNS pathologies, such as amyotrophic lateral sclerosis, that involve glutamate excitotoxicity. This may be due to the ability of these potent antibiotics to stimulate strong upregulation of EAAT2 expression *(134)*. This is the first identification of a useful pharmaceutical regulator of glutamate transporter activity.

Anesthetics

Anesthetics are another class of drugs that modulate glutamate transporter activity in various experimental systems. The local anesthetic lidocaine, acting through the PKC and PI3K pathways, stimulates EAAT3-mediated glutamate transport *(135)*. The volatile general anesthetic, isoflurane, stimulates the activity of EAAT3, by triggering redistribution to the cell surface, in a PKC-dependent manner *(136)*. Conversely, the intravenous anesthetic etomidate suppresses EAAT3-mediated glutamate uptake by modulating PKC activity *(137)*.

Other Agents

Diverse agents modulate EAAT activity experimentally. These include WAY-855 and WAY-209429 (selective inhibitors of EAAT2 and EAAT3 respectively) *(138)*, the fatty acid docosahexaenoic acid (DHA) (increases EAAT2 and EAAT3-mediated uptake, but suppresses EAAT1 transport) *(139)*, ethanol (stimulates EAAT3 via kinase signaling) *(140)*, the antiepileptic benzodiazepine clobazam (increases EAAT2 protein levels) *(141)* and neuroimmunophilins (modulate EAAT2 expression) *(142)*.

Conclusion

It is now apparent that glutamate transporter dysfunction contributes to the etiology of many diseases, including retinal ischemia and glaucoma. An understanding of the mechanisms that control glutamate homeostasis and the manipulation of specific glutamate transporters in the retina may provide a framework for the development of pharmacological strategies necessary for the successful clinical management of excitotoxic retinal injury.

ACKNOWLEDGMENTS

We would like to thank Amanda Barnett, Kei Takamoto, and Rowan Tweedale for their invaluable contributions to this chapter.

REFERENCES

1. Thoreson WB and Witkovsky P. Glutamate receptors and circuits in the vertebrate retina. Prog Retin Eye Res 1999;18:765–810.
2. Choi DW. Glutamate receptors and the induction of excitotoxic neuronal death. Prog Brain Res 1994;100:47–51.
3. Danbolt NC. Glutamate uptake. Prog Neurobiol 2001;65:1–105.
4. Kanai Y and Hediger MA. Primary structure and functional characterization of a high-affinity glutamate transporter. Nature 1992;360:467–71.
5. Pines G, Danbolt NC, Bjoras,M, Zhang Y, Bendahan A, Eide L, Koepsell H, Storm MJ, Seeberg E, Kanner BI. Cloning and expression of a rat brain L-glutamate transporter. Nature 1992;360:464–7.
6. Storck T, Schulte S, Hofmann K, Stoffel W. Structure, expression, and functional analysis of a Na^+-dependent glutamate/aspartate transporter from rat brain. Proc Natl Acad Sci USA 1992;89:10955–9.
7. Vandenberg RJ. Molecular pharmacology and physiology of glutamate transporters in the central nervous system. Clin Exp Pharmacol Physiol 1998;25:393–400.
8. Seal RP and Amara SG. Excitatory amino acid transporters: a family in flux. Annu Rev Pharmacol Toxicol 1999;39:431–56.
9. Eliasof S and Jahr CE. Retinal glial cell glutamate transporter is coupled to an anionic conductance. Proc Natl Acad Sci USA 1996;93:4153–8.
10. Fyk-Kolodziej B, Qin P, Dzhagaryan A, Pourcho RG. Differential cellular and subcellular distribution of glutamate transporters in the cat retina. Vis Neurosci 2004;21:551–65.
11. Pow DV and Barnett NL. Changing patterns of spatial buffering of glutamate in developing rat retinae are mediated by the Müller cell glutamate transporter GLAST. Cell Tissue Res 1999;297:57–66.
12. Pow DV and Barnett NL. Developmental expression of excitatory amino acid transporter 5: a photoreceptor and bipolar cell glutamate transporter in rat retina. Neurosci Lett 2000;280:21–4.
13. Rauen T, Rothstein JD, Wassle H. Differential expression of three glutamate transporter subtypes in the rat retina. Cell Tissue Res 1996;286:325–36.
14. Ward MM, Jobling AI, Puthussery T, Foster LE, Fletcher EL. Localization and expression of the glutamate transporter, excitatory amino acid transporter 4, within astrocytes of the rat retina. Cell Tissue Res 2004;315:305–10.
15. Sarthy VP, Pignataro L, Pannicke T, Weick M, Reichenbach A, Harada T, Tanaka K, Marc R. Glutamate transport by retinal Müller cells in glutamate/aspartate transporter-knockout mice. Glia 2005;49:184–96.
16. Rauen T, Taylor WR, Kuhlbrodt K, Wiessner M. High-affinity glutamate transporters in the rat retina: a major role of the glial glutamate transporter GLAST-1 in transmitter clearance. Cell Tissue Res 1998;291:19–31.
17. Harada T, Harada C, Watanabe M, Inoue Y, Sakagawa T, Nakayama N, Sasaki S, Okuyama S, Watase K, Wada K, Tanaka K. Functions of the two glutamate transporters GLAST and GLT-1 in the retina. Proc Natl Acad Sci USA 1998;95:4663–6.
18. Barnett NL and Pow DV. Antisense knockdown of GLAST, a glial glutamate transporter, compromises retinal function. Invest Ophthalmol Vis Sci 2000;41:585–91.
19. Ehinger B. Glial and neuronal uptake of GABA, glutamic acid, glutamine and glutathione in the rabbit retina. Exp Eye Res 1977;25:221–34.
20. White RD and Neal MJ. The uptake of L-glutamate by the retina. Brain Res 1976;111:79–93.

21. Takayasu Y, Iino M, Kakegawa W, Maeno H, Watase K, Wada K, Yanagihara D, Miyazaki T, Komine O, Watanabe M, Tanaka K, Ozawa S. Differential roles of glial and neuronal glutamate transporters in Purkinje cell synapses. J Neurosci 2005;25:8788–93.
22. Beretta S, Mattavelli L, Sala G, Tremolizzo L, Schapira AH, Martinuzzi A, Carelli V, Ferrarese C. Leber hereditary optic neuropathy mtDNA mutations disrupt glutamate transport in cybrid cell lines. Brain 2004;1272183–92.
23. Taylor S, Srinivasan B, Wordinger RJ, Roque RS. Glutamate stimulates neurotrophin expression in cultured Müller cells. Brain Res Mol Brain Res 2003;111:189–97.
24. Barnett NL, Pow DV, Robinson SR. Inhibition of Müller cell glutamine synthetase rapidly impairs the retinal response to light. Glia 2000;30:64–73.
25. Pow DV and Robinson SR. Glutamate in some retinal neurons is derived solely from glia. Neuroscience 1994;60:355–66.
26. Derouiche A and Rauen T. Coincidence of L-glutamate/L-aspartate transporter (GLAST) and glutamine synthetase (GS) immunoreactions in retinal glia: evidence for coupling of GLAST and GS in transmitter clearance. J Neurosci Res 1995;42:131–43.
27. Shaked I, Ben-Dror I, Vardimon L. Glutamine synthetase enhances the clearance of extracellular glutamate by the neural retina. J Neurochem 2002;83:574–80.
28. Minelli A, Barbaresi P, Reimer RJ, Edwards RH, Conti F. The glial glutamate transporter GLT-1 is localized both in the vicinity of and at distance from axon terminals in the rat cerebral cortex. Neuroscience 2001;108:51–9.
29. Rozyczka J and Engele J. Multiple 5′-splice variants of the rat glutamate transporter-1. Brain Res Mol Brain Res 2005;133:157–61.
30. Reye P, Sullivan R, Fletcher EL, Pow DV. Distribution of two splice variants of the glutamate transporter GLT1 in the retinas of humans, monkeys, rabbits, rats, cats, and chickens. J Comp Neurol 2002;445:1–12.
31. Sullivan R, Rauen T, Fischer F, Wiessner M, Grewer C, Bicho A, Pow DV. Cloning, transport properties, and differential localization of two splice variants of GLT-1 in the rat CNS: implications for CNS glutamate homeostasis. Glia 2004;45:155–69.
32. Rauen T, Wiessner M, Sullivan R, Lee A, Pow DV. A new GLT1 splice variant: cloning and immunolocalization of GLT1c in the mammalian retina and brain. Neurochem Int 2004;45:1095–106.
33. Davis KE, Straff DJ, Weinstein EA, Bannerman PG, Correale DM, Rothstein JD, Robinson MB. Multiple signaling pathways regulate cell surface expression and activity of the excitatory amino acid carrier 1 subtype of Glu transporter in C6 glioma. J Neurosci 1998;18:2475–85.
34. Wiessner M, Fletcher EL, Fischer F, Rauen T. Localization and possible function of the glutamate transporter, EAAC1, in the rat retina. Cell Tissue Res 2002;310:31–40.
35. Maenpaa H, Gegelashvili G, Tahti H. Expression of glutamate transporter subtypes in cultured retinal pigment epithelial and retinoblastoma cells. Curr Eye Res 2004;28:159–65.
36. Chen Y and Swanson RA. The glutamate transporters EAAT2 and EAAT3 mediate cysteine uptake in cortical neuron cultures. J Neurochem 2003;84:1332–9.
37. Otis TS, Brasnjo G, Dzubay JA, Pratap M. Interactions between glutamate transporters and metabotropic glutamate receptors at excitatory synapses in the cerebellar cortex. Neurochem Int 2004;45:537–44.
38. Pignataro L, Sitaramayya A, Finnemann SC, Sarthy VP. Nonsynaptic localization of the excitatory amino acid transporter 4 in photoreceptors. Mol Cell Neurosci 2005;28:440–51.
39. Mim C, Balani P, Rauen T, Grewer C. The glutamate transporter subtypes EAAT4 and EAATs 1–3 transport glutamate with dramatically different kinetics and voltage dependence but share a common uptake mechanism. J Gen Physiol 2005;126:571–89.

40. Eliasof S and Werblin F. Characterization of the glutamate transporter in retinal cones of the tiger salamander. J Neurosci 1993;13:402–11.
41. Arriza JL, Eliasof S, Kavanaugh MP, Amara SG. Excitatory amino acid transporter 5, a retinal glutamate transporter coupled to a chloride conductance. Proc Natl Acad Sci USA 1997;94:4155–60.
42. Eliasof S, Arriza JL, Leighton BH, Kavanaugh MP, Amara SG. Excitatory amino acid transporters of the salamander retina: identification, localization, and function. J Neurosci 1998;18:698–712.
43. Rabl K, Bryson EJ, Thoreson WB. Activation of glutamate transporters in rods inhibits presynaptic calcium currents. Vis Neurosci 2003;20:557–66.
44. Osborne NN, Casson RJ, Wood JP, Chidlow G, Graham M, Melena J. Retinal ischemia: mechanisms of damage and potential therapeutic strategies. Prog Retin Eye Res 2004;23:91–147.
45. Mattson MP. Apoptosis in neurodegenerative disorders. Nat Rev Mol Cell Biol 2000;1:120–9.
46. Louzada-Junior P, Dias JJ, Santos WF, Lachat JJ, Bradford HF, Coutinho-Netto J. Glutamate release in experimental ischaemia of the retina: An approach using microdialysis. J Neurochem 1992;59:358–363.
47. Billups B and Attwell D. Modulation of non-vesicular glutamate release by pH. Nature 1996;379:171–4.
48. Phillis JW, Ren J, O'Regan MH. Transporter reversal as a mechanism of glutamate release from the ischemic rat cerebral cortex: studies with DL-threo-beta-benzyloxyaspartate. Brain Res 2000;868:105–12.
49. Rossi DJ, Oshima T, Attwell D. Glutamate release in severe brain ischaemia is mainly by reversed uptake. Nature 2000;403:316–21.
50. Barnett NL, Pow DV, Bull ND. Differential perturbation of neuronal and glial glutamate transport systems in retinal ischaemia. Neurochem Int 2001;39:291–9.
51. Bull ND and Barnett NL. Retinal glutamate transporter activity persists under simulated ischemic conditions. J Neurosci Res 2004;78:590–9.
52. Napper GA, Pianta MJ, Kalloniatis M. Reduced glutamate uptake by retinal glial cells under ischemic/hypoxic conditions. Vis Neurosci 1999;16:149–58.
53. Swanson RA, Farrell K, Simon RP. Acidosis causes failure of astrocyte glutamate uptake during hypoxia. J Cereb Blood Flow Metab 1995;15:417–24.
54. Boston-Howes W, Gibb SL, Williams EO, Pasinelli P, Brown RH Jr, Trotti D. Caspase-3 cleaves and inactivates the glutamate transporter EAAT2. J Biol Chem 2006;281:14076–84.
55. Bull ND and Barnett NL. Antagonists of protein kinase C inhibit rat retinal glutamate transport activity in situ. J Neurochem 2002;81:472–80.
56. Conradt M and Stoffel W. Inhibition of the high-affinity brain glutamate transporter GLAST-1 via direct phosphorylation. J Neurochem 1997;68:1244–51.
57. Barnett NL and Grozdanic SD. Glutamate transporter localization does not correspond to the temporary functional recovery and late degeneration after acute ocular ischemia in rats. Exp Eye Res 2004;79:513–24.
58. Otori Y, Shimada S, Tanaka K, Ishimoto I, Tano Y, Tohyama M. Marked increase in glutamate-aspartate transporter (GLAST/GluT-1) mRNA following transient retinal ischemia. Brain Res Mol Brain Res 1994;27:310–4.
59. Izumi Y, Hammerman SB, Kirby CO, Benz AM, Olney JW, Zorumski CF. Involvement of glutamate in ischemic neurodegeneration in isolated retina. Vis Neurosci 2003;20:97–107.
60. Payet O, Maurin L, Bonne C, Muller A. Hypoxia stimulates glutamate uptake in whole rat retinal cells in vitro. Neurosci Lett 2004;356:148–50.

61. Farkas RH and Grosskreutz CL. Apoptosis, neuroprotection, and retinal ganglion cell death: an overview. Int Ophthalmol Clin 2001;41:111–30.
62. Dreyer EB, Zurakowski D, Schumer RA, Podos SM, Lipton SA. Elevated glutamate levels in the vitreous body of humans and monkeys with glaucoma. Arch Ophthalmol 1996;114:299–305.
63. Dalton R. Private investigations. Nature 2001;411:129–30.
64. Levkovitch-Verbin H, Martin KR, Quigley HA, Baumrind LA, Pease ME, Valenta D. Measurement of amino acid levels in the vitreous humor of rats after chronic intraocular pressure elevation or optic nerve transection. J Glaucoma 2002;11:396–405.
65. Hare W, WoldeMussie E, Lai R, Ton H, Ruiz G, Feldmann B, Wijono M, Chun T, Wheeler L. Efficacy and safety of memantine, an NMDA-type open-channel blocker, for reduction of retinal injury associated with experimental glaucoma in rat and monkey. Surv Ophthalmol 2001;45(Suppl 3):S284–9.
66. Hare WA, WoldeMussie E, Lai RK, Ton H, Ruiz G, Chun T, Wheeler L. Efficacy and safety of memantine treatment for reduction of changes associated with experimental glaucoma in monkey, I: Functional measures. Invest Ophthalmol Vis Sci 2004;45:2625–39.
67. Lipton SA. Possible role for memantine in protecting retinal ganglion cells from glaucomatous damage. Surv Ophthalmol 2003;48(Suppl 1):S38–46.
68. Schroder A and Erb C. Use of memantine in progressive glaucoma. Klin Monatsbl Augenheilkd 2002;219:533–6.
69. Rejdak R, Kohler K, Kocki T, Shenk Y, Turski WA, Okuno E, Lehaci C, Zagorski Z, Zrenner E, Schuettauf F. Age-dependent decrease of retinal kynurenate and kynurenine aminotransferases in DBA/2J mice, a model of ocular hypertension. Vision Res 2004;44:655–60.
70. Dyka FM, May CA, Enz R. Metabotropic glutamate receptors are differentially regulated under elevated intraocular pressure. J Neurochem 2004;90:190–202.
71. Vorwerk CK, Naskar R, Schuettauf F, Quinto K, Zurakowski D, Gochenauer G, Robinson MB, Mackler SA, Dreyer EB. Depression of retinal glutamate transporter function leads to elevated intravitreal glutamate levels and ganglion cell death. Invest Ophthalmol Vis Sci 2000;41:3615–21.
72. Carter-Dawson L, Crawford ML, Harwerth RS, Smith EL 3rd, Feldman R, Shen FF, Mitchell CK, Whitetree A. Vitreal glutamate concentration in monkeys with experimental glaucoma. Invest Ophthalmol Vis Sci 2002;43:2633–7.
73. Woldemussie E, Wijono M, Ruiz G. Müller cell response to laser-induced increase in intraocular pressure in rats. Glia 2004;47:109–19.
74. Martin KR, Levkovitch-Verbin H, Valenta D, Baumrind L, Pease ME, Quigley HA. Retinal glutamate transporter changes in experimental glaucoma and after optic nerve transection in the rat. Invest Ophthalmol Vis Sci 2002;43:2236–43.
75. Naskar R, Vorwerk CK, Dreyer EB. Concurrent downregulation of a glutamate transporter and receptor in glaucoma. Invest Ophthalmol Vis Sci 2000;41:1940–4.
76. Mawrin C, Pap T, Pallas M, Dietzmann K, Behrens-Baumann W, Vorwerk CK. Changes of retinal glutamate transporter GLT-1 mRNA levels following optic nerve damage. Mol Vis 2003;9:10–3.
77. Prager TC, Garcia CA, Mincher CA, Mishra J, Chu HH. The pattern electroretinogram in diabetes. Am J Ophthalmol 1990;109:279–84.
78. Holopigian K, Seiple W, Lorenzo M, Carr R. A comparison of photopic and scotopic electroretinographic changes in early diabetic retinopathy. Invest Ophthalmol Vis Sci 1992;33:2773–80.

79. Sokol S, Moskowitz A, Skarf B, Evans R, Molitch M, Senior B. Contrast sensitivity in diabetics with and without background retinopathy. Arch Ophthalmol 1985;103:51–4.
80. Barber AJ, Lieth E, Khin SA, Antonetti DA, Buchanan AG, Gardner TW. Neural apoptosis in the retina during experimental and human diabetes. Early onset and effect of insulin. J Clin Invest 1998;102:783–91.
81. Lieth E, Barber AJ, Xu B, Dice C, Ratz MJ, Tanase D, Strother JM. Glial reactivity and impaired glutamate metabolism in short-term experimental diabetic retinopathy. Penn State Retina Research Group. Diabetes 1998;47:815–20.
82. Ambati J, Chalam KV, Chawla DK, D'Angio CT, Guillet EG, Rose SJ, Vanderlinde RE, Ambati BK. Elevated γ-aminobutyric acid, glutamate, and vascular endothelial growth factor levels in the vitreous of patients with proliferative diabetic retinopathy. Arch Ophthalmol 1997;115:1161–6.
83. Santiago AR, Pereira TS, Garrido MJ, Cristovao AJ, Santos PF, Ambrosio AF. High glucose and diabetes increase the release of [3H]-D-aspartate in retinal cell cultures and in rat retinas. Neurochem Int 2006;48:453–8.
84. Lieth E, LaNoue KF, Antonetti DA, Ratz M. Diabetes reduces glutamate oxidation and glutamine synthesis in the retina. The Penn State Retina Research Group. Exp Eye Res 2000;70:723–30.
85. Li Q and Puro DG. Diabetes-induced dysfunction of the glutamate transporter in retinal Muller cells. Invest Ophthalmol Vis Sci 2002;43:3109–16.
86. Ward MM, Jobling AI, Kalloniatis M, Fletcher EL. Glutamate uptake in retinal glial cells during diabetes. Diabetologia 2005;48:351–60.
87. Fletcher EL and Kalloniatis M. Neurochemical architecture of the normal and degenerating rat retina. J Comp Neurol 1996;376:343–60.
88. Fletcher EL and Kalloniatis M. Neurochemical development of the degenerating rat retina. J Comp Neurol 1997;388:1–22.
89. Delyfer MN, Forster V, Neveux N, Picaud S, Leveillard T, Sahel JA. Evidence for glutamate-mediated excitotoxic mechanisms during photoreceptor degeneration in the *rd1* mouse retina. Mol Vis 2005;11:688–96.
90. Jaken S and Parker PJ. Protein kinase C binding partners. Bioessays 2000;22:245–54.
91. Ron D and Kazanietz MG. New insights into the regulation of protein kinase C and novel phorbol ester receptors. FASEB J 1999;13:1658–76.
92. Castagna M, Takai Y, Kaibuchi K, Sano K, Kikkawa U, Nishizuka Y. Direct activation of calcium-activated, phospholipid-dependent protein kinase by tumor-promoting phorbol esters. J Biol Chem 1982;257:7847–51.
93. Frey MR, Leontieva O, Watters DJ, Black JD. Stimulation of protein kinase C-dependent and -independent signaling pathways by bistratene A in intestinal epithelial cells. Biochem Pharmacol 2001;61:1093–100.
94. Griffiths G, Garrone B, Deacon E, Owen P, Pongracz J, Mead G, Bradwell A, Watters D, Lord J. The polyether bistratene A activates protein kinase C-δ and induces growth arrest in HL60 cells. Biochem Biophys Res Commun 1996;222:802–8.
95. Herbert JM, Augereau JM, Gleye J, Maffrand JP. Chelerythrine is a potent and specific inhibitor of protein kinase C. Biochem Biophys Res Commun 1990;172:993–9.
96. Gonzalez MI, Lopez-Colom AM, Ortega A. Sodium-dependent glutamate transport in Müller glial cells: regulation by phorbol esters. Brain Res 1999;831:140–5.
97. Gonzalez MI and Ortega A. Regulation of the Na^+-dependent high affinity glutamate/aspartate transporter in cultured Bergmann glia by phorbol esters. J Neurosci Res 1997;50:585–90.

98. Espinoza-Rojo M, Lopez-Bayghen E, Ortega A. GLAST: gene expression regulation by phorbol esters. Neuroreport 2000;11:2827–32.
99. Casado M, Bendahan A, Zafra F, Danbolt NC, Aragon C, Gimenez C, Kanner BI. Phosphorylation and modulation of brain glutamate transporters by protein kinase C. J Biol Chem 1993;268:27313–7.
100. Ganel R and Crosson CE. Modulation of human glutamate transporter activity by phorbol ester. J Neurochem 1998;70:993–1000.
101. Gonzalez MI, Susarla BT, Robinson MB. Evidence that protein kinase C α interacts with and regulates the glial glutamate transporter GLT-1. J Neurochem 2005;94:1180–8.
102. Tan J, Zelenaia O, Correale D, Rothstein JD, Robinson MB. Expression of the GLT-1 subtype of Na$^+$-dependent glutamate transporter: pharmacological characterization and lack of regulation by protein kinase C. J Pharmacol Exp Ther 1999;289:1600–10.
103. Dowd LA and Robinson MB. Rapid stimulation of EAAC1-mediated Na$^+$-dependent L-glutamate transport activity in C6 glioma cells by phorbol ester. J Neurochem 1996;67:508–16.
104. Gonzalez MI, Bannerman PG, Robinson MB. Phorbol myristate acetate-dependent interaction of protein kinase C α and the neuronal glutamate transporter EAAC1. J Neurosci 2003;23:5589–93.
105. Dunlop J, Lou Z, McIlvain HB. Properties of excitatory amino acid transport in the human U373 astrocytoma cell line. Brain Res 1999;839:235–42.
106. Trotti D, Peng JB, Dunlop J, Hediger MA. Inhibition of the glutamate transporter EAAC1 expressed in *Xenopus* oocytes by phorbol esters. Brain Res 2001;914:196–203.
107. Fang H, Huang Y, Zuo Z. Enhancement of substrate-gated Cl$^-$ currents via rat glutamate transporter EAAT4 by PMA. Am J Physiol Cell Physiol 2006;290:C1334–40.
108. Schlag BD, Vondrasek JR, Munir M, Kalandadze A, Zelenaia OA, Rothstein JD, Robinson MB. Regulation of the glial Na$^+$-dependent glutamate transporters by cyclic AMP analogs and neurons. Mol Pharmacol 1998;53:355–69.
109. Hughes EG, Maguire JL, McMinn MT, Scholz RE, Sutherland ML. Loss of glial fibrillary acidic protein results in decreased glutamate transport and inhibition of PKA-induced EAAT2 cell surface trafficking. Brain Res Mol Brain Res 2004;124:114–23.
110. Barnett NL, Takamoto K, Bull ND. Glutamate transport modulation: a possible role in retinal neuroprotection. In: Hollyfield JG, Anderson RE, La Vail MM, editors. Retinal Degenerative Diseases. New York: Springer; 2006. p. 327–332.
111. Lee G, Huang Y, Washington JM, Briggs NW, Zuo Z. Carbamazepine enhances the activity of glutamate transporter type 3 via phosphatidylinositol 3-kinase. Epilepsy Res 2005;66:145–53.
112. Li LB, Toan SV, Zelenaia O, Watson DJ, Wolfe JH, Rothstein JD, Robinson MB. Regulation of astrocytic glutamate transporter expression by Akt: evidence for a selective transcriptional effect on the GLT-1/EAAT2 subtype. J Neurochem 2006;97:759–71.
113. Guillet BA, Velly LJ, Canolle B, Masmejean FM, Nieoullon AL, Pisano P. Differential regulation by protein kinases of activity and cell surface expression of glutamate transporters in neuron-enriched cultures. Neurochem Int 2005;46:337–46.
114. Lin CI, Orlov I, Ruggiero AM, Dykes-Hoberg M, Lee A, Jackson M, Rothstein JD. Modulation of the neuronal glutamate transporter EAAC1 by the interacting protein GTRAP3-18. Nature 2001;410:84–8.
115. Ikemoto MJ, Inoue K, Akiduki S, Osugi T, Imamura T, Ishida N, Ohtomi M. Identification of addicsin/GTRAP3-18 as a chronic morphine-augmented gene in amygdala. Neuroreport 2002;13:2079–84.
116. Lim G, Wang S, Mao J. cAMP and protein kinase A contribute to the downregulation of spinal glutamate transporters after chronic morphine. Neurosci Lett 2005;376:9–13.

117. Jackson M, Song W, Liu MY, Jin L, Dykes-Hoberg M, Lin CI, Bowers WJ, Federoff HJ, Sternweis PC, Rothstein JD. Modulation of the neuronal glutamate transporter EAAT4 by two interacting proteins. Nature 2001;410:89–93.
118. Gegelashvili G, Dehnes Y, Danbolt NC, Schousboe A. The high-affinity glutamate transporters GLT1, GLAST, and EAAT4 are regulated via different signalling mechanisms. Neurochem Int 2000;37:163–70.
119. Figiel M, Maucher T, Rozyczka J, Bayatti N, Engele J. Regulation of glial glutamate transporter expression by growth factors. Exp Neurol 2003;183:124–35.
120. Schluter K, Figiel M, Rozyczka J, Engele J. CNS region-specific regulation of glial glutamate transporter expression. Eur J Neurosci 2002;16:836–42.
121. Zelenaia O, Schlag BD, Gochenauer GE, Ganel R, Song W, Beesley JS, Grinspan JB, Rothstein JD, Robinson MB. Epidermal growth factor receptor agonists increase expression of glutamate transporter GLT-1 in astrocytes through pathways dependent on phosphatidylinositol 3-kinase and transcription factor NF-kappaB. Mol Pharmacol 2000;57:667–78.
122. Fournier KM, Gonzalez MI, Robinson MB. Rapid trafficking of the neuronal glutamate transporter, EAAC1: evidence for distinct trafficking pathways differentially regulated by protein kinase C and platelet-derived growth factor. J Biol Chem 2004;279:34505–13.
123. Rodriguez-Kern A, Gegelashvili M, Schousboe A, Zhang J, Sung L, Gegelashvili G. Beta-amyloid and brain-derived neurotrophic factor, BDNF, up-regulate the expression of glutamate transporter GLT-1/EAAT2 via different signaling pathways utilizing transcription factor NF-kappaB. Neurochem Int 2003;43:363–70.
124. Bonde C, Sarup A, Schousboe A, Gegelashvili G, Noraberg J, Zimmer J. GDNF pre-treatment aggravates neuronal cell loss in oxygen-glucose deprived hippocampal slice cultures: a possible effect of glutamate transporter up-regulation. Neurochem Int 2003;43:381–8.
125. Zou JY and Crews FT. TNF alpha potentiates glutamate neurotoxicity by inhibiting glutamate uptake in organotypic brain slice cultures: neuroprotection by NF kappa B inhibition. Brain Res 2005;1034:11–24.
126. Pawlak J, Brito V, Kuppers E, Beyer C. Regulation of glutamate transporter GLAST and GLT-1 expression in astrocytes by estrogen. Brain Res Mol Brain Res 2005;138:1–7.
127. Maenpaa H, Saransaari P, Tahti H. Kinetics of inhibition of glutamate uptake by antioestrogens. Pharmacol Toxicol 2003;93:174–9.
128. Gegelashvili G, Civenni G, Racagni G, Danbolt NC, Schousboe I, Schousboe A. Glutamate receptor agonists up-regulate glutamate transporter GLAST in astrocytes. Neuroreport 1996;8:261–5.
129. Poitry-Yamate CL, Vutskits L, Rauen T. Neuronal-induced and glutamate-dependent activation of glial glutamate transporter function. J Neurochem 2002;82:987–97.
130. Lopez-Bayghen E and Ortega A. Glutamate-dependent transcriptional regulation of GLAST: role of PKC. J Neurochem 2004;91:200–9.
131. Vermeiren C, Najimi M, Vanhoutte N, Tilleux S, de Hemptinne I, Maloteaux JM, Hermans E. Acute up-regulation of glutamate uptake mediated by mGluR5a in reactive astrocytes. J Neurochem 2005;94:405–16.
132. O'Shea RD. Roles and regulation of glutamate transporters in the central nervous system. Clin Exp Pharmacol Physiol 2002;29:1018–23.
133. Shimamoto K, Sakai R, Takaoka K, Yumoto N, Nakajima T, Amara SG, Shigeri Y. Characterization of novel L-threo-β-benzyloxyaspartate derivatives, potent blockers of the glutamate transporters. Mol Pharmacol 2004;65:1008–15.
134. Rothstein JD, Patel S, Regan MR, Haenggeli C, Huang YH, Bergles DE, Jin L, Dykes Hoberg M, Vidensky S, Chung DS, Toan SV, Bruijn LI, Su ZZ, Gupta P, Fisher PB.

β-lactam antibiotics offer neuroprotection by increasing glutamate transporter expression. Nature 2005;433:73–7.
135. Do SH, Fang HY, Ham BM, Zuo Z. The effects of lidocaine on the activity of glutamate transporter EAAT3: the role of protein kinase C and phosphatidylinositol 3-kinase. Anesth Analg 2002;95:1263–8.
136. Huang Y and Zuo Z. Isoflurane induces a protein kinase C α-dependent increase in cell-surface protein level and activity of glutamate transporter type 3. Mol Pharmacol 2005;67:1522–33.
137. Yun JY, Kim JH, Kim HK, Lim YJ, Do SH, Zuo Z. Effects of intravenous anesthetics on the activity of glutamate transporter EAAT3 expressed in *Xenopus* oocytes: evidence for protein kinase C involvement. Eur J Pharmacol 2006;531:133–9.
138. Dunlop J, Eliasof S, Stack G, McIlvain HB, Greenfield A, Kowal D, Petroski R, Carrick T. WAY-855 (3-amino-tricyclo[2.2.1.02.6]heptane-1,3-dicarboxylic acid): a novel, EAAT2-preferring, nonsubstrate inhibitor of high-affinity glutamate uptake. Br J Pharmacol 2003;140:839–46.
139. Berry CB, Hayes D, Murphy A, Wiessner M, Rauen T, McBean GJ. Differential modulation of the glutamate transporters GLT1, GLAST and EAAC1 by docosahexaenoic acid. Brain Res 2005;1037:123–33.
140. Kim JH, Lim YJ, Ro YJ, Min SW, Kim CS, Do SH, Kim YL, Zuo Z. Effects of ethanol on the rat glutamate excitatory amino acid transporter type 3 expressed in Xenopus oocytes: role of protein kinase C and phosphatidylinositol 3-kinase. Alcohol Clin Exp Res 2003;27:1548–53.
141. Doi T, Ueda Y, Tokumaru J, Willmore LJ. Molecular regulation of glutamate and GABA transporter proteins by clobazam during epileptogenesis in Fe^{3+}-induced epileptic rats. Brain Res Mol Brain Res 2005;142:91–6.
142. Ganel R, Ho T, Maragakis NJ, Jackson M, Steiner JP, Rothstein JD. Selective up-regulation of the glial Na^+-dependent glutamate transporter GLT1 by a neuroimmunophilin ligand results in neuroprotection. Neurobiol Dis 2006;21:556–67.

19
Glutamate Transport in Retinal Glial Cells during Diabetes

Erica L. Fletcher and Michelle M. Ward

CONTENTS

INTRODUCTION
THE IMPORTANCE OF GLUTAMATE TRANSPORT
CHANGES IN GLUTAMATE HOMEOSTASIS IN THE RETINA DURING DIABETES
MECHANISMS INVOLVED IN ALTERED GLUTAMATE FUNCTION
SIGNIFICANCE OF GLUTAMATE TRANSPORTER FUNCTION
CONCLUSION
ACKNOWLEDGMENTS
REFERENCES

INTRODUCTION

Diabetic retinopathy is the leading cause of blindness in those of working age. It develops in approximately 80% of those with type I diabetes after 10 years, and more than 90% after 20 years of diabetes *(1–3)*. Proliferative diabetic retinopathy, the vision-threatening form of the disease occurs in 25% of people with type I diabetes following 15 years *(4)*.

Diabetic retinopathy is characterized by alterations in the retinal vasculature. In particular, diabetes induces the loss of pericytes, breakdown of the blood–retinal barrier, thickening of the basement membrane and the development of microaneurysms in the early stages. More serious changes include hemorrhages, intraretinal microvascular abnormalities, and finally angiogenesis of retinal blood vessels, which can lead to scarring and retinal detachment *(5)*.

Although clinical care of those with diabetes involves assessment of the retinal vasculature and treatment is directed towards the prevention of retinal neovascularization, there is considerable evidence that both retinal neurons and glia are also affected during diabetes. Evidence from psychophysical, electrophysiological, and anatomical studies indicates that neurons are affected during diabetes, and several studies have shown that neural dysfunction occurs before the earliest vascular changes are observed *(6–8)*.

It is well known that the integrity and function of the retina is intricately linked to the normal functioning of retinal glial cells. The two principal macroglial cells within the retina are Müller cells and astrocytes. Müller cells are located within the inner nuclear

From: *Ophthalmology Research: Ocular Transporters in Ophthalmic Diseases and Drug Delivery*
Edited by: J. Tombran-Tink and C. J. Barnstable © Humana Press, Totowa, NJ

layer, and have processes that envelop all neurons and synapses extending from the inner limiting membrane to the outer limiting membrane. Astrocytes are located within the nerve fiber layer. Both glial cell types have processes that wrap around retinal blood vessels forming a *glia limitans*.

Müller cells play an essential role in the normal function of the retina *(9)*. They are intricately involved in uptake and degradation of the neurotransmitters, glutamate and GABA, the shuttling of energy metabolites to neurons, and they also act as a siphon for the uptake of extracellular potassium *(9)*. Of relevance to diabetic retinopathy, Müller cells and astrocytes have been linked with the formation and maintenance of the blood–retinal barrier *(10)*. Indeed, recently, retinal glial cells have been shown to regulate blood-vessel diameter, implying that these cells regulate retinal blood flow in the rat retina *(11)*. Therefore Müller-cell abnormalities could cause both neuronal dysfunction, as well as vascular alterations. The aim of this chapter is to examine the evidence that diabetes alters the ability of retinal glia to metabolize glutamate.

THE IMPORTANCE OF GLUTAMATE TRANSPORT

Glutamate is the major excitatory neurotransmitter within the retina, where it mediates neurotransmission between neurons of the retinal through pathway (photoreceptors, bipolar cells and ganglion cells) *(12, 13)*. The level of glutamate within the retina is approximately 1.55 mM, with the vast majority of this amino acid located intracellularly *(14)*. The high intracellular content of glutamate can be visualized immunocytochemically as shown in Fig. 1. shows a vertical section of a rat retina that has been labeled with an antibody directed against glutamate conjugated to bovine serum albumin. Every neuron within the retina labels intensely for glutamate.

In order for the retina to operate in a normal fashion, it is imperative that the extracellular concentration of glutamate is held at a low level, normally around 3–4 µM. Glutamate elicits a neuronal response by binding to one or more glutamate receptors. Under conditions where the extracellular level of glutamate is raised, glutamate receptors are stimulated and when the level of glutamate in the synaptic cleft is abnormally high, glutamate receptors are excessively stimulated leading to neuronal cell death. Glutamate transporters play a vital role in removing extracellular glutamate, to maintain a low glutamate concentration within the extracellular space.

Within the retina, retinal Müller cells play a central role in removing extracellular glutamate *(13)*. Following release of glutamate from neurons, it must be taken up by Müller cells via a high-affinity glutamate transporter. It is then rapidly converted to glutamine and shuttled back to neurons where it is the main precursor for the formation of the neurotransmitters glutamate and gamma-aminobutyric acid (GABA) (Fig. 2). If glutamate recycling is altered by disease, the function of neurons is affected.

From the discussion above, it is clear that glutamate found extracellularly can be toxic at high levels, and therefore removal of glutamate is important for the health of neurons. Glutamate removal from the extracellular space is also important to maintain the neurotransmitter pools within presynaptic terminals. Glutamate is the principal precursor for the formation of GABA, and therefore adequate glutamate must be present in GABAergic neurons. In addition, glutamate can be transaminated to form α-ketoglutarate, an

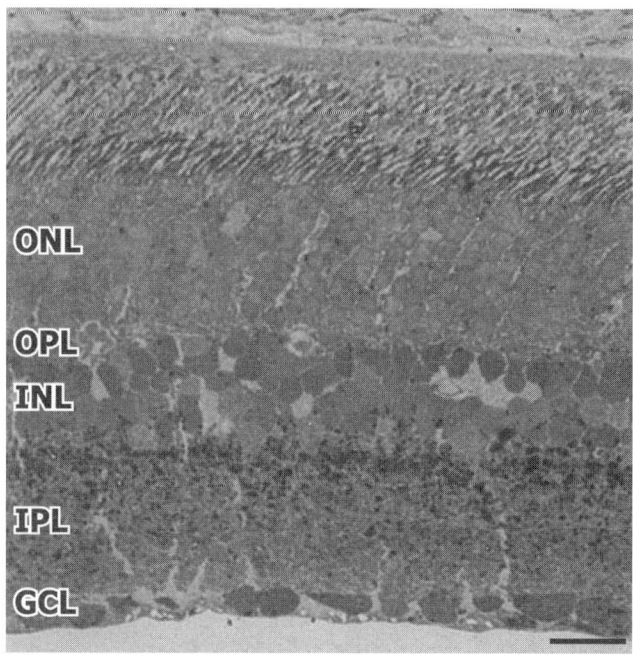

Fig. 1. Vertical section of rat retina labeled with an antibody directed against glutamate. Photoreceptors, bipolar cells, and ganglion cells are labeled, most likely because these neurons utilize glutamate as their neurotransmitter. In addition, horizontal and amacrine cells are labeled for glutamate, indicative of the precursor role that glutamate plays in these cells. Müller-cell somata and processes are devoid of glutamate because they rapidly degrade glutamate to glutamine. ONL, outer nuclear layer; OPL, outer plexiform layer; INL, inner nuclear layer; IPL, inner plexiform layer; GCL, ganglion cell layer; NFL, nerve fiber layer. Scale bar = 20 µM.

intermediary of the tricarboxylic cycle. Therefore, intracellular glutamate can be useful as a substrate for metabolism, especially in situations of metabolic stress *(13,15)*.

In summary, for normal function of the retina, it is important that there are adequate systems for the removal of extracellular glutamate. Removal of extracellular glutamate is important for the prevention glutamate-induced cell death, for the adequate supply of neurotransmitter precursors and to provide an intermediary for the tricarboxylic acid cycle in times of need.

Excitatory Amino Acid Transporters in the Retina

Removal of synaptically released glutamate occurs via glutamate transporters. At the molecular level five excitatory amino acid transporters (EAATs) have been cloned from human tissue: EAAT1–5 *(16,17)*. Rat homologs of the transporters EAAT1–4 have also been cloned, including EAAT1 (GLAST), EAAT2 (GLT1), EAAT3 (EAAC1), and EAAT4 (see table 1) *(18–21)*. These transporters have variable affinity for glutamate (1–100 µM) and are driven by the gradients of both sodium and potassium ions. All five EAATs transport L-glutamate, as well as L- and D-aspartate, and also cotransport

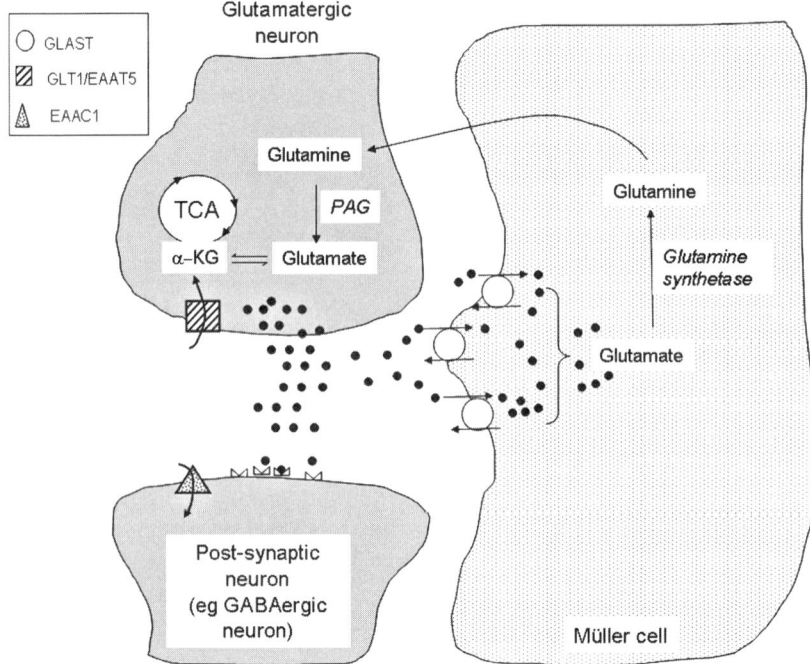

Fig. 2. Schematic diagram showing the uptake and turnover of glutamate by retinal Müller cells. Glutamate released by glutamatergic neurons within the retina is taken up via the high-affinity electrogenic glutamate transporter GLAST. Within Müller cells, glutamate is rapidly degraded to glutamine via the enzyme glutamine, and glutamine is shuttled back into neurons to act as a precursor for the formation of the neurotransmitters glutamate and GABA. Glutamate transporters are located either on glial cell membranes [glutamate-aspartate transporter (GLAST)], or on presynaptic [glutamate transporter 1 (GLT1), excitatory amino acid transporter 5 (EAAT5)] or postsynaptic [excitatory amino acid carrier 1 (EAAC1)] terminals, as indicated.

Table 1 Summary of the common names and abbreviations of glutamate transporters and their cellular expression pattern within the mammalian retina

Glutamate transporter	Alternative name in rat retina.	Cellular expression pattern
EAAT1	Glutamate–aspartate transporter (GLAST)	Müller cells and astrocytes
EAAT2	Glutamate transporter 1 (GLT1)	Bipolar cells
EAAT3	Excitatory amino acid carrier 1 (EAAC1)	Amacrine and horizontal cells
EAAT4	-	Astrocytes
EAAT5	-	Photoreceptors, Müller cells (?)
Cystine–glutamate exchanger	-	Müller cells

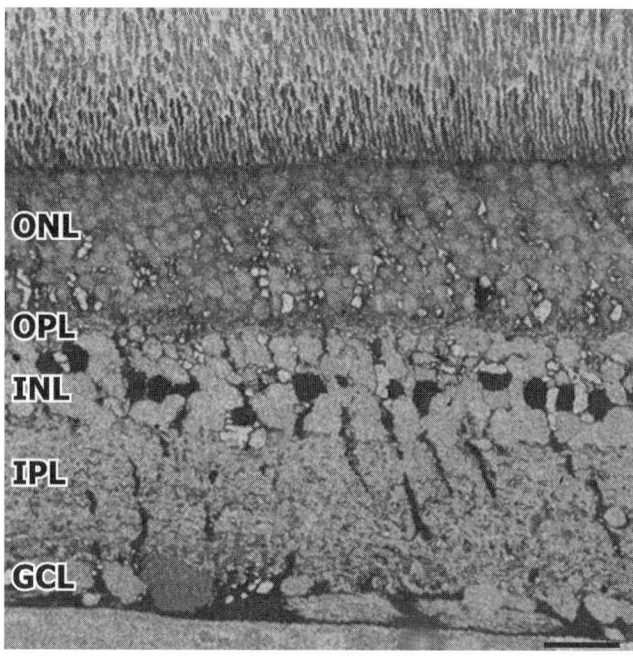

Fig. 3. A vertical section of an adult rat retina incubated in D-aspartate, fixed and immunolabeled for D-aspartate. D-aspartate labeling is seen labeling Müller cell somata, and processes. Abbreviations as in Fig. 1. Scale bar = 20 μM.

sodium into the cell, while potassium is transported out. In addition to the transport of glutamate, these proteins also function as channels for anions, a property that is particular prominent in EAAT4 and EAAT5 *(16)*.

All five EAATs are expressed within the retina. GLAST is localized within Müller cells and astrocytes (see Fig. 4) *(22,23)*, whilst GLT1, EAAC1 and EAAT5 are all localized within specific neuronal classes including bipolar (GLT1), horizontal (EAAC1), amacrine cells (EAAC1), and photoreceptors (EAAT5, GLT1) *(22–26)*. The neuronal expression of EAAT2 is shown in Fig. 5. Immunolabeling of two splice variants of GLT1, GLT1A, and GLTB, shows labeling of bipolar cells and processes in the inner plexiform layer. EAAT5 is also shown labeling cones and bipolar cell processes in the rat retina. It has been suggested that these neuronal glutamate transporters are important for maintaining the transmitter pools within presynaptic terminals. For example, EAAC1 is localized within GABAergic neurons in the retina and brain, and is thought to maintain the levels glutamate in these terminals so that formation of GABA can occur rapidly *(27,28)*.

EAAT4 was originally localized within Purkinje cells of the cerebellum *(21)*. More recently, EAAT4 has been identified in astrocytes from the spinal cord *(29)* and retina (see Fig. 4) *(30)*. This implies that astrocytes within the retina express two glutamate transporters, GLAST and EAAT4. These glutamate transporters display different affinities for glutamate, and distinct chloride permeabilities. It is possible that expres-

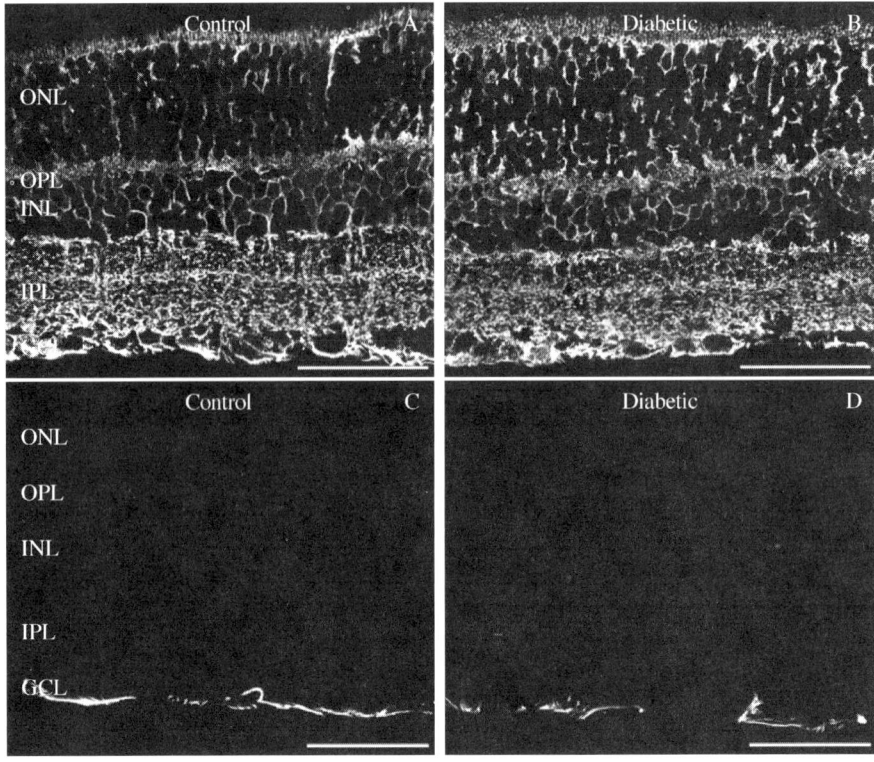

Fig. 4. Vertical sections of control, **(A)** and **(C)**, and diabetic, **(B)** and **(D)**, Sprague–Dawley rat retinas labeled for the glutamate–aspartate transporter (GLAST) (A, B) and excitatory amino acid transporter 4 (EAAT4) (C, D). GLAST is highly expressed by Müller cells and astrocytes in control and diabetic retina. No difference in expression level of GLAST can be noted in the diabetic rat retina, compared with the control. EAAT4 is expressed by astrocytes in the rat retina, which are located within the nerve fibre layer. No difference in expression of EAAT4 could be detected. Abbreviations as in Fig. 1 Scale bar = 50 µM.

sion of two glutamate transporters by the same cell occurs because of the different functions of these transporters.

Although the function of specific transporter types is contentious, it is clear that GLAST is the predominant transporter for removal of glutamate within the retina and is localized in Müller cells and astrocytes (22,23). Fig. 3 illustrates an experiment that reveals the importance of Müller cells in the removal of glutamate within the retina. D-aspartate is a nonhydrolyzable glutamate analog that is transported by all five EAATs with similar efficiency. Because D-aspartate is an amino acid, it can be fixed within tissue using conventional aldehyde fixatives and then immunocytochemically detected. Fig. 2 shows a vertical section of rat retina that has been incubated with D-aspartate, and then labeled with an antibody directed against D-aspartate. D-aspartate heavily labels Müller-cell somata and their processes. Although a variety of EAATs are known to be expressed by neurons, antibodies to D-aspartate do not label these other neurons, suggesting that Müller cells are the primary site for the removal of glutamate within the retina.

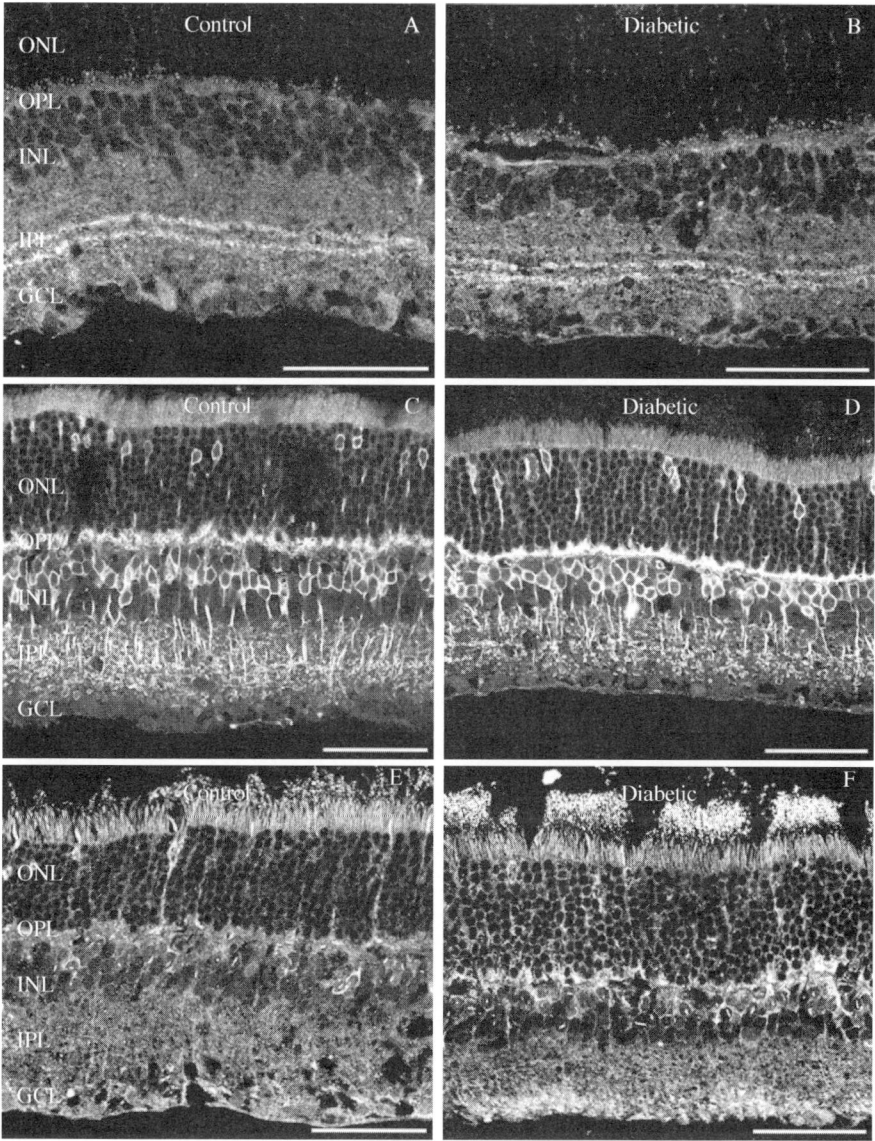

Fig. 5. Vertical sections of control (**A, C, E**) and diabetic (**B, D, F**) Sprague–Dawley rat retinae labeled for glutamate transporter (GLT)A (A, B), GLT1B (C, D) and excitatory amino acid transporter 5 (EAAT5) (E, F). GLT1A labels two prominent bands within the IPL. GLT1B, a different splice variant of GLT1, labels cone bipolar cells and their processes, as well as cones. EAAT5 labels cones on the rat retina. No differences are noted between control and diabetic retinas for GLT1A or GLT1B. However, EAAT5 did show more-prominent labeling within the OPL in the diabetic rat retina. Abbreviations as in Fig. 1. Scale bar = 50 µM.

Several studies have shown that GLAST is the most important glutamate transporter in the retina accounting for around 50% of all glutamate uptake *(23,31,32)*. Moreover, in mice where the gene for GLAST has been knocked out, visual function, as measured by the electroretinogram, is substantially affected, although not completely lost *(31)*. Therefore, glutamate uptake by Müller cells is central to maintaining the normal function of the retina. However, it should also be noted that GLAST alone does not account for the total removal of glutamate in the retina.

The five EAATs constitute the high-affinity glutamate uptake systems in the retina, but other proteins present within the retina can also transport glutamate in a sodium-independent manner. Recently, a cystine–glutamate exchanger has been identified in the retina *(33)*. This transporter is an exchanger with 1:1 stoichiometry, transporting cystine into a cell in exchange for glutamate. It is electroneutral, and uses the transmembrane gradient of glutamate as the driving force. When the extracellular level of glutamate is elevated, this exchanger can release cystine whilst taking up glutamate.

In summary, glutamate uptake into retinal Müller cells and astrocytes is dependent on the activity of GLAST and/or EAAT4. In the next section we will consider the evidence that glutamate transport is abnormal during diabetes.

CHANGES IN GLUTAMATE HOMEOSTASIS IN THE RETINA DURING DIABETES

There is considerable evidence that Müller cells are functionally abnormal in diabetes. The most obvious change that has been reported in Müller cells during diabetes is reactive gliosis *(34–38)*. However, functional abnormalities have also been observed, including anomalies in glutamate and GABA turnover *(39,40)*. In support of the notion that glutamate turnover is abnormal in diabetes is the observation that glutamate levels are elevated in the vitreous of diabetic rats and human patients *(37,41,42)*. In addition, the concentration of glutamate within the rat retina following three months of diabetes is elevated and the rate of glutamine formation is reduced, suggesting that the conversion of glutamate to glutamine by Müller cells might be abnormal *(37,42)*. As noted earlier, showing that the glutamate concentration in the whole retina is altered during diabetes is not evidence per se of an abnormality in glutamate transport, but rather of a defect or defects in the glutamate–glutamine cycle as a whole. An elevation in glutamate within the retina could be explained by; (i) a change in glutamate transport, (ii) a reduction in the release of glutamate from glutamatergic neurons, or (iii) a reduction in the activity of glutamine synthetase. In relation to these three possible explanation, the activity of glutamine synthetase, the principal enzyme involved in converting glutamate to glutamine within Müller cells was found to be reduced following as little as two months duration of diabetes in the rat retina *(42)*. In addition, there is evidence for a reduction in neuronal function from an early stage of diabetes *(43,44)*. This change in neuronal function could cause a reduction in the release of glutamate from glutamatergic neurons.

Two studies have considered, in detail, glutamate uptake by Müller cells during diabetes *(45,46)*. Li and Puro *(45)* have shown that glutamate uptake was decreased in isolated rat

Müller cells, possibly through oxidative processes. These authors used patch-clamping techniques to measure currents in isolated Müller cells, following application of the glutamate analog 1-trans-pyrrolidine-2,4-dicarboxylate. Glutamate uptake into Müller cells via GLAST is electrogenic, because sodium ions are cotransported into Müller cells with glutamate, while potassium is transported out of the cell. This movement of sodium and potassium ions across Müller cell membranes can be measured as current using patch-clamping electrodes. Li and Puro *(45)* measured GLAST function by measuring the current induced in freshly isolated control and diabetic rat Müller cells. They observed that the current induced following application of the glutamate analog, 1-trans-pyrrolidine-2,4-dicarboxylate, was reduced by 36% following as little as four weeks of diabetes, suggesting that GLAST activity is reduced early in the course of diabetes.

The second study, by Ward et al. *(46)*, examined the expression of GLAST and EAAT4 in the rat retina and measured glutamate uptake during diabetes in vivo. These authors used semiquantitative polymerase chain reaction (PCR) and immunocytochemistry to examine whether the expression of EAAT4 or GLAST was altered following 12 weeks of diabetes. No differences in messenger RNA or protein expression were noted. In order to probe the function of GLAST in vivo, these authors quantified the level of uptake of D-aspartate into Müller cells following 1, 4, and 12 weeks of diabetes. D-aspartate is taken up into Müller cells via GLAST, but is not degraded by glutamine synthetase and can therefore be visualized using immunocytochemical techniques. These authors showed that D-aspartate uptake was greater in diabetic rat retina following as little as one week of diabetes. Results from a similar experiment are shown in Fig. 6. Control and diabetic rats were injected intravitreally with D-aspartate. Rats were then sacrificed, and their eyes were removed, fixed and prepared for postembedding immunocytochemistry. Fig. 6 shows the ganglion cell layer, and nerve fiber layer, of a control and diabetic rat retina that has been labeled with an antibody directed against D-aspartate. The end-feet of Müller cells are prominently labeled in both the control and diabetic retina, and appear a little darker in the diabetic rat retina. Fig. 6C shows the intensity of D-aspartate labeling in control and diabetic Müller cell end-feet following 1, 4, or 12 weeks of diabetes. D-aspartate labeling was greater in the end-feet of Müller cells at all stages of diabetes (analysis of variance, $p < 0.05$).

How can the results of Li and Puro *(45)* be reconciled with the apparently opposite finding of Ward et al. *(46)*? First, experimentation on isolated cells may not always reflect function in vivo, because the external environment is artificially controlled. This could be particularly significant in studying the changes in cellular function during diabetes, because the external milieu can be altered by breakdown of the blood–retinal barrier *(47)*.

It has been recently recognized that measuring the function of GLAST using patch-clamping electrophysiological tools can lead to different results compared to those generated by measuring D-aspartate uptake. A recent study by Sarthy et al. *(32)* showed that Müller cells isolated from GLAST-knockout mice displayed no electrophysiologically recordable current when glutamate was applied, verifying that GLAST function did not exist in these cells. However, the retinas of these same animals showed extensive D-aspartate uptake into Müller cells. Therefore, it is likely that glutamate uptake into Müller cells does not rely solely on the function of GLAST, but requires that other nonelectrogenic transporters are present.

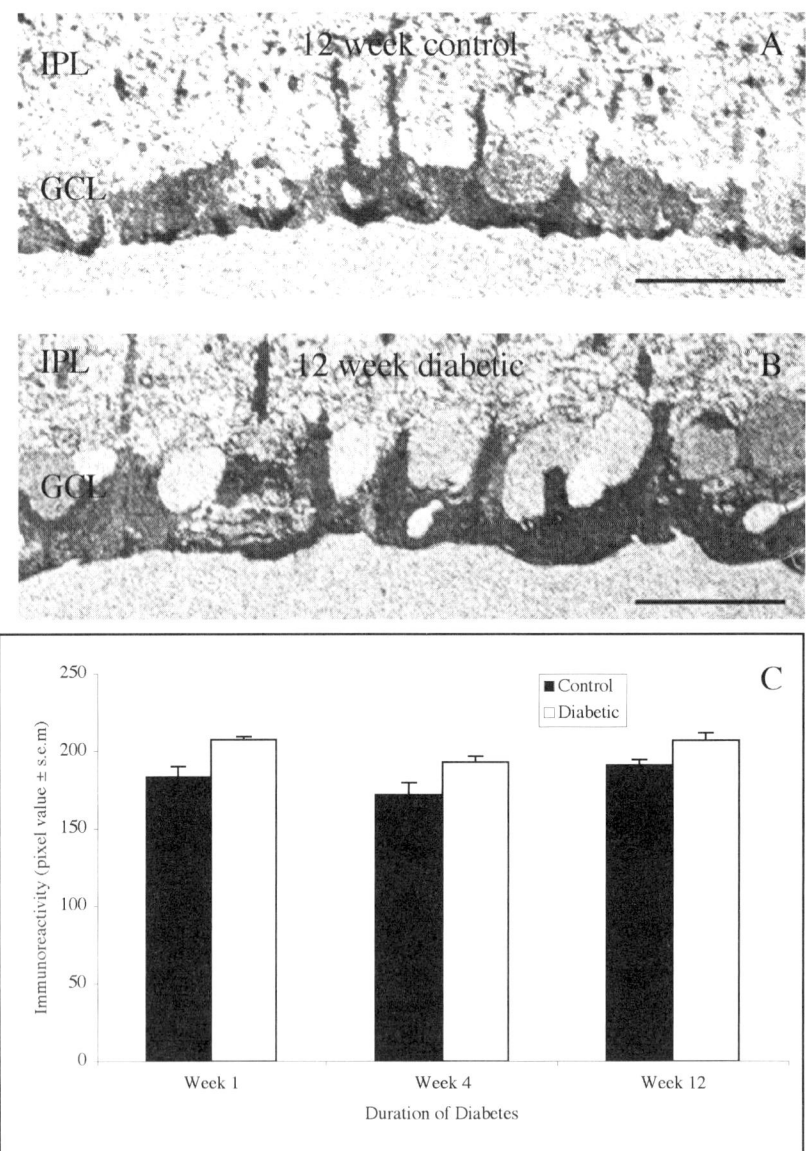

Fig. 6. Vertical sections of the inner retina of (**A**) a control retina, and (**B**) a diabetic rat retina, that were intravitreally injected with D-aspartate and then immunolabeled for D-aspartate. Müller cell end-feet within the nerve fiber layer are heavily labeled for D-aspartate. The level of D-aspartate immunoreactivity within Müller cell end-feet is higher during diabetes than in control retina, even as early as one week following the onset of diabetes (**C**) Graph shows the mean pixel intensity, plus or minus the standard error of the mean (±SEM), of D-aspartate labeling within Müller cell end-feet following 1, 4, and 12 weeks of diabetes. Uptake of D-aspartate was significantly greater in diabetic end-feet compared with control retinas, at all stages of diabetes examined. Abbreviations as in Fig. 1. Scale bar = 20 μM.

Therefore, when one considers the studies of Li and Puro *(45)* and Ward et al. *(46)* together, it is possible to conclude that GLAST function is reduced in diabetes (as shown by Li and Puro), but that another nonelectrogenic glutamate transporter is upregulated in diabetes. More importantly, overall, glutamate uptake by Müller cells during diabetes is greater than in controls and this may have a bearing on the function of retinal neurons during diabetes.

MECHANISMS INVOLVED IN ALTERED GLUTAMATE FUNCTION

There are a number of factors that can cause an alteration in glutamate transporter expression and function. Of particular relevance to diabetes are the roles that ischemia, glutamate protein kinase C, and lipid peroxidation play in regulating transporter expression and function *(44)*. In this section, we will consider possible ways that glutamate transporter expression and function could be altered during diabetes.

Diabetic retinopathy is a disease of the inner retinal vasculature, and although the precise etiology is not clear, hypoxia within the inner retina is likely to play an essential role. Therefore, it is important to ask whether hypoxia or ischemia could cause the changes in glutamate transporter function observed during diabetes. Indeed, Barnett et al. *(48)* showed that uptake of D-aspartate into Müller cells was impaired following occlusion of the central retinal artery, but expression of GLAST was unaffected *(48)*. By contrast, Stanimirovic et al. *(49)* and Payet et al. *(50)* showed that glutamate uptake was enhanced in cultured astrocytes following a short period of ischemia or hypoxia, and suggested that enhanced glutamate uptake by glial cells might act as a protective mechanism in the early stages of ischemia. Only after sustained ischemia is glutamate uptake reduced, and neuronal function substantially affected. There are some very interesting parallels between changes in the retina during short-term ischemia and during diabetes. The elevation in glutamate uptake, as shown by D-aspartate uptake, could be a protective mechanism within the retina.

It is not absolutely clear why glutamate uptake is elevated and then later reduced during ischemia *(49)*. GLAST is known to cotransport sodium ions with glutamate, at the expense of potassium. The levels of these ions are, in turn, controlled by energy-dependent Na^+,K^+-ATPase pumps. Indeed, Stanimirovic et al. *(49)* showed that, following short-term ischemia, Na^+,K^+-ATPase activity was elevated and could provide the driving force for glutamate uptake. These findings suggest that during early ischemia, glial cells are functionally altered so as to prevent excessive neuronal death secondary to glutamate excitotoxicity. Similar mechanisms could operate in the retina during diabetes, and would explain why glutamate uptake in the retina is enhanced.

Diabetes is known to be accompanied by an elevation in the activity of protein kinase C (PKC), and compounds that inhibit the beta form of PKC prevent vascular change during diabetes *(51,52)*. PKC is a potent regulator of glutamate transport, however, the PKC isoform that regulates GLAST activity in the retina is the PKC delta isoform *(53)*. Further work will be necessary to define whether isoforms other than PKC beta are altered during diabetes, and whether this could have a bearing on glutamate transport.

The retina during diabetes is prone to oxidative stress *(54)*. Moreover, glutamate levels have been shown to be ameliorated by treatment with antioxidants *(55)*. More recently, Li and Puro *(45)* examined whether the reduction in GLAST function could be returned to normal with chemical reductants. They found that treatment of isolated Müller cells with the reductant disulfide-dithreitol (DTT) restored the size of the current induced by the glutamate analog 1-trans-pyrrolidine-2,4-dicarboxylate. These studies suggest that oxidative stress can alter glutamate levels within the retina, and that the function of GLAST may be altered by oxidative stress.

In summary, a variety of biochemical alterations that are known to be important in diabetes can also directly affect the expression of glutamate transporters, including ischemia, PKC and oxidative stress. The following section will consider the significance of a change in glutamate transport.

SIGNIFICANCE OF GLUTAMATE TRANSPORTER FUNCTION

Glutamate transport plays at least three essential roles in the CNS, including, (i) the removal of extracellular glutamate so as to prevent excitotoxic damage to neurons, (ii) uptake of glutamate so as to replenish neurotransmitter pools of both glutamate and GABA, and (iii) to provide a supply of energy metabolites for the tricarboxylic acid cycle. The consequence of an alteration in glutamate transport will manifest as abnormalities in any one or more of these three functions.

Neuronal Dysfunction and Cell Death

There is a great deal of evidence that suggests that retinal neurons are affected by diabetes. Moreover, several studies have suggested that neuronal dysfunction occurs prior to the onset of vascular changes *(7,8)*. The types of neurons affected by diabetes include photoreceptors, amacrine cells, and ganglion cells; photoreceptors are known to become dysfunctional from as early as day two of diabetes and show signs of apoptosis within 12 weeks *(43,56–59)*. Amacrine and ganglion cells are known to undergo programmed cell death in a number of animal models of diabetes *(56,59–61)*.

The factors that lead to apoptosis of retinal neurons during diabetes are not known, but could be related to abnormalities in glutamate homeostasis, namely, either an elevation in glutamate uptake or a decrease in GLAST function. During diabetes, glutamate-induced neurotoxicity could arise in at least two ways. Firstly, glutamate could diffuse into the retina from the plasma following breakdown of the blood retinal barrier. The concentration of glutamate within plasma is approximately 100–300 μM, which is far in excess of levels of glutamate known to cause neuronal damage (more than 5 μM) *(62,63)*. Alternatively, because glutamate transport is abnormal from an early stage in diabetes, the extracellular concentration of glutamate within the retina could reach the threshold for neuronal damage.

Increases in the level of extracellular glutamate could lead to an increase in expression of glutamate transporters, and also an increase in functional uptake. Indeed, GLAST and GLT1 expression is elevated in the optic nerve during multiple sclerosis, and glutamate uptake in this disease has been shown to be increased *(64)*. The increase in expression and function is thought to be a protective effect, that is, a way of limiting neurodegeneration in this disease.

Hyperactivity of glutamate transport can also be deleterious to the function of neurons, as has been shown in schizophrenia *(16)*. Glutamate transporter expression is enhanced in the thalamus of patients with schizophrenia *(65)*. Smith et al. *(65)* suggest that the hyperfunction of GLAST may be linked with a decrease in N-methyl D-aspartate (NMDA) receptor expression in schizophrenia *(66)*. An increase in glutamate uptake capacity would lower the concentrations of glutamate within synapses to levels below that normally found. This could have downstream effects, such as a reduction in glutamate receptor expression and function. There are no reports of changes in glutamate receptor expression in the retina during diabetes. However, it is possible that some of the neuronal dysfunction reported to occur during diabetes could arise because of an abnormality of glutamate transport, and concomitant glutamate receptor expression changes.

Reduced Threshold for Neuronal Damage Secondary to Ischemia

A downstream effect of abnormalities in the turnover of glutamate is the susceptibility that neurons may have to ischemic damage. The threshold for damage due to ischemia could be lower during diabetes than in the normal rat retina. Although the levels of glutamate in the vitreous and retina are elevated during diabetes, they remain below the threshold for neurodegeneration *(41)*. However, it is possible that in a case of even mild ischemia, where the glutamate levels rise only subtly, the threshold for neuronal damage might be more easily attained in diabetes compared to controls. In effect, diabetes might reduce the 'functional reserve' of neurons, so that even a slight change in glutamate homeostasis is enough to ellicit neuronal apoptosis. Harada et al. *(31)* showed that the retinas of GLAST-knockout mice were more sensitive to ischemia than wild-type mice. Moreover, Hartwick et al. *(67)* showed that ganglion cells were more sensitive to glutamate-induced excitotoxic death following raised intraocular pressure than controls, suggesting that subtle changes in glutamate homeostasis render the retina more susceptible to further insult.

Abnormalities in Glutamate Transport as a Generalized Indicator of Müller Cell Dysfunction

The finding that glutamate homeostasis is altered during diabetes is evidence that Müller cells are functionally altered at an early stage in the disease process. Müller cells within the retina have a number of essential roles, some or all of which could be impaired during diabetes *(9,68)*. The significance of an alteration in glutamate turnover may be an incidental finding on its own, or may be evidence for a much larger problem with Müller-cell function generally. In this section, we will consider briefly the evidence that other functions of retinal Müller cells are abnormal during diabetes.

In addition to uptake and degradation of glutamate, retinal Müller cells are also important for the uptake and turnover of the inhibitory neurotransmitter GABA *(69)*. GABA released into the extracellular space by inhibitory neurons is removed by high-affinity GABA transporters that are expressed by Müller cells. GABA is then converted to succinate and shunted into the tricarboxylic acid cycle. Conditions that affect aerobic metabolism and the tricarboxylic acid cycle can lead to an impairment in GABA degradation. GABA accumulation has been noted in diabetic rats 12 weeks after the onset of diabetes, suggesting that GABA uptake or degradation is altered during diabetes(39,40).

Müller cells play a key role in the siphoning of extracellular potassium, and therefore express inwardly rectifying potassium channels in abundance. Recently, Pannicke et al. (70) showed that the function of a number of different types of potassium channels was altered during diabetes, which could have important implications in the swelling that occurs within the retina during macular oedema.

As noted in the above paragraphs, several Müller-cell functions are altered during diabetes. This is an important consideration in view of the observations that Müller cells express vascular endothelial growth factor (VEGF) and is thought to play an important role in regulating the blood–retinal barrier (10). Recently, Metea et al. (11) have shown that Müller cells may also regulate blood flow in the normal retina. Clearly, more work is needed to examine the possible role that Müller cells play in altering the retinal vasculature.

CONCLUSION

Glutamate uptake and removal from the synaptic cleft is a crucial function of retinal Müller cells, and failure of this system can have deleterious effects on neurons. Glutamate uptake by GLAST is altered in rats with diabetes. However, glutamate uptake overall within the retina is increased from an early stage of diabetes. This change, in turn, could lead to subtle changes in neuronal function, and an oversensitivity of neurons to glutamate-induced toxicity, or could be a sign that other aspects of Müller-cell function are abnormal.

ACKNOWLEDGMENTS

The authors wish to thank Ms Lisa Foster for her technical assistance, and Dr Theresa Puthussery and Dr Joanna Phipps for reading the manuscript. The work described in this chapter was supported by grants from the National Health and Medical Research Council (grants #350434 and #208950 to E. L. F.).

REFERENCES

1. Aiello LP, et al. Diabetic retinopathy. Diabetes Care 1998;21:143–56.
2. Klein R, et al. The Wisconsin epidemiologic study of diabetic retinopathy. III. Prevalence and risk of diabetic retinopathy when age at diagnosis is 30 or more years. Arch Ophthalmol 1984;102:527–32.
3. Klein R, et al. The Wisconsin epidemiologic study of diabetic retinopathy. II. Prevalence and risk of diabetic retinopathy when age at diagnosis is less than 30 years. Arch Ophthalmol 1984;102:520–6.
4. Diabetes Control and Complications Trial. Effect of intensive diabetes treatment on nerve conduction in the Diabetes Control and Complications Trial. Ann Neurol 1995;38:869–80.
5. National Health and Medical Research Council (NHRMC), Management of diabetic retinopathy. Clinical Practice Guidelines. Canberra: Arawang; 1997. p. 94.

6. Han Y, et al. Multifocal electroretinogram and short-wavelength automated perimetry measures in diabetic eyes with little or no retinopathy. Arch Ophthalmol 2004;122:1809–1815.
7. Han Y, et al. Multifocal electroretinogram delays predict sites of subsequent diabetic retinopathy. Invest Ophthalmol Vis Sci 2004;45:948–54.
8. Fortune B, Schneck ME, Adams AJ. Multifocal electroretinogram delays reveal local retinal dysfunction in early diabetic retinopathy. Invest Ophthalmol Vis Sci 1999;40:2638–51.
9. Ripps H and Witkovsky P. Neuron-Glia Interaction in the Brain and Retina. In: G.G. Chader and N.N. Osborne, editors. Progress in Retinal Eye Research. Oxford: Pergamon; 1985. p. 181–219.
10. Tout S, et al. The role of Muller cells in the formation of the blood-retinal barrier. Neuroscience 1993;55:291–301.
11. Metea MR and Newman EA. Glial cells dilate and constrict blood vessels: a mechanism of neurovascular coupling. J Neurosci 2006;26:2862–70.
12. Massey SC. Cell types using glutamate as a neurotransmitter in the vertebrate retina. Prog Retin Eye Res 1990;399–425.
13. Kalloniatis M and Tomisich G. Amino acid neurochemistry of the vertebrate retina. Prog Retin Eye Res 1999;18:811–66.
14. Marc RE, Murry RF, Basinger SF. Pattern recognition of amino acid signatures in retinal neurons. J Neurosci 1995;15:5106–29.
15. Kalloniatis M and Napper GA. Glutamate metabolic pathways in displaced ganglion cells of the chicken retina. J Comp Neurol 1996;367:518–36.
16. Danbolt NC. Glutamate uptake. Prog Neurobiol 2001;65:1–105.
17. Danbolt NC, et al. Properties and localization of glutamate transporters. Prog Brain Res 1998;116:23–43.
18. Kanai Y and Hediger MA. Primary structure and functional characterization of a high-affinity glutamate transporter. Nature 1992;360:467–71.
19. Pines G, et al. Cloning and expression of a rat brain L-glutamate transporter. Nature 1992;360:464–7.
20. Storck T, et al. Structure, expression, and functional analysis of a Na(+)-dependent glutamate/aspartate transporter from rat brain. Proc Natl Acad Sci USA 1992;89:10955–9.
21. Fairman WA, et al. An excitatory amino-acid transporter with properties of a ligand-gated chloride channel. Nature 1995;375:599–603.
22. Rauen T, Rothstein JD, Wassle H. Differential expression of three glutamate transporter subtypes in the rat retina. Cell Tissue Res 1996;286:325–36.
23. Rauen T, et al. High-affinity glutamate transporters in the rat retina: a major role of the glial glutamate transporter GLAST-1 in transmitter clearance. Cell Tissue Res 1998;291:19–31.
24. Eliasof S, et al. Localization and function of five glutamate transporters cloned from the salamander retina. Vision Res 1998;38:1443–54.
25. Arriza JL, et al. Excitatory amino acid transporter 5, a retinal glutamate transporter coupled to a chloride conductance. Proc Natl Acad Sci USA 1997;94:4155–60.
26. Pow DV, Barnett NL, Penfold P. Are neuronal transporters relevant in retinal glutamate homeostasis? Neurochem Int 2000;37:191–8.
27. Wiessner M, et al. Localization and possible function of the glutamate transporter, EAAC1, in the rat retina. Cell Tissue Res 2002;310:31–40.
28. Sepkuty JP, et al. A neuronal glutamate transporter contributes to neurotransmitter GABA synthesis and epilepsy. J Neurosci 2002;22:6372–9.
29. Hu WH, et al. Neuronal glutamate transporter EAAT4 is expressed in astrocytes. Glia 2003;44:13–25.

30. Ward MM, et al. Localization and expression of the glutamate transporter, excitatory amino acid transporter 4, within astrocytes of the rat retina. Cell Tissue Res 2004;315:305–10.
31. Harada T, et al. Functions of the two glutamate transporters GLAST and GLT-1 in the retina. Proc Natl Acad Sci USA 1998;95:4663–6.
32. Sarthy VP, et al. Glutamate transport by retinal Muller cells in glutamate/aspartate transporter-knockout mice. Glia 2005;49:184–96.
33. Pow DV. Visualising the activity of the cystine-glutamate antiporter in glial cells using antibodies to aminoadipic acid, a selectively transported substrate. Glia 2001;34:27–38.
34. Nork TM, et al. Muller's cell involvement in proliferative diabetic retinopathy. Arch Ophthalmol 1987;105:1424–9.
35. Mizutani M, Gerhardinger C, Lorenzi M. Muller cell changes in human diabetic retinopathy. Diabetes 1998;47:445–9.
36. Rungger-Brandle E, Dosso AA, Leuenberger PM. Glial reactivity, an early feature of diabetic retinopathy. Invest Ophthalmol Vis Sci 2000;41:1971–80.
37. Lieth E, et al. Glial reactivity and impaired glutamate metabolism in short-term experimental diabetic retinopathy. Penn State Retina Research Group. Diabetes 1998;47:815–20.
38. Lo TC, Klunder L, Fletcher EL. Increased Muller cell density during diabetes is ameliorated by aminoguanidine and ramipril. Clin Exp Optom 2001;84:276–281.
39. Ishikawa A, Ishiguro S, Tamai M. Accumulation of gamma-aminobutyric acid in diabetic rat retinal Muller cells evidenced by electron microscopic immunocytochemistry. Curr Eye Res 1996;15:958–64.
40. Ishikawa A, Ishiguro S, Tamai M. Changes in GABA metabolism in streptozotocin-induced diabetic rat retinas. Curr Eye Res 1996;15:63–71.
41. Ambati J, et al. Elevated gamma-aminobutyric acid, glutamate, and vascular endothelial growth factor levels in the vitreous of patients with proliferative diabetic retinopathy. Arch Ophthalmol 1997;115:1161–6.
42. Lieth E, et al. Diabetes reduces glutamate oxidation and glutamine synthesis in the retina. The Penn State Retina Research Group. Exp Eye Res 2000;70:723–30.
43. Phipps JA, Fletcher EL, Vingrys AJ. Paired-flash identification of rod and cone dysfunction in the diabetic rat. Invest Ophthalmol Vis Sci 2004;45:4592–600.
44. Fletcher EL, Phipps JA, Wilkinson-Berka JL. Dysfunction of retinal neurons and glia during diabetes. Clin Exp Optom 2005;88:132–45.
45. Li Q and Puro DG. Diabetes-induced dysfunction of the glutamate transporter in retinal Muller cells. Invest Ophthalmol Vis Sci 2002;43:3109–16.
46. Ward MM, et al. Glutamate uptake in retinal glial cells during diabetes. Diabetologia 2005;48:351–60.
47. Kusaka S, et al. Serum-induced changes in the physiology of mammalian retinal glial cells: role of lysophosphatidic acid. J Physiol 1998;506(2):445–58.
48. Barnett NL, Pow DV, Bull ND. Differential perturbation of neuronal and glial glutamate transport systems in retinal ischaemia. Neurochem Int 2001;39:291–9.
49. Stanimirovic DB, Ball R, Durkin JP. Stimulation of glutamate uptake and Na,K-ATPase activity in rat astrocytes exposed to ischemia-like insults. Glia 1997;19:123–34.
50. Payet O, et al. Hypoxia stimulates glutamate uptake in whole rat retinal cells in vitro. Neurosci Lett 2004;356:148–50.
51. Ishii H, et al. Amelioration of vascular dysfunctions in diabetic rats by an oral PKC beta inhibitor. Science 1996;272:728–31.
52. Ishii H, Koya D, King GL. Protein kinase C activation and its role in the development of vascular complications in diabetes mellitus. J Mol Med 1998;76:21–31.

53. Bull ND and Barnett NL. Antagonists of protein kinase C inhibit rat retinal glutamate transport activity in situ. J Neurochem 2002;81:472–80.
54. Kowluru RA. Diabetes-induced elevations in retinal oxidative stress, protein kinase C and nitric oxide are interrelated. Acta Diabetol 2001;38:179–85.
55. Kowluru RA, et al. Retinal glutamate in diabetes and effect of antioxidants. Neurochem Int 2001;38:385–90.
56. Barber AJ, et al. Neural apoptosis in the retina during experimental and human diabetes. Early onset and effect of insulin. J Clin Invest 1998;102:783–91.
57. Aizu Y, et al. Degeneration of retinal neuronal processes and pigment epithelium in the early stage of the streptozotocin-diabetic rats. Neuropathology 2002;22:161–70.
58. Park SH, et al. Apoptotic death of photoreceptors in the streptozotocin-induced diabetic rat retina. Diabetologia 2003;46:1260–8.
59. Martin PM, et al. Death of retinal neurons in streptozotocin-induced diabetic mice. Invest Ophthalmol Vis Sci 2004;45:3330–6.
60. Feit-Leichman RA, et al. Vascular damage in a mouse model of diabetic retinopathy: relation to neuronal and glial changes. Invest Ophthalmol Vis Sci 2005;46:4281–7.
61. Barber AJ, et al. The Ins2Akita mouse as a model of early retinal complications in diabetes. Invest Ophthalmol Vis Sci 2005;46:2210–8.
62. Castillo J, Davalos A, Noya M. Progression of ischaemic stroke and excitotoxic aminoacids. Lancet 1997;349:79–83.
63. Lipton SA and Rosenberg PA. Excitatory amino acids as a final common pathway for neurologic disorders. N Engl J Med 1994;330:613–22.
64. Vallejo-Illarramendi A, et al. Increased expression and function of glutamate transporters in multiple sclerosis. Neurobiol Dis 2006;21:154–64.
65. Smith RE, et al. Expression of excitatory amino acid transporter transcripts in the thalamus of subjects with schizophrenia. Am J Psychiatry 2001;158:1393–9.
66. Ibrahim HM, et al. Ionotropic glutamate receptor binding and subunit mRNA expression in thalamic nuclei in schizophrenia. Am J Psychiatry 2000;157:1811–23.
67. Hartwick AT, et al. Functional assessment of glutamate clearance mechanisms in a chronic rat glaucoma model using retinal ganglion cell calcium imaging. J Neurochem 2005;94:794–807.
68. Newman E and Reichenbach A. The Muller cell: a functional element of the retina. Trends Neurosci 1996;19:307–12.
69. Ehinger B. Glial and neuronal uptake of GABA, glutamic acid, glutamine and glutathione in the rabbit retina. Exp Eye Res 1977;25:221–34.
70. Pannicke T, et al. Diabetes alters osmotic swelling characteristics and membrane conductance of glial cells in rat retina. Diabetes 2006;55:633–9.

VIII
Ocular Drug Delivery

20
The Emerging Significance of Drug Transporters and Metabolizing Enzymes to Ophthalmic Drug Design

Mayssa Attar and Jie Shen

CONTENTS

SUMMARY
INTRODUCTION
TRANSPORTER EXPRESSION IN THE EYE
ENZYME DISTRIBUTION IN OCULAR TISSUES
FACTORS IMPACTING OCULAR TRANSPORTERS AND METABOLIZING ENZYMES
DISEASE
DRUG TRANSPORTER–METABOLIC ENZYME INTERPLAY
CONCLUSION
REFERENCES

SUMMARY

Ophthalmic drugs typically achieve less than 10% ocular bioavailability. A drug applied to the surface of the eye may cross ocular–blood barriers, where it may encounter cellular transporters and metabolizing enzymes, before it is distributed to the site of action. Characterization of ocular membrane transporters, enzyme systems and their respective substrate selectivity has provided new insight into the roles these proteins may play in ocular drug delivery and distribution. Altered drug transport and metabolism have been proposed to contribute to a number of ocular disease processes including inflammation, glaucoma, cataract, dry eye and neurodegeneration. With the development of more-sensitive technologies, ocular transport and enzyme systems will continue to be better characterized, and their properties will become an integral consideration in drug design and development.

INTRODUCTION

Eye disease can profoundly impact a patient's quality of life. As the population of the United States ages, the number of people with age-related eye disease is expected to

From: *Ophthalmology Research: Ocular Transporters in Ophthalmic Diseases and Drug Delivery*
Edited by: J. Tombran-Tink and C. J. Barnstable © Humana Press, Totowa, NJ

double within the next 30 years (1). Currently, it is estimated that blindness and visual impairment cost the United States federal government more than four billion dollars each year (1). Thus, a need exists to develop more effective drug treatments to prevent the progression of ocular disease.

A shift in the treatment paradigm for many ocular disorders has been observed in recent years. In the past, a disease such as glaucoma was managed by lowering intraocular pressure. New treatment strategies now involve preventing neurodegeneration and maintaining retinal function. While ocular therapies have evolved, this area is still largely ignored as compared to the much more established systemic therapeutics. However, many therapeutic principles applicable to systemic disorders are applicable to ocular disorders.

The pharmacokinetic processes of absorption, distribution, metabolism and excretion determine the concentration of drug delivered to the site of action. The individual or combined activities of drug-metabolizing enzymes and transporters play an integral role in these processes. The need for information and understanding regarding metabolic pathways in ocular tissues has long been recognized (2,3). There is growing recognition of the role metabolic enzymes may play in governing the rate and extent of drug delivery to various ocular tissues. Similar to ocular metabolism, expression, and function of membrane transporters at various layers of ocular epithelial and endothelial cells may significantly influence ocular drug efficacy by means of absorption, distribution, and excretion. These transporters have not yet been studied as extensively or thoroughly as those expressed in other organs, such as liver, intestine and kidney. The recent discoveries pertaining to ocular transporter systems and their impact on drug therapy, in association with ocular metabolizing enzymes, will be discussed.

TRANSPORTER EXPRESSION IN THE EYE

The Ocular Surface

Treatment of many eye diseases, such as corneal keratitis, conjunctivitis, dry eye, eye allergies and glaucoma, relies on topically applied medication. Some of these drugs exert their effect at the ocular surface, while others may need to penetrate across the epithelial lining of the cornea and/or conjunctiva to reach their target sites within the eye. Expression and function of corneal and conjunctiva transporters, in particular peptide and amino acid transporters, have been thoroughly reviewed in 2003 by Dey et al. (4). Table 1 lists many transporter proteins identified as existing in cornea and conjunctiva, as well as other tissues in the eye.

Besides peptide and amino acid transporters, Ito et al. (5) reported expression of the organic anion transporting polypeptide-E (oatp-E), at both messenger RNA (mRNA) and protein levels, in rat corneal epithelium. The oatp family of proteins, as plasma membrane proteins, are responsible for cellular uptake of various anionic and neutral molecules.

A functional study of organic cation transporters was carried out by Ueda et al. in rabbit conjunctival epithelial tissue (6). In this study, guanidine transport in the absorptive direction was significantly greater than in the secretive direction and was temperature and concentration dependent, suggesting involvement of transporter proteins. This study also found that guanidine transport in the conjunctiva was significantly inhibited

Table 1 Ocular distribution of known transporter proteins

Transporters	Substrates	Species	Ocular Expression Site	References
LAT1, LAT2	Large neutral amino acids	Rabbit, human	Cornea, ARPE cells	(69,68)
ASCT1	Neutral amino acids	Rabbit	Cornea	(112)
B(0,+)	Neutral and cationic amino acids	Rabbit, human	Cornea	(113)
EAAC1	Glutamate	Pig, human	Pigment retinal epithelium	(114)
EAAT4	Glutamate	Human	Pigment retinal epithelium	(114)
PepT1	Dipeptides	Rabbit	Cornea, conjunctiva	(115,116)
PEPT2	Dipeptides	Rat	Retina	(117)
GLUT1	Glucose	Rat, cow, human,	Cornea, conjunctiva, retina, iris-ciliary body	(118,119, 120,121)
oatp-E	Organic anion	Rat	Cornea, retina, ciliary body-iris, pigment retinal epithelium	(5)
MCTs	Monocarboxylate	Rat, human	Retina, pigment retinal epithelium, iris and ciliary body, lens, cornea	(122)
NaDC3	Dicarboxylate	Mouse	Optic nerve, most layers of the retina, retinal pigment epithelium, ciliary body, iris, and lens	(111)
Organic cation transporter, OCT3	Organic cations	Rabbit, mouse, human	Conjunctiva, pigment retinal epithelium	(6,17)
P-glycoprotein	Large neutral or cationic compounds	Rabbit, human	Cornea, conjunctiva, retina, pigment retinal epithelium	(9,10,11,12, 13,19,20)
MRP proteins	Large neutral or anionic compounds, glucuronide, glutathione, or sulfate conjugates	Human	Pigment retinal epithelium	(21)
ABCG2/ BCRP1	Wide range of substrates from chemotherapeutic agents to organic anion conjugates	Rabbit, human	Conjunctival epithelia basal cells	(14)

by dipivefrine, brimonidine, and carbachol *(6)*. Therefore, several cationic amine-type drugs might share this transport system, using it to cross conjunctiva and gain access to the underlying ocular tissue.

Corneal expression of nucleoside/nucleobase transporters was studied by Majumdar et al. *(7)*. The antiviral nucleoside analogues acyclovir and ganciclovir appeared to permeate across the excised corneal tissues by simple passive diffusion, suggesting an absence of interaction between these drugs and nucleoside/nucleobase transporters. This may explain the low corneal permeability of both acyclovir and ganciclovir, which greatly limits clinical application of these antiviral agents *(8)*.

The efflux pump transporter P-glycoprotein (P-gp), well known and well characterized in systemic tissues, has been reported to exist in both cornea and conjunctiva at functional and molecular levels *(9–13)*. In both tissues, P-gp was localized to the mucosal membrane, facilitating efflux of substrates into the tear side of the eye. This expression is beneficial as it removes harmful toxins from the ocular epithelial lining, but at the same time could restrict topical absorption and lower the ocular bioavailability of lipophilic drugs such as cyclosporine A and erythromycin. Coadministration of P-gp inhibitors at nontoxic doses could facilitate drug absorption across cornea. A new addition to the family of efflux pump proteins, breast-cancer-resistant protein (BCRP), which was first identified in human cancer cell lines, was recently found in human and rabbit conjunctival epithelial basal cells, but not in corneal epithelial cells *(14)*. This efflux pump, due to its wide substrate selectivity, could also potentially hamper ocular bioavailability of topically administered therapeutics.

The Back of the Eye

The blood–retinal barrier (BRB) consists of an outer barrier, the retinal pigmented epithelium, and an inner barrier, the retinal endothelium. The former prevents free passage of molecules from the well-perfused choroid into the retina, while the latter prevents free passage of molecules from the blood into the retina. The existence of these barriers poses a special challenge for drug delivery to the posterior segment of the eye to treat retinal disorders such as age-related macular degeneration, diabetic retinopathy, and proliferative vitreoretinopathy, which in severe cases lead to blindness. Conventional topical application of ocular drugs has not proven to be a useful route to treat these disorders. Expression of transporter proteins at the BRB, therefore, provides an opportunity for drug delivery to the retina and other tissues of the posterior eye.

The expression and function of BRB transporters, including amino acid, monocarboxylic acid, and folate transporters, have been reviewed in 2003 by Duvvuri et al. *(15)* and are listed in Table 1. These transporters were usually reported to exist in both the inner and outer BRB. However, Ocheltree et al. found that bovine and human RPE cell membranes did not actively transport the model dipeptide substrate GlySar, providing evidence against the expression of dipeptide transporter proteins in the RPE cells *(16)*. Expression and function of an organic cation transporter, OCT3, in the eye, and in particular the RPE, was reported by Rajan et al. *(17)*. The same study that identified the existence of the organic anion transporter oatp-E in cornea, also reported expression in RPE *(5)*. As for the nucleoside/nucleobase transporters, Majumdar et al. *(18)* reported functional evidence for their expression in the retina.

In BRB, P-gp serves as a biological barrier for gate keeping, in addition to the physical barrier of tight junctions in RPE and retinal endothelium that restricts paracellular diffusion (19). In human RPE cells, P-gp is expressed at both the apical and basolateral cell surfaces, possibly serving a protective function for the neural retina by clearing away unwanted substances from subretinal space (20). In addition to P-gp, the multidrug-resistance-associated protein (MRP) was also found in both cell lines and primary culture of human RPE cells (21). MRP proteins export organic anions and glucuronide, glutathione, and sulfate conjugated compounds.

It is essential to realize that in many cases, transporter studies in the eye have been limited to confirming expression at the transcriptional and translational levels, whereas further, detailed studies of various ocular tissues, e.g., studies of transporter expression level and localization, substrate selectivity, transporter activity kinetics, species differences, etc., will be necessary to effectively utilize this route to allow site-specific carrier-mediated drug delivery.

ENZYME DISTRIBUTION IN OCULAR TISSUES

Many enzyme systems that exist in systemic tissues have been identified in the various tissues of the eye (Table 2). The majority of published work has focused on enzyme systems active at the ocular surface. However, there is overlap of specific enzymes characterized at both the ocular surface and the back of the eye. Nonetheless, different

Table 2 Drug metabolizing enzymes characterized in ocular tissues

Enzyme	Substrates	References
Oxidoreductase		
Aldehyde oxidase	Brimonidine	(38)
Ketone reductase	Levobunolol and Ketanserin	(39,40,41)
Cyclooxygenase	Arachidonic acid	(123,124)
Monoamine oxidase	Serotonin	(125)
Cytochrome P450	Broad array of commercial drugs and endogenous substrates	See Table 3
Hydrolytic		
Aminopeptidase	L-leucine, L-alanine and L-arginine-4-methoxy-2-naphthylamide	(126,127)
Acetylcholinesterase	Naphthyl esters	(44)
Butyrylcholinestesase	Butyl esters	(44)
Carboxylesterase	Flestolol	(45)
Phosphatase	Phosphorylated amino acid residues	(43,128,129,130)
Aryl sulfatase	Sulfate esters	(43,131)
N-acetyl-β-glucosaminidase	Glucosamidated compounds	(43,132)
β-glucuronidase	Glucuronidated compounds	(132,133,134)
Conjugating		
Arylamine acetyltransferase	Aminozolamide and p-aminobenzoic acid	(48,49)
Glutathione S-transferase	Polyunsaturated fatty acids i.e., 4-hydroxynonenal	(50,51)

enzyme systems and different enzyme isoforms are expressed in the different tissues. These differences likely arise from the unique functions served by each tissue. The observation of region- and gender-specific gene expression patterns in drug-metabolizing enzymes in rat ocular tissues supports the notion that these enzymes play selective roles in order to maintain normal eye function *(22)*.

The Cytochrome P450 Monooxygenase System

Cytochrome P450 enzymes constitute a superfamily of heme-containing proteins that are of paramount importance to commercial drug metabolism *(23)*. In addition to xenobiotic metabolism, cytochrome P450 enzymes play crucial roles in the metabolism, either biosynthesis or degradation, of endogenous substrates such as steroids, fatty acids, vitamins and other compounds *(24)*.

Cytochrome P450 Enzymes in the Eye

Shichi first identified the presence of a microsomal electron transfer system in the bovine retinal epithelium in 1969 *(25)*. Following the first identification of the cytochrome P450 monooxygenase system in ocular tissues, specific isoforms have been identified with specific ocular tissue distributions (Table 3).

The expression of CYP1A1 and CYP1A2, isoforms important to polyaromatic hydrocarbon metabolism, were detected in the mouse ciliary and iris epithelium using in situ hybridization and immunohistochemistry *(26,27)*. The rat lens was found to express CYP2B1/2, which is phenobarbital inducible, and CYP2C11, which is an isoform specific to male rats, at the gene and protein levels *(26)*. The gene and protein expression of CYP2J, an enzyme involved in endogenous arachidonic acid and retinoic acid metabolism, and the expression of CYP2C, an enzyme involved in the metabolism of drugs

Table 3 Cytochrome P450 enzymes identified in ocular tissues

Isozyme	Species	Ocular Tissue	References
CYP1A	Bovine, Rabbit and Mouse	Cornea, choroid-retina and iris-ciliary body	*(27,135,99)*
CYP1B1	Human and Mouse	Ciliary epithelium and trabecular meshwork	*(34,33)*
CYP2B	Rabbit and Rat	Conjunctiva, cornea, ciliary epithelium and lens	*(78,77)*
CYP2C	Mouse	Cornea, ciliary-body, lens and retina	*(29)*
CYP2J	Mouse	Specific tissue not reported.	*(28)*
CYP3A	Rabbit, Dog and Human	Conjunctiva, cornea, choroid-retina, irisciliary body and lacrimal gland	*(31,136,137)*
CYP4B1	Human, Bovine and Rabbit	Conjunctiva, cornea, choroid-retina and iris-ciliary body	*(102,101)*
CYP39A1	Bovine	Ciliary epithelium	*(138)*
NADPH-reductase	Human and Bovine	Corneal epithelium	*(139)*

such as diclofenac and propranolol, were reported in the mouse eye *(28,29)*. We recently detected mRNA transcript copies of CYP3A, an enzyme with broad substrate specificity that is involved in the metabolism of more than 50% of commercial drugs, in the rabbit lacrimal gland and conjunctiva *(30)*. We confirmed protein expression of CYP3A in the rabbit lacrimal gland and conjunctiva, in addition to the iris-ciliary body *(31)*. Furthermore, we detected the protein expression of CYP1A, CYP2D and NADPH reductase in the rabbit lacrimal gland, conjunctiva, iris ciliary body and choroid-retina *(31)*. The mRNA and protein of CYP4B1, an enzyme involved in arachidonic and retinoic acid metabolism, were detected in the rabbit corneal epithelium *(32)*. Finally, gene expression of CYP1B1, an enzyme involved in retinoic acid biosynthesis and linked to primary congenital glaucoma *(33)*, was detected in the human ciliary body, iris, a nonpigmented ciliary epithelial cell line, and at lower levels in the cornea, retinal pigment epithelium and retina *(34)*.

Other Oxidoreductase Systems

Aldehyde Oxidase

Aldehyde oxidase is a molybdenum-containing enzyme involved in the oxidative metabolism of nicotine *(35)*, and retinoic acid synthesis *(36)*. Aldehyde oxidase gene expression has been detected in rabbit ocular tissues *(36)*, and activity has been measured in bovine and rabbit ocular tissues including the ciliary body, retinal pigment epithelium-choroid, iris, retina and cornea, but not the lens *(37)*. Brimonidine, a selective α_2-adrenoreceptor agonist used to lower intraocular pressure, undergoes aldehyde-oxidase-mediated metabolism in the rabbit conjunctiva, cornea, and iris-ciliary body *(38)*.

Ketone Reductase

NADPH-dependent ketone-reductase activity has been characterized in the rabbit corneal epithelium, iris-ciliary epithelium, conjunctiva and the lens *(39)*. Levobunolol and ketanserin both act to decrease intraocular pressure, and undergo ketone-reductase-mediated metabolism in ocular tissues *(40,41)*. Following topical administration in rabbits, levobunolol is rapidly absorbed and undergoes reductive metabolism to dihydrolevobunolol, an equipotent metabolite, in the corneal epithelium and iris-ciliary body *(40)*. In fact, two-thirds of the ocularly bioavailable dose was represented by dihydrolevobunolol. Examination of concentration–time profiles following topical levobunolol administration to rabbits revealed that the area under the curve was greater for dihydrolevobunolol versus levobunolol, in the cornea, iris-ciliary body and aqueous humor *(40,42)*. Furthermore, the terminal elimination half life of dihydrolevobunolol was longer than parent drug half life.

Hydrolytic Enzymes

Several hydrolytic enzymes are active in the retina, including acid phosphatase, aryl sulfatase, N-acetyl-β-glucosamidase, and esterase *(43)*. Various esterases, such as acetylcholinesterase, butyrylcholinesterase, and carboxylesterase, have been identified as being ubiquitously expressed not only in the retina, but in several ocular tissues *(44,45)*. There is an abundance of literature describing the expression and activity of various esterases in ocular tissues, in particular, relating to the activation of prodrugs. It has long been recognized that the distribution of esterases within different cell types can influence the rate at which, and extent to which, ester linkages are hydrolyzed *(46)*. For example, esterase activity towards naphthyl esters was greater in iris-ciliary body tissue homogenates, as compared to corneal tissue homogenates, for both bovine and rabbit eyes *(46,47)*. Furthermore, there was differential esterase expression within the different cell types of the rabbit cornea, such that greater esterase activity was detected in the corneal epithelium as opposed to the stroma-endothelium *(47)*.

Conjugating Enzyme Systems

Arylamine acetyltransferase activity in the metabolism of p-aminobenzoic acid and aminozolamide, a drug developed to treat ocular hypertension that failed clinical trials, has been characterized in rabbit ocular tissues, namely the iris-ciliary body, corneal epithelium and stroma-endothelium *(48,49)*. Glutathione S-transferase has been identified in several ocular tissues including the cornea, retina, iris-ciliary body and sclera *(50)*. This enzyme system plays an important role in the adaptive response to oxidative stress, particularly in the retina *(51)*.

FACTORS IMPACTING OCULAR TRANSPORTERS AND METABOLIZING ENZYMES

When considering ocular drug transport and metabolism, similarly to at the systemic level, there are several factors of particular importance, namely polymorphism, inhibition, induction, and excipients in drug formulations. As we learn more about the expression and activity of drug transporters and metabolizing enzymes in ocular tissues, the influence of these factors can be taken into consideration in order to improve the design of ophthalmic drug treatment.

Polymorphism

Variability in transporter and cytochrome P450 activity is a critical issue in drug therapy, since it affects all aspects of drug absorption, distribution, metabolism, and elimination *(52,53)*. Phillips et al. reported that drugs frequently cited in adverse drug reaction studies were more likely metabolized by at least one enzyme with a variant allele known to cause poor metabolism than any randomly selected drug *(54)*. Historically, the major polymorphisms that have clinical implications at the systemic level are those of P-gp, MRP, CYP2D6 and CYP2C19 *(23,52,53,55,56)*. At the ocular level, no data to date have been generated to suggest transporter polymorphism alters drug disposition, although conceivably it could happen.

Timolol undergoes CYP2D6-mediated hepatic metabolism following oral administration. In ophthalmology, timolol is used to treat ocular hypertension. Adverse events, specifically excessive β-blockade, are associated with ophthalmic timolol therapy and relate to the patient's CYP2D6 genotype *(57,58)*. The mechanism proposed to underlie this event involves systemic absorption of topically applied timolol, such that exposure levels exceed a critical point. These high exposure levels may result from a poorly metabolizing patient phenotype and/or competitive inhibition of CYP2D6-mediated metabolism by oral drugs. However, it has been argued that due to the theoretical route of ophthalmic timolol into the systemic circulation via the nasolacrimal duct, with absorption through the nasal mucosa with venous delivery to the heart, the impact of a first-pass hepatic effect should be minimal *(59)*. Thus, an alternative explanation may be related to the impact of ocular tissue CYP2D6-mediated metabolism. In the past, it has been assumed that timolol is unlikely to undergo ocular metabolism that achieves clinical significance in the eye or body. This assumption is likely based on early metabolic studies conducted in rabbits that demonstrated minimal metabolism in ocular tissues *(60)*. However, these studies were conducted 20 years ago and significant advancements have been made in the methodology and technology used to study extrahepatic metabolism. In fact, we have observed CYP2D expression and activity in rabbit ocular tissues *(31)*. Potentially, polymorphic CYP2D expressed in ocular tissues may affect ophthalmic timolol therapy. Recently, a frameshift mutation that is thought to impact local codeine-to-morphine metabolism was identified in human brain CYP2D7 *(61)*. Similar polymorphism in ocular CYP2D enzymes may alter the exposure levels of timolol and other substrate drugs.

The effect of hepatic polymorphic enzyme phenotype has been investigated for arylamine acetylation in the rabbit eye. Aminozolamide undergoes arylamine acetyltransferase metabolism in the corneal epithelium and iris-ciliary body of rabbits *(48)*. However, in vitro and in vivo ocular acetylation activity do not correlate with hepatic acetylation phenotype *(49)*. Flestolol metabolism by carboxylesterase has been studied in rabbit eyes and appears to be polymorphic. The results from this study demonstrated that carboxylesterase slow- and fast-metabolizer phenotypes measured in the blood, correlated with the metabolizer phenotype measured in cornea but not other ocular tissues *(45)*. Together, these findings suggest that further studies are required to understand the impact of polymorphism on ocular metabolism better. In particular, does polymorphism impact ophthalmic drug bioavailability, and why are there differences between ocular and systemic expression? Perhaps, this differential expression in polymorphism arises from particular local requirements to serve ocular needs. For example, polymorphism in CYP1B1, which is expressed in human ocular tissues, is linked to primary congenital glaucoma *(33,62,63)*. Also, polymorphism in glutathione S-transferase has been linked as a genetic risk factor in the development of cataracts in Japanese and Estonian patients *(64,65)*.

Inhibition

The inhibition of drug transporters and/or metabolizing enzymes is an important issue in drug pharmacokinetics. Identifying potential inhibitors can aid prediction of drug–drug interactions and/or unfavorable pharmacokinetic profiles produced upon drug coadministration. Competitive inhibition of drug transport or metabolism may alter drug or metabolite concentrations to cause therapeutically and toxicologically significant consequences.

In vivo topical ocular absorption of erythromycin, a P-gp substrate, in the absence and presence of P-gp inhibitors, was carried out by Dey et al. *(9)* in rabbits to determine the role played by P-gp in ocular pharmacokinetics of erythromycin. Cyclosporin A, testosterone and quinidine, well-known inhibitors for P-gp, all significantly increased erythromycin peak concentration (C_{max}), as well as its area under the concentration–time curve (AUC) in aqueous humor following coadministration, as a result of inhibition of active corneal efflux of erythromycin by P-gp. In the interior chamber of the eye, expression and activity of P-gp at the blood–retinal barrier, as well as the blood aqueous barrier, is thought to be the mechanism behind increased vitreal quinidine C_{max} and AUC, following systemic administration of quinidine, when the P-gp inhibitor verapamil is present in the vitreous in an in vivo rabbit study *(66)*. More examples of such transporter-based drug–drug interaction via inhibition have been reported in vitro in ocular cell lines *(67–70)*, and are sometimes used as tools to identify, on a functional level, the expression and activity of specific transporter systems depending on specificity of the inhibitors used.

CYP3A activity in the lacrimal gland can be inhibited by ketoconazole *(30)*. Ketoconazole is given orally for the treatment of ocular fungal infections, such as fungal keratitis *(71–73)*. An interesting question is raised as to what effect these azole antifungals might exert on CYP3A-mediated metabolism in ocular tissues. Systemic drug interactions with ketoconazole have been extensive and the clinical consequences are severe. Therefore, azoles will most definitely affect the pharmacokinetics of drugs metabolized by CYP3A in the eye. Although, to our knowledge there has been no clinical report of this type of drug–drug interaction, it is still important to consider this possibility in ocular drug therapy. Furthermore, could inhibition of cytochrome P450 mediated metabolism in ocular tissues be targeted in drug design strategies? The importance of cytochrome P450- and myeloperoxidase-mediated generation of reactive oxygen species in the retina to the development of drug-induced retinopathy is recognized *(74)*. Understanding these metabolic pathways, and eventually being able to target them for therapeutic intervention, may offer new opportunities for drug design.

Induction

Induction of transporter proteins in ocular tissues has not been studied extensively. Ohtsuki et al. *(75)* recently found in a comparison study between brain and retinal capillary endothelial cells that, despite structural and functional similarity between the two cell types, there is a difference in expression of organic anion transporter 3 (oat3) mRNA in response to treatment with dihydrotestosterone (DHT), an androgen receptor ligand. In brain capillary endothelial cells, which selectively express the androgen receptor gene, OCT3 mRNA, as well as OCT3 transport activity, was induced by DHT and this induction was suppressed by flutamide. In contrast, oat3 mRNA expression in retinal capillary endothelial cells was not affected by DHT treatment.

Zhao and Shichi were the first to demonstrate specific cytochrome P450 isoform induction in a specific ocular tissue *(27)*. Following intraperitoneal β-naphthoflavone treatment in mice, CYP1A1/1A2 expression was induced in the ciliary non-pigmented epithelium, choroid, retinal pigmented epithelium, cornea epithelium, and iris epithelium. The authors proposed that, since the ciliary epithelium is involved in aqueous humor formation by acting as a metabolic ultrafiltration system for blood plasma, then drug-metabolizing enzymes in these tissues may play a critical role in metabolic detoxification of plasma prior to its secretion as aqueous humor.

Dexamethasone induces CYP3A in ocular tissues following both systemic and topical administration. When rabbits received intraperitoneal injections of dexamethasone, CYP3A activity increased fourfold in the cornea, as measured by testosterone 6β-hydroxylation *(76)*. Topical treatment with dexamethasone increased CYP3A activity two- to fivefold, in the lacrimal gland and conjunctiva of rabbits, as measured by benzyloxyquinoline dealkylation and testosterone hydroxylation *(30)*. Intraperitoneal phenobarbital treatment in rats and rabbits induces expression and activity of various cytochrome P450 enzymes in the cornea, conjunctiva, ciliary epithelium and lens *(76–78)*. Induction of glutathione S-transferase has been observed in human retinal pigment epithelial cell lines in response to the chemopreventative agent oltipraz and in response to oxidative stress *(51,79)*.

Excipients

Common excipients used in various formulations have been shown to impact function of many transporter proteins. For example, the block copolymer Pluronic P85 decreased maximal reaction rates (V_{max}) and increased apparent Michaelis constants (K_m) for P-gp, MRP1, and MRP2 drug-efflux transport proteins, and the extent of alterations was increased in the order MRP1, MRP2, P-gp from smallest to greatest effect *(80)*. The same copolymer was not found to affect activity of the glucose transporter GLUT1, but significantly decreased transport of lactate, a monocarboxylate transporter 1 (MCT1) substrate, across the primary culture of bovine brain microvessel endothelial cells *(81)*. Among the transporters studied for excipient effect, P-gp appears to be the most sensitive, being affected by many, including Labrasol, Imwitor 742, Acconon E, Softigen 767, Cremophor EL, Miglyol, Solutol HS 15, Sucrose monolaurate, Polysorbate 20, TPGS, and Polysorbate 80 in addition to Pluronic P85 *(82)*. Among these, cremophor and pluronic P85 have been used in ophthalmic formulations for topical administration *(83)*. It is conceivable that if a P-gp substrate, for example cyclosporine A, is formulated with an excipient that happens to modulate P-gp activity, its ocular bioavailability could be changed. Therefore, when transporters are involved in facilitating the permeation, selection of ophthalmic formulation excipients for topical administration could be utilized as a tool to improve ocular bioavailability of compounds with limited passive permeability across ocular surface.

DISEASE

Transporters

Mutations occurring in transporter proteins may be the cause of certain ocular diseases. Pseudoxanthoma elasticum (PXE), an autosomally inherited disease, is caused by mutations in MRP6, encoding a multidrug-resistance-associated protein that is a member of the ABC transporter gene family *(84,85)*. PXE is characterized by progressive dystrophic calcification of the elastic structures in the skin, eyes, and cardiovascular system. The ocular abnormalities include retinitis pigmentosa, comet-like streaks, pinpoint white lesions of the choroid, and angioid streaks.

In addition to being the underlying cause of ocular diseases, ocular transporter expression and function may be altered in disease states, with diabetic retinopathy being the most frequently studied. Using human retinal pigment epithelial cells, it was demonstrated that expression and activity of the taurine transporter is upregulated under oxidative stress, but overexpression of aldose reductase and high glucose, as happens in diabetic retinopathy, impairs this response *(86)*. In contrast, expression and activity of the MCT1 remain unchanged in ocular tissues in diabetic rats *(87)*. The glucose transporter GLUT1 mediates glucose entry into the endothelial cells of the inner BRB. Retinal GLUT1 abundance decreases in experimental diabetes with exposure of retinal endothelial cells to elevated glucose concentrations. Ubiquitinylation of GLUT1 is proposed to be the mechanism targeting GLUT1 for degradation in diabetes *(88)*. Similarly, it was found that, early in the course of diabetic retinopathy, the function of the glutamate transporter in retinal Muller cells is decreased *(89)*.

Metabolism and Ocular Disease

Altered metabolism can underlie the pathogenesis of ocular disease. The production of reactive metabolites, or changes in the metabolism rates of endogenous gene regulators such as steroids and retinoids, may elicit undesirable effects.

Cataract

When ocular metabolism at a specific site becomes saturated, reactive and/or toxic metabolites could be formed locally and cause tissue damage. This pathophysiological mechanism may explain the observation of cataract formation following an overdose of acetaminophen in mice *(90)*. When mice received an overdose of acetaminophen in the presence of the CYP1A1/1A2 inhibitor, diallyl sulfide, cataract development was prevented *(91)*. The protective effect of diallyl sulfide presumably results from inhibition of biotransformation of acetaminophen to the reactive metabolite N-acetyl-p-benzoquinone (NAPQI) by CYP1A1/1A2. Mice injected intracamerally with NAPQI develop cataracts *(92)*.

3-hydroxy-3-methyl-glutaryl- (HMG-)CoA reductase inhibitors, commonly referred to as statins, are widely prescribed as oral drugs to lower plasma low-density lipoprotein cholesterol levels. A strong relationship between systemic exposure levels and the cataractogenic potential of this class of compounds has been observed in dogs *(93)*. Lovastatin treatment induced opacity in organ-cultured rat lens and adversely affected cell structure and proliferation of cultured epithelial cells from human and rabbit lenses *(94)*. The mechanism underlying these observations is not yet clear, however, it has been suggested that the exposure of the outer cortical region of the lens to statins is somehow associated with cataractogenesis *(93)*. The exposure levels achieved by statins at clinically therapeutic doses have not been associated with the development of cataract in man *(95–97)*. However, the concomitant administration of a drug that inhibits cytochrome P450 metabolism may be significant if this results in increased exposure levels of statins and associated adverse effects *(95,98)*. The role of ocular cytochrome P450 metabolism in chronic statin treatment and the association with cataract formation still need to be confirmed.

Neurodegeneration

CYP1A1 expression and activity were examined in bovine retinal pigment epithelial cells following exposure to the cigarette smoke constituent, benzo[a]pyrene (BaP) *(99)*. This compound was found to induce CYP1A1 mRNA levels and protein content in retinal epithelial cells. CYP1A1 is known to convert polyaromatic hydrocarbon compounds to toxic metabolites, including the metabolism of BaP to the mutagen, benzo[a]pyrene-7,8-diol 9,10-epoxide (BPDE), which forms covalent adducts with DNA in bovine retinal epithelial cells. This study data suggests a mechanism by which exposure to cigarette smoke may underlie degenerative disorders of the retina.

Ocular Lens Dislocation

Molybdenum deficiency is associated with ocular lens dislocation. Aldehyde oxidase activity requires molybdenum as a necessary cofactor. Altered retinoic acid synthesis resulting from the absence of aldehyde oxidase activity may underlie this developmental disorder.

Primary Congenital Glaucoma

Mutation in the CYP1B1 gene is associated with primary congenital glaucoma (PCG) *(33)*. CYP1B1 is expressed in the human ciliary body and iris, and human cells of nonpigmented ciliary epithelium and trabecular meshwork *(33,34)*. Libby et al. 2003 reported that CYP1B1 knockout mice present histological abnormalities resembling humans with PCG in terms of abnormal ocular drainage structures *(100)*. Altered CYP1B1 metabolism of all-trans retinoic acid, affecting cell proliferation in neural cells, was proposed to contribute to abnormal anterior segment development.

Inflammation and Angiogenesis

Schwartzman and colleagues were the first to describe a novel metabolic pathway for arachidonic acid (AA) that was cytochrome P450 dependent, in the corneal epithelium of human, bovine, and rabbit eyes *(101,102)*. AA is an important precursor to a number of cell mediators. Proposed activities of the CYP4B1-mediated AA metabolites include inhibition of the Na^+,K^+-ATPase pump in the cornea, which helps to maintain the electrochemical gradient across the epithelium *(103)*, action as a potent vasodilator, stimulation of protein influx into the aqueous humor of the eye, and action as a potent angiogenic factor *(104–106)*. Human tear film collected from subjects with inflamed eyes contains increased levels of these AA metabolites *(107)*. Furthermore, it was found that the levels of these endogenous metabolites increased following contact-lens-induced hypoxic stress in rabbits *(108)*. CYP4B1 mRNA levels were found to be increased in hypoxia-treated corneal epithelial cells as compared to control cells.

In addition, CYP4B1 activity in the cornea may be involved in wound healing through metabolism of retinoids *(109)*. More recently, CYP4B1 in the cornea has been identified as a potential new target in preventing corneal neovascularization *(105,106)*.

DRUG TRANSPORTER–METABOLIC ENZYME INTERPLAY

Membrane transporters and metabolizing enzymes are endogenously expressed in the eye and serve as defense mechanisms that regulate the entry, exit, and exposure of drugs and endogenous substrates (Fig. 1). It is not fully understood whether these systems coexist in ocular tissues to work in concert or are individually expressed to serve unique functions.

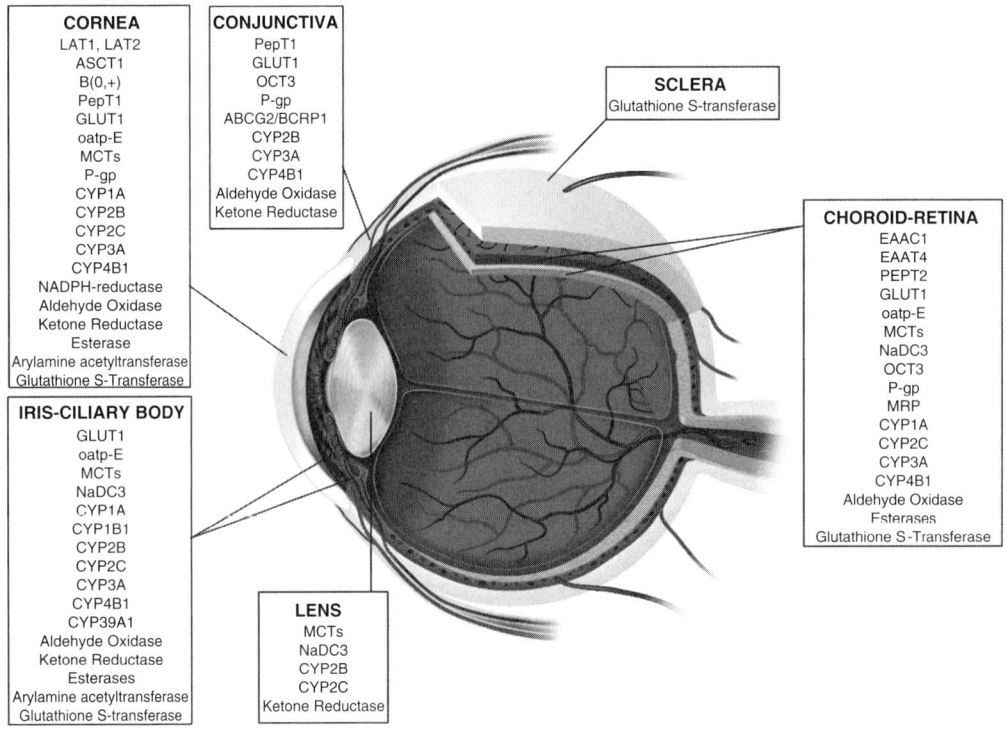

Fig. 1. Illustration of the eye with lists of drug transporters and metabolizing enzymes identified in the various ocular tissues.

Drug therapy may modulate drug transporters and/or metabolizing enzymes. When trying to assess the contribution and impact of transport and metabolism in ophthalmic tissues we must consider the vectorial nature of drug traffic in relation to the locations of transporters and enzymes. For example, the interplay of P-gp transport and CYP3A metabolism will affect drug exposure in the conjunctiva, depending on whether the substrate is delivered systemically or topically. The same applies to drug delivery to the posterior retina. Fig. 2 is adapted from a model first proposed and then reviewed by Benet et al. describing metabolic and transport considerations at the small intestine versus the liver *(110)*. Our adapted figure illustrates the vectorial nature of the conjunctival epithelium and depicts the situations where a drug may be delivered topically or systemically. Drugs delivered topically will encounter efflux prior to metabolism, whereas drugs or endogenous substrates delivered systemically will encounter metabolism prior to efflux. The interplay of P-gp and CYP3A in the conjunctiva will generate scenarios where drug absorption could be increased or decreased, or alternatively a drug may undergo recycling. These effects would be based on the vectorial nature of the drug traffic and either overlapping substrate specificity or lack of substrate specificity. Together, these effects may influence which chemical stimulants are present in the cell to elicit an effect.

Topical Delivery

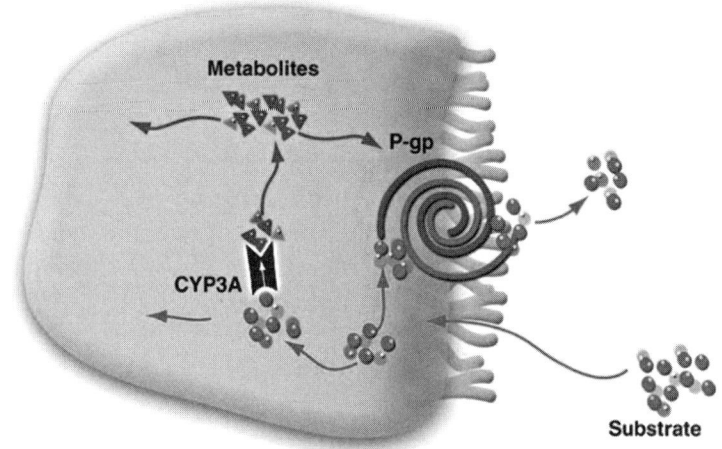

Systemic Delivery

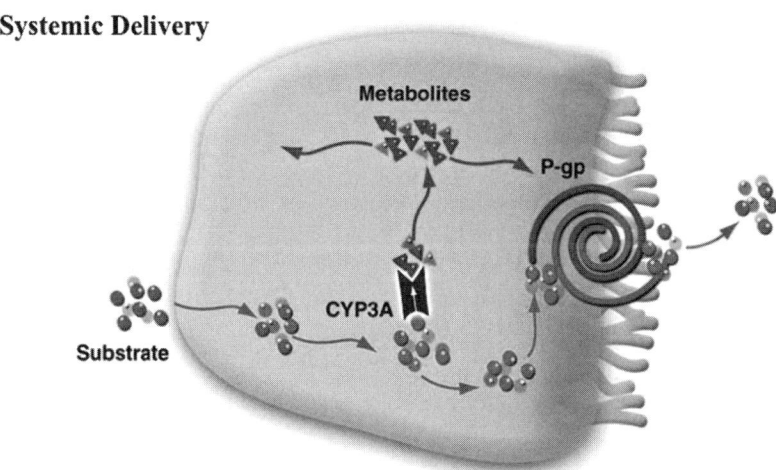

Fig. 2. Illustration demonstrating the vectorial considerations that impact topical drug bioavailability versus systemic drug or endogenous substrate bioavailability.

An example of transporter and enzyme function in concert in intraocular tissues was proposed by George et al. *(111)*. Canavan disease is a genetic disorder associated with optic neuropathy, and the metabolism of N-acetylaspartate is defective in this disorder due to mutations in the gene coding for the enzyme aspartoacylase II. The plasma membrane transporter NaDC3, a Na$^+$-coupled transporter for dicarboxylates, is able to transport N-acetylaspartate. The expression pattern of NaDC3 and aspartoacylase II in ocular tissues was studied in mouse. These studies showed that NaDC3 mRNA is expressed in the optic nerve, most layers of the retina, the retinal pigment epithelium, ciliary body, iris, and lens, while aspartoacylase II mRNA is coexpressed in most of these cell types. Therefore, it is most likely that transport of N-acetylaspartate into ocular tissues via NaDC3 and its subsequent hydrolysis by aspartoacylase II play essential roles in the maintenance of visual function.

CONCLUSION

Drug metabolism, and recently, transporters, have emerged as important considerations in ophthalmic drug design. The development of newer compounds such as brimonidine and latanoprost has benefited from the availability of modern laboratory tools and therefore their ocular metabolic fate has been fully characterized as an integral part of drug development. In the coming years we will broaden our knowledge of transport and metabolic systems in ocular tissues. We will strengthen our understanding as to how these proteins can affect pharmacokinetics and pharmacodynamics and how they may underlie disease processes. Together, this information will result in the development of more effective drug therapies to treat and prevent the progression of eye disease.

REFERENCES

1. Vision problems in the US: Prevalence of adult vision impairment and age-related eye disease in America. (2002) Prevent Blindness America. www.nei.nih.gov
2. Zimmerman TJ, Leader B, Kaufman HE. Advances in ocular pharmacology. Annu Rev Pharmacol Toxicol 1980;20:415–428.
3. Lee VHL. Precorneal, corneal and postcorneal factors. In: Mitra, editor. Ophthalmic Drug Delivery Systems. New York: Marcel Dekker, 1993. p. 59–81.
4. Dey S, Anand BS, Patel J, et al. Transporters/receptors in the anterior chamber: pathways to explore ocular drug delivery strategies. Expert Opin Biol Ther 2003;3:23–44.
5. Ito A, Yamaguchi K, Tomita H, et al. Distribution of rat organic anion transporting polypeptide-E (oatp-E) in the rat eye. Invest Ophthalmol Vis Sci 2003;44:4877–4884.
6. Ueda H, Horibe Y, Kim KJ, et al. Functional characterization of organic cation drug transport in the pigmented rabbit conjunctiva. Invest Ophthalmol Vis Sci 2000;41:870–876.
7. Majumdar S, Tirucherai GS, Pal D, et al. Functional differences in nucleoside and nucleobase transporters expressed on the rabbit corneal epithelial cell line (SIRC) and isolated rabbit cornea. AAPS PharmSci 2003;5:E15.
8. Majumdar S, Nashed YE, Patel K, et al. Dipeptide monoester ganciclovir prodrugs for treating HSV-1-induced corneal epithelial and stromal keratitis: in vitro and in vivo evaluations. J Ocul Pharmacol Ther 2005;21:463–474.
9. Dey S, Gunda S, Mitra AK. Pharmacokinetics of erythromycin in rabbit corneas after single-dose infusion: role of P-glycoprotein as a barrier to in vivo ocular drug absorption. J Pharmacol Exp Ther 2004;311:246–255.
10. Dey S, Patel J, Anand BS, et al. Molecular evidence and functional expression of P-glycoprotein (MDR1) in human and rabbit cornea and corneal epithelial cell lines. Invest Ophthalmol Vis Sci 2003;44:2909–2918.
11. Kawazu K, Yamada K, Nakamura M, et al. Characterization of cyclosporin A transport in cultured rabbit corneal epithelial cells: P-glycoprotein transport activity and binding to cyclophilin. Invest Ophthalmol Vis Sci 1999;40:1738–1744.
12. Yang JJ, Kim KJ, Lee VH. Role of P-glycoprotein in restricting propranolol transport in cultured rabbit conjunctival epithelial cell layers. Pharm Res 2000;17:533–538.

13. Saha P, Yang JJ, Lee VH. Existence of a p-glycoprotein drug efflux pump in cultured rabbit conjunctival epithelial cells. Invest Ophthalmol Vis Sci 1998;39:1221–1226.
14. Budak MT, Alpdogan OS, Zhou M, et al. Ocular surface epithelia contain ABCG2-dependent side population cells exhibiting features associated with stem cells. J Cell Sci 2005;118:1715–1724.
15. Duvvuri S, Majumdar S, Mitra AK. Drug delivery to the retina: challenges and opportunities. Expert Opin Biol Ther 2003;3:45–56.
16. Ocheltree SM, Keep RF, Shen H, et al. Preliminary investigation into the expression of proton-coupled oligopeptide transporters in neural retina and retinal pigment epithelium (RPE): lack of functional activity in RPE plasma membranes. Pharm Res 2003;20:1364–1372.
17. Rajan PD, Kekuda R, Chancy CD, et al. Expression of the extraneuronal monoamine transporter in RPE and neural retina. Curr Eye Res 2000;20:195–204.
18. Majumdar S, Macha S, Pal D, et al. Mechanism of ganciclovir uptake by rabbit retina and human retinal pigmented epithelium cell line ARPE-19. Curr Eye Res 2004;29:127–136.
19. Greenwood J. Characterization of a rat retinal endothelial cell culture and the expression of P-glycoprotein in brain and retinal endothelium in vitro. J Neuroimmunol 1992;39:123–132.
20. Kennedy BG and Mangini NJ. P-glycoprotein expression in human retinal pigment epithelium. Mol Vis 2002;8:422–430.
21. Aukunuru JV, Sunkara G, Bandi N, et al. Expression of multidrug resistance-associated protein (MRP) in human retinal pigment epithelial cells and its interaction with BAPSG, a novel aldose reductase inhibitor. Pharm Res 2001;18:565–572.
22. Nakamura K, Fujiki T, Tamura HO. Age, gender and region-specific differences in drug metabolising enzymes in rat ocular tissues. Exp Eye Res 2005;81:710–715.
23. Parkinson A. An overview of current cytochrome P450 technology for assessing the safety and efficacy of new materials. Toxicol Pathol 1996;24:48–57.
24. Wilkinson GR. Pharmacokinetics. In: Hardman AJG, Limbird LE, Goodman Gilman A, editors. The pharmacological basis of therapeutics. New York: McGraw-Hill; 2001. p. 3–30.
25. Shichi H. Microsomal electron transfer system of bovine retinal pigment epithelium. Exp Eye Res 1969;8:60–68.
26. McAvoy M, Singh AK, Shichi H. In situ hybridization of Cyp1a1, Cyp1a2 and Ah receptor mRNAs expressed in murine ocular tissues. Exp Eye Res 1996;62:449–452.
27. Zhao C and Shichi H. Immunocytochemical study of cytochrome P450 (1A1/1A2) induction in murine ocular tissues. Exp Eye Res 1995;60:143–152.
28. Xie Q, Zhang QY, Zhang Y, et al. Induction of mouse CYP2J by pyrazole in the eye, kidney, liver, lung, olfactory mucosa, and small intestine, but not in the heart. Drug Metab Dispos 2000;28:1311–1316.
29. Tsao CC, Coulter SJ, Chien A, et al. Identification and localization of five CYP2Cs in murine extrahepatic tissues and their metabolism of arachidonic acid to regio- and stereoselective products. J Pharmacol Exp Ther 2001;299:39–47.
30. Attar M, Ling KH, Tang-Liu DD, et al. Cytochrome P450 3A expression and activity in the rabbit lacrimal gland: glucocorticoid modulation and the impact on androgen metabolism. Invest Ophthalmol Vis Sci 2005;46:4697–4706.
31. Attar M, Lee VHL, Tang-Liu DS, et al. Characterization of cytochrome P450 1A, 2D and 3A in the rabbit eye. Presented at the AOPT 2003.
32. Mastyugin V, Aversa E, Bonazzi A, et al. Hypoxia-induced production of 12-hydroxyeicosanoids in the corneal epithelium: involvement of a cytochrome P-4504B1 isoform. J Pharmacol Exp Ther 1999;289:1611–1619.
33. Stoilov I, Akarsu AN, Sarfarazi M. Identification of three different truncating mutations in cytochrome P4501B1 (CYP1B1) as the principal cause of primary congenital glaucoma

(Buphthalmos) in families linked to the GLC3A locus on chromosome 2p21. Hum Mol Genet 1997;6:641–647.

34. Stoilov I, Akarsu AN, Alozie I, et al. Sequence analysis and homology modeling suggest that primary congenital glaucoma on 2p21 results from mutations disrupting either the hinge region or the conserved core structures of cytochrome P4501B1. Am J Hum Genet 1998;62:573–584.

35. Berkman CE, Park SB, Wrighton SA, et al. In vitro-in vivo correlations of human (S)-nicotine metabolism. Biochem Pharmacol 1995;50:565–570.

36. Huang DY, Furukawa A, Ichikawa Y. Molecular cloning of retinal oxidase/aldehyde oxidase cDNAs from rabbit and mouse livers and functional expression of recombinant mouse retinal oxidase cDNA in Escherichia coli. Arch Biochem Biophys 1999;364:264–272.

37. Shimada S, Mishima H, Kitamura S, et al. Nicotinamide N-oxide reductase activity in bovine and rabbit eyes. Invest Ophthalmol Vis Sci 1987;28:1204–1206.

38. Acheampong AA, Shackleton M, Tang-Liu DD. Comparative ocular pharmacokinetics of brimonidine after a single dose application to the eyes of albino and pigmented rabbits. Drug Metab Dispos 1995;23:708–712.

39. Lee VH, Chien DS, Sasaki H. Ocular ketone reductase distribution and its role in the metabolism of ocularly applied levobunolol in the pigmented rabbit. J Pharmacol Exp Ther 1988;246:871–878.

40. Tang-Liu DD, Liu S, Neff J, et al. Disposition of levobunolol after an ophthalmic dose to rabbits. J Pharm Sci 1987;76:780–783.

41. Schoenwald RD and Zhu J. The ocular pharmacokinetics of ketanserin and its metabolite, ketanserinol, in albino rabbits. J Ocul Pharmacol Ther 2000;16:481–495.

42. Tang-Liu DD, Shackleton M, Richman JB. Ocular metabolism of levobunolol. J Ocul Pharmacol 1988;4:269–278.

43. Essner E, Gorrin GM, Griewski RA. Localization of lysosomal enzymes in retinal pigment epithelium of rats with inherited retinal dystrophy. Invest Ophthalmol Vis Sci 1978;17:278–288.

44. Lee VH, Chang SC, Oshiro CM, et al. Ocular esterase composition in albino and pigmented rabbits: possible implications in ocular prodrug design and evaluation. Curr Eye Res 1985;4:1117–1125.

45. Stampfli HF and Quon CY. Polymorphic metabolism of flestolol and other ester containing compounds by a carboxylesterase in New Zealand white rabbit blood and cornea. Res Commun Mol Pathol Pharmacol 1995;88:87–97.

46. Lee VH, Iimoto DS, Takemoto KA. Subcellular distribution of esterases in the bovine eye. Curr Eye Res 1982;2:869–876.

47. Lee VH, Morimoto KW, Stratford RE Jr. Esterase distribution in the rabbit cornea and its implications in ocular drug bioavailability. Biopharm Drug Dispos 1982;3:291–300.

48. Putnam ML, Schoenwald RD, Duffel MW, et al. Ocular disposition of aminozolamide in the rabbit eye. Invest Ophthalmol Vis Sci 1987;28:1373–1382.

49. Campbell DA, Schoenwald RD, Duffel MW, et al. Characterization of arylamine acetyltransferase in the rabbit eye. Invest Ophthalmol Vis Sci 1991;32:2190–2200.

50. Srivastava SK, Singhal SS, Bajpai KK, et al. A group of novel glutathione S-transferase isozymes showing high activity towards 4-hydroxy-2-nonenal are present in bovine ocular tissues. Exp Eye Res 1994;59:151–159.

51. Singhal SS, Godley BF, Chandra A, et al. Induction of glutathione S-transferase hGST 5.8 is an early response to oxidative stress in RPE cells. Invest Ophthalmol Vis Sci 1999;40:2652–2659.

52. Liu Y and Hu M. P-glycoprotein and bioavailability-implication of polymorphism. Clin Chem Lab Med 2000;38:877–881.
53. Lin JH and Lu AY. Inhibition and induction of cytochrome P450 and the clinical implications. Clin Pharmacokinet 1998;35:361–390.
54. Phillips KA, Veenstra DL, Oren E, et al. Potential role of pharmacogenomics in reducing adverse drug reactions: a systematic review. JAMA 2001;286:2270–2279.
55. Beringer PM and Slaughter RL. Transporters and their impact on drug disposition. Ann Pharmacother 2005;39:1097–1108.
56. Lin JH and Lu AY. Role of pharmacokinetics and metabolism in drug discovery and development. Pharmacol Rev 1997;49:403–449.
57. Edeki TI, He H, Wood AJ. Pharmacogenetic explanation for excessive beta-blockade following timolol eye drops. Potential for oral-ophthalmic drug interaction. JAMA 1995;274:1611–1613.
58. Ishii Y, Nakamura K, Tsutsumi K, et al. Drug interaction between cimetidine and timolol ophthalmic solution: effect on heart rate and intraocular pressure in healthy Japanese volunteers. J Clin Pharmacol 2000;40:193–199.
59. Novack GD. Excessive beta-blockade with timolol eye drops. JAMA 1996;275:985.
60. Putterman GJ, Davidson J, Albert J. Lack of metabolism of timolol by ocular tissues. J Ocul Pharmacol 1985;1:287–296.
61. Pai HV, Kommaddi RP, Chinta SJ, et al. A frameshift mutation and alternate splicing in human brain generate a functional form of the pseudogene cytochrome P4502D7 that demethylates codeine to morphine. J Biol Chem 2004;279:27383–27389.
62. Melki R, Lefort N, Brezin AP, et al. Association of a common coding polymorphism (N453S) of the cytochrome P450 1B1 (CYP1B1) gene with optic disc cupping and visual field alteration in French patients with primary open-angle glaucoma. Mol Vis 2005;11:1012–1017.
63. Doshi M, Marcus C, Bejjani BA, et al. Immunolocalization of CYP1B1 in normal, human, fetal and adult eyes. Exp Eye Res 2006;82:24–32.
64. Sekine Y, Hommura S, Harada S. Frequency of glutathione-S-transferase 1 gene deletion and its possible correlation with cataract formation. Exp Eye Res 1995;60:159–163.
65. Juronen E, Tasa G, Veromann S, et al. Polymorphic glutathione S-transferases as genetic risk factors for senile cortical cataract in Estonians. Invest Ophthalmol Vis Sci 2000;41:2262–2267.
66. Duvvuri S, Gandhi MD, Mitra AK. Effect of P-glycoprotein on the ocular disposition of a model substrate, quinidine. Curr Eye Res 2003;27:345–353.
67. Majumdar S, Gunda S, Pal D, et al. Functional activity of a monocarboxylate transporter, MCT1, in the human retinal pigmented epithelium cell line, ARPE-19. Mol Pharm. 2005;2:109–117.
68. Gandhi MD, Pal D, Mitra AK. Identification and functional characterization of a Na(+)-independent large neutral amino acid transporter (LAT2) on ARPE-19 cells. Int J Pharm 2004;275:189–200.
69. Jain-Vakkalagadda B, Dey S, Pal D, et al. Identification and functional characterization of a Na+-independent large neutral amino acid transporter, LAT1, in human and rabbit cornea. Invest Ophthalmol Vis Sci 2003;44:2919–2927.
70. Hosoya K, Kondo T, Tomi M, et al. MCT1-mediated transport of L-lactic acid at the inner blood-retinal barrier: a possible route for delivery of monocarboxylic acid drugs to the retina. Pharm Res 2001;18:1669–1676.
71. Scott IU, Flynn HW Jr, Miller D, et al. Exogenous endophthalmitis caused by amphotericin B-resistant Paecilomyces lilacinus: treatment options and visual outcomes. Arch Ophthalmol 2001;119:916–919.

72. Thomas PA. Fungal infections of the cornea. Eye 2003;17:852–862.
73. Arthur RR, Drew RH, Perfect JR. Novel modes of antifungal drug administration. Expert Opin Investig Drugs 2004;13:903–932.
74. Toler SM. Oxidative stress plays an important role in the pathogenesis of drug-induced retinopathy. Exp Biol Med (Maywood) 2004;229:607–615.
75. Ohtsuki S, Tomi M, Hata T, et al. Dominant expression of androgen receptors and their functional regulation of organic anion transporter 3 in rat brain capillary endothelial cells; comparison of gene expression between the blood–brain and –retinal barriers. J Cell Physiol 2005;204:896–900.
76. Madhu C, Dinh V, Babusis D, et al. Cytochrome P450 isozyme activities in rabbit ocular tissues. ISSX Proceedings 1998;13.
77. Matsumoto K, Kishida K, Manabe R, et al. Induction of cytochrome P-450 in the rabbit eye by phenobarbital, as detected immunohistochemically. Curr Eye Res 1987;6:847–854.
78. Tanaka H, Hirayama I, Takehana M, et al. Cytochrome P450 expression in rat ocular tissues and its induction by phenobarbital. Journal of Health Science 2002;48:346–349.
79. Nelson KC, Armstrong JS, Moriarty S, et al. Protection of retinal pigment epithelial cells from oxidative damage by oltipraz, a cancer chemopreventive agent. Invest Ophthalmol Vis Sci 2002;43:3550–3554.
80. Batrakova EV, Li S, Li Y, et al. Effect of pluronic P85 on ATPase activity of drug efflux transporters. Pharm Res 2004;21:2226–2233.
81. Batrakova EV, Zhang Y, Li Y, et al. Effects of pluronic P85 on GLUT1 and MCT1 transporters in the blood-brain barrier. Pharm Res 2004;21:1993–2000.
82. Cornaire G, Woodley J, Hermann P, et al. Impact of excipients on the absorption of P-glycoprotein substrates in vitro and in vivo. Int J Pharm 2004;278:119–131.
83. Carmignani C, Rossi S, Saettone MF, et al. Ophthalmic vehicles containing polymer-solubilized tropicamide: "in vitro/in vivo" evaluation. Drug Dev Ind Pharm 2002;28:101–105.
84. Yoshida S, Honda M, Yoshida A, et al. Novel mutation in ABCC6 gene in a Japanese pedigree with pseudoxanthoma elasticum and retinitis pigmentosa. Eye 2005;19:215–217.
85. Ringpfeil F, Lebwohl MG, Christiano AM, et al. Pseudoxanthoma elasticum: mutations in the MRP6 gene encoding a transmembrane ATP-binding cassette (ABC) transporter. Proc Natl Acad Sci USA 2000;97:6001–6006.
86. Nakashima E, Pop-Busui R, Towns R, et al. Regulation of the human taurine transporter by oxidative stress in retinal pigment epithelial cells stably transformed to overexpress aldose reductase. Antioxid Redox Signal 2005;7:1530–1542.
87. Layton CJ, Chidlow G, Casson RJ, et al. Monocarboxylate transporter expression remains unchanged during the development of diabetic retinal neuropathy in the rat. Invest Ophthalmol Vis Sci 2005;46:2878–2885.
88. Fernandes R, Carvalho AL, Kumagai A, et al. Downregulation of retinal GLUT1 in diabetes by ubiquitinylation. Mol Vis 2004;10:618–628.
89. Li Q and Puro DG. Diabetes-induced dysfunction of the glutamate transporter in retinal Muller cells. Invest Ophthalmol Vis Sci 2002;43:3109–3116.
90. Shichi H, Gaasterland DE, Jensen NM, et al. Ah locus: genetic differences in susceptibility to cataracts induced by acetaminophen. Science 1978;200:539–541.
91. Zhao C and Shichi H. Prevention of acetaminophen-induced cataract by a combination of diallyl disulfide and N-acetylcysteine. J Ocul Pharmacol Ther 1998;14:345–355.
92. Qian W, Amin RH, Shichi H. Cytotoxic metabolite of acetaminophen, N-acetyl-p-benzoquinone imine, produces cataract in DBA2 mice. J Ocul Pharmacol Ther 1999;15:537–545.
93. Gerson RJ, MacDonald JS, Alberts AW, et al. On the etiology of subcapsular lenticular opacities produced in dogs receiving HMG-CoA reductase inhibitors. Exp Eye Res 1990;50:65–78.

94. Rao PV, Robison WG Jr, Bettelheim F, et al. Role of small GTP-binding proteins in lovastatin-induced cataracts. Invest Ophthalmol Vis Sci 1997;38:2313–2321.
95. Smeeth L, Hubbard R, Fletcher AE. Cataract and the use of statins: a case-control study. QJM 2003;96:337–343.
96. Boccuzzi SJ, Bocanegra TS, Walker JF, et al. Long-term safety and efficacy profile of simvastatin. Am J Cardiol 1991;68:1127–1131.
97. Harris ML, Bron AJ, Brown NA, et al. Absence of effect of simvastatin on the progression of lens opacities in a randomised placebo controlled study. Oxford Cholesterol Study Group. Br J Ophthalmol 1995;79:996–1002.
98. Einarson TR, Metge CJ, Iskedjian M, et al. An examination of the effect of cytochrome P450 drug interactions of hydroxymethylglutaryl-coenzyme A reductase inhibitors on health care utilization: a Canadian population-based study. Clin Ther 2002;24:2126–2136.
99. Patton WP, Routledge MN, Jones GD, et al. Retinal pigment epithelial cell DNA is damaged by exposure to benzo[a]pyrene, a constituent of cigarette smoke. Exp Eye Res 2002;74:513–522.
100. Libby RT, Smith RS, Savinova OV, et al. Modification of ocular defects in mouse developmental glaucoma models by tyrosinase. Science 2003;299:1578–1581.
101. Schwartzman ML, Masferrer J, Dunn MW, et al. Cytochrome P450, drug metabolizing enzymes and arachidonic acid metabolism in bovine ocular tissues. Curr Eye Res 1987;6:623–630.
102. Schwartzman ML, Davis KL, Nishimura M, et al. The cytochrome P450 metabolic pathway of arachidonic acid in the cornea. Adv Prostaglandin Thromboxane Leukot Res 1991;21A:185–192.
103. Schwartzman ML, Balazy M, Masferrer J, et al. 12(R)-hydroxyicosatetraenoic acid: a cytochrome-P450-dependent arachidonate metabolite that inhibits Na+,K+-ATPase in the cornea. Proc Natl Acad Sci USA 1987;84:8125–8129.
104. Murphy RC, Falck JR, Lumin S, et al. 12(R)-hydroxyeicosatrienoic acid: a vasodilator cytochrome P-450-dependent arachidonate metabolite from the bovine corneal epithelium. J Biol Chem 1988;263:17197–17202.
105. Chen P, Guo M, Wygle D, et al. Inhibitors of cytochrome P450 4A suppress angiogenic responses. Am J Pathol 2005;166:615–624.
106. Mezentsev A, Mastyugin V, Seta F, et al. Transfection of cytochrome P4504B1 into the cornea increases angiogenic activity of the limbal vessels. J Pharmacol Exp Ther 2005;315:42–50.
107. Mieyal PA, Dunn MW, Schwartzman ML. Detection of endogenous 12-hydroxyeicosatrienoic acid in human tear film. Invest Ophthalmol Vis Sci 2001;42:328–332.
108. Davis KL, Conners MS, Dunn MW, et al. Induction of corneal epithelial cytochrome P-450 arachidonate metabolism by contact lens wear. Invest Ophthalmol Vis Sci 1992;33:291–297.
109. Ashkar S, Mesentsev A, Zhang WX, et al. Retinoic acid induces corneal epithelial CYP4B1 gene expression and stimulates the synthesis of inflammatory 12-hydroxyeicosanoids. J Ocul Pharmacol Ther 2004;20:65–74.
110. Benet LZ, Cummins CL, Wu CY. Unmasking the dynamic interplay between efflux transporters and metabolic enzymes. Int J Pharm 2004;277:3–9.
111. George RL, Huang W, Naggar HA, et al. Transport of N-acetylaspartate via murine sodium/dicarboxylate cotransporter NaDC3 and expression of this transporter and aspartoacylase II in ocular tissues in mouse. Biochim Biophys Acta 2004;1690:63–69.
112. Katragadda S, Talluri RS, Pal D, et al. Identification and characterization of a Na+-dependent neutral amino acid transporter, ASCT1, in rabbit corneal epithelial cell culture and rabbit cornea. Curr Eye Res 2005;30:989–1002.

113. Jain-Vakkalagadda B, Pal D, Gunda S, et al. Identification of a Na+-dependent cationic and neutral amino acid transporter, B(0,+), in human and rabbit cornea. Mol Pharm 2004;1:338–346.
114. Maenpaa H, Gegelashvili G, Tahti H. Expression of glutamate transporter subtypes in cultured retinal pigment epithelial and retinoblastoma cells. Curr Eye Res 2004;28:159–165.
115. Anand BS and Mitra AK. Mechanism of corneal permeation of L-valyl ester of acyclovir: targeting the oligopeptide transporter on the rabbit cornea. Pharm Res 2002;19:1194–1202.
116. Basu SK, Haworth IS, Bolger MB, et al. Proton-driven dipeptide uptake in primary cultured rabbit conjunctival epithelial cells. Invest Ophthalmol Vis Sci 1998;39:2365–2373.
117. Berger UV and Hediger MA. Distribution of peptide transporter PEPT2 mRNA in the rat nervous system. Anat Embryol (Berl) 1999;199:439–449.
118. Bildin VN, Iserovich P, Fischbarg J, et al. Differential expression of Na:K:2Cl cotransporter, glucose transporter 1, and aquaporin 1 in freshly isolated and cultured bovine corneal tissues. Exp Biol Med (Maywood) 2001;226:919–926.
119. Gherzi R, Melioli G, De Luca M, et al. High expression levels of the "erythroid/brain" type glucose transporter (GLUT1) in the basal cells of human eye conjunctiva and oral mucosa reconstituted in culture. Exp Cell Res 1991;195:230–236.
120. Takata K, Kasahara T, Kasahara M, et al. Ultracytochemical localization of the erythrocyte/HepG2-type glucose transporter (GLUT1) in cells of the blood-retinal barrier in the rat. Invest Ophthalmol Vis Sci 1992;33:377–383.
121. Takata K, Kasahara T, Kasahara M, et al. Ultracytochemical localization of the erythrocyte/HepG2-type glucose transporter (GLUT1) in the ciliary body and iris of the rat eye. Invest Ophthalmol Vis Sci 1991;32:1659–1666.
122. Philp NJ, Wang D, Yoon H, et al. Polarized expression of monocarboxylate transporters in human retinal pigment epithelium and ARPE-19 cells. Invest Ophthalmol Vis Sci 2003;44:1716–1721.
123. Damm J, Rau T, Maihofner C, et al. Constitutive expression and localization of COX-1 and COX-2 in rabbit iris and ciliary body. Exp Eye Res 2001;72:611–621.
124. Kulkarni PS and Srinivasan BD. Cyclooxygenase and lipoxygenase pathways in anterior uvea and conjunctiva. Prog Clin Biol Res 1989;312:39–52.
125. Waltman A and Sears M. Catechol-O-methyl transferase and monoamine oxidase activity in the ocular tissues of albino rabbits. Invest Ophthalmol 1964;3:601.
126. Stratford RE Jr and Lee VH. Ocular aminopeptidase activity and distribution in the albino rabbit. Curr Eye Res 1985;4:995–999.
127. Stratford RE Jr and Lee VH. Aminopeptidase activity in albino rabbit extraocular tissues relative to the small intestine. J Pharm Sci 1985;74:731–734.
128. Azzarolo AM, Bjerrum K, Maves CA, et al. Hypophysectomy-induced regression of female rat lacrimal glands: partial restoration and maintenance by dihydrotestosterone and prolactin. Invest Ophthalmol Vis Sci 1995;36:216–226.
129. Mircheff AK, Miller SS, Farber DB, et al. Isolation and provisional identification of plasma membrane populations from cultured human retinal pigment epithelium. Invest Ophthalmol Vis Sci 1990;31:863–878.
130. Li DW-C, Xiang H, Fass U, et al. Analysis of Expression Patterns of Protein Phosphatase-1 and Phosphatase-2A in Rat and Bovine Lenses. Invest Ophthalmol Vis Sci 2001;42:2603–2609.
131. Feeney L. Lipofuscin and melanin of human retinal pigment epithelium. Fluorescence, enzyme cytochemical, and ultrastructural studies. Invest Ophthalmol Vis Sci 1978;17:583–600.

132. Cabral L, Unger W, Boulton M, et al. Regional distribution of lysosomal enzymes in the canine retinal pigment epithelium. Invest Ophthalmol Vis Sci 1990;31:670–676.
133. Hayasaka S and Sears ML. Distribution of acid phosphatase, beta-glucuronidase, and lysosomal hyaluronidase in the anterior segment of the rabbit eye. Invest Ophthalmol Vis Sci 1978;17:982–987.
134. Robertson MJ, Erwig LP, Liversidge J, et al. Retinal Microenvironment Controls Resident and Infiltrating Macrophage Function during Uveoretinitis. Invest Ophthalmol Vis Sci 2002;43:2250–2257.
135. Shichi H, Atlas SA, Nebert DW. Genetically regulated aryl hydrocarbon hydroxylase induction in the eye: possible significance of the drug-metabolizing enzyme system for the retinal pigmented epithelium-choroid. Exp Eye Res 1975;21:557–567.
136. Madhu C, Dinh V, Babusis D, et al. Tissue specific distribution of cytochrome P450 isozymes in the monkey eye. ARVO Proceedings 1999.
137. Madhu C, Dinh V, Babusis D, et al. Expression of cytochrome P450 isoenzymes in dog and human eye. ISSX Proceedings 1999.
138. Ikeda H, Ueda M, Ikeda M, et al. Oxysterol 7alpha-hydroxylase (CYP39A1) in the ciliary nonpigmented epithelium of bovine eye. Lab Invest 2003;83:349–355.
139. Abraham NG, Lin JH, Dunn MW, et al. Presence of heme oxygenase and NADPH cytochrome P-450 (c) reductase in human corneal epithelium. Invest Ophthalmol Vis Sci 1987;28:1464–1472.

21
Barriers in Ocular Drug Delivery

Sriram Gunda, Sudharshan Hariharan,
Nanda Mandava, and Ashim K. Mitra

CONTENTS

INTRODUCTION
BARRIERS IN OCULAR DRUG DELIVERY
PHYSICOCHEMICAL PROPERTIES OF DRUGS AFFECTING PERMEABILITY
   ACROSS OCULAR BARRIERS
EFFLUX PUMPS
STRATEGIES TO OVERCOME OCULAR BARRIERS
SUMMARY
ACKNOWLEDGEMENTS
REFERENCES

INTRODUCTION

Ocular drug delivery, particularly to the posterior segment, is one of the most challenging tasks currently facing ocular drug-delivery scientists. Ophthalmic drugs are primarily administered by the topical route to treat ocular viral and bacterial infections, glaucoma, inflammations, and immunosuppression. Other routes of ocular drug delivery would be by systemic, intraocular and periocular injections, trans-scleral and subconjunctival pathways. Research on the posterior segment diseases, such as age-related macular degeneration (AMD) and diabetic macular edema (DME), is of high clinical significance and there is an urgent requirement for efficient drug-delivery systems. Delivery of therapeutic agents to the retina remains a challenging task because of the barrier properties of retinal pigment epithelium (RPE) and the endothelium lining the inner side of the retinal blood vessels. Drug delivery to the anterior segment is equally important, since a large fraction (up to 90%) of the administered dose is washed away into the nasolacrimal ducts. In order for a drug to reach therapeutic concentrations in the anterior segment, it has to pass through the anterior segment barriers, such as the corneal and conjunctival epithelia. Various formulation strategies, such as oil solutions, emulsions, nanoemulsions, ointments and gels, have been employed to increase the residence time of the drug on the corneal surface. A conjunctival sac can accommodate a maximal

From: *Ophthalmology Research: Ocular Transporters in Ophthalmic Diseases and Drug Delivery*
Edited by: J. Tombran-Tink and C. J. Barnstable © Humana Press, Totowa, NJ

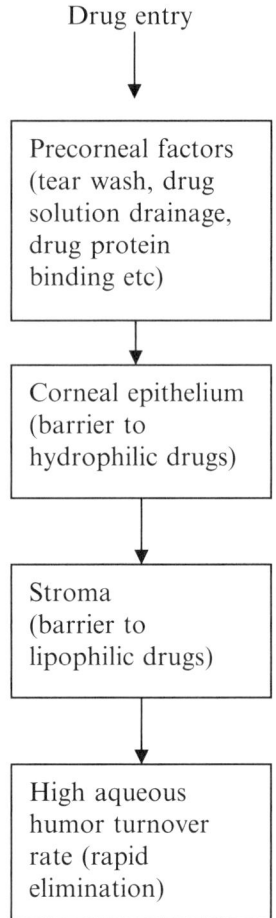

Fig. 1. Flow chart depicting the factors/barriers influencing the drug absorption after topical administration.

fluid volume of 30 µl. When a 50 µl volume is instilled topically, about 20 µl of drug will be drained away. In order to generate high drug concentrations in the precorneal area, a large dose (about 2–3% weight/volume) is usually required in formulations to achieve 500–600 µM of drug in the precorneal area. Fifty percent of the normal human tear film is replaced every 2–20 minutes. Such a high rate of tear turnover also reduces drug residence in the pre-corneal and pre-conjunctival areas. Therefore, the tear film also acts as a barrier to topical drug absorption. (Fig. 1 and Table 1)

Drugs administered topically have a low probability of reaching the posterior segment in significant amounts, as they have to pass through the corneal and conjunctival epithelia, aqueous humor, and lens, to reach the retina. However, a highly potent drug may achieve sufficient concentrations to exert therapeutic response. Currently, intravitreal bolus injection is the primary mode of therapy, but repetition of these can lead to

Table 1 Factors influencing drug absorption after topical administration *(53–55)*

Physiological factors	
Tear volume	7–30 µL
Tear turnover rate	0.5–2.2 µL/min
Spontaneous blinking rate	6–15/min
pH of lacrimal fluids	7.3
Turnover rate for aqueous humor	2–3 µL/min
Volume of aqueous humor	100–250 µL
Anatomical factors	
Thickness of the cornea	0.52 mm
Diameter of the cornea	12 mm
Surface area of the cornea	1.04 cm^2
Pore size of the epithelium	60 Å

retinal puncture, peeling, and/or detachment. Such bolus delivery of injections can also lead to higher concentrations of drugs, which in turn may cause toxicity, as in the case of gentamycin. There are several challenges to efficient ocular drug delivery. Some of the major challenges and hurdles are discussed in this chapter.

BARRIERS IN OCULAR DRUG DELIVERY

Anterior Segment

Cornea

The cornea is the outer coat (air–tissue interface) of the anterior segment. It is one of the few transparent tissues in the body, and it needs to maintain a high degree of clarity to transmit light to the retina without causing significant scattering. The transparency of the cornea is attributed to its avascularity, regularly arranged epithelium, and stroma. The central thickness of the cornea is about 0.52 mm and the central radius is 7.8 mm. It is composed of five layers: the columnar epithelium, Bowman's layer, stroma, Descemet's membrane, and endothelium.

Urtti et al. studied the permeation of polyethylene glycols (PEGs) across rabbit cornea, conjunctiva, and sclera. Conjunctival and scleral tissues were 15 to 25 times more permeable than the cornea, and molecular size affected the conjunctival permeability less than for cornea. The palpebral and bulbar conjunctiva exhibit similar permeabilities. However, the scleral permeability was approximately half of that in the conjunctiva, and approximately 10 times more than in the cornea. The conjunctival epithelia had approximately two times larger pores, and 16 times higher pore density, than the cornea. The total paracellular space in the conjunctiva was estimated to be 230 times more than in the cornea.

Corneal Epithelium Epithelium is the outermost layer of the cornea, which contributes more than half of the corneal resistance. It is composed of five or six layers of stratified,

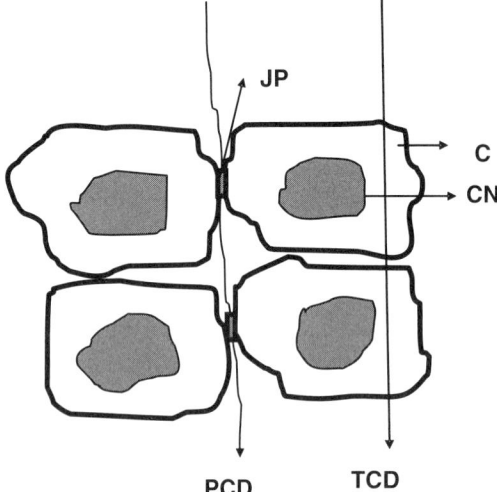

Fig. 2. Depicts drugs diffusing across the model ocular barrier by para- and transcellular routes.

C, cell; CN, cell nucleus; JP, junction proteins; PCD, paracellular drug diffusion; TCD, transcellular drug diffusion.

squamous, nonkeratinized cells of 50–60 μm total thickness. This layer contains 90% of the total cells in the cornea, which makes it the most lipophilic membrane in the cornea. Epithelial cells are tightly bound by cell adhesion proteins to form tight junctions. These tight junctions are mainly occludin, ZO-1, and ZO-2 *(1)*.

Tight junctions provide a continuous seal around the epithelial cells, thereby preventing the entry of polar drug molecules into the cornea. Drug molecules traverse the membrane by a number of mechanisms: paracellular and/or transcellular passive diffusion, active transport, and carrier-and receptor-mediated transport. Paracellular diffusion of polar drugs across the cornea is hindered by epithelial tight junctions. Lipophilic molecules (nonpolar) can diffuse passively through the lipid membranes of the epithelium (Fig. 2). Once the molecules cross the epithelial barrier, passage through the stroma and endothelium to enter the aqueous humor is not hindered. Molecular size does not play a significant role in the rate of permeation across epithelium, but the ionization may decrease the transcellular transport. Thus, the lipophilicity of a drug molecule, its ionization constant, and the pH of the eye drop formulation play major roles in determining the rate of drug permeation across the epithelium. Permeation enhancers can be used to improve the permeation of hydrophilic drugs across the cornea. Poly-L-arginine, with a molecular weight of 38 kDa, was found to increase the permeability of FITC-Dextran (molecular weight 3800) across cornea *(2)*. A detailed description of how lipophilicity affects the drug permeability across ocular barriers is described later in this chapter.

Bowman's Layer   Bowman's Layer is a modified, acellular part of the stroma. The layer mostly contains collagen fibrils. It separates the epithelium and stroma. This layer does not play any role in restricting drug permeation across cornea.

Stroma Stroma constitutes a major part of the cornea. Ninety percent of stroma is made up of water. It also consists of thick collagenous lamellae parallel to the surface of the cornea. Between the lamellae, fibroblastic keratocytes are sparsely dispersed. Stroma is considered the hydrophilic component of the cornea. Lipophilic drugs which enter epithelium, may not readily partition into the stroma. For such drugs the epithelium can act as depot. Stroma can act as a depot for relatively hydrophilic compounds.

Descemet's Membrane This is a thin, collagen-rich layer between the posterior stroma and endothelium, with a thickness of 8–12 µm. This layer does not act as a barrier to drug absorption.

Corneal Endothelium This constitutes a single layer of squamous epithelium. The membrane plays a critical role in corneal hydration and transparency. It is leaky in nature, relative to epithelium, and is not found to act as a transport barrier.

Conjunctiva

Drug absorption into the anterior chamber can take place through corneal or noncorneal routes. The noncorneal route includes permeation across conjunctiva. Delivery of drugs to posterior segment through the conjunctiva is one of the potential routes of drug delivery. The conjunctiva is the conduit for topically applied drugs to reach the posterior segment of eye *(3)*. The permeability of the conjunctiva to peptides and proteins is usually larger than that of the cornea, for example, the permeability to insulin (molecular weight 5800) and p-aminoclonidine (molecular weight 245.1) *(4,5)*. The reasons for the high conjunctival permeability are the increased surface area and leaky blood vessels of the conjunctiva.

The conjunctiva is a mucus membrane that lines the inner surfaces of eyelids and folds back to cover the front surface of the eyeball, except for the central portion of the outer eye (the cornea). It consists of two to five layers of columnar cells resting on a continuous basal lamina. In rabbits, the conjunctiva occupies nine times larger a surface area, and in humans, it occupies 17 times larger a surface area than the cornea *(6)*.

The conjunctiva is composed of three sections: palpebral, bulbar, and fornix. Palpebral conjunctiva covers the posterior surface of eyelid covered with thick, opaque, red muscle. Bulbar conjunctiva coats the anterior portion of the globe. It is a thin, semitransparent and colorless tissue. Blood vessels in bulbar conjunctiva can be differentiated from those underlying the sclera by their movement, as conjunctival vessels move, whereas scleral capillaries are fixed. These blood vessels are sensitive to irritation and can easily become congested. The white sclera and Tenon's capsule can be seen deep within the bulbar conjunctiva. Although the palpebral conjunctiva is moderately thick, the bulbar conjunctiva is very thin and movable. Fornix is the transition portion that forms the junction between the posterior eyelid and the globe (where palpebral conjunctiva reflects to become the bulbar conjunctiva); it is portion of conjunctiva covering cornea. This tissue consists of pseudostratified columnar epithelium rich in goblet cells, and contains accessory lacrimal glands, ductules of main lacrimal gland, and lymphoid follicles. The space between the palpebral and bulbar conjunctiva is called the conjunctival cul-de-sac. The conjunctiva (within the bulbar conjunctiva) contains goblet cells and melanocytes.

Goblet cells secret mucin, which is an important component of the precorneal tear film that protects and nourishes the cornea.

The conjunctiva can be divided into three layers, the outer epithelium, substantia propias, and submucosa. Outer epithelium acts as a permeability barrier that restricts the entry of particles into the ocular structures as a part of protective function. Submucosa contains nerves, lymphatic tissue, and blood vessels. Submucosa provides a loose attachment to the underlying sclera.

Conjunctival epithelium is similar to corneal epithelium with respect to the organization of epithelial cells. Both epithelia contain superficial, wing and basal cells. Conjunctival epithelium is two to three cell layers thick with non-keratinized, stratified, and squamous cells at the eyelids and columnar cells towards the cornea. Tight junctions at the apical pole of the conjunctival epithelium render it a relatively impermeable barrier. Conjunctival epithelial cells are attached together by desmosomes and are attached to the basal lamina through hemidesmosomes. Such junctional complexes seal the intercellular space between conjunctival epithelial cells restricting the transport of hydrophilic compounds such as peptides *(7,8)*. The rate-limiting barrier for hydrophilic compounds is the thin, apical layer and anatomically, the intercellular space is much smaller than the lipid membrane area *(9)*.

Other than cellular tight-junctional complexes in epithelium, the expression of efflux transporters is another important factor that contributes to the barrier properties of conjunctiva. Recent studies have shown the existence of a P-glycoprotein efflux pump (P-gp) in cultured rabbit conjunctival epithelial cells *(10)*. P-glycoprotein is a 170 kDa membrane protein encoded by multidrug resistance gene (MDR1) that functions as energy-dependent efflux pump *(11)*.

This energy-dependent efflux pump appears to be predominantly located on the apical plasma membrane of conjunctival epithelium and may play an important role in restricting the conjunctival absorption of various lipophilic compounds *(12)*. A recent report has indicated that the apical uptake of 50 nM propranolol increased from 43% to 66% in the presence of cyclosporine A, progesterone, rhodamine 123, verapamil, 4E3 monoclonal antibody and 2,4-dinitrophenol (2,4-DNP). This result indicates the functional expression of P-gp on conjunctival epithelium *(12)*.

The apparent permeability coefficient of cyclosporine A in the basal-to-apical direction (efflux) is 9.3 times higher than in the apical-to-basal direction (influx). Verapamil and progesterone increase cyclosporine A influx by 300%, while reducing efflux by 50% to 70%. Verapamil and progesterone inhibit cyclosporine A efflux in a concentration-dependent manner. These results indicate the localization and functional activity of P-gp in the apical membrane of the conjunctiva *(10)*.

Posterior Segment

The posterior segment is comprised of retina, choroid, vitreous humor, sclera, and the optic nerve. Many sight-threatening diseases that affect the posterior segment of the eye, with the retina being the primary target, are increasing in prevalence. Therapeutic concentrations of drugs in the retina and vitreous are needed to treat such diseases as cytomegaloviral (CMV) retinitis, age-related macular degeneration (ARMD), proliferative and diabetic vitreoretinopathy, and endophthalmitis. The topical mode of administering drugs is the preferred route when it comes to treating disorders of the eye. Cornea is the primary barrier to drug delivery and precorneal constraints such as solution drainage, tear turnover, and conjunctival absorption together limit drug absorption into the retina, choroid and vitreous significantly. Therapeutic agents can also be administered systemically to treat disease at the back of the eye. Choroid is richly perfused, with plentiful blood vessels, and allows easy exchange of molecules. Limited permeation into the retina primarily occurs due to the blood–aqueous barrier (BAB) and blood–retinal barrier (BRB). These two barriers together comprise the blood–ocular barrier. This chapter discusses the blood–retinal barrier in detail, and how it affects the transport of molecules into the retina. Before understanding the barriers, a closer examination of retinal physiology is essential.

Retina

The retina is a multilayered membrane of neuro-ectodermal origin that lines the internal space of the posterior segment. The following layers comprise a fully developed retina: inner limiting membrane, optic fiber layer, ganglion cell layer, inner plexiform layer, inner nuclear layer, outer plexiform layer, outer nuclear layer, outer limiting membrane, photoreceptor layer, and the retinal pigment epithelium (RPE). In short, a retina can be broadly divided into the neural retina and the RPE. The former is involved in signal transduction, leading to vision. Light enters the retina through the ganglion cell layer and penetrates all cell layers before reaching the rods and cones. The transduced signal is then relayed out of the retina by various neuronal cells to the optic nerve. The latter subsequently conducts the signal to the brain, where it is registered and an image is formed. RPE, on the other hand, is a single cell layer that separates the outer surface of the neural retina from the choroid and appears as a uniform and continuous layer extending through the entire retina, playing a vital role in supporting, and maintaining the viability of, the neural retina.

The neural elements of the retina are separated from the blood at two levels. An outer level, where the RPE covers the entire interface between the neural retina and the vascular layers of the choroid, and an inner level, where the endothelial cells of the retinal blood vessels separate the neural retina from the retinal blood supply *(13)*. Together, these two components constitute the BRB, which plays a critical role in homeostasis of the neural retina by limiting the entry of xenobiotics and by preventing the loss of essential solutes. Disruption of the BRB is associated with many of the disorders of the posterior segment.

Blood–Retinal Barrier

The barrier properties of the BRB are similar to the blood–brain barrier (BBB). The BRB protects the retina and vitreous from the entry of toxic substances and maintains the homeostatic control that underpins the physiology of the retina.

Retinal Vessels Morphological comparisons between the continuous endothelium of the retinal vessels (of the BRB) and the BBB were drawn to examine any marked differences in cellular organization. Results revealed that the interendothelial junctions of the retinal vessels were slightly different from those in BBB. The intercellular spaces of the retinal endothelium were found to be extensively sealed with zonulae occludens, limiting the permeability of molecules across the membrane. The retinal endothelium is also referred to as inner blood–retinal barrier (iBRB). The endothelial cells, along with their junctional complexes, comprise the bulk of the BRB with respect to substances like thorium dioxide, trypan blue, and fluorescein *(14)*. BRB has also been shown to restrict the entry of small protein tracers like microperoxidase (1.9 kDa, hydrodynamic radius around 2 nm), which is capable of entering the perioxonal space. Obviously large-molecular-weight protein tracers such as horseradish peroxidase, with a molecular weight of 40 kDa and a hydrodynamic radius of 5 nm, are no exceptions *(15)*. The transport of these tracers are not only restricted from the capillary lumen to the vitreous side, but also from the interstitial vitreous spaces towards the lumenal side of the retina, showing the bidirectional effectiveness of the barrier *(16)*. Excessive amounts of zonulae occludens in the intercellular junctions of the retinal endothelium appears to contribute to the BRB's barrier properties. Hydration and dehydration of the cat retina did not cause swelling or shrinkage of the endothelial cells, which remained firmly closed *(13)*. The capillary permeability surface area products (PSs) for mannitol and for sucrose in BRB were similar to that achieved across the BBB *(17)*. BAB, on the other hand, is not as effective as BRB in restricting molecular diffusion. The non-pigmented epithelium of the ciliary body is leaky with respect to small non-electrolytes such as sucrose. Multicompartmental analyses, as well as in vitro measurements, suggest that the ciliary epithelium of the rabbit eye is a relatively leaky layer with loose tight junctions, whereas the retina expresses very tight junctions, like those of brain capillaries *(18)*. Higher amounts of the test substance were found in the anterior side of the vitreous humor after intravenous injection *(13,19)*. Such a phenomenon was observed with proteins, urea, sodium, potassium, chloride, phosphate, inulin, sucrose, and several antibiotics, which gain entry into the anterior vitreous from the anterior chamber or ciliary circulation rather than across the BRB *(13,20–23)*.

Retinal Pigment Epithelium The RPE lies at the interface between the neural retina and the choroid. The microvilli of its apical surface interdigitate with the outer segments of photoreceptors, whereas its highly infolded basal membrane interacts with Bruch's membrane, serum components that cross the fenestrated choriocapillaris, and the secretions of the various choroidal cell types. RPE plays a central role in regulating the microenvironment surrounding the photoreceptors in the distal retina, where phototransduction takes place *(24)*. The outer segments of the rods and cones are closely associated with the RPE through villous and pseudopodial attachments. It is also responsible for the phagocytosis of the distal portions of the rods and cones outer segment *(24)*.

RPE forms the outer blood–retinal barrier (oBRB) by regulating transport between the neural retina and the fenestrated capillaries of the choroid. This layer was first discovered when organic anions were shown to migrate by an active transport mechanism at the retinal endothelium, as well as at the RPE. Light microscopy studies showed that the penetration of tryphan blue and fluorescein from choroid to retina stops at the level of RPE. Adherens junctions and associated actin filaments are characteristic features of this tight junction that retards diffusion through the paracellular spaces. Such tight junctions form an apical junctional complex that completely encircles each cell near the apical end of the lateral membranes. Increase in ZO-1 protein content has been correlated with decreased permeability across biological membranes *(25,26)*, but there are some conflicting results regarding the correlation of higher levels of ZO-1 with lower permeabilities across RPE and retinal endothelium *(27,28)*. The barrier properties of the RPE develop gradually. Studies with the chick RPE show that the epithelium is highly permeable in early stages of development. Tight junctions are evident by embryonic day 7, but are not effective in restricting the passage of protein markers like horseradish peroxidase. Freeze-fracture studies on embryonic days 15 to 19 show the complexity of the tight junctions, which suggests a dramatic decrease in the permeability during the later stage *(29)*.

Pore radius is one of the functions considered when evaluating the barrier properties of a biological membrane. The pore dimensions are probed by examining the permeation characteristics of solutes with various shapes, charges, molecular weights, and solubilities. Studies conducted by Vargas et al. suggest that compounds like inulin and sucrose, with molecular radii of 14 Å and 5.3 Å respectively, do not penetrate into the vitreous, whereas small molecules like glycerol (radius 3 Å) and sodium (radius 2.5 Å) permeate very slowly *(13)*. The permeability coefficient of fluorescein across BRB, with a molecular radius of 5.5 Å, is at least 30 times less than across other vessels, for molecules like sucrose (radius 5.3 Å) and raffinose (radius 6 Å) with similar molecular radii *(13,30)*. However, at the same time, the permeability coefficient of fluorescein across BRB is comparable to BBB.

Though the BRB retards the entry of most water-soluble molecules into the vitreous, several active carrier-mediated processes are present that regulate the transport of nutrients and metabolites such as glucose (via GLUT1 and GLUT3), amino acids, nucleosides (purine nucleoside transporter), folic acid (reduced-folate transporter [RFT]1 and FRα), lactic acid (monocarboxylate transporter [MCT]1–4), ascorbic acid and retinoids (cellular retinol-binding protein and interphotoreceptor retinoid-binding protein) across the retina *(31–38)*. In addition to these nutrient transporters, several active ion transporters are also present in order to maintain cell homeostasis.

PHYSICOCHEMICAL PROPERTIES OF DRUGS AFFECTING PERMEABILITY ACROSS OCULAR BARRIERS

Drug permeability prediction across the ocular tissues is important in the development of new drugs and drug-delivery strategies. To develop models that broadly predict permeability, it is necessary to understand the effects of various physicochemical properties of a drug, such as lipophilicity, charge, molecular radius and size, on its transport across ocular barriers.

The cornea is composed of five or six layers of columnar epithelium with tight junction proteins. On the other hand, passive diffusion is the primary route for hydrophilic drugs to permeate the cornea. Thus the lipoidal nature of the corneal epithelium presents a major barrier to the entry of hydrophilic drugs like acyclovir, ganciclovir, epinephrine, and pilocarpine. In humans, ocular bioavailability is predicted to be 1–5% for lipophilic molecules (octanol-to-water distribution coefficient greater than 1) and to be less than 0.5% for hydrophilic molecules (octanol-to-water distribution coefficient less than 0.01) *(39)*. Chemical modifications have been carried out to improve the partition coefficient of hydrophilic drugs by acyl ester prodrug design, and the results showed improved permeability across corneal epithelium *(40,41)*, but one of the major disadvantages is that achieving the desired lipophilicity often requires compromising on the aqueous solubility of the molecule. For a compound to be effective topically and to be formulated into eye drops, it must possess sufficient hydrophilicity, and at the same time exhibit sufficient permeability across the cornea. The cornea is an effective barrier to compounds larger than 10Å, which cannot cross the membrane at any significant rate *(42)*. Also, there is no apparent dependence on corneal permeability for compounds with small molecular radius, but for macromolecular peptides like thyrotropin-releasing hormone (TRH), p-nitrophenyl beta-cellopentaoside (PNP), and luteinizing hormone-releasing hormone (LHRH), a trend is observed with increasing molecular size *(43)*. The ionization state of the drug molecule can also affect its permeability across corneal epithelium. In vitro corneal transport studies suggest that the permeability of the unionized pilocarpine species is twice that of the ionized species *(44)*. Thus, the lipoidal epithelial layer of the corneal membrane appears to be a predominant barrier to the transport of polar species. The conjunctiva offers an attractive route for drug delivery to the posterior segment of the eye because of its large surface area compared to cornea, but the conjunctival epithelium, like corneal epithelium, restricts the entry of molecules across the membrane. The limited data available regarding conjunctiva show no clear dependence on distribution coefficient, but some dependence on molecular size. However, one would expect conjunctival permeability to show a preference for lipophilic molecules *(45)*. Also, in general, the conjunctiva appears to possess similar or higher permeability than cornea *(45)*. Considering the posterior segment of the eye, the RPE functions as a protective barrier against the entry of xenobiotics into the retina and vitreous from the systemic circulation. Pitkanen et al. studied the effects of solute molecular weight and lipophilicity on the permeability across a RPE-choroid preparation from bovine eyes *(46)*. Carboxyfluoresceins, a group of fluorescein isothiocyanate (FITC)-labeled dextrans with molecular weights of 4–80 kDa, and β-blockers exhibiting a wide range of lipophilicity, were chosen as permeation markers. Results showed that the RPE-choroid preparation was 35 times more permeable to carboxyfluorescein (376 Da) than to FITC-dextran (80 kDa). The permeabilities of lipophilic β-blockers were about 8- and 20-fold higher than those of hydrophilic atenolol and carboxyfluorescein, respectively *(46)*. Indeed, with ascending partition coefficients, vitreal concentrations, as well as the rate of vitreal penetration of antibiotics, were elevated.

Hence, knowledge of the effect of physicochemical properties on permeation across barriers can lead to the development of theoretical models, which in turn can be broadly applied to predict permeability of new drugs, leading to novel drug-delivery strategies.

EFFLUX PUMPS

Multidrug resistance (MDR) can be partially caused by efflux pumps belonging to the adenosine triphosphate (ATP)-binding cassette (ABC) superfamily. There are two major families of efflux proteins, P-glycoprotein (P-gp) and multidrug-resistance-associated proteins (MRPs).

P-Glycoprotein

P-gp is a versatile xenobiotic efflux pump. Several classes of drugs that are structurally and pharmacologically unrelated are known to be substrates for P-gp, including anticancer agents, steroids, protease inhibitors, and cardiovascular drugs. P-gp is expressed in various tissues in eye, including cornea, conjunctiva and retina. It has been reported that P-gp is expressed in the retinal capillary endothelial cells, RPE cells, ciliary non-pigmented epithelium, conjunctival epithelial cells, and iris and ciliary muscle cells *(10,47,48)*. A study was performed in our laboratory using quinidine as a model P-gp substrate and verapamil as inhibitor of P-gp. Intravitreal concentration of quinidine increased twofold in the presence of Verapamil, indicating the importance of P-gp in ocular disposition of drugs *(49)*. Reverse-transcriptase polymerase chain reaction (RTPCR) and Western blot analysis showed that P-gp belonging to the MDR1 family is expressed in rabbit corneal epithelium. RTPCR experiments also suggest the presence

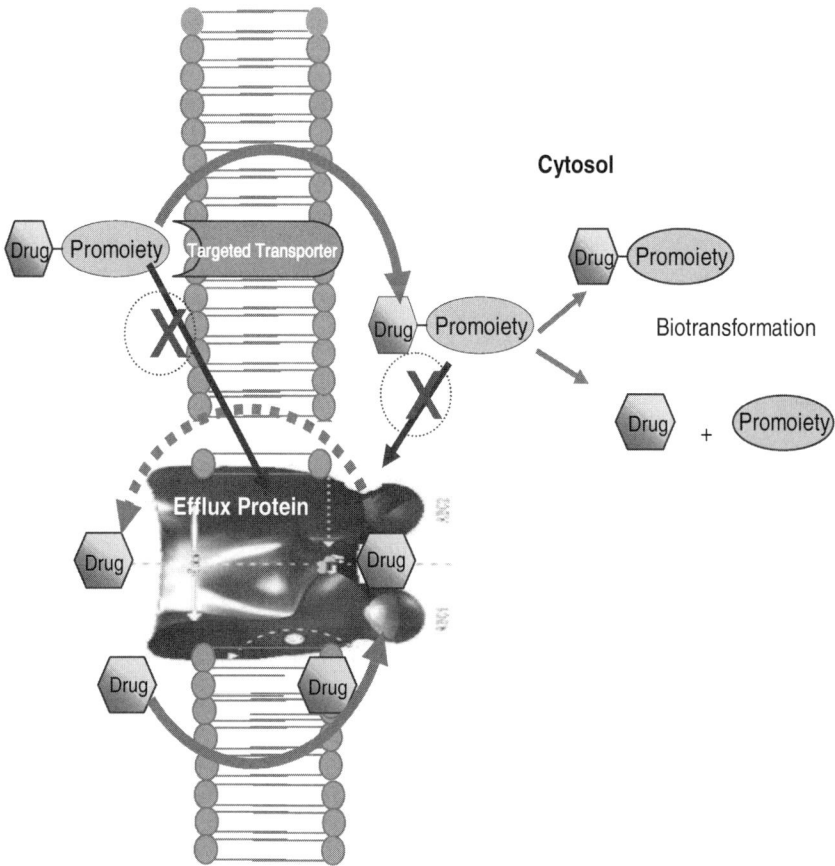

Fig. 3. Strategy to overcome efflux pumps through transporter-targeted prodrug derivatization.

of P-gp messenger RNA (mRNA) in human cornea *(50)*. Studies from our laboratory have also shown that ocular bioavailability of erythromycin is enhanced in the presence of P-gp inhibitors such as cyclosporine A, quinidine, and verapamil *(51)*.

Multi-Drug-Resistance Protein

MRP is an integral membrane glycophosphoprotein belonging to the superfamily of ATP-binding cassette transmembrane transporter proteins. The MRP family contains at least seven members: MRP1 and its six homologs, known as MRP2–7. A functional, biochemically active MRP has been identified in human RPE cells *(52)*. MRP1 has been found in rabbit conjunctival epithelial cells. In our laboratory, MRP2 has been identified and its role in drug efflux from rabbit primary corneal epithelial cells has been studied. MRP2 mRNA has also been identified in the human cornea. Other MRPs in ocular tissues have not yet been identified.

STRATEGIES TO OVERCOME OCULAR BARRIERS

A prodrug strategy has been employed successfully in our laboratory to circumvent P-glycoprotein-mediated efflux. In this strategy a P-gp substrate is attached to a promoiety so that it is no longer identified by the efflux proteins. However, the conjugate is recognized by an influx transporter, for which the promoiety acts as a substrate. By such derivatization, the drug molecule is transported into the cytosol in an active manner by the influx transporter, and consequently avoids interaction with the efflux protein. Specific influx transport systems can be targeted by prodrug derivatization for a specific substrate. Such a strategy has previously resulted in a threefold increase in corneal penetration of quinidine permeation, a model P-gp/MRP substrate. The strategy is depicted in Fig. 3. Success of the above strategy depends mainly on the properties of the influx transporters, i.e., capacity, affinity, and substrate specificity.

SUMMARY

Ocular barriers pose a challenge to the delivery of effective therapeutic concentrations of drugs to the eye, especially to the posterior segment. Depending on the delivery target, various factors such as size, lipophilicity, solubility, charge on the molecule, pH and viscosity of the formulation, drug release rates, etc., can be manipulated to effectively deliver drugs to the eye. Various novel routes of administration, such as subconjuctival, subtenon, etc., can be effectively used to overcome the ocular barriers.

ACKNOWLEDGEMENTS

This work was supported by NIH grants R01EY09171-12 and R01EY10659-11.

REFERENCES

1. Yi X, Wang Y, Yu FS. Corneal epithelial tight junctions and their response to lipopolysaccharide challenge. Invest Ophthalmol Vis Sci 2000;41:4093–100.
2. Nemoto E, et al. Effects of poly-L-arginine on the permeation of hydrophilic compounds through surface ocular tissues. Biol Pharm Bull 2006;29:155–60.

3. Agarwal S, et al. Functional characterization of peptide transporters in MDCKII-MDR1 cell line as a model for oral absorption studies. Int J Pharm 2007;332:147–52.
4. Chien DS, et al. Role of enzymatic lability in the corneal and conjunctival penetration of timolol ester prodrugs in the pigmented rabbit. Pharm Res 1991;8:728–33.
5. Gunda S, Hariharan S, Mitra AK. Corneal absorption and anterior chamber pharmacokinetics of dipeptide monoester prodrugs of ganciclovir (GCV): in vivo comparative evaluation of these prodrugs with Val-GCV and GCV in rabbits. J Ocul Pharmacol Ther 2006;22:465–76.
6. Watsky MA, Jablonski MM, Edelhauser HF. Comparison of conjunctival and corneal surface areas in rabbit and human. Curr Eye Res 1988;7:483–6.
7. Grass GM, Robinson JR. Mechanisms of corneal drug penetration. I: In vivo and in vitro kinetics. J Pharm Sci 1988;77:3–14.
8. Schoenwald RD, et al. Tear film stability of protein extracts from dry eye patients administered a sigma agonist. J Ocul Pharmacol Ther 1997;13:151–61.
9. Pfister RR. The normal surface of conjunctiva epithelium. A scanning electron microscopic study. Invest Ophthalmol 1975;14:267–79.
10. Saha P, Yang JJ, Lee VH. Existence of a p-glycoprotein drug efflux pump in cultured rabbit conjunctival epithelial cells. Invest Ophthalmol Vis Sci 1998;39:1221–6.
11. Bellamy WT. P-glycoproteins and multidrug resistance. Annu Rev Pharmacol Toxicol 1996; 36:161–83.
12. Yang JJ, Kim KJ, Lee VH. Role of P-glycoprotein in restricting propranolol transport in cultured rabbit conjunctival epithelial cell layers. Pharm Res 2000;17:533–8.
13. Cunha-Vaz JG. The blood-retinal barriers. Doc Ophthalmol 1976;41:287–327.
14. Cunha-Vaz J. The blood-ocular barriers. Surv Ophthalmol 1979;23:279–96.
15. Smith RS, Rudt LA. Ocular vascular and epithelial barriers to microperoxidase. Invest Ophthalmol 1975;14:556–60.
16. Peyman GA, Bok D. Peroxidase diffusion in the normal and laser-coagulated primate retina. Invest Ophthalmol 1972;11:35–45.
17. Lightman S, et al. Assessment of the permeability of the blood-retinal barrier in hypertensive rats. Hypertension 1987;10:390–5.
18. Green K, Pederson JE. Effect of 1-tetrahydrocannabinol on aqueous dynamics and ciliary body permeability in the rabbit. Exp Eye Res 1973;15:499–507.
19. Hariharan S, et al. Identification and functional expression of a carrier-mediated riboflavin transport system on rabbit corneal epithelium. Curr Eye Res 2006;31:811–24.
20. Bleeker GM and Maas EH. Penetration of penethamate, a penicillin ester, into the tissues of the eye. AMA Arch Ophthalmol 1958;60:1013–20.
21. Davson H, et al. The penetration of some electrolytes and non-electrolytes into the aqueous humour and vitreous body of the cat. J Physiol 1949;108:203–17.
22. Jain R, et al. Evasion of P-gp mediated cellular efflux and permeability enhancement of HIV-protease inhibitor saquinavir by prodrug modification. Int J Pharm 2005;303:8–19.
23. Kinsey VE. Ion movement in the eye. Circulation 1960;21:968–87.
24. Luo S, et al. Functional characterization of sodium-dependent multivitamin transporter in MDCK-MDR1 cells and its utilization as a target for drug delivery. Mol Pharm 2006;3:329–39.
25. Buse P, et al. Glucocorticoid-induced functional polarity of growth factor responsiveness regulates tight junction dynamics in transformed mammary epithelial tumor cells. J Biol Chem 1995;270:28223–7.
26. Gardner TW. Histamine, ZO-1 and increased blood-retinal barrier permeability in diabetic retinopathy. Trans Am Ophthalmol Soc 1995;93:583–621.
27. Rubin LL, et al. A cell culture model of the blood-brain barrier. J Cell Biol 1991;115:1725–35.

28. Stevenson BR, et al. Tight junction structure and ZO-1 content are identical in two strains of Madin-Darby canine kidney cells which differ in transepithelial resistance. J Cell Biol 1988;107:2401–8.
29. Ban Y and Rizzolo LJ. A culture model of development reveals multiple properties of RPE tight junctions. Mol Vis 1997;3:18.
30. Vargas F and Johnson JA. Permeability of rabbit heart capillaries to nonelectrolytes. Am J Physiol 1967;213:87–93.
31. Blazynski C. The accumulation of [3H]phenylisopropyl adenosine ([3H]PIA) and [3H]adenosine into rabbit retinal neurons is inhibited by nitrobenzylthioinosine (NBI). Neurosci Lett 1991;121:1–4.
32. Bok D, Ong DE, Chytil F. Immunocytochemical localization of cellular retinol binding protein in the rat retina. Invest Ophthalmol Vis Sci 1984;25:877–83.
33. Chancy CD, et al. Expression and differential polarization of the reduced-folate transporter-1 and the folate receptor alpha in mammalian retinal pigment epithelium. J Biol Chem 2000;275:20676–84.
34. Mantych GJ, Hageman GS, Devaskar SU. Characterization of glucose transporter isoforms in the adult and developing human eye. Endocrinology 1993;133:600–7.
35. Mitscherlich A. [Mass-psychology and analysis of the ego–one life-time later]. Psyche (Stuttg) 1977;31:516–39.
36. Philp NJ, Yoon H, Grollman EF. Monocarboxylate transporter MCT1 is located in the apical membrane and MCT3 in the basal membrane of rat RPE. Am J Physiol 1998;274:R1824–8.
37. Tornquist P and Alm A. Carrier-mediated transport of amino acids through the blood-retinal and the blood-brain barriers. Graefes Arch Clin Exp Ophthalmol 1986;224:21–5.
38. Yoon H, et al. Identification of a unique monocarboxylate transporter (MCT3) in retinal pigment epithelium. Biochem Biophys Res Commun 1997;234:90–4.
39. Zhang W, Prausnitz MR, Edwards A. Model of transient drug diffusion across cornea. J Control Release 2004;99:241–58.
40. Hughes PM and Mitra AK. Effect of acylation on the ocular disposition of acyclovir. II: Corneal permeability and anti-HSV 1 activity of 2′-esters in rabbit epithelial keratitis. J Ocul Pharmacol 1993;9:299–309.
41. Tirucherai GS, Dias C, Mitra AK. Corneal permeation of ganciclovir: mechanism of ganciclovir permeation enhancement by acyl ester prodrug design. J Ocul Pharmacol Ther 2002;18:535–48.
42. Janoria KG, et al. Biotin uptake by rabbit corneal epithelial cells: role of sodium-dependent multivitamin transporter (SMVT). Curr Eye Res 2006;31:797–809.
43. Sasaki H, et al. Ocular membrane permeability of hydrophilic drugs for ocular peptide delivery. J Pharm Pharmacol 1997;49:135–9.
44. Mitra AK, Mikkelson TJ. Mechanism of transcorneal permeation of pilocarpine. J Pharm Sci 1988;77:771–5.
45. Prausnitz MR and Noonan JS. Permeability of cornea, sclera, and conjunctiva: a literature analysis for drug delivery to the eye. J Pharm Sci 1998;87:1479–88.
46. Pitkanen L, et al. Permeability of retinal pigment epithelium: effects of permeant molecular weight and lipophilicity. Invest Ophthalmol Vis Sci 2005;46:641–6.
47. Holash JA and Stewart PA. The relationship of astrocyte-like cells to the vessels that contribute to the blood-ocular barriers. Brain Res 1993;629:218–24.
48. Schlingemann RO, et al. Ciliary muscle capillaries have blood-tissue barrier characteristics. Exp Eye Res 1998;66:747–54.
49. Duvvuri S, Gandhi MD, Mitra AK. Effect of P-glycoprotein on the ocular disposition of a model substrate, quinidine. Curr Eye Res 2003;27:345–53.

50. Dey S, et al. Molecular evidence and functional expression of P-glycoprotein (MDR1) in human and rabbit cornea and corneal epithelial cell lines. Invest Ophthalmol Vis Sci 2003;44:2909–18.
51. Dey S, Gunda S, Mitra AK. Pharmacokinetics of erythromycin in rabbit corneas after single-dose infusion: role of P-glycoprotein as a barrier to in vivo ocular drug absorption. J Pharmacol Exp Ther 2004;311:246–55.
52. Aukunuru JV, et al. Expression of multidrug resistance-associated protein (MRP) in human retinal pigment epithelial cells and its interaction with BAPSG, a novel aldose reductase inhibitor. Pharm Res 2001;18:565–72.
53. Schoenwald R.D. Ocular pharmacokinetics/pharmacokinetics. In: AK Mitra, editor. Ophthalmic drug delivery systems. New York: Marcel Dekker; 1993. p. 83–110.
54. Worakul N and Robinson JR. Ocular pharmacokinetics/pharmacodynamics. Eur J Pharm Biopharm 1997;44:71.
55. Lee VH. Mechanism and facilitation of corneal drug penetration. J Control Rel 1990;11:79.

22
Ophthalmic Applications of Nanotechnology

Swita Raghava, Gaurav Goel, and Uday B. Kompella

CONTENTS

> INTRODUCTION
> NANOSYSTEMS AND FUNDAMENTALS OF NANOTECHNOLOGY
> MANUFACTURING METHODS FOR NANOPARTICLES
> NANOSURGERY
> NANOTECHNOLOGY IN RETINAL PROSTHESES
> NANOTECHNOLOGY IN OPHTHALMIC DIAGNOSTICS
> NANOTECHNOLOGY FOR GENE DELIVERY TO THE EYE
> NANOTECHNOLOGY FOR OCULAR DRUG DELIVERY
> TOXICITY CONCERNS WITH NANOTECHNOLOGY
> CONCLUSIONS
> ACKNOWLEDGEMENTS
> REFERENCES

INTRODUCTION

Nanotechnology, the current catchphrase in applied technology, broadly refers to the design, fabrication, evaluation, or application of materials with at least one of the dimensions in the nanometer range. The enthusiasm for nanotechnology in biomedical sciences is due to several unique properties of nanomaterials. These unique properties include small size, large surface area, easy suspendability in liquids, access to cells and organelles, and tunable physicochemical characteristics including optical and magnetic properties (1). Several nanosystems are currently in use for biomedical applications including disease therapy and diagnosis. Some of these systems are described in Table 1. This chapter reviews nanosystems that can potentially be used for ophthalmic applications and provides some examples of nanotechnology applications in the areas of ophthalmic devices, diagnostics, and gene- or drug-delivery systems.

NANOSYSTEMS AND FUNDAMENTALS OF NANOTECHNOLOGY

Nanosystems

Nanosystems include (i) soluble macromolecules, (ii) dendrimers, (iii) micelles, (iv) nanoplexes, (v) nanoparticles (nanospheres and nanocapsules), (vi) liposomes, and (vii) nanorods, tubes, and fibers, among others. Some of these systems are shown in Fig. 1.

From: *Ophthalmology Research: Ocular Transporters in Ophthalmic Diseases and Drug Delivery*
Edited by: J. Tombran-Tink and C. J. Barnstable © Humana Press, Totowa, NJ

Table 1 Biomedical nanosystems marketed for human use

Product name	Company	Drug	Description
Abraxane®	Abraxis Biosciences Inc.	Paclitaxel	Albumin nanoparticles (~130 nm) for breast cancer
Combidex®	Advanced Magnetics Inc.	Iron oxide	Ultrasmall (< 50 nm) superparamagnetic iron oxide nanoparticles for the detection of metastatic disease in lymph nodes by magnetic resonance imaging (MRI)
Doxil®	Ortho Biotech	Doxorubicin	Liposomes for ovarian carcinoma and acquired immune deficiency syndrome-related Kaposi's sarcoma
Emend®	Merck	Aprepitant	Nanocrystals of aprepitant in a capsule for prevention of emesis during cancer chemotherapy
Ferridex®	Advanced Magnetics Inc.	Iron oxide	Superparamagnetic dextran coated iron oxide nanoparticles as MRI contrast agents for diagnosis of liver lesions
GastroMARK®	Advanced Magnetics Inc.	Iron oxide	Superparamagnetic silicone coated iron oxide nanoparticles as MRI contrast agents for abdominal imaging
Macugen®	Pfizer Inc.	VEGF Aptamer	Nanoconjugates of aptamer with polyethylene glycol (PEG) for wet age-related macular degeneration
Rapamune®	Wyeth-Ayerst Labs.	Sirolimus	Nanocrytals of sirolimus in tablets for immunosuppresion following kidney transplant

Soluble Macromolecules

Virtually any drug, including macromolecules, is in the molecular or nanometer dimensions when present in solution. Some macromolecular preparations such as soluble drug conjugates can offer unique properties to the drug molecule. For instance, Macugen®, a marketed treatment for neovascular age-related macular degeneration is a covalent conjugate of an oligonucleotide with two 20 kDa monomethoxy polyethylene glycol (PEG) units *(2)*. The molecular weight of the conjugate is around 50 kDa. This conjugate prolongs the half life of the drug in the vitreous humor. Depending on the concentration of the conjugate, it can be present in solution or as a colloidal dispersion.

Dendrimers

Dendrimers (from the Greek *dendron*, meaning tree, and *meros*, meaning part) consist of a central core molecule, from which a number of highly branched, tree-like arms originate in an ordered and symmetric fashion *(3)*. Dendrimers have several unique physicochemical properties that make them attractive for biomedical applications. Dendrimers are made by stepwise attachment of layers of a second branched molecule on a central core. Addition of one layer on top of the core molecule results in generation zero or G0 of a dendrimer. Addition of subsequent layers results in G1, G2, …., Gn generations of

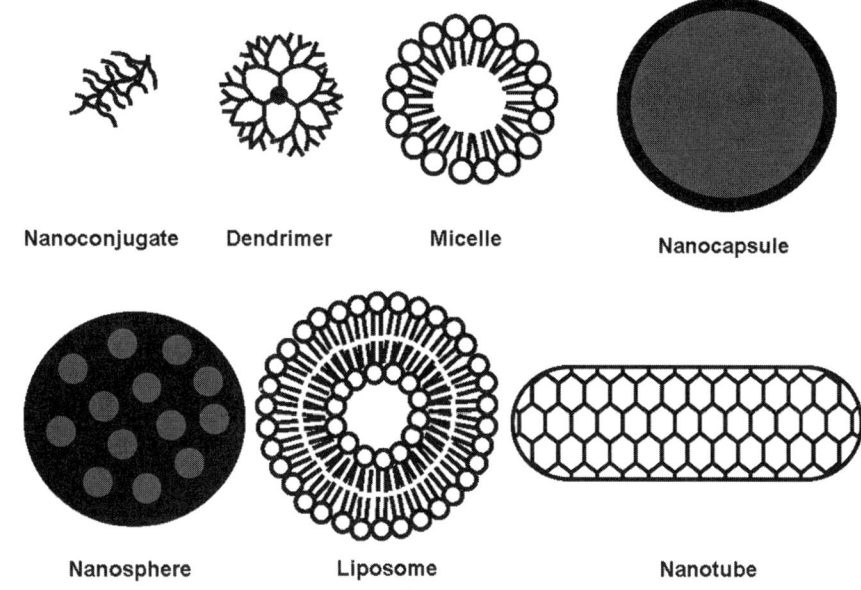

Fig. 1. Nanosystems for ophthalmic applications.

dendrimers. In practice, dendrimers up to G6 are used. With dendrimers, a precise size control is possible because of the stepwise chemical synthetic processes used in generating the uniform annular monolayer of each generation. Each dendrimer preparation is chemically more homogeneous, resulting in a low polydispersity index as compared to conventional linear or branched polymers *(4)*. Dendrimers are similar in size to a number of biological structures. For instance, G5 polyamidoamine (PAMAM) dendrimers have a diameter of 5.5 nm, which is similar to the size of a hemoglobin molecule *(5)*.

With increasing generations, the number of surface functional groups of a dendrimer increases. Large number of surface groups on a dendrimer can easily be utilized to fabricate systems for applications in drug and gene delivery, imaging, boron-neutron capture therapy, and biotechnological diagnostics and sensing. Cationic dendrimers based on PAMAM (Superfect®) for experimental gene delivery are commercially available *(4)*. Since net positive surface charge results in nonspecific cell-surface adsorption and membrane perturbation, cationic dendrimers cause in vivo toxicity, which increases with the number of dendrimer generation *(6)*. Other materials that have been employed for dendrimer synthesis include polylysine, polyethylene oxide, polyethyleneglycol monomethacrylate, dihydroxybenzoic acid, and dihydroxy gallic acid *(6,7)*.

Lysine-based drug-containing dendrimers inhibit choroidal neovascularization in vivo. An intravitreally administered lysine-based dendrimer anti-vascular endothelial growth factor (VEGF) oligodeoxynucleotide inhibits the severity of laser-induced choroid neovascularization (CNV) significantly better than plain oligonucleotide, for at least four months, in rats *(7)*. The severity score in eyes with oligonucleotide alone at two and four months post-injection was 2.86 ± 0.09 and 2.15 ± 0.17, as compared to 0.18 ± 0.12 and 1.39 ± 0.17, respectively, with the dendrimer preparation.

Unlike the widely investigated positively charged dendrimers, anionic dendrimers are expected to be safer. Indeed, an anionic dendrimer drug-delivery system was shown to be useful in preventing scar tissue formation after glaucoma filtration surgery. Anionic G3.5 PAMAM dendrimers conjugated to glucosamine (an immunomodulator) or glucosamine-6-sulphate (an antiangiogenic molecule), when coadministered subconjunctivally, increase the long-term clinical success of glaucoma surgery *(8)*. In this study, which employed a rabbit model, the day 30 bleb survival increased from 30% to 80% with the use of dendrimer formulations.

Micelles

Micelles are colloidal aggregates of amphiphilic molecules. Micellization is a self-assembling property exhibited by amphiphilic molecules in aqueous solution above a specific concentration, referred to as critical micelle concentration (CMC). Although any surfactant molecule is suitable for forming a micelle, micelles based on block copolymers made up of repeating hydrophilic and lipophilic blocks are being extensively investigated as therapeutic carriers for drug and gene delivery. Micelles may be spherical, rod shaped, or lamellar, depending on the nature of the monomer units. An example of such a block copolymer is Pluronic®, which consists of ethylene oxide (EO) and propylene oxide (PO) blocks arranged in the order EO_x-PO_y-EO_x *(9)*. A doxorubicin formulation in Pluronic® micelles (SP1049C, Supratek Pharma, Canada) is under clinical investigation for the treatment of drug-resistant esophageal adenocarcinoma. SP1049C consists of doxorubicin-containing Pluronic® micelles that are 10 to 20 nm in size *(10)*. The partial response rate to SP1049C in esophageal adenocarcinoma is 47%, and significantly higher compared to 20% with doxorubicin alone. One of the major drawbacks of micellar formation is the demicellization that occurs as a result of dilution upon injection in vivo.

Ocular bioavailability of topical pilocarpine increases upon instillation of a micellar formulation of pilocarpine as compared to pilocarpine hydrochloride solution *(11)*. The area under the curve (AUC) and miotic response duration, in rabbits, increase by 64% and 1.5-fold, respectively, with Pluronic F127, which uses pilocarpine micelles as opposed to pilocarpine hydrochloride solution. Enhanced drug delivery with the micellar formulation is a likely explanation for this finding.

Nanoplexes

Nanoplexes are complexes of a therapeutic agent with one or more carrier materials. For instance, cationic polymers such as polyethyleneimine (PEI) are routinely used to complex and condense anionic agents such as deoxyribonucleic acid (DNA). The positively charged groups on the polymer interact electrostatically with the negatively charged phosphate groups in the backbone of DNA molecule, allowing the compaction of DNA and hence, better transfection. A reason for the high transfection efficiency of PEI is considered to be its ability to buffer endosomes *(12)*. The presence of unprotonated amino groups in PEI structure at physiological pH is responsible for this characteristic. In the acidic pH of endosome, PEI captures protons, thereby lowering ion concentration and osmotic pressure within the endosome. This results in water influx, swelling of endosome, and finally, burst release of endosomal contents including DNA into cytosol. Cationic carriers for the formation of nanoplexes can be chosen from a variety of materials including lipids, polymers, and peptides. Practical application of these cationic polyplexes is limited due to the cytotoxicity caused by most of the cationic polymers, especially after repeated use. The cytotoxicity can potentially be reduced by optimally titrating the charge groups in the polycation with other anionic carriers. Indeed, use of anionic dextran sulfate as a co-carrier reduces the cytotoxicity of cationic polyallylamine and resulted in a higher transfection efficiency compared to lipofectin®, a cationic lipid formulation *(13)*.

Nanoparticles

Nanoparticles are spherical particles with diameters in the nanometer size range that can be prepared using lipids, proteins, polysaccharides, or polymers *(14)*. These can be broadly classified into nanospheres and nanocapsules. Nanospheres are nanoparticles with drug or gene molecules dispersed in a carrier matrix. Nanocapsules are composed of a reservoir of drug solution, or solid, coated by a rate-limiting layer *(15)*. Thus, nanocapsules have a distinct core containing the drug and an outer shell made up of a membrane. Table 2 summarizes various materials employed for the preparation of nanoparticles for ophthalmic purposes.

Liposomes

Liposomes are vesicles consisting of hydrated phospholipid bilayers, designed to entrap drug either in the core or bilayer. Based on structural parameters such as size, number and position of lamellae, liposomes can be classified into multilamellar large vesicles (MLV) (greater than 0.5 µm), oligolamellar vesicles (OLV) (0.1 to 1 µm), unilamellar vesicles (UV) (all sizes), small unilamellar vesicles (SUV) (20 to 100 nm), large unilamellar vesicles (LUV) (greater than 100 nm), giant unilamellar vesicles (GUV) (greater than 1 µm) and multivesicular vesicles (MVV) (greater than 1 µm) *(31)*. Liposomal drug formulations can reduce toxicity, prolong circulation, and drug delivery, and also target a particular tissue. Indeed, light-targeted drug delivery to posterior segment

Table 2 Materials used for ophthalmic nanoparticles

Chemical class	Carrier material assessed	References
	Albumin	*(16,17)*
Proteins	Gelatin	*(18)*
	Mucin	*(19)*
Lipids	Stearic acid	*(20)*
Carbohydrates	Chitosan	*(21)*
	Poly(alkyl) cyanoacrylates	*(22)*
	Poly(lactide) or poly(lactide-co-glycolide)	*(23,24)*
	Poly(epsilon) caprolactone	*(25)*
	Polyacrylic acid	*(26)*
Polymers	Cellulose acetate phthalate	*(27)*
	Poly(methylmethacrylate)	*(28)*
	Eudragit	*(29)*
	Poly(methylmethacrylate) sulfopropylmethacrylate	*(30)*

tissues like choriocapillaris and RPE can be achieved using thermosensitive liposomes *(32)*. Light-targeted liposomal delivery can be employed for diagnosis and therapy of CNV in age-related macular degeneration (ARMD). In this approach, after the systemic administration of thermosensitive liposomes, drug release in CNV and surrounding tissues is triggered by local warming of the tissue and liposomes using a light beam at a wavelength strongly absorbed by blood.

Nanorods, Nanotubes, and Nanofibers

These are cylindrical structures, with diameters typically in the nanometer range. Unlike nanotubes, which have a hollow interior, nanorods have a solid interior. Compared to nanorods, nanofibers are longer and more flexible. Nanofibers can have a solid or hollow interior. These nanosystems can be applied to a broad range of areas including biosensors, catalysis, drug delivery, and optoelectronics. Carbon nanotubes are one of the most investigated nanotube-based systems. Carbon nanotubes conduct electric signals, allowing their application in nanoelectronics *(33,34)*. More recently, carbon nanotubes have been employed for several biological applications. Carbon nanotubes can be used as cytotoxic nanobombs for inducing cancer cell death *(35)*, for highly selective detection of cancer biomarkers *(36)*, and as tissue-engineering platforms for neuronal growth *(37,38)* and bone repair *(39,40)*. Potent nanobombs can be created by adsorbing water molecules on single-wall carbon nanotube sheets. These nanotubes can be made to explode by causing rapid evaporation of adsorbed water molecules, with the generation of enormous pressure in nanotubes, by exposure to a 800 nm laser of 50–200 mW/cm^2 intensity. Neurons cultured on these nanotubes exhibit extensive branching neurites *(38)*. Embryonic hippocampal neurons cultured on carbon nanotubes coupled to 4-hydroxynonenal (4-HNE), elaborated four to six neurites as compared to just one or two neurites of neurons cultured on untreated nanotubes. Total neurite length also increased more than twofold with 4-HNE-treated nanotubes. This, in conjunction with their ability to support neuronal growth and conduct electric signals, can be potentially employed in the development of retinal prostheses.

MANUFACTURING METHODS FOR NANOPARTICLES

The approaches for engineering nanoparticles include emulsion polymerization, desolvation of macromolecules, solvent evaporation, ionic gelation, nanoprecipitation, milling, self assembly, nanolithography, and supercritical fluid technology. The engineering methods differ depending on the type of carrier, drug materials utilized, and intended application. Using the emulsion solvent evaporation method and a probe sonicator, nanoparticles of budesonide *(24)* and VEGF antisense oligonucleotide *(41)* have been prepared. In this method, controlling the energy input from probe sonicator, the duration of sonication, and the choice of emulsification system allows particle size to be controlled.

A key problem with solvent evaporation methods is the presence of residual organic solvent in the particles prepared. We have utilized supercritical fluid technology extraction-based methods to remove the residual organic solvents from the particles *(42)*. A 30-minute exposure of supercritical CO_2 to PLGA particles formed using the emulsion solvent evaporation method reduces the residual organic solvent (dichloromethane) content from

4,500 ppm to less than 25 ppm. In addition, supercritical fluid extraction-based methods can be utilized for the preparation of drug-containing nano- and microparticles of PLGA *(43)*. In a supercritical fluid process, several parameters, including the concentration of the drug/polymer, the flow rates of solute-containing liquids or supercritical CO_2, the temperature and pressure conditions of supercritical fluid, and the diameters and type of capillary nozzles used for spraying, can be varied to optimize the particle properties *(44)*. Using a desolvation method, we prepared albumin nanoparticles encapsulating Cu, Zn superoxide dismutase, and VEGF intraceptor plasmid-loaded albumin nanoparticles *(17,45)*.

Nanoelectromechanical systems (NEMS) and microelectromechanical systems (MEMS) are emerging technologies for generating electrical or mechanical devices in the nanometer to micrometer size range. Use of biocompatible polymers like polymethylmethacrylate (PMMA) and polydimethylsiloxane (PDMS) as alternatives to silicon in NEMS/MEMS has made these technologies available for biological applications such as controlled drug delivery *(46)*. NEMS and nanotechnology in general have laid the foundation for nanosurgery, biological prostheses, imaging, biological tagging, biotechnology, and drug delivery. NEMS has the potential to address the unmet need in the programmed delivery of biological macromolecules. Indeed, slow zero-order release of radiolabeled bovine serum albumin has been achieved with a 13 nm nanopore membrane loaded with the protein *(47)*. The release follows non-Fickian diffusion kinetics as pore diameter approaches the hydrodynamic diameter of the solute. Further, a biodegradable polymer-chip multireservoir drug-delivery device has been fabricated using the above technologies to obtain sequential release of macromolecules in vitro and in vivo *(48)*. PLA was used as the polymer to fabricate such a device by compression molding. The device had 36 individual 120–130 nL reservoirs loaded with human growth hormone, dextran, or heparin. The release control was achieved by capping reservoirs with PLGA membranes ranging in molecular weight from 4,400 to 64,000 Da. The activation of reservoirs in sequence was illustrated in vitro and in vivo. The sequential release of the encapsulated drug from reservoirs was achieved by using reservoir capping membranes with different degradation rates.

In addition to the above examples of nanosystems, several additional examples of nanosystems in the areas of nanosurgery, retinal prostheses, diagnostics, gene delivery, and drug delivery are provided in the following sections.

NANOSURGERY

Although the dimensions of retinal artery, arteriole and choriocapillary are 120 μm, 8–15 μm, and 10–15 μm, the smallest dimension of surgical tools such as tanoforceps used in retinal surgery is 1300 μm (personal communications with Dr. Marco A. Zarbin). The diameter of a 25-gauge needle is 500 μm. Thus, there is a need to develop better surgical devices and needles to accomplish micro- and nano-manipulations of blood vessels, and potentially nerve tissue in the eye. This forms the basis for the field of nanosurgery, which is aimed at developing tools such as nanoscalpels, nanotweezers, and nanoneedles *(49)*. These tools are expected to allow small-vessel, cellular, and subcellular manipulations.

Nanoscalpels are based on femtosecond laser pulses. Nanoscalpels allow subcellular dissections, without collateral damage, using laser beams near the infrared region of the visual spectrum.

Nanotweezers are essentially two nanotubes arranged together to form a tweezer-like structure *(50)*. Nanotweezers are commonly made of silicon and carbon nanoelectrodes. The principle of their action involves voltage-triggered to and fro motion of the free end of the tweezers. The application of voltage causes the two tips to bend closer, while removal of the applied voltage makes the tips to get back to the original position. Nanotweezers can potentially be used in the manipulation of subcellular components. These are currently being investigated as a means of enhancing the design of scanning probe microscopes, such as scanning tunneling microscopes (STM) and atomic force microscopes (AFM). The current tip design cannot grab an object for manipulation. It is believed that tweezers could overcome this problem.

Nanoneedles as thin as 100 nm have been developed and these might be helpful in performing nanoscale surgery. Nanoneedles are currently used for modification of AFM tips *(51)*. AFM nanoneedles have the advantage of being able to enter the cells with relatively little force and high accuracy. Obataya et al. *(52)* developed ultrathin (200–300 nm) needle-like tips for AFM microscopes. These could potentially be used for protein, nucleic acid, and drug delivery in target cells. Although these approaches are exciting in terms of understanding events at the cellular level in experimental medicine, the practical applicability of such approaches in therapeutic intervention is yet to be established.

NANOTECHNOLOGY IN RETINAL PROSTHESES

Prostheses for vision restoration in visually impaired patients with degenerative disorders like ARMD, retinitis pigmentosa and glaucoma are based on one of three techniques: stimulation of the visual cortex, stimulation of optic nerve, and stimulation of retinal cells *(53)*. The visual prostheses can be broadly classified into cortical and retinal implants, which are further discussed below.

Cortical Implants

These consist of electrodes embedded in the visual cortex. For localized intracortical stimulation with minimal pain, small electrodes and low electrical currents are desired. Nanosize electrodes could potentially be employed in such devices to enhance visual signals. The main advantage of cortical stimulation for vision restoration is that the technique bypasses disease proximal to the primary visual cortex, making it potentially useful for treating diseases of optic nerve and retina.

Retinal Implants

These serve as artificial photoreceptors and consist of an array of microphotodiodes attached to microelectrodes. Depending on the layer of retina receiving the device, retinal prostheses can be classified into epiretinal, when the implant is placed on the surface of the retina between vitreous and inner limiting membrane, and subretinal, when the implant is placed in the subretinal space between outer retina and RPE. Retinal implants are limited in application to disorders in which the visual pathway distal to retina is intact and functional.

During the past decade, tremendous progress has been made towards electronic retinal prostheses; however, it is generally believed that a fully functional long-lasting device is not on the immediate horizon. One of the challenges with retinal prostheses is long-term biocompatibility. For an electronic implant, in addition to ensuring chemical, biophysical and immunological compatibility, the reaction to external electrical stimuli and heating of the tissue need to be considered. To this effect, carbon nanotube (CNT)-based retinal prostheses are being developed to avoid the incompatibility issues with metallic electrodes *(54)*. Carbon nanotubes offer unique mechanical and electrical properties for use in retinal prostheses, such as tunable electrode height to allow fabrication of penetrating microelectrode arrays with controllable stimulation depth, easy penetrability in retina, and good biocompatibility, as suggested by their capability to support retinal ganglion cell growth. Thus, nanotechnology advancement is benefiting the development of smaller, more-efficient, and biocompatible retinal prostheses.

NANOTECHNOLOGY IN OPHTHALMIC DIAGNOSTICS

Nanostructures are attractive probes for biological detection in vivo, ex vivo, or in vitro, because of their high surface-to-volume ratio, tunable physicochemical properties, and structural stability. The advantages offered by nanomaterials as diagnostics are increased sensitivity, selectivity, and practicality as compared to conventional diagnostic systems *(55)*.

In Vivo Diagnosis

Various imaging approaches can be employed for the diagnosis of ophthalmic disorders. Such techniques include positron emission tomography (PET), computed tomography (CT), magnetic resonance imaging (MRI), optical coherence tomography (OCT), fundus photography, fluorescein angiography, scanning laser ophthalmoscopy, in vivo confocal imaging, and ultrasonography. This imaging technique armamentarium offers options with a wide range of sensitivity, spatial resolution, and cost effectiveness. Nuclear imaging techniques such as PET have a high sensitivity, but lack spatial resolution. MRI, on the other hand, has good spatial resolution and provides images with good anatomical definition, but has inherently low sensitivity. Nanotechnology provides multiple options to overcome issues of low sensitivity.

Nanotechnology has revolutionized magnetic resonance imaging by providing highly sensitive and targetable contrast agents. MRI contrast agents are classified into positive contrast agents, based on paramagnetic complexes of Gd^{3+} or Mn^{2+} ions, and negative contrast agents, based on superparamagnetic particles based on iron oxide *(56)*. Different generations of Gd(III) DTPA–terminated poly(propyleneimine) dendrimers enabled delineation of sub-millimeter-sized blood vessels after intravenous injection into a colon-cancer tumor model *(57)*. The minimum detectable dendrimer concentration depends on the dendrimer generation, with G5 dendrimers having two orders of magnitude lower minimum detectable concentration (8.1 ± 10^{-8} M) than for G0 dendrimers (3.1 ± 10^{-5} M).

In vivo imaging of neovascularization in the eye can help in monitoring the prognosis of ocular disorders like diabetic retinopathy and ARMD. For instance, the MRI signal intensity, as well as angiogenic vessel- targeting in vivo, can potentially be enhanced using Gd-perfluorocarbon nanoparticles directed towards angiogenic blood vessels using a biotinylated anti-$\alpha_v\beta_3$ monoclonal antibody *(58)*. A 25% increase in average MRI signal was observed in a rabbit corneal neovascularization model receiving the targeted contrast agent as early as 90 minutes post-contrast. This contrast enhancement can be attributed to the pooling of targeted contrast agent through the openings or gaps in angiogenic blood vessels.

Similarly, superparamagnetic iron oxide nanoparticles have higher relaxivities as compared to paramagnetic ions. Since iron oxide nanoparticles passively target the reticuloendothelial system, these nanoparticles have been employed as contrast agents for MRI of reticuloendothelial system *(59,60)*. Furthermore, magnetic nanoparticles can be fabricated for active targeting as well. Indeed, PEG-coated magnetic nanoparticles covalently conjugated to a monoclonal antibody specific for a human glioma cell surface antigen accumulated in the tumor tissue 24 or 48 hours after intravenous injection *(61)*. The clinical diagnosis of malignant tumors can be envisioned in near future, with the development of the magneto impedance sensor for sensing of tumor tissue that incorporates magnetic nanoparticles. This technique has been successfully tested in rodent models of glioma and a malignant brain tumor *(62)*.

Gold nanoparticles, nanoshells *(63)*, and nanocages *(64)* are promising nanotechnology-based contrast agents for OCT. OCT produces real-time cross-sectional images through biological tissues by detecting the reflections of a low-coherence light source directed at the tissue, and determining the depth at which the reflection occurred. Gold nanoparticles are particularly attractive for OCT contrast enhancement as their optical resonance wavelengths can be precisely tuned over a broad range by controlling their sizes and shapes. Nanoshells are composed of a dielectric core such as silica coated with an ultrathin metallic layer, typically gold. The optical response of nanoshells can be varied over a broad range by varying the relative size of the core and the shell thickness. The optical spectrum of nanoshells spans from visible to near-infrared spectral region. The absorption cross section of nanoshells in the near-infrared region is around 4×10^{-14} m^2, a million-fold higher than that of the conventional near-infrared dye indocyanine green at 800 nm, around 10^{-20} m^2 *(63)*. Similarly, gold nanocages of 35 nm edge length have an absorption cross section approximately five orders of magnitude larger than conventional dyes (indocyanine green) upon OCT imaging of phantom samples in the near-infrared spectral region *(64)*.

Quantum dots are bright photostable fluorophores made up of semiconductor crystals that have a broad excitation spectrum, but a narrow emission at wavelengths controllable by the size and composition of a core *(65)*. The cause of these remarkable properties is the nanoscale size that leads to a quantum confinement effect, resulting from physical confinement of excitons in a semiconductor crystal. Similar to other nanostructures, quantum dots can be functionalized to facilitate tissue targeting. CdSe quantum dots functionalized with peptides binding to dipeptidase on the endothelial cells in lung or tumor blood vessels, accumulated in lungs and tumors, respectively, after intravenous injection in BALB/c mice *(66)*. Potentially, enhanced fluorophores such as quantum dots can be used for improving fluorescence imaging of the eye.

Ex Vivo Diagnosis

In addition to the in vivo imaging techniques that can be employed for imaging in isolated eyes, confocal imaging and fluorescence microscopy are routinely used for imaging excised ocular tissues. Isolated tissue preparations provide invaluable ex vivo alternatives to study uptake, and transport of therapeutic molecules and delivery systems, in preclinical studies. Understanding the interactions of various retinal cell types during normal and pathological conditions is essential for the development of therapeutic strategies. In isolated retinas, cell–cell interactions can be assessed using various dye-loading approaches or electrophysiological approaches that are not readily amenable for human applications.

The challenge of bulk dye loading with minimal damage to cells in whole retina ex vivo has recently been addressed using a biolistic delivery of silver nanoparticles coated with the calcium indicator, rhodamine dextran, in mouse retina *(67)*. After pneumatic delivery of nanoparticles coated with the red calcium indicator, rods in the photoreceptor layer were stained. This bulk loading was achieved with minimal damage to the retinal cells. The average amount of collateral damage using nanoparticle protocol was found to be 2%, compared to 25% using microparticles loaded in the same biolistic gun.

NANOTECHNOLOGY FOR GENE DELIVERY TO THE EYE

The various nanosystems discussed in this chapter can potentially be used for gene therapy. Gene therapy is a promising therapeutic approach for the treatment of a wide array of inherited and acquired disorders. Clinical use of gene therapy is limited by effective gene delivery in vivo. An ideal gene-delivery vector should be nontoxic, efficiently taken up by target cells, should protect the plasmid cargo against enzymatic degradation, and be conducive to gene expression. A number of viral vectors on the nanometer lengthscale have been shown to result in successful intraocular gene expression. The viral vectors have the drawback of limited carrying capacity, immunogenicity, and toxicity. Therefore, a number of nonviral nanosystems are currently being investigated. In the eye, gene delivery has potential applications in treating various disorders including neovascular and non-neovascular retinal degenerative disorders, glaucoma, and corneal graft rejection among others.

Evidence to date indicates the suitability of nanosystems in enhancing the delivery of nucleic acids, including oligonucleotides and genes, to the tissues of the eye. We are investigating the use of albumin and PLGA/PLA-based nanoparticles for facilitating gene delivery to the eye. With the United States Food and Drug Administration approval of albumin nanoparticles (130 nm) loaded with paclitaxel (Abraxane®) in 2005, albumin serves as an acceptable carrier for preparing nanoparticles. Albumin-based nanoparticles, as a gene-delivery system, offer the advantages of an established safety profile, biodegradability, and potential clinical viability *(17)*. We encapsulated a plasmid coding for Cu, Zn superoxide dismutase plasmid (SOD1) and enhanced yellow fluorescent protein, and demonstrated that these particles with a diameter of about 150 nm sustain the release of the plasmid, protect it against serum and nuclease degradation, and allow in vitro transfection efficiencies better than lipofectamine, a positively charged commercial reagent *(17)*. Furthermore, intravitreal injection of the albumin nanoparticles in mice followed by Western blot analysis indicated retinal expression of enhanced yellow fluorescent protein in the retina within two days. Albumin nanoparticles are also useful in enhancing corneal gene expression. We observed that corneal gene expression and effects of VEGF intraceptor (Flt-23K) can be sustained for at least a few weeks in a mouse model of corneal injury *(45)*. Intrastromal injection of Flt-23K-loaded nanoparticles sustained corneal gene expression for four weeks. Corneal neovascularization development post mechanical-alkali injury was reduced by Flt-23K plasmid-loaded nanoparticles administered three weeks prior to injury. The corneal neovascularization areas after injection of intraceptor plasmid and naked plasmid were 35.0 ± 6.0 % and 58.3 ± 8.7 %, respectively.

Biodegradable polymers such as PLGA and PLA offer an additional platform for preparing nanoparticles for nucleic acid delivery. Several pharmaceutical products containing these polymers are approved for parenteral administration. Currently, Posurdex®, an intravitreal implant system based on these polymers, is undergoing clinical trials. Nanoparticles based on these polymers are useful for enhancing the cellular uptake and efficacy of oligonucleotides. For instance, we observed that PLGA (50:50) nanoparticles (252 ± 3.4 nm) enhance the delivery and activity of a VEGF antisense oligonucleotide in cultured human retinal pigment epithelial cell line (ARPE 19) *(41)*. The cellular uptake of VEGF antisense oligonucleotide encapsulated in PLGA nanoparticles was enhanced by 4.3-fold as compared to unencapsulated oligonucleotide. Nanoparticles containing the oligonucleotide significantly reduced VEGF mRNA levels, as well as VEGF secretion, from ARPE 19 cells, suggesting their potential usefulness for the treatment of neovascular disorders of the eye. An increase in VEGF secretion has been at least partly implicated in the retinal neovascularization seen in diabetic retinopathy, retinopathy of prematurity, and ARMD. We have recently demonstrated the suitability of PLGA-based nanoparticle systems for retinal gene delivery of plasminogen kringle 5 peptide (unpublished data). Other investigators have shown that PLA and PLGA nanoparticles loaded with plasmids encoding green fluorescent protein or red nuclear fluorescent protein (643 ± 74 nm), successfully transfect cultured bovine and human retinal pigment epithelial cells (RPE) in vivo upon intravitreal injection in male Lewis rats *(68)*.

Nanosystems of oligonucleotides can be used for improving the outcomes of glaucoma filtration surgery, which is complicated by excessive scarring during wound healing, which can eventually lead to obliteration of surgically created subconjunctival filtration space, or bleb, for the passage of aqueous humor. Inhibition of transforming growth factor–β2 (TGF-β2) by administration of nanosize complexes of polyethyleneimine and anti-TGF-β2 oligonucleotide significantly improves the outcome of glaucoma surgery *(69)*. Nanocomplexes (220 ± 40 nm) administered following encapsulation in porous particles sustained the release of oligonucleotide for 15 days and enhanced the intracellular penetration of the oligonucleotide upon subconjunctival administration of particle suspension, in pigmented Fauve de Bourgogne female rabbits. The clinical evaluation of these rabbits was based on overall inflammatory state of the eye, and the time to bleb filtering failure. Microspheres containing antisense-TGFβ2–PEI nanosize complexes prolonged 100% bleb survival to 42 days, as compared to 28 days with microspheres containing uncomplexed oligonucleotide. No significant hyperemia was noted in the six weeks following the procedure.

Even inorganic nanoparticles, such as those made of iron oxide, have the potential for enhancing gene transfection. Streptavidin-coated superparamagnetic iron oxide nanoparticles can be conjugated to biotin-labeled DNA fragments to enhance gene transfection

(70). Using these layered particles of 100 nm, gene transfection could be enhanced in an immortalized human hepatoma cell line (Huh-7) and adult retinal endothelial cell lines from dog and human sources.

In comparison to viral vectors, these nonviral nanoparticle vectors offer greater flexibility in customizing the system for the purpose of targeting or enhancing the in vivo circulation time.

NANOTECHNOLOGY FOR OCULAR DRUG DELIVERY

Nanoparticles can enhance ocular drug delivery by increasing cellular uptake, sustaining drug release, preventing drug degradation, and targeting the drug intraocularly *(14)*. The delivery benefit accrued by using nanoparticles is achievable by all three local ocular routes, namely, topical, intraocular, and periocular.

Topical ocular delivery for the treatment of anterior chamber disorders is limited by rapid precorneal clearance. Nanoparticles can potentially be employed for enhanced corneal and conjunctival interaction. Even for very small 20 nm fluospheres, the corneal uptake following eye drop administration is only 2.4% in five minutes (Table 3) *(71)*. The uptake is not higher at 60 minutes (Fig. 2). The five-minute uptake can be enhanced up to 23% by denuding the epithelium, suggesting that epithelium is a formidable barrier even for nanoparticles. Functionalizing the particle surface with an LHRH agonist (deslorelin) or transferrin increases corneal uptake to 9% and 16% in five minutes (Table 3 and Fig. 2), indicating the usefulness of such approaches in enhancing bioavailability of pharmacological agents present in nanoparticle formulations.

Use of mucoadhesive polymers to enhance drug residence in the corneal, conjunctival, and cul-de-sac areas is another approach to enhance drug-residence time. Chitosan is one such polymer. Cyclosporine A-loaded chitosan nanoparticles increase the levels of drug in cornea and conjunctiva by two- and sixfold when compared to cyclosporine A suspension *(72)*. The corneal and conjunctival levels at the end of 24 hours with cyclosporine A suspension are subtherapeutic. On the other hand, those with cyclosporine A chitosan nanoparticles are therapeutic.

Polymers assessed in topical nanoparticles include poly(alkyl cyanoacrylate), cellulose acetate phthalate, Eudragit®, calcium phosphate, poly(lactic acid), and poly(lactide-co-glycolide) *(14)*. The nanoparticles tested ranged in size from 50 to 600 nm, with most of the studies done with particles ranging from 150 to 300 nm *(14)*. Some of

Table 3 Nanoparticle corneal uptake enhancement of surface functionalizations with deslorelin and transferrin *(71)*

Particle type	Effective diameter (nm)	Zeta potential (mV)	Corneal epithelial uptake in 5 min (%)	Corneal stromal uptake in 5 min (%)
NP	85.2 ± 0.6	−57.93 ± 4.55	1.05	1.32
Deslorelin-NP	98.2 ± 0.6	−35.76 ± 1.73	3.38	5.54
Transferrin-NP	84.7 ± 0.8	−29.24 ± 4.33	4.96	10.98

NP, plain nanoparticle; Deslorelin-NP, deslorelin-conjugated nanoparticles; Transferrin-NP, transferrin-conjugated nanoparticles. The data are expressed as mean ± standard deviation for $n = 3$

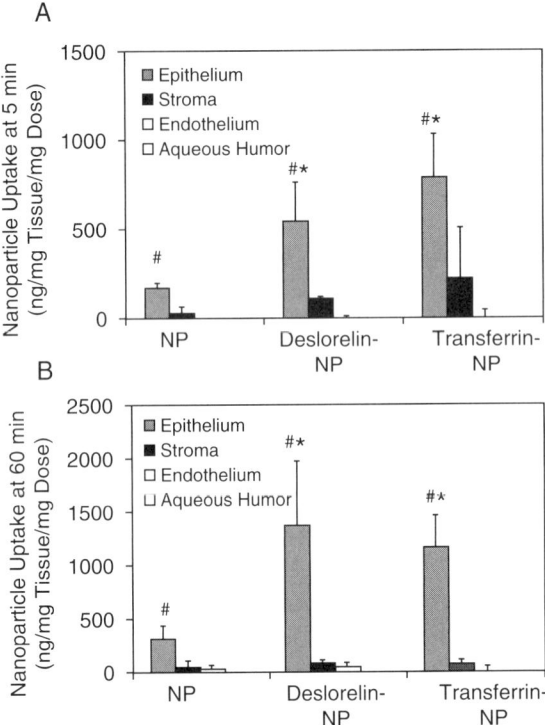

Fig. 2. Deslorelin and transferrin functionalizations enhance corneal nanoparticle uptake in bovine ex vivo eye model. A 50 μl drop of 1 mg/ml nanoparticle suspension was topically instilled on an isolated bovine eye maintained at 37°C. After 5 (**A**) and 60 (**B**) minutes, particle uptake in corneal epithelium, stroma, endothelium, and aqueous humor was quantified. The data are expressed as mean ± standard deviation for $n = 3$. The asterisk indicates $p < 0.05$ compared to NP group and the hash marks indicate $p < 0.05$ compared to stroma, endothelium or aqueous humor. NP, plain nanoparticles; Deslorelin-NP, deslorelin-conjugated nanoparticles; Transferrin-NP, transferrin-conjugated nanoparticles.

these formulations have been shown to sustain the miotic and IOP-lowering effect of the encapsulated drugs, reduce the systemic effects of β-blockers, and increase intraocular drug bioavailability.

Intravitreal nanoparticles can sustain drug levels in the posterior segment of the eye. Polymers tested for intravitreal nanoparticle administration include bovine serum albumin, poly(hexadecyl cyanoacrylate), poly(ethyl cyanoacrylate) and poly(lactide-co-glycolide) *(14)*. Indeed, intravitreal administration of acyclovir- and ganciclovir-loaded poly(ethyl cyanoacrylate) nanoparticles in rabbits sustains drug levels higher than MIC (0.25–1.22 μg/ml) in retina for at least 10 days *(73)*. Nanoparticles also reduced the plasma drug levels as compared to drug solution.

Intravitreal injection of PLGA nanospheres loaded with neuroprotective pigment epithelial-derived growth factor (PEDF) peptides in a mouse model of retinal ischemia protects the retinal ganglion cell layer from toxicity for seven days, as compared to

48 hours with the peptide alone *(74)*. PLA nanoparticles loaded with rhodamine-6G localized preferentially in RPE layer and persisted in retinal pigment epithelium layer for four months following single intravitreal injection in rats *(23)*. Transretinal movement and persistence of nanoparticles suggests that these nanoparticles can be employed for a steady and continuous retinal drug delivery.

Following intravitreal injection, significant amounts of albumin nanoparticles loaded with ganciclovir remain in a thin layer overlying retina up to two weeks and are well tolerated *(75)*. The retina and other ocular tissues maintain their cytoarchitecture. Histological evaluation does not show infiltration or vascular inflammation in the eye. Further, no alterations in expression and localization of ocular antigens, namely S-antigen and rhodopsin, were observed as compared to controls, suggesting the absence of organ-specific autoimmune phenomena following intravitreal ganciclovir-loaded albumin nanoparticles.

Periocular injections, a less-invasive approach compared to intravitreal injections, can be classified based on the exact site of injection into subtenon, subconjunctival, peribulbar, retrobulbar, and posterior juxtascleral injections *(76)*. Indeed, subconjunctival administration of poly(lactic acid) nanoparticles loaded with budesonide equivalent to 50 µg drug results in higher levels of the drug in retina, vitreous, lens and cornea, as compared to a solution form of budesonide (Fig. 3) *(24,77,78)*. However, prolonged drug delivery is better achieved with microparticles, which have a lower surface-to-volume ratio and a lower burst release of encapsulated drug. We observed that microparticles of budesonide, as well as celecoxib, exhibit low burst release and sustain quantifiable drug levels for longer periods.

For sustained drug release and prevention of systemic side-effects, it is desired that the nanoparticles are retained at the site of injection without causing any toxicity. The retention of nanoparticles at a periocular site of administration is governed by the size of nanoparticles *(79)*. Nanoparticles 200 nm or larger remain almost completely at the periocular site of injection for at least two months. The nanoparticle retention at periocular site of injection is not dependent on dose, for these particles. Carboxylate- (hydrophilic, negatively charged), amine- (hydrophilic, positively charged), and aldehyde sulfate- (hydrophobic) modified nanoparticles (200 nm) are all retained at the site of administration to the same extent, suggesting that retention is not dependent on the surface properties for these particles. On the other hand, 20 nm particles are rapidly cleared from the site of administration with 15% and 8% remaining after day 1 and day 9, respectively. Even 20 nm particles do not enter the retina significantly, suggesting the inadequacy of even nanoparticles in reaching retina via the trans-scleral pathway following periocular administration. The more-rapid clearance of 20 nm particles possibly occurs via the needle track and/or blood/lymphatic circulation. Thus, drug-particulate systems greater than 200 nm in size, formulated using biodegradable polymers, can potentially be utilized for sustained drug delivery to retina by the trans-scleral approach. Consistent with this hypothesis, we observed that periocularly administered 1140 nm PLGA particles encapsulating celecoxib sustain drug delivery for two months in the retina and inhibit retinal prostaglandin E_2 secretion, VEGF expression, and vascular leakage for at least two months *(80)*.

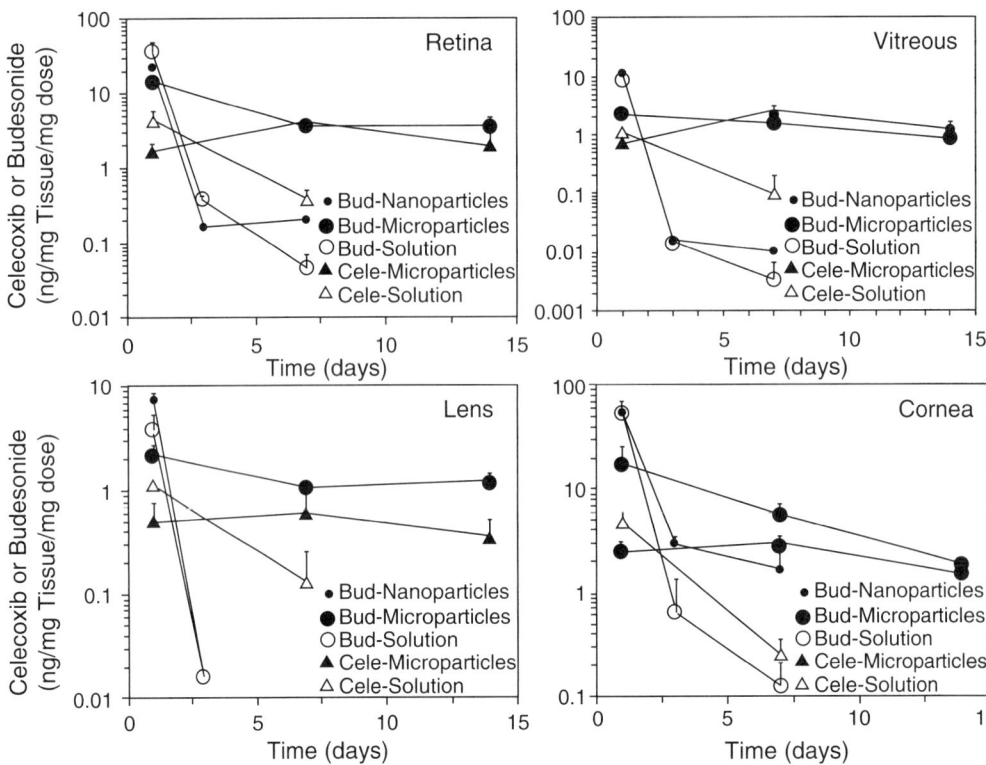

Fig. 3. Nanoparticles and microparticles sustain ocular tissue drug levels after periocular administration in a rat model. Budesonide (Bud) or celecoxib (Cele) was administered in the eyes of rats, either in the form of a solution, nanoparticles, or microparticles, and drug levels were estimated in retina, vitreous, cornea and lens. Data are expressed as mean ± standard deviation of results in four experiments. Data are shown for the ipsilateral eye. All formulations were administered at a 75 μg drug dose, except budesonide nanoparticles, which contained a 50 μg drug dose. At the doses used, the drug levels for all formulations were below detection limits in the contralateral eye. Also, budesonide levels were below detection limit on day 14 in solution and nanoparticle groups.

TOXICITY CONCERNS WITH NANOTECHNOLOGY

The ocular drug/gene delivery benefit with nanoparticles made using biodegradable polymers is reasonably well established. However, caution needs to be exercised while using new nanotechnology products, including polymers and inorganic materials, in humans (81). All new materials should be thoroughly assessed for their safety before human use. The toxicity evaluation of drug-delivery systems must take into account the toxicity of the drug, polymer, and entire system in vivo.

PLA and PLGA are very commonly used for fabrication of particulate delivery systems. For ophthalmic purposes, budesonide-PLA microparticles administered periocularly to New Zealand white male rabbits do not cause elevation in IOP. Furthermore, no change in lens opacity, fundus photography or blood chemistry was observed up to two months after dosing. The particles remained at the periocular site of injection at the end of two months, and even in periocular surrounding tissue no histological abnormalities are seen *(82)*. Periocularly administered celecoxib-PLGA particles sustained drug levels and drug effects in a rat model for diabetic retinopathy without inducing any fibrotic reaction at the site of administration when assessed in normal rats. Furthermore, these particles maintained the retinal architecture, as well as the thickness of the various retinal layers. In addition, these particles did not induce any changes in blood clinical chemistry involving parameters related to kidney function, liver function, or blood cell counts *(80)*.

A recent report compared the toxicity of nanoparticles for gene delivery by intravitreal or subretinal injections in rabbits *(83)*. The nanoparticles investigated include chitosan, poly-[(cholesteryl oxocarbonylamido ethyl) methyl bis(ethylene) ammonium iodide]-ethyl phosphate (PCEP), and magnetic nanoparticles (MNP). All of these nanoparticles successfully transfect cells in vitro and in vivo, but do show varying degrees of toxicity in vivo. Following chitosan nanoparticle injection, the study reported ocular inflammation, membranous opacity, and hazy vitreous in 92%, 62% and 32% of eyes, respectively. This might be related to the concentration of chitosan used and the positive charge density of the nanoparticles. Subretinal injection of PCEP nanoparticles induced RPE dysfunction in 11% and retinal degeneration in 15% of the eyes. The in vitro transfection efficiency of nanoparticles correlated with their in vivo toxicity, with the most efficient transfection agents causing the most toxicity in vivo. Surface coating of nanoparticles reduces their toxicity. However, caution should be exercised since the toxicity may reemerge following removal of the coat in vivo.

CONCLUSIONS

Nanotechnology offers several modern tools including; delivery systems for diagnostic agents, genes, and drugs; NEMS-based electrodes and devices for retinal prostheteses; and tools for nanosurgery. Of particular interest are nanosized colloidal delivery systems, broadly referred to as nanoparticles made of lipids, proteins, polysaccharides, or polymers that can carry a variety of chemicals. These systems can be administered by various routes including topical, periocular, intravitreous, intravenous, and oral routes. These systems enhance intracellular delivery, especially for poorly permeable macromolecules such as peptides, proteins, oligonucleotides, and genes. They can also sustain the delivery, although not as well as the currently available microparticulate systems. Furthermore, the nanoparticles can be functionalized with ligands to enhance their residence time, targeting, and cell entry in the body. Nanoparticles, like all other drugs come with their own risk–benefit ratio, depending on the specific materials used, system configuration, route of administration, frequency and dose of administration, and patient factors. A careful balancing of risks and benefits is critical to the development of these systems. A material safe in one compartment of the body may not be tolerated, even at a fraction of a dose, in another compartment. For this reason, rigorous procedures for drug and device development should be followed in the development of nanotechnology products for human use. With such diligent efforts, significant nanotechnology-based advances are anticipated that will be useful in preventing, diagnosing, and treating various ocular disorders.

ACKNOWLEDGEMENTS

This work was supported by NIH grants DK064172, EY013842, and EY017045 (through Emory University).

REFERENCES

1. Gupta RB. Fundamentals of drug nanoparticles. In: Gupta RB and Kompella UB, editors. New York: Taylor and Francis; 2006).
2. Katz B and Goldbaum M. Macugen (pegaptanib sodium), a novel ocular therapeutic that targets vascular endothelial growth factor (VEGF). Int Ophthalmol Clin 2006;46:141–54.
3. Niederhafner P, Sebestik J, Jezek J. Peptide dendrimers. J Pept Sci 2005;11:757–88.
4. Dufes C, Uchegbu IF, Schatzlein AG. Dendrimers in gene delivery. Adv Drug Deliv Rev 2005;57:2177–202.
5. D'Emanuele A and Attwood D. Dendrimer–drug interactions. Adv Drug Deliv Rev 2005;57:2147–62.
6. Duncan R and Izzo L. Dendrimer biocompatibility and toxicity. Adv Drug Deliv Rev 2005;57:2215–37.
7. Marano RJ, Toth I, Wimmer N, Brankov M, Rakoczy PE. Dendrimer delivery of an anti-VEGF oligonucleotide into the eye: a long-term study into inhibition of laser-induced CNV, distribution, uptake and toxicity. Gene Ther 2005;12:1544–50.
8. Shaunak S. et al. Polyvalent dendrimer glucosamine conjugates prevent scar tissue formation. Nat Biotechnol 2004;22:977–84.
9. Kabanov AV, Batrakova EV, Alakhov VY. Pluronic block copolymers for overcoming drug resistance in cancer. Adv Drug Deliv Rev 2002;54:759–79.
10. Danson S, et al. Phase I dose escalation and pharmacokinetic study of pluronic polymer-bound doxorubicin (SP1049C) in patients with advanced cancer. Br J Cancer 2004;90:2085–91.
11. Pepic I, Jalsenjak N, Jalsenjak I. Micellar solutions of triblock copolymer surfactants with pilocarpine. Int J Pharm 2004;272:57–64.
12. Kichler A. Gene transfer with modified polyethylenimines. J Gene Med 2004;6(Suppl 1): S3–10.
13. Nimesh S, Kumar R, Chandra R. Novel polyallylamine-dextran sulfate-DNA nanoplexes: Highly efficient non-viral vector for gene delivery. Int J Pharm 2006;320:143–9.
14. Amrite AC and Kompella UB. Nanoparticles for ocular drug delivery. Gupta RB and Kompella UB, editors. New York: Taylor and Francis; 2006).
15. Kreuter J. Nanoparticles. New York: Marcel Dekker; 1994.
16. Merodio M, Arnedo A, Renedo MJ, Irache JM. Ganciclovir-loaded albumin nanoparticles: characterization and in vitro release properties. Eur J Pharm Sci 2001;12:251 9.
17. Yun M, Barnett ME, Raghava S, Takemoto D, Kompella UB. In: The 33rd Annual Conference of Controlled Release Society. Vienna, Austria, 2006.
18. Kaul G and Amiji M. Cellular interactions and in vitro DNA transfection studies with poly(ethylene glycol)-modified gelatin nanoparticles. J Pharm Sci 2005;94:184–98.
19. Zimmer A, et al. Microspheres and nanoparticles used in ocular drug delivery systems. Adv Drug Deliv Rev 1995;16:61–73.
20. Cavalli R, Gasco MR, Chetoni P, Burgalassi S, Saettone MF. Solid lipid nanoparticles (SLN) as ocular delivery system for tobramycin. Int J Pharm 2002;238:241–5.
21. Alonso MJ and Sanchez A. The potential of chitosan in ocular drug delivery. J Pharm Pharmacol 2003;55:1451–63.
22. Marchal-Heussler L, Maincent P, Hoffman M, Sirbat D. [Value of the new drug carriers in ophthalmology: liposomes and nanoparticles]. J Fr Ophtalmol 1990;13:575–82.

23. Bourges JL, et al. Ocular drug delivery targeting the retina and retinal pigment epithelium using polylactide nanoparticles. Invest Ophthalmol Vis Sci 2003;44:3562–9.
24. Kompella UB, Bandi N, Ayalasomayajula SP. Subconjunctival nano- and microparticles sustain retinal delivery of budesonide, a corticosteroid capable of inhibiting VEGF expression. Invest Ophthalmol Vis Sci 2003;44:1192–201.
25. Marchal-Heussler L, Sirbat D, Hoffman M, Maincent P. Poly(epsilon-caprolactone) nanocapsules in carteolol ophthalmic delivery. Pharm Res 1993;10:386–90.
26. De TK, Rodman DJ, Holm BA, Prasad PN, Bergey EJ. Brimonidine formulation in polyacrylic acid nanoparticles for ophthalmic delivery. J Microencapsul 2003;20:361–74.
27. Gurny R. Preliminary study of prolonged acting drug delivery system for the treatment of glaucoma. Pharm Acta Helv 1981;56:130–2.
28. Harmia T, Speiser P, Kreuter J. Nanoparticles as drug carriers in ophthalmology. Pharm Acta Helv 1987;62:322–31.
29. Pignatello R, et al. Preparation and characterization of eudragit retard nanosuspensions for the ocular delivery of cloricromene. AAPS PharmSciTech 2006;7:E27.
30. Langer K, M E, Lambrecht G, Mayer D, Troschau G, Stieneker F, Kreuter J. Methymethacrylate sulfopropylmethacrylate copolymer nanoparticles for drug delivery Part III: Evaluation as drug delivery system for ophthalmic application. Int J Pharm 1997;158:219–231.
31. Crommelin DJA and Schreier H. Liposomes. In: Kreuter J, editor. New York: Marcel Dekker; 1994).
32. Zeimer R and Goldberg MF. Novel ophthalmic therapeutic modalities based on noninvasive light-targeted drug delivery to the posterior pole of the eye. Adv Drug Deliv Rev 2001;52:49–61.
33. Tans SJ, Vershueren ARM, Dekker C. Room temperature transistor based on a single carbon nanotube. Nature 1998;393:49–52.
34. Wong EW, Sheehan PE, Lieber CM. Nanobeam mechanics: elasticity, strength, and toughness of nanorods and nanotubes. Science 1997;277:1971–1975.
35. Panchapakesan B, L S, Shivakumar K, Teker K, Cesarone G, Wickstrom E. Single wall carbon nanotube nanobomb agents for killing breast cancer cells. NanoBiotechnology 2005;1:133–139.
36. Yu X, et al. Carbon nanotube amplification strategies for highly sensitive immunodetection of cancer biomarkers. J Am Chem Soc 2006;128:11199–205.
37. Hu H, et al. Polyethyleneimine functionalized single-walled carbon nanotubes as a substrate for neuronal growth. J Phys Chem B Condens Matter Mater Surf Interfaces Biophys 2005;109:4285–9.
38. Mattson MP, Haddon RC, Rao AM. Molecular functionalization of carbon nanotubes and use as substrates for neuronal growth. J Mol Neurosci 2000;14:175–82.
39. Shi X, et al. Injectable nanocomposites of single-walled carbon nanotubes and biodegradable polymers for bone tissue engineering. Biomacromolecules 2006;7:2237–42.
40. Zanello LP, Zhao B, Hu H, Haddon RC. Bone cell proliferation on carbon nanotubes. Nano Lett 2006;6:562–7.
41. Aukunuru JV, Ayalasomayajula SP, Kompella UB. Nanoparticle formulation enhances the delivery and activity of a vascular endothelial growth factor antisense oligonucleotide in human retinal pigment epithelial cells. J Pharm Pharmacol 2003;55:1199–206.
42. Koushik K and Kompella UB. Preparation of large porous deslorelin-PLGA microparticles with reduced residual solvent and cellular uptake using a supercritical carbon dioxide process. Pharm Res 2004;21:524–35.
43. Martin TM, Bandi N, Shulz R, Roberts CB, Kompella UB. Preparation of budesonide and budesonide-PLA microparticles using supercritical fluid precipitation technology. AAPS PharmSciTech 2002;3:E18.

44. Kompella UB and Koushik K. Preparation of drug delivery systems using supercritical fluid technology. Crit Rev Ther Drug Carrier Syst 2001;18:173–99.
45. Jani PD, et al. Nanoparticles sustain the release of Flt intraceptors and inhibit injury-induced corneal angiogenesis. Invest Ophthalmol Vis Sci, submitted 2007.
46. Staples M, Daniel K, Cima MJ, Langer R. Application of micro- and nano-electromechanical devices to drug delivery. Pharm Res 2006; 23:847–63.
47. Martin F, et al. Tailoring width of microfabricated nanochannels to solute size can be used to control diffusion kinetics. J Control Release 2005;102:123–33.
48. Grayson AC, et al. Differential degradation rates in vivo and in vitro of biocompatible poly(lactic acid) and poly(glycolic acid) homo- and co-polymers for a polymeric drug-delivery microchip. J Biomater Sci Polym Ed 2004;15:1281–304.
49. Leary SP, Liu CY, Yu C, Apuzzo ML. Toward the emergence of nanoneurosurgery: part I–progress in nanoscience, nanotechnology, and the comprehension of events in the mesoscale realm. Neurosurgery 2005;57:606–34; discussion 606–34.
50. Kim P and Lieber CM. Nanotube nanotweezers. Science 1999;286:2148–50.
51. Obataya I, Nakamura C, Han S, Nakamura N, Miyake J. Mechanical sensing of the penetration of various nanoneedles into a living cell using atomic force microscopy. Biosens Bioelectron 2005;20:1652–5.
52. Obataya I, Nakamura C, Han S, Nakamura N, Miyake J. Nanoscale operation of a living cell using an atomic force microscope with a nanoneedle. Nano Lett 2005;5:27–30.
53. Hossain P, Seetho IW, Bowning AC, Amoaku WM. Artificial means of restoring vision. Brit Medical Journal 2006;330:30–33.
54. Wang K, Lofus D, Leng T, Harris JS, Fishman H. Carbon nanotubes as microelectrodes for retinal prosthesis. Invest Ophthalmol Vis Sci 2003;5054:B713.
55. Leary SP, Liu CY, Apuzzo ML. Toward the emergence of nanoneurosurgery: part II–nanomedicine: diagnostics and imaging at the nanoscale level. Neurosurgery 2006;58:805–23; discussion 805–23.
56. Mulder WJ, Strijkers GJ, van Tilborg GA, Griffioen AW, Nicolay K. Lipid-based nanoparticles for contrast-enhanced MRI and molecular imaging. NMR Biomed 2006;19:142–64.
57. Langereis S, et al. Evaluation of Gd(III)DTPA-terminated poly(propylene imine) dendrimers as contrast agents for MR imaging. NMR Biomed 2006;19:133–41.
58. Anderson SA, et al. Magnetic resonance contrast enhancement of neovasculature with alpha(v)beta(3)-targeted nanoparticles. Magn Reson Med 2000;44:433–9.
59. Saini S, et al. Ferrite particles: a superparamagnetic MR contrast agent for the reticuloendothelial system. Radiology 1987;162:211–6.
60. Stark DD, et al. Superparamagnetic iron oxide: clinical application as a contrast agent for MR imaging of the liver. Radiology 1988;168:297–301.
61. Suzuki M, et al. Development of a target-directed magnetic resonance contrast agent using monoclonal antibody-conjugated magnetic particles. Noshuyo Byori 1996;13:127–32.
62. Ito A, Shinkai M, Honda H, Kobayashi T. Medical application of functionalized magnetic nanoparticles. J Biosci Bioeng 2005;100:1–11.
63. Loo C, et al. Nanoshell-enabled photonics-based imaging and therapy of cancer. Technol Cancer Res Treat 2004;3:33–40.
64. Cang H, et al. Gold nanocages as contrast agents for spectroscopic optical coherence tomography. Opt Lett 2005;30:3048–50.
65. Arya H, et al. Quantum dots in bio-imaging: Revolution by the small. Biochem Biophys Res Commun 2005;329:1173–7.

66. Akerman ME, Chan WC, Laakkonen P, Bhatia SN, Ruoslahti E. Nanocrystal targeting in vivo. Proc Natl Acad Sci U S A 2002;99:12617–21.
67. Roizenblatt R, et al. Nanobiolistic delivery of indicators to the living mouse retina. J Neurosci Methods 2006;153:154–61.
68. Bejjani RA, et al. Nanoparticles for gene delivery to retinal pigment epithelial cells. Mol Vis 2005;11:124–32.
69. Gomes dos Santos AL, et al. Sustained release of nanosized complexes of polyethylenimine and anti-TGF-beta 2 oligonucleotide improves the outcome of glaucoma surgery. J Control Release 2006;112:369–81.
70. Prow T, et al. Construction, gene delivery, and expression of DNA tethered nanoparticles. Mol Vis 2006;12:606–15.
71. Kompella UB, Sundaram S, Raghava S, Escobar ER. Luteinizing hormone-releasing hormone agonist and transferrin functionalizations enhance nanoparticle delivery in a novel bovine ex vivo eye model. Mol Vis 2006;12:1185–98.
72. De Campos AM, Sanchez A, Alonso MJ. Chitosan nanoparticles: a new vehicle for the improvement of the delivery of drugs to the ocular surface. Application to cyclosporine A. Int J Pharm 2001;224:159–168.
73. El-Samaligy MS, R Y, Charlton JF, Weinstein GW, Lim JK. Ocular disposition of nanoencapsulated acyclovir and ganciclovir via intravitreal injection in rabbit's eye. Drug Deliv 1996;3:93–97.
74. Li H, et al. A PEDF N-terminal peptide protects the retina from ischemic injury when delivered in PLGA nanospheres. Exp Eye Res 2006;83:824–33.
75. Merodio M, Irache JM, Valamanesh F, Mirshahi M. Ocular disposition and tolerance of ganciclovir-loaded albumin nanoparticles after intravitreal injection in rats. Biomaterials 2002;23:1587–94.
76. Raghava S, Hammond M, Kompella UB. Periocular routes for retinal drug delivery. Expert Opin Drug Deliv 2004;1:99–114.
77. Amrite AC, Ayalasomayajula SP, Kompella UB. Sustained transscleral delivery of budesonide and celecoxib for treating diabetic retinopathy. The Proceedings of the 33rd International Conference of the Controlled Release Society, 2006.
78. Ayalasomayajula SP and Kompella UB. Subconjunctivally administered celecoxib-PLGA microparticles sustain retinal drug levels and alleviate diabetes-induced oxidative stress in a rat model. Eur J Pharmacol 2005;511:191–8.
79. Amrite AC and Kompella UB. Size-dependent disposition of nanoparticles and microparticles following subconjunctival administration. J Pharm Pharmacol 2005;57:1555–63.
80. Amrite AC, Ayalasomayajula SP, Cheruvu NP, Kompella UB. Single periocular injection of celecoxib-PLGA microparticles inhibits diabetes-induced elevations in retinal PGE2, VEGF, and vascular leakage. Invest Ophthalmol Vis Sci 2006;47:1149–60.
81. Hammond M and Kompella UB. In: Gupta RB and Kompella UB, editors. Nanoparticle Technology for Drug Delivery. New York: Taylor and Francis; 2006. p. 381–395.
82. Escobar ER, NPC, Zhan G, Toris CB, Kompella UB. Subconjunctival budesonide and budesonide poly-(lactide) microparticles do not elevate intraocular pressure or induce lens opacities in rabbit model. Invest Ophthalmol Vis Sci 2006;47:E-abstract 4493.
83. Lutty GA, et al. In: XVII International Congress of Eye Research. Puerto Madero, Buenos Aires, Argentina, 2006.

23
Vitamin C Transporters in the Retina

Vadivel Ganapathy, Sudha Ananth, Sylvia B. Smith,
and Pamela M. Martin

CONTENTS

CHEMISTRY AND BIOLOGIC FUNCTIONS OF ASCORBIC ACID (VITAMIN C)
VITAMIN C AND THE RETINA
DELIVERY OF VITAMIN C TO THE RETINA
MOLECULAR IDENTITY AND FUNCTIONAL FEATURES OF VITAMIN C
   TRANSPORTERS
VITAMIN C TRANSPORTERS IN THE RETINA
DIABETES AND RETINAL VITAMIN C STATUS
POTENTIAL OF VITAMIN C TRANSPORTERS IN DRUG DELIVERY
CONCLUSIONS
REFERENCES

CHEMISTRY AND BIOLOGIC FUNCTIONS OF ASCORBIC ACID (VITAMIN C)

Ascorbic acid, also known as vitamin C, is the γ-lactone of an L-hexanoic acid with an enediol structure at carbon atoms 2 and 3 (Fig. 1A). It is water soluble. The two hydroxyl groups located at carbon atoms 2 and 3 are ionizable with pKa values of 11.8 and 4.2, respectively. At physiologic pH (around 7.4), more than 99.9% of ascorbic acid exists as a monovalent anion, ascorbate, with ionization of the hydroxyl group at carbon atom 3 (Fig. 1B). L-ascorbic acid can be oxidized, enzymatically or non-enzymatically, with the removal of two hydrogen atoms to form dehydro-L-ascorbic acid (DHAA) (Fig. 1C). The oxidized form is a weak acid and does not ionize under physiologic conditions; it exists predominantly as an unionized and uncharged molecule. In aqueous solutions and in plasma, DHAA is present in a hydrated form (Fig. 1D). The concentration of the reduced form (L-ascorbate) in plasma is in the range 50–100 μM *(1–4)*. In contrast, the oxidized form, DHAA, is present in the circulation at much lower levels (about 10 μM) *(1–4)*. Ascorbic acid is synthesized de novo in the liver, from glucose, in most mammalian species. However, primates (including humans) and guinea pigs are not capable of endogenous synthesis of ascorbic acid due to inactivation of one of the key enzymes

From: *Ophthalmology Research: Ocular Transporters in Ophthalmic Diseases and Drug Delivery*
Edited by: J. Tombran-Tink and C. J. Barnstable © Humana Press, Totowa, NJ

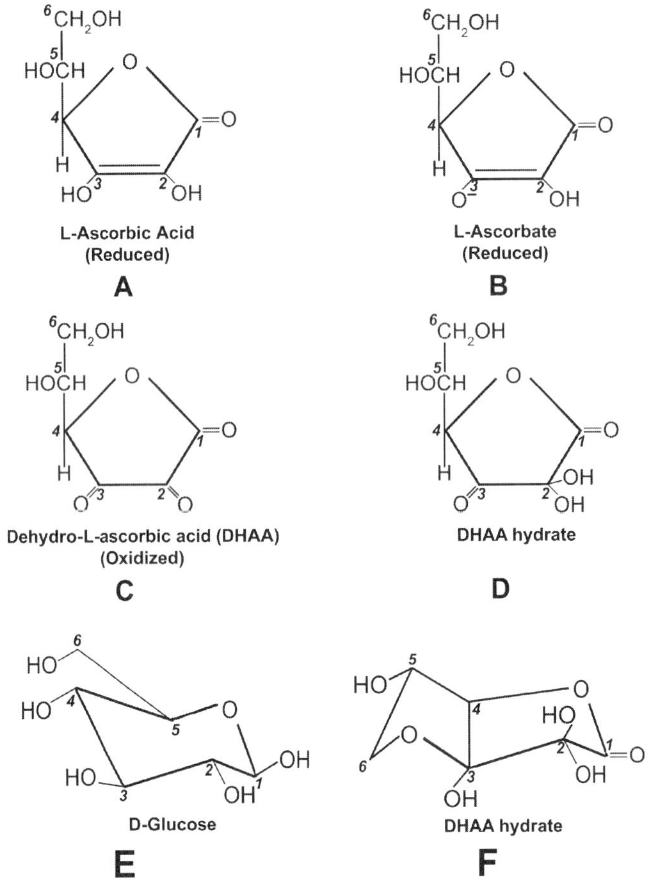

Fig. 1. Structures of ascorbic acid, ascorbate, dehydroascorbic acid (DHAA) and DHAA hydrate, and the structural similarity between D-glucose and DHAA hydrate.

(gulono-γ-lactone oxidase) in the biosynthetic pathway, and hence depend obligatorily on dietary sources to meet their daily requirements of ascorbic acid. Thus, ascorbic acid is an essential nutrient (a vitamin) for humans, but not for the widely used laboratory animals such as the rats and mice. This vitamin is an essential cofactor for mixed-function oxidases involved in the hydroxylation of lysine and proline found in protocollagen (lysyl hydroxylase and prolyl hydroxylase), in the synthesis of norepinephrine and epinephrine (dopamine β-hydroxylase) and carnitine (trimethyllysine hydroxylase and γ-butyrobetaine hydroxylase), and in the processing of neuropeptides (peptidylglycine hydroxylase) *(5,6)*. Deficiency of this vitamin leads to a clinical condition known as scurvy, which is associated with skin alterations, gum decay, loss of teeth, fragile blood capillaries, and bone abnormalities. Most of these clinical manifestations seem to arise from defective collagen synthesis due to decreased activities of vitamin C-dependent lysyl hydroxylase and prolyl hydroxylase. Another important biologic role of vitamin C is as an antioxidant *(5–7)*. Ascorbate is capable of interacting with a variety of free radicals (e.g., hydroxyl radical, peroxyl radical) and facilitating their detoxification *(8,9)*. It also helps to preserve vitamin E (α-tocopherol) by converting tocopheroxyl radical, which is produced during oxidation of vitamin E, into the native vitamin E necessary for biologic functions. In the process of detoxifying free radicals, ascorbate itself is converted into a stable ascorbyl free radical, which can subsequently dismutate, in the presence of protons, into ascorbate and dehydroascorbic acid. Ascorbyl free radical is not highly reactive and does not damage cellular components. In addition to the dismutation mechanism for the recycling of the ascorbyl free radical, enzymes such as the cytosolic thioredoxin reductase and membrane-bound NADH-dependent reductases also mediate the conversion of ascorbyl free radical into ascorbate and thus help in the reutilization of the ascorbyl free radical.

Interconversion between Ascorbate and Dehydroascorbic Acid

During participation in enzymatic and antioxidant reactions, ascorbate is converted into DHAA. Ascorbate is regenerated from DHAA inside cells by several mechanisms. Firstly, there is the non-enzymatic mechanism in which DHAA can be reduced to ascorbate directly by glutathione (γ-glutamyl cysteinyl glycine) *(10)*. Secondly, there are several glutathione-dependent enzymes that catalyze the reduction of DHAA to ascorbate. This includes thiol transferases (glutaredoxins), protein disulfide isomerases, and the omega class of glutathione transferases *(11–13)*. The significance of glutathione in the regeneration of ascorbate from DHAA is evident from animal studies which have shown that glutathione deficiency leads to decreased ascorbate levels and increased DHAA/ascorbate ratio *(14)*. DHAA can also be reduced to ascorbate by glutathione-independent enzymatic processes. This includes NADPH-dependent enzymes such as 3α-hydroxysteroid dehydrogenase *(15)* and thioredoxin reductase *(16)*. Recent studies have shown that ebselen, a seleno-organic compound, possesses dehydroascorbate reductase activity, contributing to the compound's significant antioxidative actions in vivo *(17,18)*. The physiologic significance of DHAA reduction inside the cells is not just related to the regeneration of the biologically useful ascorbate; it also serves to maintain high intracellular concentrations of ascorbate. DHAA, being uncharged, can exit cells either by simple diffusion or by transporter-mediated processes. When DHAA is reduced, the resultant ascorbate becomes charged and cannot exit the cell. Thus, ascorbate produced by reduction of DHAA gets trapped inside cells, leading to high intracellular concentrations of this antioxidant.

VITAMIN C AND THE RETINA

Vitamin C is present throughout the eye in humans and other animals at concentrations several-fold higher than found in most other tissues *(19,20)*. Concentrations of ascorbate within different regions of the retina vary in the range of 0.2–1.5 mM, the highest levels being in the neural retina *(21–23)*. The intracellular concentration of ascorbate in the neural retina is approximately 15 times higher than in plasma. There is a unique need for ascorbate in the retina to provide protection against light-induced oxidative damage. Photo-oxidation generates damaging free radicals and other reactive oxygen species, which, if not detoxified, would disrupt the function of molecular components in the retina. In addition, the generation of reactive oxygen species within the retina increases significantly under various pathologic conditions such as diabetes, ocular inflammation, and age-related macular degeneration *(24–27)*. Ascorbate, being a potent antioxidant, plays an essential role in the maintenance of retinal function, not only under normal physiologic conditions but also under various pathologic conditions.

DELIVERY OF VITAMIN C TO THE RETINA

Primates, including humans, and guinea pigs do not have the ability to synthesize ascorbic acid and hence have to obtain this essential nutrient from the diet. The needs of the retina for this vitamin have to be met by ascorbate and DHAA present in the plasma. Even in those animal species capable of endogenous synthesis of ascorbic acid, synthesis occurs primarily in the liver; hence, the retina in these species also relies upon plasma ascorbate and DHAA as a source of this biologically important nutrient. Neither the reduced form (ascorbate), nor the oxidized form (DHAA) of vitamin C is easily permeable across biologic membranes, necessitating involvement of specific transport processes for the entry of this vitamin from the circulation into the retina. The retina is separated from the peripheral circulation by the blood–retinal barrier, composed of the retinal capillary endothelial cells (inner blood–retinal barrier) *(28)* and the tight junctions of the retinal pigment epithelial cells (outer blood–retinal barrier) *(29)*. Ascorbate and DHAA present in the circulation have to be transported across these barriers for the cells of the neural retina to have access to them. Moreover, individual cell types within the neural retina must possess transport processes to take up these compounds from the extracellular space. Distinct transporters operate in the retinal capillary endothelial cells, retinal pigment epithelial cells, and various cell types of the neural retina to facilitate the entry of ascorbate and DHAA from the blood into neural retina and retinal cells.

MOLECULAR IDENTITY AND FUNCTIONAL FEATURES OF VITAMIN C TRANSPORTERS

Transporters for Ascorbate

The reduced form of vitamin C, ascorbate, is a monovalent anion at physiologic pH; hence, it cannot diffuse across the hydrophobic lipid bilayer of the plasma membrane. Furthermore, the inside-negative membrane potential, which exists across the plasma membrane of cells, acts as a barrier for diffusion of ascorbate. This necessitates involvement of specific transport systems for the entry of ascorbate. Two different transporters have been identified at the molecular level for the transport of ascorbate in mammalian cells *(30)*. Both of them are active transporters, driven by a transmembrane electrochemical Na$^+$ gradient. These transporters, known as **S**odium-dependent **V**itamin **C** **T**ransporters 1 and 2 (SVCT1 and SVCT2), are members of the solute-linked carrier gene family SLC23. According to the Human Genome Organization nomenclature, SVCT1 is referred to as SLC23A1 and SVCT2 as SLC23A2 *(31)*. Human SVCT1 and SVCT2 have been cloned and functionally characterized *(32–35)*. The gene coding for SVCT1 is located on human chromosome 5q31.2–31.3 and the gene coding for SVCT2 on human chromosome 20p12.2–12.3. Both transporters are specific for the reduced form of vitamin C, namely ascorbate; they do not recognize the oxidized form, DHAA, as a substrate. Glucose, at physiologic concentrations (around 5 mM), has little effect on the transport of ascorbate via SVCT1 or SVCT2. The transport function of SVCT1 and SVCT2 is obligatorily dependent on the presence of Na$^+$, and the transport process exhibits a Na$^+$:ascorbate stoichiometry of 2:1. This stoichiometry renders the transport process electrogenic, resulting in the transfer of a net positive charge into the cell per transport cycle. Therefore, the energy source for the transport process comes not only from the inwardly directed Na$^+$ gradient, but also from the inside-negative membrane potential. The Michaelis constant for ascorbate is in the range 10–100 µM for both transporters; these values are within the physiologic concentrations of ascorbate in plasma. Though functionally very similar, SVCT1 and SVCT2 differ in tissue distribution pattern. The expression of SVCT1 is primarily restricted to epithelial cells in transport organs such as the small intestine, kidney, and liver. In contrast, SVCT2 is much more widely expressed, its expression being evident in the brain, placenta, ocular tissues (ciliary body, cornea, lacrimal gland, and retina), adrenal gland, chondrocytes, and in many endocrine, exocrine, and endothelial tissues *(30)*. Mammalian cells express an isoform of SVCT2 in which 115 amino acids are deleted in the region encompassing the putative transmembrane domains four to six; this splice variant is nonfunctional, but when coexpressed with wild-type SVCT2, it behaves as a dominant-negative inhibitor of the wild-type transporter *(36)*. The transport function of wild-type SVCT1 is also affected in a similar manner by coexpression of the SVCT2 splice variant. There is also evidence for a nonfunctional splice variant of SVCT1 *(32)*, but it is not known whether this isoform functions as a dominant-negative inhibitor of wild-type SVCTs. There are no specific inhibitors of SVCTs. Though phloretin and flavanoids inhibit SVCT1 at micromolar concentrations, the effects are not specific as other transport systems are also susceptible to inhibition by these compounds.

Transporters for DHAA

Uptake of DHAA in mammalian cells occurs via an Na$^+$-independent, facilitative process. The glucose transporters belonging to the solute-linked carrier gene family SLC2 are responsible for this process *(12)*. DHAA exists in solution as a hydrate and this hydrated form has a structure very similar to that of glucose (Fig. 1E and F). This provides the molecular basis for the recognition of DHAA as a substrate by members of SLC2 gene family. This gene family consists of at least 13 members; of these, only GLUT1 (SLC2A1), GLUT3 (SLC2A3), and GLUT4 (SLC2A4) have thus far been shown to possess the ability to transport DHAA *(37–39)*. Since glucose is also a substrate for these transporters, there is competition between glucose and DHAA for transport via these transporters under normal physiologic conditions. The Michaelis constants (K_m) for GLUT1, GLUT3, and GLUT4 to interact with glucose and DHAA are similar. In the case of glucose, the K_m values for GLUT1, GLUT3, and GLUT4 are around 3 mM, 1.5 mM, and 5 mM, respectively. For DHAA, the corresponding K_m values are around 1.1 mM, 1.7 mM, and 1 mM, respectively. It has to be noted that, even though the affinities of these transporters for DHAA and glucose are in the same range, the plasma concentrations are very different (3–5 mM for glucose and about 10 µM for DHAA). This would suggest that the transport of DHAA via any of these transporters would be minimal under normal conditions. However, most cells possess active DHAA reductase activity, which effectively reduces DHAA into ascorbate intracellularly as soon as DHAA enters the cells. This rapid metabolic conversion generates a steep inwardly directed concentration gradient for DHAA which facilitates the entry of DHAA via GLUTs even in the presence of normal concentrations of glucose. Therefore, the involvement of GLUT1, GLUT3, and GLUT4 in the cellular uptake of DHAA is physiologically relevant.

VITAMIN C TRANSPORTERS IN THE RETINA

The Blood–Retinal Barrier

The blood–retinal barrier consists of retinal capillary endothelial cells (inner blood–retinal barrier) and the retinal pigment epithelium (outer blood–retinal barrier). The inner two-thirds of the neural retina gets its nutrient supply via transfer across the inner blood–retinal barrier. The remainder of the retina is nourished via transfer across the outer blood–retinal barrier. The endothelial cells, which constitute the inner blood–retinal barrier, form tight junctions and thus are polarized. These cells have distinct luminal and abluminal membranes. GLUT1 is expressed on both membranes *(40,41)*. Cultured endothelial cells from retinal capillaries express both GLUT1 and GLUT3 *(42,43)*, but it seems that GLUT3 expression may be specific only to the cells maintained in culture, as there is no evidence for the expression of this isoform in retinal capillary endothelial cells in intact tissue. Elegant studies by Hosoya et al. *(43)* have shown that, in intact animals, transfer of vitamin C across the inner blood–retinal barrier occurs predominantly in the form of DHAA, implicating a facilitative glucose transporter in the transport process. Even though these investigators also have found evidence for expression of GLUT3 in a cell-line model system for the inner blood–retinal barrier, the expression levels of GLUT1 are several times greater than those of GLUT3, suggesting that GLUT1 is the likely transporter responsible for the transfer of DHAA across this barrier. There is no evidence in the literature for the expression of the Na^+-dependent vitamin C transporters SVCT1 and SVCT2 in retinal capillary endothelial cells. The cell line used by Hosoya et al. *(43)* expresses SVCT2 messenger RNA (mRNA) to a small extent and no SVCT1 mRNA at all. Notably, functional studies in this model system show that uptake of ascorbate, the substrate for the Na^+-dependent transporters, is negligible compared to the uptake of DHAA. Therefore, it seems that transfer of vitamin C across the inner blood–retinal barrier occurs almost exclusively in the form of DHAA via GLUT1.

The outer blood–retinal barrier consists of the retinal pigment epithelium (RPE). Small molecules are permeable across the endothelial cells of the highly fenestrated choroidal blood vessels. The RPE cells possess tight junctions and are polarized, with distinct basolateral membranes facing the choroidal circulation and apical membranes facing the subretinal space. Studies with intact retinal tissue have provided evidence for the expression of GLUT1 in this cell layer, the transporter being found both on the apical membrane as well as on the basolateral membrane *(41)*. These findings have been validated with cultured human RPE cells *(44)*. GLUT3 may be expressed in these cells, but only as a minor component *(45)*. Therefore, it is likely that DHAA is transferred across the RPE cell layer primarily via GLUT1. The predominant mechanism of vitamin C transport in this cell layer, however, is the transfer of ascorbate, the reduced form of vitamin C, and not DHAA. This is in contrast to the transport of vitamin C in retinal capillary endothelial cells, which take up primarily DHAA as the form of vitamin C.

Studies with cultured RPE cells or RPE cell lines have shown that ascorbate is taken up into the cells by a Na$^+$-dependent, high-affinity transport system (Michaelis constant, 20–70 µM) *(32,33,46–50)*. DHAA does not interact with this transport process. Na$^+$-activation kinetics of ascorbate uptake in a human RPE cell line exhibits a sigmoidal pattern, suggesting involvement of more than one Na$^+$ in the transport process *(32,33)*. The Na$^+$-dependent transport activity seems to reside principally on the apical membrane of the RPE *(48)*. Even though these functional studies have demonstrated the expression of a Na$^+$-coupled active transport system for ascorbate in RPE cells, virtually nothing is known about the molecular identity of the transporter responsible for this process. To date, the only two transporters that are known at the molecular level and are capable of mediating Na$^+$-dependent, high-affinity ascorbate transport are SVCT1 and SVCT2. Therefore, it would be of interest to know the expression pattern of these two transporters in RPE. We recently carried out reverse-transcriptase polymerase chain reaction (RT-PCR) with RNA isolated from mouse RPE-eyecup and the human RPE cell line ARPE-19 to analyze the expression of SVCT1 mRNA and SVCT2 mRNA in this tissue. These studies showed that RPE-eyecup, as well as ARPE-19, cells express mRNA for both isoforms, SVCT2 mRNA being much more abundant than SVCT1 mRNA (Fig. 2). The presence of SVCT1 mRNA in RPE cells, though at lower levels than SVCT2 mRNA, is interesting because it is widely believed at present that retina expresses only SVCT2 *(31)*. There is no information available, however, regarding the expression of SVCT proteins in RPE. Based on the current information available in the literature, it seems that vitamin C enters RPE cells from the choroidal circulation mainly in the form of DHAA, via GLUT1 expressed in the basolateral membrane. Since the apical membrane of RPE expresses SVCT2, it is likely that reduced ascorbate is taken up actively into the cells

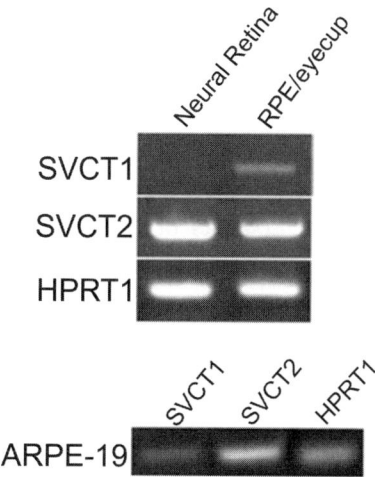

Fig. 2. Reverse-transcriptase polymerase chain reaction (RT-PCR) analysis of expression of messenger RNAs (mRNAs) for **S**odium-dependent **V**itamin **C** **T**ransporters 1 and 2 (SVCT1 and SVCT2, respectively) in neural retina, retinal pigment epithelium (RPE)/eyecup, and ARPE-19 cell line. Hypoxanthine phosphoribosyl transferase 1 (HPRT1) was used as an internal control.

from the subretinal space via this Na$^+$-coupled transporter. RPE cells may accumulate reduced ascorbate for use as a protective agent against oxidative damage. It is interesting that these cells express GLUT1 not only on the basolateral membrane, but also on the apical membrane. This raises the possibility that DHAA, which comes into the cells from the choroidal circulation via GLUT1 in the basolateral membrane, and which gets generated inside the cells from reduced ascorbate during the detoxification reactions, may exit the cells via GLUT1 in the apical membrane to enter the subretinal space. Thus, RPE cells may serve as a supplier of DHAA to the neural retina.

Neural Retina

Neural retina consists of several neuronal cell types (photoreceptor cells, ganglion cells, amacrine cells, bipolar cells, and horizontal cells) and the glial Müller cells. Mantych et al. *(51)* have performed detailed immunolocalization studies to delineate the expression of GLUT isoforms in the retina. Their studies showed that GLUT1 is the main isoform expressed in neural retina. Its expression is evident in Müller cells and, to a minor extent, in photoreceptor cells. Interestingly, other neuronal cell types in the retina are negative for GLUT1 expression. In contrast, the GLUT3 isoform is specifically localized to the inner synaptic layer, a region enriched with neuronal connections. These data suggest that Müller cells are capable of taking up the oxidized form of vitamin C, DHAA, via GLUT1. Analysis of the expression of SVCT1 and SVCT2 has revealed that only the latter is expressed in the neural retina *(30)*. This is corroborated with our recent RT-PCR data (Fig. 2). However, we do not know whether this Na$^+$-coupled transporter for reduced ascorbate is expressed widely in various cell types within the neural retina, or whether the expression is restricted to specific neuronal cell types. It has been proposed that, in the brain, ascorbate is recycled between astrocytes and neurons, in a process involving GLUT1, GLUT3, and SVCT2 *(31)*. Neurons express GLUT3 and SVCT2, whereas astrocytes express only GLUT1. According to this proposal, neurons take up reduced ascorbate from the extracellular space via SVCT2 for use as an antioxidant. When ascorbate participates in the detoxification of free radicals, it is converted into DHAA which is exported out of neurons via GLUT3. This is possible because GLUT3 is a facilitative transporter with no driving force; the direction of transport is dictated solely by the concentration of gradient of the substrate. Astrocytes then take up DHAA via GLUT1, another facilitative transporter. Inside astrocytes, DHAA is reduced to ascorbate for subsequent release into the extracellular space for reuse by neurons. The molecular mechanism responsible for the release of ascorbate from these cells remains to be established. A similar phenomenon may occur in the neural retina. Müller cells, the major glial cells within the retina, express GLUT1 while neuronal cells express GLUT3. We postulate that SVCT2 expression in the neural retina is restricted to neuronal cells. Such an expression pattern would be analogous to that found in the central nervous system. With this kind of cell-type-specific expression pattern of the transporters for DHAA and ascorbate in the neural retina, it is possible that ascorbate is recycled between Müller cells and neuronal cells to protect the neurons against oxidative damage (Fig. 3). As in the central nervous system, the molecular identity of the transport process responsible for the release of ascorbate from Müller cells remains unknown.

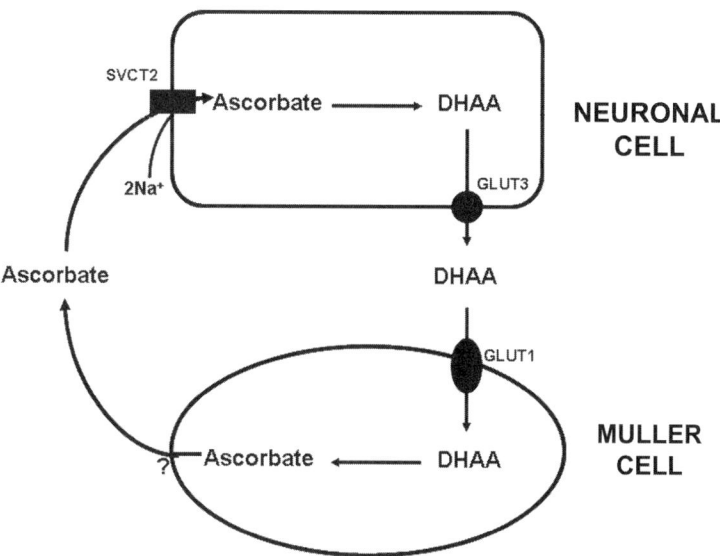

Fig. 3. A model for ascorbate recycling between neuronal cells and Müller cells in neural retina. The question mark (?) indicates that the molecular identity of the transport system responsible for the release of ascorbate from Müller cells is not known. DHAA, dehydroascorbic acid; SVCT2, Sodium-dependent Vitamin C Transporter 2.

DIABETES AND RETINAL VITAMIN C STATUS

Diabetes is associated with hyperglycemia and glucose-induced generation of reactive oxygen species. Both of these factors are directly relevant to the transport and biologic functions of vitamin C. The neural retina relies on the transfer of DHAA from the peripheral circulation across the inner blood–retinal barrier, as well as across the outer blood–retinal barrier, to meet its requirements for ascorbate. In both cases, the transfer process is mediated by GLUT1. Since this transporter handles glucose as well as DHAA, diabetes-associated hyperglycemia is likely to interfere with the transfer of DHAA into neural retina, due to increased competition with DHAA by glucose for the transport process. Therefore, it is likely that the delivery of vitamin C to the neural retina is impaired in diabetes. At the same time, glucose-induced production of reactive oxygen species in diabetes increases the oxidative burden on the retina, and consequently enhances the requirements for ascorbate for effective prevention of oxidative damage. Thus, the neural retina becomes increasingly susceptible to oxidative damage under diabetic conditions, aided by the increased generation of reactive oxygen species and the decreased transfer of vitamin C across the blood–retinal barriers. Considering the values for normal plasma glucose levels as 3–5 mM and those for plasma levels in poorly controlled diabetic patients as 8–15 mM, it has been calculated theoretically that DHAA transfer via GLUT1 will be decreased in diabetic patients by 30–80% (52). This has recently been confirmed experimentally in intact rats (53). In streptozotocin-induced diabetes in rats with the occurrence of hyperglycemia for three weeks, the transfer of DHAA across the blood-retinal barriers is reduced by around 65%. A reduction of this magnitude in the transfer of vitamin C into neural retina will have important implications for the pathogenesis of diabetic retinopathy. Ascorbate deficiency has been linked to occurrence of retinopathy and accumulation of melanin in retinal cells *(54–56)*. Diabetes is not expected to cause systemic ascorbate deficiency in humans because intestinal and renal absorption of this vitamin is mediated primarily via SVCT1, whose transport function is not affected by glucose. However, we hypothesize that diabetes may cause ascorbate deficiency within the neural retina because of the involvement of GLUT1 in the transfer of the vitamin across the blood-retinal barriers. Ascorbate deficiency in the retina will lead not only to increased oxidative stress but also to disturbed catecholamine metabolism since this vitamin is an essential cofactor for dopamine β-hydroxylase. Both synthesis and catabolism of catecholamines occur in the retina. Melanin pigment is produced in RPE. Decreased activity of dopamine β-hydroxylase due to ascorbate deficiency, as occurs in diabetes, may lead to disruption of catecholamine metabolism and accumulation of catecholamine metabolites and their polymerized byproducts in RPE and neural retina. Such events are likely to play a significant role in the pathogenesis of retinal complications associated with diabetes.

POTENTIAL OF VITAMIN C TRANSPORTERS IN DRUG DELIVERY

There is some information available in the literature supporting the potential use of vitamin C transporters in drug delivery. The facilitative glucose transporter GLUT1 can transport glucosamine, an essential sugar derivative necessary for the synthesis of glycoproteins and glycosaminoglycans *(57)*. Since this transporter operates at the inner blood–retinal barrier and the outer blood–retinal barrier, transfer of glucosamine from peripheral blood into neural retina may be facilitated by this transporter. Glucosamine and its derivatives have significant modulatory effects on insulin resistance and diabetes-associated complications *(58)*. At the same time, glucosamine is widely used as a nutraceutical agent, with an apparent benefit in the treatment of osteoarthritis *(59)*. The expression of GLUT1 at the blood–retinal barriers suggests that glucosamine can have easy access to the neural retina via the transporter. Pharmacological doses of glucosamine are likely to increase the blood levels of this sugar derivative which might compete with DHAA for transfer across the blood–retinal barriers. SVCT2, the primary Na^+-dependent ascorbate transporter expressed in the retina, also has potential for exploitation as a drug delivery system. Several studies has demonstrated that neuroactive drugs such nipocotic acid and kynurenic acid can be conjugated to ascorbic acid, and that the resultant conjugates are recognized by SVCT2 as transportable substrates *(50,60,61)*. Thus, SVCT2 may be useful for the delivery of certain drugs into neuronal cells of the retina in the form of ascorbyl prodrugs.

CONCLUSIONS

Vitamin C is essential for the maintenance of retinal health. Transporters exist in the retina for the handling of reduced ascorbate as well as oxidized ascorbic acid (DHAA). Neural retina obtains this vitamin, mostly in the form of DHAA, across the inner and outer blood–retinal barriers, and the transfer process is mediated by the facilitative glucose transporter GLUT1. It is likely that the transfer of DHAA via this mechanism is inhibited to a significant extent in diabetes, though this has not been investigated. Neurons and RPE express Na⁺-dependent transporters for reduced ascorbate, primarily SVCT2, to enable these cells to take up ascorbate actively. RPE, which constitutes the outer blood-retinal barrier, expresses not only GLUT1 (on apical and basolateral membrane), but also SVCT2 (on apical membrane). Thus, while RPE uses GLUT1 to transfer DHAA from the choroidal circulation into the neural retina, it also takes up reduced ascorbate from the subretinal space for its own use for detoxification of reactive oxygen species. Within the neural retina, there seems to be an effective crosstalk between Müller cells and neuronal cells in the handling of ascorbate, in which Müller cells function as a supplier of reduced ascorbate for the neuronal cells.

REFERENCES

1. Moeslinger T, Brunner M, Volf I, Spieckermann PG. Spectrophotometric determination of ascorbic acid and dehydroascorbic acid. Clin Chem 1995;41:1177–1181.
2. Jacob RA. Assessment of human vitamin C status. J Nutr 1990;120(Suppl. 11):1480–1485.
3. Tessier F, Birlouez-Aragon I, Tjani C, Guilland JC. Validation of a micromethod for determining oxidized and reduced vitamin C in plasma by HPLC-fluorescence. Int J Vitam Nutr Res 1996;66:166–170.
4. Chatterjee IB and Banerjee A. Estimation of dehydroascorbic acid in blood of diabetic patients. Anal Biochem 1979;98:368–374.
5. Englard S and Seifter S. The biochemical functions of ascorbic acid. Annu Rev Nutr 1986;6:365–406.
6. Padh H. Cellular functions of ascorbic acid. Biochem Cell Biol 1990;68:1166–1173.
7. Padayatty SJ, Katz A, Wang Y, Eck P, Kwon O, Lee JH, Chen S, Corpe C, Dutta A, Dutta SK, Levine M. Vitamin C as an antioxidant: evaluation of its role in disease in prevention. J Am Coll Nutr 2003;22:18–35.
8. Beyer RE. The role of ascorbate in antioxidant protection of biomembranes: interaction with vitamin E and coenzyme Q. J Bioenerg Biomembr 1994;26:349–358.
9. May JM. Is ascorbic acid an antioxidant for the plasma membrane? FASEB J 1999;13: 995–1006.
10. Winkler BS, Orselli SM, Rex TS. The redox couple between glutathione and ascorbic acid: a chemical and physiological perspective. Free Radic Biol Med 1994;17:333–349.
11. Wells WW and Xu DP. Dehydroascorbate reduction. J Bioenerg Biomembr 1994;26:369–377.
12. Wilson JX. The physiological role of dehydroascorbic acid. FEBS Lett 2002;527:5–9.
13. Whitbread AK, Masoumi A, Tetlow N, Schmuck E, Coggan M, Board PG. Characterization of the omega class of glutathione transferases. Methods Enzymol 2005;401:78–99.
14. Martensson J and Meister A. Glutathione deficiency decreases tissue ascorbate levels in newborn rats: ascorbate spares glutathione and protects. Proc Natl Acad Sci USA 1991;88:4656–4660.
15. Del Bello B, Maellaro E, Sugherini L, Santucci A, Comporti M, Casini AF. Purification of NADPH-dependent dehydroascorbate reductase from rat liver and its identification with 3α-hydroxysteroid dehydrogenase. Biochem J 1994;304:385–390.
16. May JM, Mendiratta S, Hill KE, Burk RF. Reduction of dehydroascorbate to ascorbate by the selenoenzyme thioredoxin reductase. J Biol Chem 1997;272:22607–22610.
17. Jung CH, Washburn MP, Wells WW. Ebselen has dehydroascorbate reductase and thioltransferase-like activities. Biochem Biophys Res Commun 2002;291:550–553.
18. Zhao R and Holmgren A. Ebselen is a dehydroascorbate reductase mimic, facilitating the recycling of ascorbate via mammalian thioredoxin systems. Antioxid Redox Signal 2004;6:99–104.
19. Garland DL. Ascorbic acid and the eye. Am J Clin Nutr 1991;54:1198S–1202S.

20. Rose RC and Bode AM. Ocular ascorbate transport and metabolism. Comp Biochem Physiol 1991;100A:273–285.
21. Woodford BJ, Tso MO, Lam KW. Reduced and oxidized ascorbates in guinea pig retina under normal and light-exposed conditions. Invest Ophthalmol Vis Sci 1983;24:862–867.
22. Lai YL, Fong D, Lam KW, Wang HM, Tsin AT. Distribution of ascorbate in the retina, subretinal fluid, and pigment epithelium. Curr Eye Res 1986;5:933–938.
23. Nielsen JC, Naash MI, Anderson RE. The regional distribution of vitamin E and C in mature and premature human retinas. Invest Ophthalmol Vis Sci 1988;29:22–26.
24. Van Reyk DM, Gillies MC, Davies MJ. The retina: oxidative stress and diabetes. Redox Rep 2003;8:187–192.
25. SanGiovanni JP and Chew EY. The role of omega-3 long-chain polyunsaturated fatty acids in health and disease of the retina. Prog Retin Eye Res 2005;24:87–138.
26. Rao NA. Role of oxygen free radicals in retinal damage associated with experimental uveitis. Trans Am Ophthalmol Soc 1990;88:797–850.
27. Beatty S, Koh H, Phil M, Henson D, Boulton M. The role of oxidative stress in the pathogenesis of age-related macular degeneration. Surv Ophthalmol 2000;45:115–134.
28. Hosoya K and Tomi M. Advances in the cell biology of transport via the inner blood-retinal barrier: establishment of cell lines and transport functions. Biol Pharm Bull 2005;28:1–8.
29. Rizzolo LJ. Polarity and the development of the outer blood-retinal barrier. Histol Histopathol 1997;12:1057–1067.
30. Tsukaguchi H, Tokui T, Mackenzie B, Berger UV, Chen XZ, Wang Y, Brubaker RF, Hediger MA. A family of mammalian Na$^+$-dependent L-ascorbic acid transporters. Nature 1999;399:70–75.
31. Takanaga H, Mackenzie B, Hediger MA. Sodium-dependent ascorbic acid transporter family SLC23. Pflugers Arch. Eur J Physiol 2004;447:677–682.
32. Wang H, Dutta B, Huang W, Devoe LD, Leibach FH, Ganapathy V, Prasad PD. Human Na$^+$-dependent vitamin C transporter 1 (hSVCT1): primary structure, functional characteristics, and evidence for a non-functional splice variant. Biochim Biophys Acta 1999;1461:1–9.
33. Rajan DP, Huang W, Dutta B, Devoe LD, Leibach FH, Ganapathy V, Prasad PD. Human placental sodium-dependent vitamin C transporter (SVCT2): molecular cloning and transport function. Biochem Biophys Res Commun 1999;262:762–768.
34. Daruwala R, Song J, Koh WS, Rumsey SC, Levine M. Cloning and functional characterization of the human sodium-dependent vitamin C transporters hSVCT1 and hSVCT2. FEBS Lett 1999;460:480–484.
35. Wang Y, Mackenzie B, Tsukaguchi H, Weremowicz S, Morton CC, Hediger MA. Human vitamin C (L-ascorbic acid) transporter SVCT1. Biochem Biophys Res Commun 2000;267:488–494.
36. Lutsenko EA, Carcamo JM, Golde DW. A human sodium-dependent vitamin C transporter 2 isoform acts as a dominant-negative inhibitor of ascorbic acid transport. Mol Cell Biol 2004;24:3150–3156.
37. Vera JC, Rivas CI, Fischbarg J, Golde DW. Mammalian facilitative hexose transporters mediate the transport of dehydroascorbic acid. Nature 1993;364:79–82.
38. Rumsey SC, Kwon O, Xu GW, Burant CF, Simpson I, Levine M. Glucose transporter isoforms GLUT1 and GLUT3 transport dehydroascorbic acid. J Biol Chem 1997;272:19882–19889.
39. Rumsey SC, Daruwala R, Al-Hasani H, Zarnowski MJ, Simpson I, Levine M. Dehydroascorbic acid transport by GLUT4 in Xenopus oocytes and isolated rat adipocytes. J Biol Chem 2000;275:28246–28253.

40. Kumagai AK. Glucose transport in brain and retina: implications in the management and complications of diabetes. Diabetes Metab Res Rev 1999;15:261–273.
41. Takata K, Kasahara T, Kasahara M, Ezaki O, Hirano H. Ultracytochemical localization of the erythrocyte/HepG2-type glucose transporter (GLUT1) in cells of the blood-retinal barrier in the rat. Invest Ophthalmol Vis Sci 1992;33:377–383.
42. Knott RM, Robertson M, Muckersie E, Forrester JV. Regulation of glucose transporters (GLUT-1 and GLUT-3) in human retinal endothelial cells. Biochem J 1996;318:313–317.
43. Hosoya KI, Minamizono A, Katayama K, Terasaki T, Tomi M. Vitamin C transport in oxidized form across the rat blood-retinal barrier. Invest Ophthalmol Vis Sci 2004;45:1232–1239.
44. Senanayake P, Calabro A, Hu JG, Bonilha VL, Darr A, Bok D, Hollyfield JG. Glucose utilization by the retinal pigment epithelium: evidence for rapid uptake and storage in glycogen, followed by glycogen utilization. Exp Eye Res 2006;83:235–246.
45. Takagi H, Tanihara H, Seino Y, Yoshimura N. Characterization of glucose transporter in cultured human retinal pigment epithelial cells: gene expression and effect of growth factors. Invest Ophthalmol Vis Sci 1994;35:170–177.
46. Khatami M, Stramm LE, Rockey JH. Ascorbate transport in cultured cat retinal pigment epithelial cells. Exp Eye Res 1986;43:607–615.
47. Khatami M. Na^+-linked active transport of ascorbate into cultured bovine retinal pigment epithelial cells: Heterologous inhibition by glucose. Memb Biochem 1988;7:115–130.
48. Dimattio J and Streitman J. Active transport of ascorbic acid across the retinal pigment epithelium of the bullfrog. Curr Eye Res 1991;10:959–965.
49. Lam KW, Yu HS, Glickman RD, Lin T. Sodium-dependent ascorbic and dehydroascorbic acid uptake by SV-40-transformed retinal pigment epithelial cells. Ophthalmic Res 1993;25:100–107.
50. Manfredini S, Vertuani S, Pavan B, Vitali F, Scaglianti M, Bortolotti F, Biondi C, Scatturin A, Prasad P, Dalpiaz A. Design, synthesis and in vitro evaluation on HRPE cells of ascorbic acid and 6-bromoascorbic acid conjugates with neuroactive molecules. Bioorg Med Chem 2004;12:5453–5463.
51. Mantych GJ, Hageman GS, Devaskar SU. Characterization of glucose transporter isoforms in the adult and developing human eye. Endocrinology 1993;133:600–607.
52. Root-Bernstein R, Busik JV, Henry DN. Are diabetic neuropathy, retinopathy and nephropathy caused by hyperglycemic exclusion of dehydroascorbate uptake by glucose transporters? J Theor Biol 2002;216:345–359.
53. Minamizono A, Tomi M, Hosoya K. Inhibition of dehydroascorbic acid transport across the rat blood–retinal and –brain barriers in experimental diabetes. Biol Pharm Bull 2006;29:2148–2150.
54. Greco AM, Fioretti F, Rimo A. Relationship between hemorrhagic ocular diseases and vitamin C deficiency: clinical and experimental data. Acta Vitaminol Enzymol 1980;2:21–25.
55. Sinclair AJ, Girling AJ, Gray L, Le Guen C, Lunec J, Barnett AH. Disturbed handling of ascorbic acid in diabetic patients with and without microangiopathy during high dose ascorbate supplementation. Diabetologia 1991;34:171–175.
56. Augsten R, Konigsdorffer E, Schweitzer D, Strobel J. Multisubstance analysis of reflection spectra before and after laser photocoagulation for proliferative diabetic retinopathy. Eur J Ophthalmol 1997;7:317–321.
57. Uldry M, Ibberson M, Hosokawa M, Thorens B. GLUT2 is a high-affinity glucosamine transporter. FEBS Lett 2002;524:199–203.

58. Buse MG. Hexosamines, insulin resistance, and the complications of diabetes: current status. Am J Physiol Endocrinol Metab 2006;290:E1–E8.
59. Barclay TS, Tsourounis C, McCart GM. Glucosamine. Ann Pharmacother 1998;32: 574–579.
60. Manfredini S, Pavan B, Vertuani S, Scaglianti M, Compagnone D, Biondi C, Scatturin A, Tanganelli S, Ferraro L, Prasad P, Dalpiaz A. Design, synthesis and activity of ascorbic acid prodrugs of nipecotic, kynurenic and diclophenamic acids, liable to increase neutrotropic activity. J Med Chem 2002;45:559–562.
61. Dalpiaz A, Pavan B, Vertuani S, Vitali F, Scaglianti M, Bortolotti F, Biondi C, Scatturin A, Tanganelli S, Ferraro L, Marzola G, Prasad P, Manfredini S. Ascorbic and 6-Br-ascorbic acid conjugates as a tool to increase the therapeutic effects of potentially central active drugs. Eur J Pharm Sci 2005;24:259–269.

24
The Plasma Membrane Transporters and Channels of Corneal Endothelium

Jorge Fischbarg

CONTENTS

SUMMARY
INTRODUCTION
ELEMENTS OF THE MODEL
RATES OF TURNOVER
SYSTEM OF EQUATIONS
CONCLUSIONS AND PREDICTIONS FROM THE MATHEMATICAL MODEL
ACKNOWLEDGEMENTS
REFERENCES

SUMMARY

The idea of developing a theoretical model of a transporting epithelium appears worthwhile. With it, physiological and pharmacological manipulations could be tested in silico to develop agents that can suitably affect function. However, there are apparently no such models readily available for use by interested researchers. Part of the problem is that it has been presumed that a large amount of experimental information is required prior to modeling, yet, that is not always the case; we have modeled the corneal endothelium using five experimental values. The remaining six parameter values are calculated algebraically assuming cellular steady-state conditions. We have used the distribution of channels and/or transporters of the endothelium, and find that the program reproduces quite well the experimental behavior of the electrical potential difference and rate of fluid transport across that layer, in the steady state, as well as after changes in ambient conditions and after addition of inhibitors. In addition, the program favors the hypothesis that translayer fluid transport originates in electro-osmosis. This tool is readily available and written in Basic, hence it is easily modifiable for the modeling of other epithelia.

INTRODUCTION

The corneal endothelium is the site the of a fairly intense fluid transport mechanism. This mechanism moves fluid out of the corneal stroma and into the aqueous, thereby maintaining the stroma's somewhat dehydrated state and providing optimal thickness

From: *Ophthalmology Research: Ocular Transporters in Ophthalmic Diseases and Drug Delivery*
Edited by: J. Tombran-Tink and C. J. Barnstable © Humana Press, Totowa, NJ

for transparency. Given the importance of the layer and its function, over the years, the molecular composition of corneal endothelium has attracted much attention. By now, many of the membrane proteins of the endothelium have been located and characterized (1–3). There are two major questions that remain unanswered; (i) what is the mode by which bicarbonate ions cross the apical membrane? (ii) what is the mechanism by which fluid is transported?

Still, the arrangement of the membrane proteins has allowed us to elaborate a descriptive model which includes eleven of them. In the model, these proteins are seen to be the site of fluxes of four ions of interest (Na^+, K^+, Cl^-, and HCO_3^-) across the plasma membrane. The key to this development is that the ionic fluxes cannot assume arbitrary values, as they are all interdependent and linked by equations of conservation of mass and electrical charge. We have solved the system of 11 equations and 11 unknowns that allow one to find values for the 11 fluxes at any given time. As a result, we are in a position to sketch how each of the transporters and channels could be contributing to the task of transporting fluid and electrolytes across the layer. In addition, we are able to evaluate differing hypotheses to explain fluid transport. The assumption of solute–fluid coupling via electro-osmosis produces the closest match to experimental data. The characteristics and consequences of the arrangement of the endothelial transporters and channels are presented in the remainder of this chapter.

ELEMENTS OF THE MODEL

The channel and transporter constituents of the model are described in Fig. 1 and Table 1. Their distribution on either side of the corneal endothelial cell corresponds to that found immunocytochemically or deduced from experimental data. They are as follows, with numbers identifying their position in Fig. 1. Their electrogenic characteristics are noted because they will be important for the model.

Apical Membrane:

Electrogenic: H^+-ATPase (5), sodium, three-bicarbonate cotransporter (2), potassium channel (3), chloride channel (4), epithelial sodium channel (1).

Basolateral Membrane:

Neutral: sodium, potassium, two-chloride cotransporter (8), sodium/hydrogen exchanger (11), chloride/bicarbonate exchanger (9).

Electrogenic: sodium pump (6), sodium, two-bicarbonate cotransporter (7), potassium channel (10).

The evidence for the presence and locations of these elements is discussed in our recent publications (4,5).

RATES OF TURNOVER

The model as detailed in Fig. 1 carries symbols that stand for the rate of molecular turnover of each element. To emphasize this meaning, the symbols begin with a *t*. These rates (see also Table 1) are:

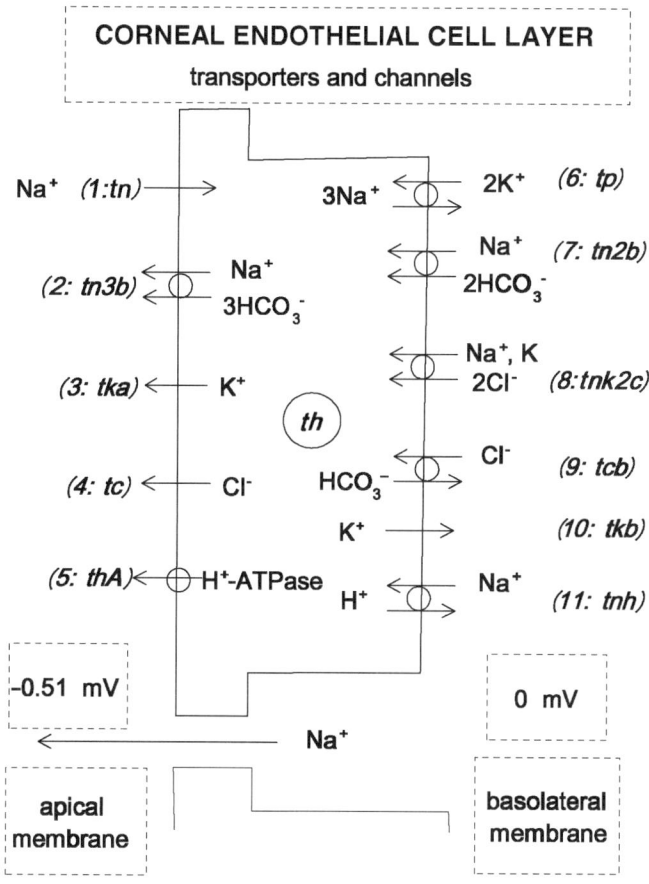

Fig. 1. Schematic cross-section across corneal endothelial cells. It includes all 11 transporters and channels that contribute to cell homeostasis and translayer transport in this model. Abbreviations for the turnover rates given in parenthesis are explained in Table 1.

Table 1 System of 11 equations and 11 unknowns to find values for all 11 rates of turnover in the model

	Equations	*Basis of the equations*
(1)	$tp = af$	Na^+ pump (af known)
(2)	$tn2b = bf$	Na^+-$2HCO_3^-$ cotransporter (bf known)
(3)	$tc = cf$	Cl^- channel (cf known)
(4)	$tn = df$	epithelial Na^+ channel (df known)
(5)	$tkb + tp + tn2b = Foc$	basolateral K^+ channel/s (Foc known)
(6)	$tn2b + tp + tkb = tn + 2^*tn3b + tc - tka$	Cellular electroneutrality
(7)	$tnk2c + 2^*tp = tka + tkb$	K^+ conservation
(8)	$tn + tnh + tn2b + tnk2c = 3^*tp + tn3b$	Na^+ conservation
(9)	$2^*tn2b = 3^*tn3b + tcb$	HCO_3^- conservation
(10)	$2^*tnk2c + tcb = tc$	Cl^- conservation
(11)	$th = tnh + thA$	H^+ conservation

The five symbols in Eqs. 1–5, *af, bf, cf, df*, and *Foc*, represent flux values obtained from experiments. The 11 parameters used in the model begin with a '*t*' because they represent rates of turnover (or element fluxes) for the given transporters and ion channels (or elements).

(1) *tn* (sodium channel); (2) *tn3b* (sodium, three-bicarbonate cotransporter); (3) *tka* (apical potassium channels); (4) *tc* (chloride channels); (5) *thA* (hydrogen-ATPase); (6) *tp* (sodium pump); (7) *tn2b* (sodium, two-bicarbonate cotransporter); (8) *tnk2c* (sodium, potassium, two-chloride cotransporter); (9) *tcb* (chloride/bicarbonate exchanger); (10) *tkb* (basolateral potassium channels); (11) *tnh* (sodium/hydrogen exchanger).

SYSTEM OF EQUATIONS

We assumed that the cells were in a steady state; given conservation of mass, this implies that, for each ion, the absolute value of the total flux through a given side of the cell is equal to that through the opposite side (one in, one out). In addition, there is conservation of electrical charge, that is, the currents through both cell sides are equal in magnitude and direction.

From this we constructed a system of 11 equations and 11 turnover rates (Table 1). The turnover rates 1–4 came from experimental values, so the rates were simply equated to the experimentally deduced numerical values (*af*, *bf*, *cf*, and *df*) of the corresponding element fluxes. The suffix *f* in this context implies an element flux. How these experimental values were determined is explained in the relevant publication *(4)*.

For turnover rate 5 (Table 1), we postulated that the potential difference existing across the endothelium results in a standing current across the low resistance that the junctions represent in this leaky preparation. This current would then reverse direction and close the circuit, returning through the cell membranes. We term this a recirculating, paracellular open-circuit current (I_{oc}). By conservation of electrical charge, this current will have to be equal to the currents across the apical and basolateral membranes in series: $I_{oc} = I_{ap} = I_{bl}$. The value of I_{oc} was calculated, given the potential difference (PD) determined across rabbit corneal endothelial preparations (around $510\,\mu V$) *(6,7)* and the transendothelial total specific electrical resistance (R_t is $20\,\Omega\cdot cm^2$) *(7,8)*. Since the resistance across the cell due to the two membranes in series is much higher than the resistance across the junctions *(9)*, it can be neglected for our purposes, so that $I_{oc} \sim PD/Rt = 25.5\,\mu A\,cm^{-2} = 0.955\,\mu Eq\,h^{-1}\,cm^{-2}$. We then equated I_{oc} to the sum of the rates of the electrogenic elements in the basolateral membrane (K^+ channels plus Na^+ pump plus Na^+-HCO_3^- cotransporter). K^+ channels and the Na^+ pump drive positive charge out of the cell, as does the Na^+-$2HCO_3^-$ cotransporter in moving two negative charges and one positive charge into the cell per cycle.

The rate of turnover 6 (Table 1) is calculated from cellular electroneutrality, for which $I_{bl} = I_{ap}$; note that for I_{bl} a positive term means positive charge going out of the cell, and for I_{ap} positive charge going into the cell. The rates of turnover 7 to 11 were obtained from the conservation for each given ion species (K^+, Na^+, HCO_3^-, Cl^-, and H^+). In equation 11, *th* represents the rate of cellular production of H^+.

The system of 11 simultaneous equations can be solved to obtain the value of each of the turnover rates. We solved it using Mathcad® (Mathsoft, Cambridge, MA; solve block function). The solutions obtained are shown in Table 2, which also shows the experimental values used for parameters *af* through *df*, and the values calculated from the algebraic solutions for the unknown parameters.

Using the turnover values of Table 2, one can calculate steady-state ionic fluxes across each one of the 11 transporters/channels in the model (Fig. 1). Transporters/channels are

Table 2 Solutions and values for the rates of turnover

	Solutions	Values obtained from experiments	Values calculated
		(μmol h^{-1} cm^{-2})	
(1)	$tn = df$	0.532	
(2)	$tn3b = -cf + 6 \cdot af - 2 \cdot df$		0.294
(3)	$tka = -Foc - 3 \cdot df - cf + 12 \cdot af$		0.327
(4)	$tc = cf$	0.162	
(5)	$thA = th$		0 (set)
(6)	$tp = af$	0.253	
(7)	$tn2b = bf$	0.456	
(8)	$tnk2c = -bf - cf + 9 \cdot af - 3 \cdot df$		0.066
(9)	$tcb = 2 \cdot bf + 3 \cdot cf - 18 \cdot af + 6 \cdot df$		0.03
(10)	$tkb = Foc - af - bf$		0.246
(11)	$tnh = 0$		

$Foc = 0.955$ (μmol h^{-1} cm^{-2})
Equation numbers correspond to the numbers identifying the elements in Fig. 1 and Table 1.

also called 'elements' in this context. Figure 2 shows these results; bars, which represent fluxes, are oriented according to their direction. As can be seen, the cell is transporting a large amount of HCO_3^- from the basolateral to the apical side (stroma to aqueous), and much smaller amounts of K^+ and Cl^- in the same direction. Interestingly, the Na^+ flux is also substantial, but it goes in the opposite direction, from apical to basolateral.

One key finding is therefore that these ionic movements do not correspond to salt transport across the cell. For salt to be transported across the layer, a cation has to move not across the cell, but instead across the junction to follow the movement of the anion (HCO_3^-) across the cell. That is precisely what happens; Na^+ moves paracellularly through the leaky tight junction and down the electrical field. This movement is passive. There is evidence for this; as the preparation is short-circuited, the HCO_3^- flux remains, but the Na^+ flux is eliminated *(8)*; hence, the HCO_3^- flux must be active and that of Na^+ passive.

Other secretory systems appear to depend heavily on the transport of Cl^-, but not HCO_3^-. Hence, Fig. 2 may suggest a hypothetical argument: could the cell operate instead by transporting less bicarbonate and more chloride? We explored this, and found that the mechanism cannot operate in such way. Fluxes cannot deviate much from the values found, otherwise the behavior of the model is compromised. Figure 3 illustrates this by showing the dependence of the unknown or computed parameters on the values of the known parameters. In this case the independent variable is the apical Cl^- channel turnover. We can try to come up with larger chloride transcellular fluxes by assuming a larger turnover rate of the apical Cl^- channel. However, that leads to an increase in intracellular Cl^- and in turn to a reversal of the flux through the Na^+-K^+-2Cl^- cotransporter to meet the steady-state conditions set by the conservation equations for the major ions and electroneutrality (Table 1). This is an unphysiological condition. Conversely, if the Cl^- channel turnover decreases even a little from its set value shown (0.162 in our standard units), that is sufficient to reverse the direction of the anion exchanger, which

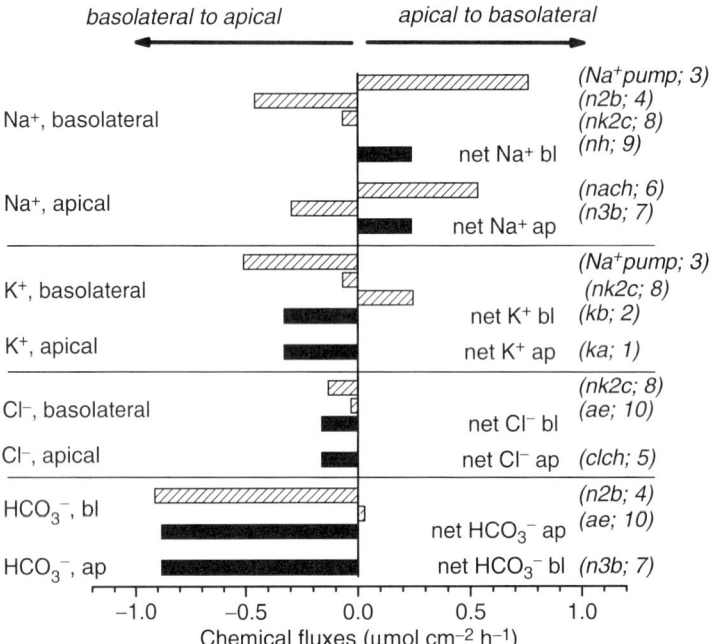

Fig. 2. Unidirectional fluxes (hatched bars) and net fluxes (solid bars) of electrolytes across each of the transporters/channels described in the model of Fig. 1 and listed here at the right margin. The values shown were obtained by solving the system of 11 linear equations listed in Table 1. For three of the fluxes (apical K^+, Cl^-, and HCO_3^-), the unidirectional fluxes are not shown, as they were equal to the net flux. Several of the 11 elements appear more than once in this figure (Na^+ pump, Na^+-HCO_3^- cotransporter, Na^+-K^+-$2Cl^-$ cotransporter, anion exchanger), because they carry more than one ion species simultaneously. ap, apical side; bl, basolateral side.

under normal conditions is thought to bring Cl^- into the cell. Such a reversal in turn is only possible if the intracellular chloride concentration ($[Cl^-]_i$) would rise to unphysiological concentrations. This stresses that the rates of turnover we find are constrained by conservation equations and cannot be varied at will.

The rest of the treatment yielded two main conclusions; (i) the model reproduced experimental data reasonably well, and (ii) the model accounted for experimental data best when it was assumed that fluid movement originated via electro-osmotic coupling of the Na^+ movement across the junctions with the water in them. The prevalent assumption that the coupling was of local osmotic nature instead, led to failure.

CONCLUSIONS AND PREDICTIONS FROM THE MATHEMATICAL MODEL

The model allows one to test the effects of specific inhibitors on the different transport pathways. It also predicts the values of a number of parameters that can be tested experimentally; these include changes in intracellular electrolytes, cell volume, cell

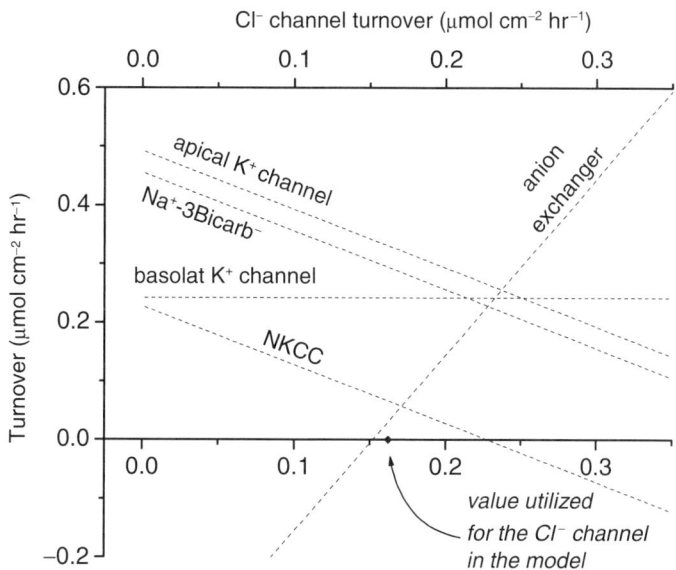

Fig. 3. This figure details how the rates of turnover initially chosen for the model are all interlinked by 11 equations corresponding to conservation of mass. Changing the value set for the turnover for the Cl⁻ channel results in changes in the turnovers of several other elements. NKCC, Na⁺-K⁺-2Cl⁻ transporter; basolat, basolateral; Na⁺-3Bicarb⁻, SVCT2, sodium, three-bicarbonate cotransporter.

pH, membrane potential, transcellular potential difference, and transendothelial fluid transport. Lastly, the model permits expansion to include the regulation of the function of individual elements by cell-signaling compounds. The model, thus, can serve as the basis for a comprehensive description and explanation of the function of the corneal endothelial fluid pump in steady and nonsteady states and during pharmacological challenges.

ACKNOWLEDGEMENTS

Our work was supported by National Institutes of Health grant EY06178, and in part by Research to Prevent Blindness, Inc.

REFERENCES

1. Fischbarg J, Hernandez JA, Liebovitch LS, Koniarek JP. The mechanism of fluid and electrolyte transport across corneal endothelium: critical revision and update of a model. Curr Eye Res 1985;4:351–360.
2. Sanchez JM, Li Y, Rubashkin A, Iserovich P, Wen Q, Ruberti JW, Smith RW, Rittenband D, Kuang K, Diecke FPJ, Fischbarg J. Evidence for a central role for electro-psmosis in fluid transport by corneal endothelium. J Membr Biol 2002;187:37–50.
3. Bonanno JA. Identity and regulation of ion transport mechanisms in the corneal endothelium. Prog Retin Eye Res 2003;22:69–94.

4. Fischbarg J and Diecke FP. A mathematical model of electrolyte and fluid transport across corneal endothelium. J Membr Biol 2005;203:41–56.
5. Fischbarg J and Diecke FP. The 9th World Multi-Conference on Ssytemics, Cybernetics and Informatics, Orlando, Florida. 2005.
6. Fischbarg J and Lim JJ. Role of cations, anions and carbonic anhydrase in fluid transport across rabbit corneal endothelium. J Physiol 1974;241:647–675.
7. Fischbarg J. Hormonal control of epithelial transport. Paris; 1979.
8. Hodson S and Miller F. The bicarbonate ion pump in the endothelium which regulates the hydration of the rabbit cornea. J Physiol 1976;263:563–577.
9. Lim JJ and Fischbarg J. Electrical properties of rabbit corneal endothelium as determined from impedance measurements. Biophys J 1981;36:677–695.

Index

12(R)-hydroxy-5,8,10,14-eicosatetraenoic acid [12(R)-HETE] 23
1-trans-pyrrolidine-2,4-dicarboxylate 363, 366
2-5-N-ethylcabamidoadenosine (NECA) 176
3α-hydroxysteroid dehydrogenase 439

A

A_3AR-selective agonists 74
ABC transporter – see *ATP-binding cassette transporter*
Abcg2 – see *ATP-binding cassette transporter G2*
ABCR – see also *ATP-binding cassette transporter* 319
   nucleotide binding domains 322
   related degenerative maculopathies 323
Acetazolamide 171
Acid phosphatase 381
Acyclovir 378
Adenosine 71, 74, 148
Adenosine receptor 73
Adenosine triphosphate (ATP)-binding cassette (ABC) 408
Adenosine triphosphate (ATP)-sensitive K^+ channels 163
Adenylyl cyclase (AC) 166
Adherens junctions 188
Adrenergic receptors 19
Advanced glycation endproducts (AGEs) 130, 132, 225
AE2 antiports 69
Age-related macular degeneration 317, 399, 404, 420, 144, 151, 159, 207, 378, 423, 426, 439
Alanine-serine-cysteine transporters (ASCT1-2) 104
Aldehyde oxidase 381

α1-adrenergic receptors 168
α-ketoglutarate 356
Amacrine cells 306
AMD – see *Age-related macular degeneration*
Amiloride 171
Amiloride-sensitive Na^+ channel 72
Aminozolamide 382
Angioid streaks 385
Anti-epileptic drugs 308
Antinatriuretic peptide 73
Anti-VEGF 207
Aphakic bullous keratopathy (ABK) 22
Apical Cl^- transporters 165
Apical junctional complex 188, 190
Apical potassium channels 454
APK/JNK 25
Aquaglyceroporins 6
Aquaporin 3, 6, 7, 10, 113, 172
Aquaporin 1 (AQP1) 7, 8
Aquaporin 3 (AQP3) 7, 12, 13
Aquaporin 4 (AQP4) 5
Aquaporin 5 (AQP5) 5, 11
Aquaporin 9 (AQP9) 172
Arachidonic acid (AA) 387
ARMD – see *Age-related macular degeneration*
Aryl sulfatase 381
Ascorbic acid (vitamin C) 31, 47, 52, 144, 145, 437, 438, 440, 445
aspartate 357
Astrocytes 355, 359
ATP-binding cassette (ABC) transporters 149, 151
ATP-binding cassette transporter G2 (Abcg2) 139, 150
ATP-sensitive channels 164
Atrial natriuretic peptide (ANP) 167
Autosomal recessive cone-rod dystrophy 317

B

Baclofen 298
Barbiturates 298
Basolateral potassium channels 454
BayK8644 204
Benzodiazepine 298, 345
Benzolamide 171
Best's vitelliform macular dystrophy 166, 207
Bestrophin 159, 166, 206
β-alanine 296
β-lactam 345
Betaine glycine transporter (BGT) 295
bFGF 202
bFGF receptor 205
Bicarbonate transporter family (SLC4 family) 165, 168
Bicuculline 298
Bipolar cell 304, 307, 356
Bistratene A 341
Blood retinal barrier 129
Blood-aqueous barrier 405
Blood-brain barrier 132, 139, 141, 150, 171, 185, 187, 218, 339, 355, 363, 368, 378, 384, 402, 405, 440
Brain-derived neurotrophic factor (BDNF) 295, 344
Breast-cancer-resistant protein (BCRP) 378
Brimonidine 378
Bruch's membrane 159, 188, 222, 406

C

Cl$^-$ channels 72, 99
Cl$^-$/HCO$_3^-$ anion exchanger (CBE) 26
Cl$^-$/HCO$_3^-$ antiport 64, 69, 72
Ca^{2+}-activated Cl$^-$ channels 132, 159, 165, 166, 176
Ca^{2+}-activated K$^+$ channels 131, 164
Ca^{2+}-activated K$^+$ channels-Intermediate conductance [IK$_{Ca}$] 163
Ca^{2+}-activated K$^+$ channels-Large conductance [BK$_{Ca}$] 163
Ca^{2+}-activated K$^+$ channels-Small conductance [SK$_{Ca}$] 163
Ca^{2+}-ATPase (PMCA) transporter 28
Ca^{2+}-calmodulin-dependent kinase 167
Ca^{2+}-dependent calmodulin (CaM) II kinase 27
Ca^{2+}-dependent Cl$^-$ channels 166, 202, 206, 208
Ca^{2+}-dependent K$^+$ channels 202
Ca^{2+} extrusion 258
Ca^{2+} homeostasis 208
Ca^{2+} ionophores 202
Ca^{2+}-pumps 162
CaBP (calbindin) 304
CaCC – see *Ca^{2+}-activated Cl$^-$ channels*
CACL – see *Ca^{2+}-activated Cl$^-$ channels*
Calcium channels 204
Calcium-permeable nonspecific cation (NSC) channels 132
Calmodulin 28
cAMP 73
cAMP-dependent Cl$^-$ transport 167
Canavan disease 389
Cannabinoids 73
Carbachol 378
Carbamazepine 342
Carbonic anhydrase (CA) 64, 66, 69, 170
Carbonic anhydrase inhibitor 67
Cardiac antiporter 167
Cardiac glycosides 119
Cardiac sarcolemmal membranes 257
Carotid artery disease 336
Catalase 51
Cataract 4, 89, 106, 120
    Age-related nuclear (ARN) 95, 101, 102, 105
    Congenital 4
    Cortical 121
    Diabetic 95, 100
    Surgery 23
Catecholamines 73
Cationic and neutral amino acid transporter, B$^{o,+}$ 30
Celecoxib-PLGA 431
Cellular phosphorylation potential 164
Cell-volume homeostasis 165
Central retinal artery or vein occlusion 336
Central retinal artery 220
CFTR – see *Cystic-fibrosis transmembrane regulator*
Cyclic guanosine monophosphate (cGMP) 73
Charybdotoxin 164
Chelerythrine 341
Chloride channel 33, 452, 454
Chloride/bicarbonate exchanger 162, 169, 452, 454
Cholinergic receptors 19
Choriocapillaris 162, 185, 220, 224, 420

Choroid 143, 189, 405
Choroidal neovascularization (CNV) 207, 417, 420
Choroidal blood vessels 442
Choroidal capillaries 186
Choroidal circulation 220, 236
Ciliary body 115, 406
Ciliary epithelia 298
Ciliary neurotrophic factor (CNTF) 158
Cis-3 aminocyclohexanecarboxylic acid (ACHC) 296
C-jun N-terminal/stress-activated protein kinase (JNK/SAPK) 21
Claudin 1 194
Claudin 2 194
Claudin 4L2 194
ClC channel 165
ClC-2 165
ClC-3 34, 165
ClC-5 165
Comet-like streaks 385
Computed tomography (CT) 423
Cone/ganglion cell Na^+/Ca^{2+}-K^+ exchanger (NCKX2) 257
Conjunctiva 403
Conjunctivitis 376
Connexins 70, 113
Cornea 3, 18, 376, 401
Corneal edema 23
Corneal endothelium 27, 403, 451
Corneal epithelium 18, 22, 24, 32, 382
Corneal epithelial ion transport 17, 19
Corneal epithelial wound healing 11
Corneal graft rejection 425
Corneal keratitis 376
Corneal transparency 8
Cortical implants 422
Creatine transporter (CRT) 139
Creatine 143
Cyclic adenosine monophosphate (cAMP) 71, 159, 340
Cyclic guanosine monophosphate (cGMP)-gated channels 174, 257
Cyclosporine A 241, 378
Cysteine sulphinic acid decarboxylase (CSD) 218
Cystic-fibrosis transmembrane regulator (CFTR) 34, 159, 165, 166, 318
Cystine/glutamate exchanger, X 95, 362

Cystoid macular edema 171
Cytochrome P450 380
Cytokine endothelin (ET) 27
Cytokine receptor 20, 23
Cytokine 17, 20, 30
Cytomegaloviral (CMV) retinitis 404
Cytosolic Ca^{2+} stores 201

D
D-aspartate 360
Dehydroascorbic acid 48, 52, 144, 437, 442
Delayed-rectifier K^+ channel 164
Dendrimers 416
Descemet's membrane 403
Desmosomes 404
Deturgescence 17, 20
Deturgescence 20
Diabetes 129, 362, 439, 445
Diabetic macular edema (DME) 399
Diabetic retinopathy 129, 133, 142, 148, 163, 275, 339, 355, 365, 386, 423, 446
Diacylglycerol (DAG) 341
Diclofenac 381
DIDS 166
Dihydroindenyloxyalkanoic acid (DIOA) 167
Dipivefrine 378
DL-threo-beta-benzyloxyaspartate (DL-TBOA) 345
Docosahexaenoic acid (DHA) 158, 345
Dopamine 147
Drug-resistant esophageal adenocarcinoma 418
Drusen deposits 159
Dry eye syndrome (DES) 20, 24, 376
Dystrophic calcification 385

E
Early receptor potential 220
Edema 8
EGF receptor 23
EGF 23, 28
Electroneutral cation-Cl^--coupled cotransporters (SLC12 family) 165
Electro-oculogram (EOG) 166
Emulsion polymerization 420
ENaC Na^+ channel 72
Endophthalmitis 404
Endothelin 73, 116
ENT2 149

Epidermal growth factor (EGF) 22, 343
Epithelial sodium channel 452
Equilibrative nucleoside transporter 2 (ENT2) 139
ERK 23
Erythromycin 378, 383
ET_A receptor 27
ET_B 28
Ethanol 298, 345
Etomidate 345
Excitatory amino acid transporter (EAAT) 95, 104, 276, 283, 333, 357
Excitatory amino acid transporter EAAT1 (GLAST) 276, 279, 334, 360, 365
Excitatory amino acid transporter EAAT1-5 95
Excitatory amino acid transporter EAAT2 (GLT-1) 276, 335
Excitatory amino acid transporter EAAT3 (EAAC1) 276, 335
Excitatory amino acid transporter EAAT4 276, 335
Excitatory amino acid transporter EAAT4/5 expression 104
Excitatory amino acid transporter EAAT4-associated protein 279
Excitatory amino acid transporter EAAT5 279, 335
Excitotoxic damage 280, 366
Extracellular matrix (ECM) 222
Extracellular signal-regulated kinase (ERK) 21
Eye allergies 376

F
Fast oscillation (FO) 166
Fatty acids 341
FGFR2 205
Fibre cell 90, 97, 104
Fibroblast growth factor (FGF) 158, 344
Fibroblast growth factor-β (βFGF) 167
Fluorescein angiography 423
Folate receptor α (FRα) 175
Fuchs' dystrophy 23
Fundus flavimaculatus (FFM) 317
Fundus photography 423

G
Gabaculine 298
GABAergic transmission 306
γ-glutamylcysteine synthetase (γGCS) 101
Gamma aminobutyric acid (GABA) transporters (GATs) 293, 295, 304
Gamma aminobutyric acid (GABA) 146, 219, 295, 304, 356, 367
Gangcyclovir 378
Ganglion cells 299, 356, 363
Gap junction proteins 162
Gap junction 70, 91, 96, 102, 130
GAT-1 (Gamma aminobutyric acid (GABA) transporter) 299, 305
  amacrine cells 300
  horizontal cells 304
  Müller cells 305
G-coupled receptors 281
Gene therapy 425
GLAST – see *Excitatory amino acid transporter EAAT1*
Glaucoma surgery 418
Glaucoma 61, 74, 336, 376, 399, 425
Glial fibrillary acidic protein (GFAP) 342
Glial glutamate transporter, GLAST 282
Glial transporters 281
Glial-derived neurotropic factor (GDNF) 344
Glucocorticoids 73
Glucose transporter (GLUT) 95, 142, 185, 441
Glucose transporter GLUT1 30, 95, 102, 139, 444
Glucose transporter GLUT3 95, 102, 141, 444
Glucose transporter GLUT4 4
Glucose transporters-RPE expression 191
Glucose 113, 131, 141
Glutamate homeostasis 362, 366
Glutamate protein kinase C 365
Glutamate transporters 275, 279, 356, 362
Glutamate transporter-associated proteins (GTRAPs) 282, 342
Glutamate transporters (SLC1 gene family) 294
Glutamate transporters-Gamma aminobutyric acid (GABA) synthesis 281
Glutamate turnover 362
Glutamate 95, 146, 275, 339, 356
Glutamate-glutamine cycle 284, 334, 362
Glutamatergic synapses 281
Glutaminase 104
Glutamine – system A transporters 284
Glutamine – system ASC transporters 286
Glutamine synthetase 279, 339, 362

Index

Glutamine transporters – disease involvement 286
Glutamine transporters 284, 285
Glutamine 356
Glutamine/glutamate transporter 2 (ASCT2) 95
Glutathione (GSH) 97, 100, 144, 439
Glutathione peroxidase 51
glutathione reductase (GR) 101
Glutathione S-transferase 382, 439
glutathione synthetase (GS) 101
Glycolysis 186
Gold nanoparticles 424
GSH-peroxidase 52
GTRAP41 279
Guanidoethane sulphonic acid (GES) 218
Gulono-γ-lactone oxidase 438
Guvacine 296
Gyrate atrophy 143

H
H^+-ATPase 452
H^+-dependent monocarboxylate transporter (MCT1) 142
Heart disease 336
Hemorrhages 339, 355
Hepatocyte growth factor 23
Horizontal cells 306
Hormones 295
Human nonpigmented epithelial (HNPE) cells 174
Hydrogen peroxide 51
Hydrogen-ATPase 454
Hypotonic cell swelling 71
Hypoxia 338, 365
Iberiotoxin 164

I
In vivo confocal imaging 423
Inner blood-retinal barrier 445
Inositol triphosphate (InsP3) Ca^{2+} system 202
Insulin-like growth factor-I (IGF-I) 26, 132, 158, 202
Interphotoreceptor matrix (IPM) 171
Intracellular chloride concentration ([Cl^-]) 159
Intracellular chloride pool 165
Intracellular free Ca^{2+} 202

Intraocular pressure (IOP) 61, 63, 70
Intra-retinal distribution 218
Intravitreal nanoparticles 428
Inwardly-rectifying potassium (K_{IR}) channels 8, 129, 163
Ion homeostasis 202
Ionic gelation 420
Ionomycin 166
Ionotropic ATP receptors 209
Ionotropic glutamate receptors 204, 336
Ischemia 336, 365
Isoflurane 345
Isoguvacine 298

K
K^+/Cl^- cotransporter 25, 159, 176
K^+-$2Cl^-$ cotransporter 168
Kainate receptors 209, 344
K_{ATP} channels 134
KCC – see *Potassium chloride cotransporter*
Keratinocyte growth factor 23
Ketone reductase 381
K_{IR} channels 129
Kynurenic acid 338

L
Lactate transport 30, 172
Lactic acid 171
Lens epithelium-derived growth factor (LEDGF) 158
Lens transparency 4, 118
Lens 47, 89, 112,
Lidocaine 345
Lipid peroxidation 365
Lipofuscin 227, 320
Liposomes 419
Low-affinity sodium-independent carriers 224
L-threo-beta-methoxyaspartate (L-TMOA) 345
L-type amino acid transporter (LAT) 30, 139, 146
L-type Ca^{2+} channels 205

M
Macular degeneration 217
Macular edema 148, 171, 339, 368
Magnetic resonance imaging (MRI) 423
MAP kinase (MAPK) 23, 73

MCT – see *H^+-dependent monocarboxylate transporter*
Melanosomes 201
Membrane spanning Ca^{2+} pore 204
Membrane-bound CA IV 171
Metabotropic ATP receptors 209
Metabotropic glutamate receptors (mGluRs) 344
Metalloproteases (MMPs) 159
Micelles 418
Microaneurysms 339, 355
Microelectromechanical systems (MEMS) 421
Monocarboxylate transporter 1 (MCT1) 139, 171
MP20 105
Müller cell 140, 146, 298, 306, 355, 362, 365, 445
Multidrug resistance gene (MDR1) 404
Multidrug-resistance-associated protein (MRP) 139, 142, 149, 236, 379, 408
Multiple sclerosis 366
Mural cells 129
Muscarinic 73
Myeloperoxidase 384

N

Na:K:2Cl cotransporter (NKCC) 22, 69, 98
Na^+-and Cl^--dependent transport 295
Na^+-Cl^- symport 72
Na^+ channels 164
Na^+-K^+-$2Cl^-$ symport 70
Na^+, K^+-ATPase 22, 63, 72, 111, 120, 159, 365, 387
Na^+/ Ca^{2+} exchanger (NCXs) 28, 162, 222, 257
Na^+/ Ca^{2+}-K^+ exchangers 259
Na^+/Ca^{2+} exchangers 133
Na^+/Cl^--dependent taurine transporter 222
Na^+/Cl^- cotransporter 168, 176
Na^+/H^+ (NH) antiporter 171, 176
Na^+/H^+ exchanger (NHE) 27, 64, 69, 133
Na^+-Cl^--coupled transporters (SLC6 gene family) 294
Na^+-dependent concentrative nucleoside transporters (CNTs) 148
Na^+-dependent organic-anion transporting protein (Oatp) 172
Na^+-dependent vitamin C transporters (SVCTs) 49, 144, 442
Na^+-driven Cl^-/HCO_3^- anion exchangers 168
Na^+-HCO_3^- cotransporters 159, 168
Na^+-independent equilibrative nucleoside transporters (ENTs) 148
Na^+-K^+-$2Cl^-$ cotransporters 68, 167
Na^+-K^+-$2Cl^-$ symport 64
Na^+-myoinositol distinct cotransporters 175
N-acetylaspartate 389
N-acetyl-β-glucosamidase 381
Nanocages 424
Nanocapsules 419
Nanoelectromechanical systems (NEMS) 421
Nanofibers 420
Nanolithography 420
Nanoneedles 422
Nanoparticles 419
Nanoparticles 426
Nanoplexes 418
Nanoprecipitation 420
Nanorods 420
Nanoshells 424
Nanospheres 419
   Nanosystems Liposomes
   Nanosystems Nanocapsules
   Nanosystems Nanoparticles
   Nanosystems Nanoplexes
   Nanosystems Nanorods, tubes, and fibers
   Nanosystems Nanospheres
Nanosystems 415, 420
Nanotechnology 415
Nanotubes 420
Nanotweezers 422
Natriuretic peptide receptor (NPR) 176
NBMPR-insensitive ENT2 148
NCKX genes Single-nucleotide polymorphism (SNP) 269
NCKX genes Cone-rod degeneration Cone dysfunction syndromes 266
NCKX protein – structure-function relationships 266
NCKX 258
NCKX1 post-translational modification 263
NCKX1-mediated Ca^{2+} fluxes 258
NCKX2 alternative splicing 263
NCKX2 Na^+/ Ca^{2+}-K^+ exchangers 258
NCKX2 post-translational modification 263

NCKX2 transcripts 262
NCX1 Na$^+$/Ca^{2+} exchanger 261
NECA – see 2-5-N-ethylcabamidoadenosine
Neovascularization 244, 339
Nerve fiber layer 363
N-ethylmaleimide (NEM) 167
Neurodegeneration 148
Neuroepithelium 188
Neuroimmunophilins 345
Neuropeptide Y (NPY) 167, 176
Neurosteroids 298
NHE-1 69
Nifedpine 204
Nipecotic acid 296, 304
Nitric oxide synthase (NOS) 75
nitric oxide 73
Nitrobenzylmercaptopurine riboside (NBMPR)-sensitive ENT1 148
NKCC – see Na:K:2Cl cotransporter
N-methy-D-aspartate (NMDA) receptors 209, 336, 367
NO-711 (GABA transport inhibitor) 304, 307
Nonpigmented ciliary epithelial (NPE) cells 7, 63
N-retinylidene-N-retinyl-ethanolamine (A2E) 320
Nutrient transporters 101

O

Occludin 159, 190, 402
Ocular drug delivery 399
Ocular inflammation 439
Ocular lens dislocation 387
Ocular surface 10
Ophthalmic timolol therapy 382
Optic nerve head astrocytes 174
Optic neuropathy 340, 389
Optic vesicle 188
Optical coherence tomography (OCT) 423
Organic anion transporter polypeptide (Oatp) 142, 149
Organic anion transporter polypeptide la4 (Oatpla4) 139
Organic anion transporters (OATs/ Slc22a) 149
Organic anion transporting peptide (OATP)-E 30, 376
Organic cation transporter 378

Osmolyte transporters 22
Osmotic gradients 3
Osmotic stress 17
Ouabain 163
Outer blood-retinal barrier 235, 445
Outer retinal transport system 221
Oxidative stress 366

P

P13K/Akt 26
P2X purinoceptors 129, 131
P2Y2 agonists 167
P$_2$Y$_2$ATP receptors 72, 176
P38 MAPK 21
P44/42 MAPK 25
p-aminobenzoic acid 382
Paracellular diffusion 402
Paracellular pathways 194
Parallel Cl$^-$/HCO$_3^-$ 69
Parkinson's disease 147
PDGF-BB 134
PE cells 64, 68
Pericyte 127, 130, 140
Peripheral fundus 221
Peripheral iridectomy 62
P-glycoprotein (P-gp) 235, 243, 318, 378, 408
  efflux pump (P-gp) 404
  inhibition 241
  substrates 238
Phagocytosis 159, 202
Phorbol esters 341
Phosphate-activated glutaminase 284
Phosphatidylinositol 3,4,5-triphosphate (PIP$_3$) 341
Phosphatidylinositol 3-kinase (P13K) 340
Phospholipase C 344
Photoreceptor dark current 174
Photoreceptor degeneration 220
Photoreceptor 141, 158, 201, 356
Picrotoxin 298
Pigment epithelium-derived growth factor (PEDF) 158, 245, 428
Pigmented ciliary epithelial (PE) layer 63
PLA (poly-lactic acid) 426
Plasma membrane transporter NaDC3 389
Platelet-derived growth factor (PDGF) 133, 158, 344

PLGA (poly –(lactide-co-glycolide)) 426
Polyethylene glycols (PEGs) 401
Positron emission tomography (PET) 423
Posterior segment 404
Potassium channel 163, 452
Potassium chloride cotransporter (KCC) 25, 97
Potassium chloride cotransporter 4 (KCC4) 95
Potassium-selective ion channels 32
Pressure-dependent trabecular pathway 62
Pressure-independent uveoscleral pathway 62
Presynaptic glutamate transporters 281
Primary congenital glaucoma 381
Primary open-angle glaucoma (POAG) 61
Probenecid 172
Pro-inflammatory cytokines 34
Proliferative and diabetic vitreoretinopathy 404
Proliferative vitreoretinopathy 242
Propranolol 381
Prostanoids 73
Protein disulfide isomerases 439
Protein kinase A (PKA) 73, 165, 295, 340
Protein kinase C (PKC) 27, 71, 130, 167, 206, 295, 340, 365
Pseudoexfoliation syndrome (PEX) 74
Pseudophakic bullous keratopathy 22
Pseudoxanthoma elasticum (PXE) 385
Purinergic receptors 209
Purinoreceptors 131

Q
Quantum dots 424

R
Reactive gliosis 362
Reactive oxygen species 143, 439
Rectifying K^+ channels (K_{ir} channels) 162
Rectifying Kir6.2 163
Reduced-folate transporter (RFT-1) 175
Regulatory volume decrease (RVD) 20, 97
Regulatory volume increase (RVI) 20, 97
Retina 133, 142, 185, 444
Retinal-11-cis 236
Retinal-all-trans 236, 319
Retinal capillaries 221
Retinal capillary endothelial cells 139, 150, 243, 440
Retinal degeneration 159, 165, 172, 244
Retinal detachment 165
Retinal edema 172
Retinal ganglion cells 174
Retinal implants 422
Retinal microvasculature 128, 405
Retinal pigment epithelium (RPE) 139, 144, 158, 201, 218, 298, 318, 399, 406, 420, 426, 440
Retinal rod exchanger 167
Retinal rod NCKX1 257
Retinal signal transduction 8
Retinal synapses 306
Retina-specific ABC transporter, ABCR 317
Retinitis pigmentosa 165, 171, 217, 385
Retinol-binding protein/transthyretin (RBP/TTR) 158
Retinopathy of prematurity 426
Retinylidene-phosphatidylethanolamine 319
Reversed transport of GABA 295
Rhodopsin 319
Riboflavin 52, 175
Rod Na^+/Ca^{2+}-K^+ exchanger 258
Rod NCKX1 protein interactions 260
Rod outer segments (ROS) 257
RPE development 191
RPE tight junctions 189

S
Scanning laser ophthalmoscopy 423
Schizophrenia 367
Scurvy 438
Serotonergic receptors 19
Src family tyrosine kinase (SFK) 117
SKF89976A 304, 308
Sodium calcium exchanger (NCX) 119
Sodium calcium exchanger (SL8 family) 167
Sodium channel 33, 454
Sodium efflux 159
Sodium pump 452
Sodium, potassium, two-chloride cotransporter 452
Sodium, three-bicarbonate cotransporter 452
Sodium, two-bicarbonate cotransporter 452
Sodium/hydrogen exchanger 171, 454
Sodium-coupled glucose transporters (SGLT1) 95, 102, 103, 186, 192
Sodium-dependent multivitamin transporter (SMVT) 31

Index

Sodium-dependent taurine transporter 227
Sodium-dependent vitamin C transporter (SVCT)1 31
Sodium-dependent **V**itamin **C** **T**ransporters 1 and 2 (SVCT1 and SVCT2) 440
Sodium-hydrogen exchanger 452
Sodium-potassium-chloride cotransporter (NKCC) 97
Solute carriers 165
Solute-carrier family 167
Spermine 130
Src family of tyrosine kinases (SFKs) 116
Stargardt disease (STGD) 317
Streptozotocin-induced diabetes 130, 142, 145, 445
Stroke 336
Stroma 403
Stroma-endothelium 382
Subretinal space 158
Succinate 367
Sulfinpyrazone 172
Sulfobromophthalein 172
Sulfonylurea SUR subunit 163
Supercritical fluid technology 420
Super-oxide dismutase 51
Swelling activated Cl⁻ channels 71, 165
Syntaxin 1A 295

T

Taurine 218
Taurine deficiency 220
Taurine transport by retinal pigment epithelium 222
Taurine transporter(TauT) 32, 139
Taurine-calcium complex 219
Taurine-zinc 219
Tetraethylammonium (TEA) 164
Tetrodotoxin 164
TGF-β receptor 190
Thiazide-sensitive Na⁺-Cl⁻ cotransporter 167
Thiol transferases (glutaredoxins) 439
Thioredoxin reductase 439
Thrombin 116
Tiagabine 296, 308
Tight junctions 18, 128, 139, 188, 402, 440
Toll-like receptors (TLRs) 34

Transcellular glucose transport 194
Transepithelial electrical resistance (TER) 163, 189
Transepithelial potential difference (PD) 65
Transepithelial transport 159, 185
Transforming growth factor-alpha (TGFα) 344
Transforming growth factor-β (TGF-β) 23, 158
Tricarboxylic acid cycle 357, 366
TRP channels 19, 37, 204, 208
Tumor necrosis factor-alpha 73

U

Ubiquitinylation 386
Ultrasonography 423
uridine triphosphate (UTP) 167
Uveitis 171

V

Vascular endothelial growth factor (VEGF) 132, 158, 202, 207, 368, 417
VDCC – see *Voltage-dependent calcium channels*
Verapamil 241
Vesicular glutamate transporters (VGLUTs) 333
Vigabatrin 308
Visual cycle of retinal 158
Vitamin B2 175
Vitamin C 47, 52, 144, 437
Vitamin C-transport 50
Vitamin E (α-tocopherol) 438
Voltage-activated K⁺ channels 162
Voltage-dependent calcium channels (VDCC) 130, 204
Voltage-gated Cl⁻ channels (ClC) 159, 165

W

WAY-209429 345
WAY-855 345

Z

ZO-1 protein 190, 402, 406
ZO-2 protein 402
Zonula occludens 188, 222, 405

Printed in the United States of America